Imperfect Inventory Systems

Ata Allah Taleizadeh

Imperfect Inventory Systems

Inventory and Production Management

Ata Allah Taleizadeh
School of Industrial Engineering
University of Tehran
Tehran, Iran

ISBN 978-3-030-56976-1   ISBN 978-3-030-56974-7 (eBook)
https://doi.org/10.1007/978-3-030-56974-7

© Springer Nature Switzerland AG 2021
This work is subject to copyright. All rights are reserved by the Publisher, whether the whole or part of the material is concerned, specifically the rights of translation, reprinting, reuse of illustrations, recitation, broadcasting, reproduction on microfilms or in any other physical way, and transmission or information storage and retrieval, electronic adaptation, computer software, or by similar or dissimilar methodology now known or hereafter developed.
The use of general descriptive names, registered names, trademarks, service marks, etc. in this publication does not imply, even in the absence of a specific statement, that such names are exempt from the relevant protective laws and regulations and therefore free for general use.
The publisher, the authors, and the editors are safe to assume that the advice and information in this book are believed to be true and accurate at the date of publication. Neither the publisher nor the authors or the editors give a warranty, expressed or implied, with respect to the material contained herein or for any errors or omissions that may have been made. The publisher remains neutral with regard to jurisdictional claims in published maps and institutional affiliations.

This Springer imprint is published by the registered company Springer Nature Switzerland AG
The registered company address is: Gewerbestrasse 11, 6330 Cham, Switzerland

Contents

1	**Introduction**	1
	1.1 Imperfect Items	2
	1.2 Scrap	2
	1.3 Rework	3
	1.4 Multi-product Single Machine	3
	1.5 Quality Considerations	4
	1.6 Maintenance	4
	References	4
2	**Imperfect EOQ System**	7
	2.1 Introduction	7
	2.2 Literature Review	7
	2.2.1 Deterioration, Perishability, and Lifetime Constraints	8
	2.2.2 Imperfect-Quality Items	9
	2.3 EOQ Model with No Shortage	10
	2.3.1 Imperfect Quality	10
	2.3.2 Maintenance Actions	16
	2.3.3 Screening Process	18
	2.3.4 Learning Effects	23
	2.3.5 EOQ Models with Imperfect-Quality Items and Sampling	37
	2.3.6 Buy and Repair Options	56
	2.3.7 Entropy EOQ	61
	2.4 EOQ Model with Backordering	70
	2.4.1 Imperfect Quality and Inspection	70
	2.4.2 Multiple Quality Characteristic Screening	74
	2.4.3 Rejection of Defective Supply Batches	82
	2.4.4 Rework and Backordered Demand	87
	2.4.5 Learning in Inspection	101
	2.4.6 EOQ Model for Imperfect-Quality Items	109

		2.5	EOQ Model with Partial Backordering	112
			2.5.1 EOQ Model of Imperfect-Quality Items	112
			2.5.2 Screening	118
			2.5.3 Reparation of Imperfect Products	126
			2.5.4 Replacement of Imperfect Products	142
		References		147
3	Scrap			153
	3.1	Introduction		153
	3.2	No Shortage		155
		3.2.1	Continuous Delivery	155
		3.2.2	Discrete Delivery	166
	3.3	Fully Backordered		186
		3.3.1	Continuous Delivery	186
	3.4	Partial Backordered		223
		3.4.1	Continuous Delivery	223
	3.5	Conclusion		232
	References			233
4	Rework			235
	4.1	Introduction		235
	4.2	Literature Review		236
	4.3	No Shortage		239
		4.3.1	Imperfect Item Sales	239
		4.3.2	Rework Policy	244
		4.3.3	Imperfect Rework	249
		4.3.4	Quality Screening	254
	4.4	Backordering		261
		4.4.1	Simple Rework	261
		4.4.2	Defective Product	268
		4.4.3	Random Defective Rate: Same Production and Rework Rates	277
		4.4.4	Rework Process and Scraps	293
		4.4.5	Rework and Preventive Maintenance	300
		4.4.6	Random Defective Rate: Different Production and Rework Rates	307
		4.4.7	Imperfect Rework Process	312
	4.5	Partial Backordering		318
		4.5.1	Immediate Rework	318
		4.5.2	Repair Failure	324
	4.6	Multi-delivery		331
		4.6.1	Multi-delivery Policy and Quality Assurance	331
		4.6.2	Multi-delivery and Partial Rework	337
		4.6.3	Multi-delivery Single Machine	340
		4.6.4	Multi-product Two Machines	345

		4.6.5	Shipment Decisions for a Multi-product	351
		4.6.6	Pricing with Rework and Multiple Shipments	357
	References			362
5	**Multi-product Single Machine**			367
	5.1	Introduction		367
	5.2	Literature Review		368
	5.3	No Shortage		371
		5.3.1	Simple Model	371
		5.3.2	Discrete Delivery	375
		5.3.3	Rework	380
		5.3.4	Auction	385
		5.3.5	Scrapped	391
	5.4	Backordering		401
		5.4.1	Defective Items	401
		5.4.2	Multidefective Types	409
		5.4.3	Interruption in Manufacturing Process	419
		5.4.4	Immediate Rework Process	433
		5.4.5	Repair Failure	443
	5.5	Partial Backordering		449
		5.5.1	Rework	449
		5.5.2	Repair Failure	460
		5.5.3	Scrapped	461
		5.5.4	Immediate Rework	470
		5.5.5	Preventive Maintenance	479
	5.6	Conclusion		494
	References			494
6	**Quality Considerations**			497
	6.1	Introduction		497
	6.2	Literature Review		500
	6.3	EOQ Model with No Return		501
		6.3.1	No Return Without Shortage	502
		6.3.2	Two Quality Levels with Backordering	502
		6.3.3	Learning in Inspection with Backordering	506
		6.3.4	Partial Backordering	506
	6.4	EOQ Model with Return		515
		6.4.1	Inspection and Sampling	515
		6.4.2	Inspection Error	522
		6.4.3	Different Defective Quality Levels and Partial Backordering	525
	6.5	EPQ Model Without Return		534
		6.5.1	Quality Assurance Without Shortage	535
		6.5.2	Quality Screening and Rework Without Shortage	536

6.6	EPQ Model with Return		537
	6.6.1	Continuous Quality Characteristic Without Shortage	537
	6.6.2	Inspections Errors Without Shortage	542
6.7	Conclusion		546
References			547
7 Maintenance			**549**
7.1	Introduction		549
7.2	No Shortage		550
	7.2.1	Preventive Maintenance	550
	7.2.2	Imperfect Preventive Maintenance	561
	7.2.3	Imperfect Maintenance and Imperfect Process	569
	7.2.4	Aggregate Production and Maintenance Planning	574
7.3	Backordering		578
	7.3.1	Preventive Maintenance	578
7.4	Partial Backordering		582
	7.4.1	Preventive Maintenance	582
7.5	Conclusion		582
References			582
Index			**585**

Chapter 1
Introduction

In today's manufacturing environment, managing inventories is one of the basic concerns of enterprises dealing with materials according to their activities, because material as the principal inventories of enterprises specially production ones composes the large portion of their assets. As a result, managing inventories influences directly financial, production, and marketing segments of enterprises so that efficient management of inventories leads to improving their profits. In addition, the effect of managing inventories on the selling prices of finished products is undeniable because more than half of production systems' revenues are spent to buy materials or production components. On the other hand, customers expect to receive their orders at a lower price apace. So, an efficient managing inventories and production planning are key managerial and operational tools to achieve the main goals, which are satisfying the customers' demand and becoming lower-cost producer, in order to increase market share.

Economic production quantity (EPQ) model is a well-known economic lot size model used in production enterprises that internally produce products. However, traditional EPQ model is utilized for perfect production process to determine the optimal production lot size so that overall production/inventory costs are minimized. In reality, a perfect production run rarely exists. Breakdown is an inevitable issue in production processes. Indeed, after a production period, a production process often shifts to out-control state owing to machine wear or corrosion which leads to generating defective items with loss cost. In order to reimburse these costs, some production strategies including reworking and repairing defective items, quality control, and maintenance planning to reduce the defective or scrape item costs are employed. So, the main prophecy of this book is to introduce all mentioned production strategies which can lessen unexpected imperfect item costs. The main focus of this book is to introduce mathematical models of imperfect inventory control systems in which at least one of imperfect items, scraped item, rework policy, quality control or maintenance planning may be used. In the following a brief introduction about each chapter is presented.

1.1 Imperfect Items

Since the introduction of the economic order quantity (EOQ) model by Harris (1913), frequent contributions have been made in the literature toward the development of alternative models that overcome the unrealistic assumptions embedded in the EOQ formulation. For example, the assumption related to the perfect quality items is technologically unattainable in most supply chain applications (Cheng 1991). In contrast, products can be categorized as "good quality," "good quality after reworking," "imperfect quality," and "scrap" (Chan et al. 2003). In practice, the presence of defective items in raw material or finished products inventories may deeply affect supply chain coordination, and, consequently, the product flows among supply chain levels may become unreliable (Roy et al. 2011). In response to this concern, the enhancement of currently available production and inventory order quantity models, which accounts for imperfect items in their mathematical formulation, has become an operational priority in supply chain management (Khan et al. 2011). This enhancement may also include the knowledge transfer between supply chain entities in order to reduce the percentage of defective items. In the second chapter of this book, the main focus is on introducing several mathematical models of EOQ inventory systems with imperfect items considering different kinds of shortages under different assumptions.

1.2 Scrap

The economic order quantity (EOQ) model was first introduced in 1913. Seeking to minimize the total cost, the model generated a balance between holding and ordering costs and determined the optimal order size. Later, the EPQ model considered items produced by machines inside a manufacturing system with a limited production rate, rather than items purchased from outside the factory. Despite their age, both models are still widely used in major industries. Their conditions and assumptions, however, rarely pertain to current real-world environments. To make the models more applicable, different assumptions have been proposed in recent years, including random machine breakdowns, generation of imperfect and scrap items, and discrete shipment orders. The assumption of discrete shipments using multiple batches can make the EPQ model more applicable to real-world problems. The EPQ inventory models assume that all the items are manufactured with high quality and defective items are not produced. However, in fact, defective items appear in the most of manufacturing systems; in this sense, researchers have been developing EPQ inventory models for defective production systems. In these production systems, defective items are of two types: scrapped items and reworkable items. Usually non-conforming products are scrapped and are removed from the systems' inventories. This strategy is employed for production enterprises in which either imperfect items cannot be repaired or both repair/reworking cost is more than their selling revenues. In turn, enterprises prefer to reject imperfect items instead of performing reworking/repair procedure. In the third chapter of this book, the EPQ model with scraped items under different kinds of shortages and both continuous and discrete delivery are introduced.

1.3 Rework

Rework is one of the key drivers of production designs applied in imperfect production systems in which their production lines face defectives. It helps producers reproduce the non-conforming items, which are detected within/after inspecting process, and sell them as healthy ones. Although a reworking process makes an additive cost for production companies, it causes the producers to profit from buying the reworked items more than their reworking costs, so they prefer to rework the imperfect items in order to reduce their unexpected expenses. In the fourth chapter of this book, rework process in imperfect EPQ model under different assumptions is introduced. Indeed, several mathematical models of EPQ problem with defective and rework process are presented.

1.4 Multi-product Single Machine

The economic production quantity (EPQ) is a commonly used production model that has been studied extensively in the past few decades. One of the considered constraints in the EPQ inventory models is producing all items by a single machine. Since all of the products are manufactured on a single machine with a limited capacity, a unique cycle length for all items is considered. It is assumed there is a real constant production capacity limitation on the single machine on which all products are produced. If the rework is placed, both the production and rework processes are accomplished using the same resource, the same cost, and the same speed. The first economic production quantity inventory model for a single-product single-stage manufacturing system was proposed by Taft (1918). Perhaps Eilon (1985) and Rogers (1958) were the first researchers that studied the multi-products single manufacturing system. Eilon (1985) proposed a multi-product lot-sizing problem classification for a system producing several items in a multi-product single-machine manufacturing system. In the fifth chapter of this book, multi-product single-machine EPQ model with defective and scraped items and also rework process under different assumptions are presented.

1.5 Quality Considerations

Traditional economic order quantity (EOQ) models offer a mathematical approach to determine the optimal number of items a buyer should order to a supplier each time. One major implicit assumption of these models is that all the items are of perfect quality (Rezaei and Salimi 2012). However, presence of defective products in manufacturing processes is inevitable. There is no production process which can guarantee that all its products would be perfect and free from defect. Hence, there is a yield for any production process. Basic and classical inventory control models usually ignore this fact. They assume all output products are perfect and with equal quality; however, due to the limitation of quality control procedures, among other factors, items of imperfect quality are often present. So it has given researchers the opportunity to relax this assumption and apply a yield to investigate and study its impact on several variables of inventory models such as order quantity and cycle time. In the sixth chapter of this book, several inventory control models under quality considerations such as sampling, inspections, return, etc. with different assumptions of inventory systems are presented.

1.6 Maintenance

The role of the equipment condition in controlling quality and quantity is well-known (Ben-Daya and Duffuaa 1995). Equipment must be maintained in top operating conditions through adequate maintenance programs. Despite the strong link between maintenance production and quality, these main aspects of any manufacturing system are traditionally modeled as separate problems. In the last chapter of this book, maintenance and inventory systems are considered together, and several mathematical models are presented.

References

Ben-Daya, M., & Duffuaa, S. O. (1995). Maintenance and quality: The missing link. *Journal of Quality in Maintenance Engineering, 1*(1), 20–26.

Chan, W. M., Ibrahim, R. N., & Lochert, P. B. (2003). A new EPQ model: Integrating lower pricing, rework and reject situations. *Production Planning & Control, 14*(7), 588–595.

Cheng, T. C. E. (1991). EPQ with process capability and quality assurance considerations. *Journal of the Operational Research Society, 42*(8), 713–720.

References

Eilon, S. (1985). Multi-product batch production on a single machine—A problem revisited. *Omega, 13*, 453–468.

Harris, F. W. (1913). What quantity to make at once. In *The library of factory management. Operation and costs* (The factory management series) (Vol. 5, pp. 47–52). Chicago, IL: A.W. Shaw Co..

Khan, M., Jaber, M. Y., & Bonney, M. (2011). An economic order quantity (EOQ) for items with imperfect quality and inspection errors. *International Journal of Production Economics, 133*(1), 113–118.

Rezaei, J., & Salimi, N. (2012). Economic order quantity and purchasing price for items with imperfect quality when inspection shifts from buyer to supplier. *International Journal of Production Economics, 137*(1), 11–18.

Rogers, J. (1958). A computational approach to the economic lot scheduling problem. *Management Science, 4*, 264–291.

Roy, M. D., Sana, S. S., & Chaudhuri, K. (2011). An economic order quantity model of imperfect quality items with partial backlogging. *International Journal of Systems Science, 42*(8), 1409–1419.

Taft, E. W. (1918). The most economical production lot. *Iron Age, 101*(18), 1410–1412.

Chapter 2
Imperfect EOQ System

2.1 Introduction

Since the introduction of the economic order quantity (EOQ) model by Harris (1913), frequent contributions have been made in the literature toward the development of alternative models that overcome the unrealistic assumptions embedded in the EOQ formulation. For example, the assumption related to the perfect-quality items is technologically unattainable in most supply chain applications. In contrast, products can be categorized as "good quality," "good quality after reworking," "imperfect quality," and "scrap" (Chan et al. 2003; Pal et al. 2013). In practice, the presence of defective items in raw material or finished product inventories may deeply affect supply chain coordination, and, consequently, the product flows among supply chain levels may become unreliable (Roy et al. 2015). In response to this concern, the enhancement of currently available production and inventory order quantity models, which accounts for imperfect items in their mathematical formulation, has become an operational priority in supply chain management (Khan et al. 2011). This enhancement may also include the knowledge transfer between supply chain entities in order to reduce the percentage of defective items (Adel et al. 2016).

Also some related works can be found in Hasanpour et al. (2019), Keshavarz et al. (2019), Taleizadeh et al. (2015, 2016a, 2018a, b), Taleizadeh and Zamani-Dehkordi (2017a, b), Salameh and Jaber (2000), Maddah and Jaber (2008), and Papachristos and Konstantaras (2006).

The EOQ models with imperfect-quality items in three categories are categorized and their subcategories are shown in Fig. 2.1.

The common notations of imperfect EOQ models are shown in Table 2.1.

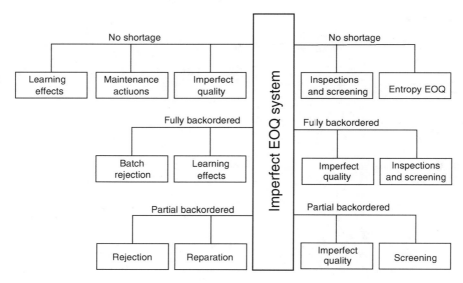

Fig. 2.1 Categories of EOQ model of imperfect-quality items

2.2 Literature Review

The academic literature related to inventory control for imperfect-quality items is multidisciplinary in nature and, for reviewing/presentation purposes in this chapter, is thematically organized around two main streams: (1) deterioration, perishability, and shelf lifetime constraints and (2) model formulations and related solution techniques that consider imperfect-quality items (Adel et al. 2016).

2.2.1 Deterioration, Perishability, and Lifetime Constraints

The terms "deterioration," "perishability," and "obsolescence" are used interchangeably in the literature and may often be perceived as ambiguous because they are linked to particular underlying assumptions regarding the physical state/fitness and behavior of items over time. Usually, deterioration refers to the process of decay, damage, or spoilage of a product, i.e., the product loses its value of characteristics and can no longer be sold/used for its original purpose (Wee 1993). In contrast, an item with a fixed lifetime perishes once it exceeds its maximum shelf lifetime and then must be discarded (Ferguson and Ketzenberg 2005). Obsolescence incurs a partial or a total loss of value of the on-hand inventory in such a way that the value for a product continuously decreases with its perceived utility (Song and Zipkin 1996; Also some related works can be found in works of Nobil, et al. (2019), Lashgary et al. (2016, 2018), Kalantary and Taleizadeh (2018), Diabat et al.

2.2 Literature Review

Table 2.1 Notations

P	Production rate (units per unit time)
D	Demand rate (units per unit time)
R	Repair rate (units per unit time)
s	Selling price for good-quality items ($/unit)
v	Selling price for imperfect or salvage value per items ($/unit)
C	Production/purchasing cost ($/unit)
C_R	Rework cost per unit ($/unit)
C_J	Reject cost per unit (including transportation, handling, and damage cost) ($/unit)
C_b	Backordering cost ($/unit/unit time)
C_T	Transportation cost per unit ($/unit)
C_d	Disposal cost per unit ($/unit)
g	Goodwill cost per unit ($/unit)
$\hat{\pi}$	Lost sale cost per unit ($/unit)
K	Fixed setup/ordering cost ($/lot)
K_S	Fixed transportation or shipment cost ($/lot)
h	Holding cost per unit per unit time ($/unit/unit time)
h_1	The holding cost for defective items per unit per unit time ($/unit/unit time)
h_R	The holding cost for reworked items per unit per unit time ($/unit/unit time)
γ	Fraction of imperfect items (percent)
C_I	The unit screening or inspection cost ($/unit)
x	Inspection rate (units per unit time)
p	Imperfect rate (units per unit time)
$E[p]$	Expected imperfect rate
$f(p)$	Probability density function of p
T	Ordering cycle duration (time)
t	Screening time (time)
$f(\gamma)$	Probability density function of imperfect products (γ)
y	Production/ordering quantity (unit)
B	Backordered level (unit)
β	Partial backordering rate (%) $0 < \beta \leq 1$
$E[.]$	Expected value of a random variable

(2017), Mohammadi et al. (2015), Tat et al. (2015), Hasanpour et al. (2019), Taleizadeh (2014), Taleizadeh and Rasouli-Baghban, (2015, 2018), Taleizadeh et al. (2013a, b, 2015, 2016, 2019), Taleizadeh and Nemattolahi (2014) and Tavakkoli and Taleizadeh (2017), Bakker et al. (2012).

2.2.2 Imperfect-Quality Items

The classical EOQ has been a widely accepted model for inventory control purposes due to its simple and intuitively appealing mathematical formulation. However, it is true to say that the operation of the model is based on a number of explicitly or implicitly made unrealistic mathematical assumptions that are never actually met in practice (Jaber et al. 2004). Salameh and Jaber (2000) developed a mathematical model that permits some of the items to drop below the quality requirements, i.e., a random proportion of defective items are assumed for each lot size shipment, with a known probability distribution. The researchers assumed that each lot is subject to a 100% screening, where defective items are kept in the same warehouse until the end of the screening process and then can be sold at a price lower than that of perfect-quality items. Huang (2004) developed a model to determine an optimal integrated vendor–buyer inventory policy for flawed items in a just-in-time (JIT) manufacturing environment. Maddah and Jaber (2008) developed a new model that rectifies a flaw in the one presented by Salameh and Jaber (2000) using renewal theory. Jaber et al. (2008) extended it by assuming that the percentage defective per lot reduces according to a learning curve.

Jaggi and Mittal (2011) investigated the effect of deterioration on a retailer's EOQ when the items are of imperfect quality. In their research, defective items were assumed to be kept in the same warehouse until the end of the screening process. Jaggi et al. (2011) and Sana (2012) presented inventory models, which account for imperfect-quality items under the condition of permissible delay in payments. Moussawi-Haidar et al. (2014) extended the work of Jaggi and Mittal (2011) to allow for shortages.

In a real manufacturing environment, the defective items are not usually stored in the same warehouse where the good items are stored. As a result, the holding cost must be different for the good items and the defective ones (e.g., Paknejad et al. 2005). With this consideration in mind, Wahab and Jaber (2010) presented the case where different holding costs for the good and defective items are assumed. They showed that if the system is subject to learning, then the lot size with the same assumed holding costs for the good and defective items is less than the one with differing holding costs. When there is no learning in the system, the lot size with differing holding costs increases with the percentage of defective items. For more details about the extensions of a modified EOQ model for imperfect-quality items, see Khan et al. (2011).

Here are some main models in literature with their mathematical model, solution procedure, and numerical examples. In the next sections, these models starting from the basic to complicated ones are presented. First, EOQ models with imperfect quality items are studied considering no shortage, back-ordering shortage, and partial back-ordering.

2.3 EOQ Model with No Shortage

2.3.1 *Imperfect Quality*

In this section, two imperfect EOQ models developed by Salameh and Jaber (2000) and Maddah and Jaber (2008) are presented. Consider the EOQ model with a

2.3 EOQ Model with No Shortage

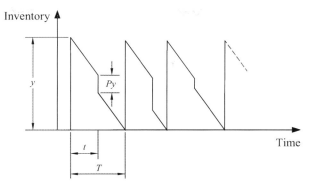

Fig. 2.2 Inventory level (Salameh and Jaber 2000; Maddah and Jaber 2008)

demand rate of D units per unit time. An order of size y is placed every time the inventory level reaches zero and is assumed to be delivered instantaneously. The fixed ordering cost is K, the fixed shipping of imperfect-quality items is K_S, the unit purchasing cost is C, and the inventory holding cost is h per unit per unit time. Each order contains a fraction P of defective items, a random variable with support in [0, 1]. Each order is subjected to a 100% inspection process at a rate of x units per unit time, $x \cdot D$. The screening cost is d per unit. Upon completion of the screening process, items of imperfect quality are sold as a single batch at a reduced price of v per unit. The price of a perfect-quality item is s per unit, $s < v$.

The behavior of the inventory level in an ordering cycle is shown in Fig. 2.2, where T is the ordering cycle duration ($T = (1 - p)y/D$, and $t = y/x$). Salameh and Jaber (2000) assumed that $(1 - p)y \geq D \cdot t$, or, equivalently, $p \leq 1 - D/x$, in order to avoid shortages. Under the above assumptions, the expected profit is presented as:

$$\mathrm{TP}(y) = \overbrace{sy(1-p) + vyp}^{\text{Revenue}} - \overbrace{K}^{\text{Fixed cost}} - \overbrace{Cy}^{\text{Purchasing cost}} - \overbrace{C_1 y}^{\text{Inspection cost}} - \overbrace{h\left(\frac{[y(1-p)]^2}{2D} + \frac{py^2}{x}\right)}^{\text{Holding cost}} \tag{2.1}$$

Then the expected profit per unit time is derived as:

$$E[\mathrm{TPU}(y)] = E\left[\frac{\mathrm{TP}(y)}{T}\right] \tag{2.2}$$

After some simplifications,

$$E[\text{TPU}(y)] = D\left(s - v + h\frac{y}{x}\right) + D\left(v - \frac{hy}{x} - C - C_1 - \frac{K}{y}\right) \times E\left[\frac{1}{1-p}\right]$$
$$- \frac{hy(1 - E[p])}{2} \tag{2.3}$$

And the optimal order quantity is derived as:

$$y^{SJ} = \sqrt{\frac{2KDE[1/(1-p)]}{h[1 - E[p] - 2D(1 - E[1/(1-p)])/x]}} \tag{2.4}$$

Then, Maddah and Jaber (2008) corrected Eq. (2.2) as:

$$E[\text{TPU}(y)] = \frac{E[\text{TP}(y)]}{E[T]} \tag{2.5}$$

and derived a new expected profit function as:

$$E[\text{TP}(y)] = sy(1 - E[p]) + vyE[p] - K - Cy - C_1 y$$
$$- h\left(\frac{y^2 E\left[(1-p)^2\right]}{2D} + \frac{E[p]y^2}{x}\right)$$

Since $E[T] = (1 - E[p])y/D$, then Eq. (2.5) is rearranged as:

$$E[\text{TPU}(y)] = \frac{[s(1 - E[p]) + vE[p] - C - C_1]D - KD/y - hy(E\left[(1-p)^2\right]/2 + E[p]D/x)}{1 - E[p]}$$
$$\tag{2.6}$$

After proofing the concavity of Eq. (2.6) with respect to y, the optimum order size is derived as:

$$y^* = \sqrt{\frac{2KD}{h\left(E\left[(1-p)^2\right] + 2E[p]D/x\right)}} \tag{2.7}$$

The expected profit in Eq. (2.6) has several terms independent of y. In subsequent analysis, these are dropped, and the objective function is redefined in terms of minimizing the expected "relevant" cost per unit time as (Maddah and Jaber 2008):

2.3 EOQ Model with No Shortage

$$\text{EC}(y) = \frac{1}{1 - E[p]} \left[KD/y + hy \left(E\left[(1-p)^2\right]/2 + E[p]D/x \right) \right] \quad (2.8)$$

Maddah and Jaber (2008) showed that for a large inspection rate, the optimum order size in Eq. (2.7) converges to:

$$y^* = \sqrt{\frac{2KD}{h\left(E\left[(1-p)^2\right]\right)}} \quad (2.9)$$

In real cases, it is not optimal to ship imperfect-quality items as a single batch in each ordering cycle (Maddah and Jaber 2008). So Maddah and Jaber assumed that shipping any number of imperfect-quality batches has a fixed cost of K_S and developed their previous model under multiple batches. Now the decision variables are order size (y) and ordering cycle number (n), and they derived the expected cost per unit time of perfect-quality items similar to Eq. (2.8) as below:

$$E[CP(y)] = \left[KD/y + hy\left(E\left[(1-p)^2\right]/2\right)\right]/(1 - E[p])$$

Figure 2.3 corresponds to the case when $n = 3$, where the imperfect-quality inventory is held for two ordering cycles and then shipped upon completing the screening of the last order. According to Maddah and Jaber (2008), let T_i be the duration of ordering period i of a shipping cycle, $i = 1, \ldots, n$. Note that $T_i = (1 - P_i)y/D$, where P_i is the fraction of imperfect-quality items in order i of a shipping cycle. Then, the expected holding cost per shipping cycle is:

$$\text{ECI}_h(y,n) = hE \left[\underbrace{\sum_{i=1}^{n-1} P_i y (1-P_i) y/D}_{\substack{\text{Imperfect quality inventory} \\ \text{cost from an order} \\ \text{carried over the ordering} \\ \text{period of the order itself}}} + \underbrace{\sum_{i=1}^{n-2} P_i y \sum_{j=i+1}^{n-1} (1-P_j)y/D}_{\substack{\text{Imperfect inventory cost from an order} \\ \text{carried through subsequent ordering} \\ \text{periods during a shipping cycle} \\ \text{excluding the nth ordering period}}} + \underbrace{\sum_{i=1}^{n} \frac{P_i y^2}{x}}_{\substack{\text{Imperfect inventory cost accumulated} \\ \text{in the nth period, which is carried} \\ \text{for a duration of } y/x \\ \text{before being shipped}}} \right]$$

$$(2.10)$$

Assuming that $P_1, \ldots, P_n$ are independent and identically distributed, the expression for $\text{ECI}_h(y, n)$ in Eq. (2.10) after some simplifications changes to:

$$\text{ECI}_h(y, n) = \frac{hy^2}{D}\left[\frac{n(n-1)}{2}\times E[p](1-E[p]) + nE[p]D\frac{y}{x} - (n-1)\text{var}[p]\right]$$

Knowing $E\left[\sum_{i=1}^{n}(1-P_i)y/D\right] = n(1-E[p])y/D$, the expected imperfect-quality item cost per unit time is:

$$E[CI(y,n)] = \frac{1}{1-E[p]}\left\{\frac{K_S}{n}\frac{D}{y}\right.$$
$$\left. + hy\left[\frac{(n-1)}{2}E[p](1-E[p]) + E[p]\frac{D}{x} - \frac{n-1}{n}\text{var}[p]\right]\right\}$$

And total cost will be:

$$E[\text{TC}(y,n)] = \frac{1}{1-E[p]}\left\{\left(K + \frac{K_S}{n}\right)\frac{D}{y} + \frac{hy}{2}\left[E\left[(1-p)^2\right] - \frac{2(n-1)}{n}\text{var}[p]\right.\right.$$
$$\left.\left. + (n-1)E[p](1-E[p]) + 2E[p]\frac{D}{x}\right]\right\}$$

(2.11)

Because of convexity of Eq. (2.11), one can easily derive that:

$$y^*(n) = \sqrt{\frac{2(K+(K_S/n))D}{\underbrace{h\left[E\left[(1-p)^2\right] - (2(n-1)/n)\text{var}[p] + (n-1)E[p](1-E[p]) + 2E[p](D/x)\right]}_{\gamma(n)}}}$$

(2.12)

Maddah and Jaber (2008) showed that optimal values of n can be found by optimizing the expected total cost presented in Eq. (2.13) which is presented in Eq. (2.14):

$$\text{ECT}_1(n) = \text{ECT}(n, y^*(n)) = \sqrt{2\kappa(n)\gamma(n)D} \qquad (2.13)$$

$$\tilde{n} = \sqrt{\frac{K_1\left[E\left[(1-p)^2\right] - 2(1-K/K_S)\text{var}[p] - E[p](1-E[p]) + 2E[p](D/x)\right]}{KE[p](1-E[p])}}$$

(2.14)

Then, the optimal value of n is one of the two integers which is closest to $\tilde{n}$, whichever leads to lower value of $\text{ECT}_1(n)$. That is, $n^* = \text{argmin}(\text{ECT}_1(n))$ where $[x]$ is the largest integer $\leq x$ and $[x]$ is the smallest integer $\geq x$. Finally, the optimal order quantity is found from Eq. (2.12) as $y^* = y^*(n^*)$ (Maddah and Jaber 2008).

2.3 EOQ Model with No Shortage

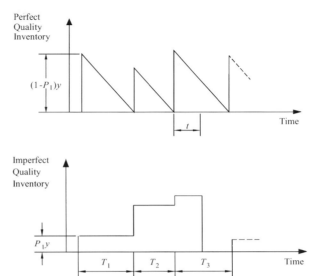

Fig. 2.3 Perfect and imperfect inventory levels when shipments are consolidated, $n = 3$ (Maddah and Jaber 2008)

Example 2.1 Maddah and Jaber (2008) developed numerical results similar to those in Salameh and Jaber (2000). This illustrates the application of their model and allows comparing their results with those of Salameh and Jaber. Consider a situation with the following parameters: demand rate, $D = 50{,}000$ units/year; ordering cost, $K = \$100$/cycle; holding cost, $h = \$5$/unit/year; screening rate, $x = 175{,}200$ units/year; screening cost, $C_I = \$0.5$/unit; purchasing cost, $C = \$25$/unit; selling price of good-quality items, $s = \$50$/unit; selling price of imperfect-quality items, $v = \$20$/unit; and the fraction of imperfect-quality item, p, uniformly distributed on (a, b), $0 < a < b < 1$, i.e., $P \sim (a, b)$. With $p \sim U(a, b)$, $E[p] = (a + b)/2$, $\mathrm{Var}[p] = (b - a)/12$ and:

$$E\left[(1-p)^2\right] = \frac{1}{b-a}\int_a^b (1-p)^2 dp = \frac{a^2 + ab + b^2}{3} + 1 - a - b \quad (2.15)$$

Assuming $a = 0$ and $b = 0.04$, then the optimal order quantity using Eq. (2.7) becomes $y^* = 1434$ units, and the related cost from Eq. (2.6) is $E[\mathrm{TPU}(y^*)] = \$1{,}212{,}274$.

Assuming shipping of imperfect-quality items has a fixed cost of $K_S = \$50$ with same values for other parameters used in the previous example, in the following, the continuous value of n that minimizes $\mathrm{ECT}_1(n)$ is $\tilde{n} = 4.93$. So, n^* is either 4 or 5 where $\mathrm{ECT}_1(4) = 7614 > \mathrm{ECT}_1(5) = 7600$. So, $n^* = 5$. The optimal order quantity is then given from Eq. (2.11) as $y^*(5) = 1447$.

2.3.2 Maintenance Actions

The objective of the analysis in this section is to determine the optimal lot size $y*$ such that the expected total cost is minimized when maintenance and reworking actions are taken into account. For describing this section, some new notations are used as presented in Table 2.2 (Porteus 1986).

Before presenting the model, first we should take notice of the remark presented by Hou et al. (2015) in Eq. (2.16). He derived the expected number of unhealthy item as below:

$$E(N) = \theta\left(y - \sum_{j=1}^{y} q^{-j}\right) \quad (2.16)$$

and

$$\Pr\{X = j\} = \begin{cases} q^{-j}q, 0 \leq j \leq y \\ q^{-y}, j = y \end{cases} \quad (2.17)$$

Then, they showed that:

$$E(X) = q\sum_{j=1}^{y-1} jq^{-j} + yq^{-y} = \sum_{j=1}^{y} q^{-j} \quad (2.18)$$

Finally, the number of defective items in y is $N = \theta(y - X)$ and the $E(N)$ is what presented in Eq. (2.16). Now based on Eq. (2.18), the expected cyclic cost of rework process will be:

$$C_R E(N) = C_R \theta\left(y - \sum_{j=1}^{y} q^{-j}\right) \quad (2.19)$$

Since the related cost to maintenance should be considered when the manufacturing process is out-of-control at the end of a production uptime for a lot of size y, the expected cyclic-related cost is:

Table 2.2 Notations of a given problem

Q	The probability that the system from in-control state shifts to out-of-control state
q	The probability that the system stays in-control state during the production of an item and $\bar{q} = 1 - q$
θ	The percentage of defective items when the process is in the out-of-control state
X	Random variable representing number of items produced in the in-control state
C_m	Maintenance cost per unit (\$/unit)

2.3 EOQ Model with No Shortage

$$C_m(1 - q^{-y}) \tag{2.20}$$

So the cyclic total cost is (Hou et al. 2015):

$$\text{TC}(y) = \underbrace{K}_{\text{Fixed cost}} + \underbrace{\frac{hy^2}{2D}}_{\text{Holding cost}} + \underbrace{C_m(1 - q^{-y})}_{\text{Maintenance cost}} + \underbrace{C_R \theta \left(y - \sum_{j=1}^{y} q^{-j} \right)}_{\text{Rework cost}} \tag{2.21}$$

And the expected cost per unit of time becomes:

$$f(y) = \text{TC}(y)/T$$

$$= \frac{DK}{y} + \frac{h}{2}y + C_R D\theta + \frac{D}{y}\left[C_m(1 - q^{-y}) - C_R \theta \sum_{j=1}^{y} q^{-j} \right] \tag{2.22}$$

It should be noticed that for $q = 0$, the production system is always in the in-control state and the produced items are healthy, and Eq. (2.22) reverses to the traditional EOQ model with healthy item. But $C_m = 0$ means all produced items are defective, and Eq. (2.22) will reduce to the approximated model (using Taylor series expansion) in Porteus (1986) as presented in Eq. (2.23):

$$f_p(y) = \frac{DK}{y} + \frac{y}{2}(h + C_R Dq) \tag{2.23}$$

and derive an approximately optimal lot size as follows (Hou et al. 2015):

$$y_p^* = \sqrt{\frac{2DK}{h + C_R Dq}} \tag{2.24}$$

Since Eq. (2.23) was not a good approximation, Hou et al. (2015) presented a comprehensive method to derive the optimal values. They provided the bounds for searching the optimal lot size y∗ that minimizes $f(y)$ of Eq. (2.22) as $\beta = C_m - \frac{C_R \theta q}{q}$ and using necessary condition for optimal points ($f'(y^*) = 0$) derived optimal values. They prove that y∗ exists and is unique when q equals 0 or 1 such that $f'(y^*) = 0$ satisfies. But for $0 < q < 1$, let:

$$g(y) = y^2 f'(y^*) = -DK + \frac{h}{2}y^2 - D\beta(1 - q^{-y} + yq^{-y} \ln q^{-}) \tag{2.25}$$

since $g(y)$ is a continuous function with $\lim_{y \to 0^+} g(y) = -D$ and $\lim_{y \to \infty} g(y) = \infty > 0$. Furthermore, the first derivative of $g(y)$ is given by Hou et al. (2015):

$$g'(y) = y\left[h - D\beta(\ln q^-)^2 q^{-y}\right] \qquad (2.26)$$

After some algebra, Hou et al. (2015) proposed the optimal lot size $y*$ when $0 < q < 1$:

$$y_1 = \sqrt{\frac{2D(K + C_m)}{h}} \quad y_2 = \sqrt{\frac{2DK}{h}} \qquad (2.27)$$

And proved that:

$$\begin{array}{l} \text{if} \quad \beta \leq 0 \rightarrow 0 < y \leq y_2^* \leq y_1^* \\ \text{if} \quad \beta > 0 \rightarrow 0 < y_2^* < y < y_1^* \end{array} \qquad (2.28)$$

Using Eqs. (2.27) and (2.28), the following algorithm is proposed to find the optimal values.

Algorithm 2.1 *Step 1*: Let $\varepsilon > 0$, and compute β, y_2, and y_1.
Step 2: If $\beta \leq 0$, set $y_L = 0$, $y_U = y_2$; otherwise, set $y_L = y_2$, $y_U = y_1$.
Step 3: Set $y_{opt} = \frac{y_L + y_U}{2}$.
Step 4: If $|g(y_{opt})| < \varepsilon$, go to Step 6; otherwise, go to Step 5.
Step 5: If $|g(y_{opt})| < 0$, set $y_L = y_{opt}$; however, if $|g(y_{opt})| > 0$, $y_U = y_{opt}$. Then, go to Step 3.
Step 6: Set $y* = y_{opt}$ and compute $f(y*)$.

Example 2.2 Consider $K = \$600$/cycle, $h = \$8$/unit/year, $D = 1000$ units/year, and $C_R = \$5$/unit, $C_m = \$200$/cycle, $\theta = 0.75$, and $\theta = 0.1$. Then, it can be verified that $\beta = 166.25$. Using Algorithm 2.1, $y* = 437.68$ units and $f(y*) = \$7251.43$ (Hou et al. 2015).

2.3.3 Screening Process

In this section, now consider a general EOQ model for items with imperfect quality under varying demand, defective items, screening process, and deterioration rates for an infinite planning horizon presented by Alamri et al. (2016). Assume that each lot is subject to a 100% screening where items that are not conforming to certain quality standards are stored in a different warehouse. Therefore, different holding costs for the good and defective items are considered. Items deteriorate while they are in storage, with demand, screening, and deterioration rates being arbitrary functions of time. The percentage of defective items per lot reduces according to a learning curve. After a 100% screening, imperfect-quality items may be sold at a discounted price as a single batch at the end of the screening process or incur a disposal penalty charge. Moreover, a general step-by-step solution procedure is provided for continuous intra-cycle periodic review applications.

2.3 EOQ Model with No Shortage

Table 2.3 Notations of a given problem

$D(t)$	Demand rate (units per unit time)
$x(t)$	Screening rate (units per unit time)
$\delta(t)$	Deterioration rate (units per unit time)
p_j	The percentage defective per lot reduces according to a learning curve
j	Cycle index ($j = 1, 2, \ldots$)
Q_j	Lot of size delivered at the beginning of each cycle j (unit)

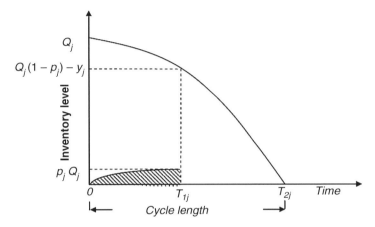

Fig. 2.4 Inventory variation of an economic order quantity (EOQ) model for one cycle (Alamri et al. 2016)

Some related notation for this problem is presented in Table 2.3.

Alamri et al. (2016) assumed that a single item held in stock lead time is zero and no restrictions exist. Moreover, any order arrives before the end of that same cycle.

In order to avoid the shortage, Alamri et al. (2016) assumed $(1 - p_j)x(t) \geq D(t)$, $\forall t \geq 0$. *Lot size* covers both deterioration and demand during both the first phase (screening) and the second phase (non-screening). Each lot is subjected to a 100% screening process that starts at the beginning of the cycle and ceases by time T_{1j}, by which point in time Q_j units have been screened and y_j units have been depleted, which is the summation of demand and deterioration. During this phase, items not conforming to certain quality standards are stored in a different warehouse. The variation in the inventory level during the first and second phase (please refer to Fig. 2.4) and the variation in the inventory level for the defective items (Alamri et al. 2016) are presented in Eq. (2.29):

$$\frac{dI_{gj}(t)}{dt} = -D(t) - p_j x(t) - \delta(t) I_{gj}(t), \quad 0 \le t \le T_{1j} \tag{2.29}$$

Using boundary condition $I_{gj}(0) = Q_j$:

$$Q_j = \int_0^{T_{1j}} x(u) du, \tag{2.30}$$

$$\frac{dI_{gj}(t)}{dt} = -D(t) - \delta(t) I_{gj}(t), \tag{2.31}$$

And with the boundary condition $I_{gj}(T_{2j}) = 0$:

$$\frac{dI_{dj}(t)}{dt} = p_j x(t), \quad 0 \le t \le T_{1j} \tag{2.32}$$

knowing $I_{dj}(0) = 0$.

After some complicated algebra, the solutions of the above differential equations are (Alamri et al. 2016):

$$I_{gj}(t) = e^{-(g(t)-g(0))} \int_0^{T_{1j}} x(u) du - e^{-g(t)} \int_0^t [D(u) + p_j x(u)] e^{g(u)} du, \quad 0 \le t$$
$$\le T_{1j} \tag{2.33}$$

$$I_{gj}(t) = e^{-g(t)} \int_t^{T_{2j}} D(u) e^{g(u)} du, \quad 0 \le t \le T_{1j} \tag{2.34}$$

$$I_{gj}(t) = \int_0^t p_j x(u) du, \quad 0 \le t \le T_{1j} \tag{2.35}$$

$$g(t) = \iota \zeta \delta(t) dt \tag{2.36}$$

The per cycle cost components for the given inventory system are as follows. The total purchasing cost during the cycle $= C \int_0^{T_{1j}} x(u) du$. Note that this cost includes the defective and deterioration costs. Holding cost $= h[I_{gj}(0, T_{1j}) + I_{gj}(T_{1j}, T_{2j})] + h_1 I_{dj}(0, T_{1j})$. Thus, the total cost per unit time of the underlying inventory system during the cycle $[0, T_{2j}]$, as a function of T_{1j} and T_{2j}, say $Z(T_{1j}, T_{2j})$ is given by:

$$G(t) = \zeta e^{g(t)} dt \tag{2.37}$$

Our objective is to find T_{1j} and T_{2j} that minimize $Z(T_{1j}, T_{2j})$. However, the variables T_{1j} and T_{2j} are related to each other as follows:

2.3 EOQ Model with No Shortage

$$0 < T_{1j} < T_{2j} \tag{2.38}$$

$$e^{g(0)} \int_0^{T_{1j}} x(u)du = \int_0^{T_{2j}} D(u)e^{g(u)}du + \int_0^{T_{1j}} p_j x(u)e^{g(u)}du \tag{2.39}$$

Thus, their goal is to solve the following optimization problem, which they shall call problem (m).

$(m) = \{$minimize $Z(T_{1j}, T_{2j})$ given by Eq. (2.37) subject to Eq. (2.39) and $h_j = 0$:

$$h^j = e^{g(0)} \int_0^{T_{1j}} x(u)du - \int_0^{T_{1j}} p_j x(u)e^{g(u)}du - \int_0^{T_{2j}} D(u)e^{g(u)}du \tag{2.40}$$

It can be noted from Eq. (2.40) that $T_{1j} = 0 \Rightarrow T_{2j} = 0$ and $T_{1j} > 0 \Rightarrow T_{1j} < T_{2j}$. Thus Eq. (2.40) implies constraint (Eq. 2.38). Consequently, if they temporarily ignore the monotony constraint (Eq. 2.38) and call the resulting problem as (m_1), then it does satisfy any solution of (m_1). Hence, (m) and (m_1) are equivalent. Moreover, $T_{1j} > 0 \Rightarrow$ RHS of Eq. (2.33) > 0, i.e., Eq. (2.39) guarantees that the number of good items is at least equal to the demand during the first phase.

First, Alamri et al. (2016) noted from Eq. (2.30) that T_{1j} can be determined as a function of Q_j, say:

$$T_{1j} = f_{1j}(Q_j) \tag{2.41}$$

Taking also into account Eq. (2.40), they found that T_{2j} can be determined as a function of T_{1j}, and thus of Q_j, say:

$$T_{2j} = f_{2j}(Q_j) \tag{2.42}$$

Thus, if they substitute Eqs. (2.40)–(2.42) in Eq. (2.36), then problem (m) will be converted to the following unconstrained problem with the variable Q_j (which they shall call problem (m_2)):

$$W(Q_j) = \frac{1}{f_{2j}} \left\{ (C+C_1) \int_0^{f_{1j}} x(u)du + h\left[-G(0)e^{g(0)} \int_0^{f_{1j}} x(u)du \right. \right.$$
$$\left. \left. + \int_0^{f_{1j}} p_j x(u)G(u)e^{g(u)}du + \int_0^{f_{2j}} D(u)G(u)e^{g(u)}du \right] + h_1 \left[\int_0^{f_{1j}} [f_{1j} - u]p_j x(u)du \right] + K \right\} \tag{2.43}$$

Now, the necessary condition for having a minimum for problem (m_2) is:

$$\frac{dW}{dQ_j} = 0 \tag{2.44}$$

Letting $W = \frac{w}{f_{2j}}$, then:

$$\frac{dW}{dQ_j} = \frac{w'_{Q_j} f_{2j} - f'_{2j,Q_j} w}{f_{2j}^2} \tag{2.45}$$

where w'_{Q_j} and f'_{2j,Q_j} are the derivatives of w and f_{2j} w.r.t. Q_j, respectively. Hence, Eq. (2.45) is equivalent to (Alamri et al. 2016):

$$w'_{Q_j} f_{2j} = f'_{2j,Q_j} w \tag{2.46}$$

Also, taking the first derivative of both sides of Eq. (2.40) w.r.t. Q_j, one obtains:

$$e^{g(0)} - p_j e^{g(f_{1j})} = f'_{2j,Q_j} D(f_{2j}) e^{g(f_{2j})} \tag{2.47}$$

From which and Eqs. (2.38)–(2.40) it can be obtained:

$$w'_{Q_j} = (C + C_1) + h\bigl[(G(f_{2j}) - G(0))e^{g(0)} + (G(f_{1j})$$
$$\times\, - G(f_{2j}))p_j e^{g(f_{1j})}\bigr] + \frac{h_1}{x(f_{1j})} \int_0^{f_{1j}} p_j x(u)\, du. \tag{2.48}$$

$$W = \frac{w}{f_{2j}} = \frac{w'_{Q_j}}{f'_{2j,Q_j}} \tag{2.49}$$

where W is given by Eq. (2.40) and w'_{Q_j} is given by Eq. (2.48). Equation (2.49) can be used to determine the optimal value of Q_j and its corresponding total minimum cost and then the optimal values of T_{1j} and T_{2j} (Alamri et al. 2016).

Example 2.3 Alamri et al. (2016) presented an example to illustrate the efficiency of their mathematical model and solution procedures. They considered $x(t) = at + b$, $D(t) = at + r$, $p_j = \frac{\tau}{C_b + e^{\eta}}$, and $\delta(t) = \frac{l}{z - \beta t}$ where $b, d, l, \tau, C_b, z > 0$; $a, r, \gamma, \beta, t \geq 0$; and $\beta t < z$.

Alamri et al. (2016) adopted the values considered in the study by Wahab and Jaber (2010), as presented in Table 2.4.

The optimal values of Q_j^*, T_{1j}^*, T_{2j}^*, and ω_j^*, the corresponding total minimum cost for ten successive cycles, are obtained, and the results are shown in Table 2.5.

2.3 EOQ Model with No Shortage

Table 2.4 Input parameters (Alamri et al. 2016)

Parameter	Value	Parameter	Value
C	100 ($/unit)	α	500 (unit/year)
C_I	0.5 ($/unit)	r	50,000 (unit/year)
h	20 ($/unit/year)	l	1 (unit/year)
h_1	5 ($/unit/year)	z	20 (unit/year)
K	3000 ($/cycle)	β	25 (unit/year)
a	1000 (unit/year)	τ	70.067 (unit/year)
b	100,200 (unit/year)	C_b	819.76 (unit/year)
γ	0.7932 (unit/year)		

Table 2.5 Optimal results for varying demand, screening, and deterioration rates with p_j (Alamri et al. 2016)

j	p_j	T_{1j}	T_{2j}	Q_j^*	$p_j \cdot Q_j^*$	ω_j^*	W_j^*	w_j^*
1	0.08524	0.035424	0.06482	3550	303	5.4	5,585,464	362,030
2	0.08497	0.035419	0.06483	3550	302	5.4	5,583,830	361,980
3	0.08436	0.035407	0.06485	3548	299	5.4	5,580,142	361,850
4	0.08305	0.035380	0.06489	3546	294	5.4	5,572,240	361,580
5	0.08030	0.035324	0.06498	3540	284	5.4	5,555,724	361,020
6	0.07482	0.035212	0.06516	3529	264	5.5	5,523,107	359,900
7	0.06502	0.035013	0.06548	3509	228	5.5	5,465,734	357,890
8	0.05042	0.034715	0.06594	3479	175	5.6	5,382,467	354,900
9	0.03369	0.034376	0.06644	3445	116	5.7	5,290,159	351,490
10	0.01944	0.034088	0.06686	3416	66	5.8	5,214,030	348,600

2.3.4 Learning Effects

2.3.4.1 Different Holding Costs

Salameh and Jaber (2000) developed a model to determine the economic lot size by maximizing the expected total profit per unit time. Each delivered lot has defective items with a known probability function and is screened completely. Then the defective items are sold as a single batch at a discounted price at the end of the screening period (Wahab and Jaber 2010).

In Salameh and Jaber's model, it is observed that they use the same holding cost for both good items and defective items. However, in the real manufacturing environment, the good items and the defective items are treated in a different way. So, the holding cost, $h = iC$, must be different for the good items and the defective items (e.g., Paknejad et al. 2005). With this consideration, they assigned holding costs h and h_1 (where $h > h_1$) for a unit of good item per period and a unit of defective item per period, respectively. In Fig. 2.5, inventory of defective items is depicted by the shaded area. In this section, the work of Wahab and Jaber (2010) based on Salameh and Jaber (2000), Maddah and Jaber (2008), and Jaber et al.

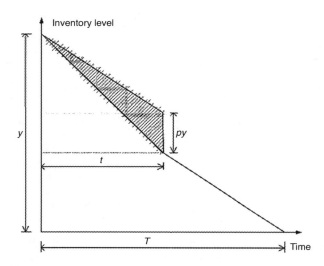

Fig. 2.5 The inventory level over time (Wahab and Jaber 2010)

(2008) with different holding costs for the good items and defective items is presented. Figure 2.5 presents the inventory level of problem on hand.

Let $N(y, p)$ be the number of good items in each lot size y where p is a random variable and is given by:

$$N(y,p) = y - py = y(1-p) \tag{2.50}$$

To avoid any shortage, good items' quantity should not be less than the demand during the screening time, t, meaning:

$$N(y,p) \geq Dt \tag{2.51}$$

Using Eqs. (2.50) and (2.51) and replacing t by $y = x$:

$$p \leq 1 - \frac{D}{x} \rightarrow E[p] \leq 1 - D/x \tag{2.52}$$

Let TR(y) and TC(y) be the total revenue and the total cost per cycle, respectively. TR(y) consists of revenues from the good and defective items and is given by TR(y) = $sy(1-p) + vyp$, and the total cost per cycle is (Wahab and Jaber 2010):

$$\mathrm{TC}(y) = \underbrace{K + Cy}_{\text{Fixed and purchasing cost}} + \underbrace{C_1 y}_{\text{Screening cost}} + \underbrace{h\left(\frac{py^2}{2x} + \frac{y(1-p)T}{2}\right)}_{\text{The holding costs of good items per cycle}} + \underbrace{h_1\left(\frac{py^2}{2x}\right)}_{\text{Holding costs of defective items}} \tag{2.53}$$

2.3 EOQ Model with No Shortage

The total profit per cycle, TP(y), is determined as the total revenue per cycle, TR (y), minus the total cost per cycle, TC(y), and given as (Wahab and Jaber 2010):

$$\text{TP}(y) = sy(1-p) + vyp \\ - \left[K + Cy + C_1 y + h\left(\frac{py^2}{2x} + \frac{y(1-p)T}{2}\right) + h_1\left(\frac{py^2}{2x}\right) \right] \quad (2.54)$$

The total profit per unit time, TPU(y), is determined by TP(y) = T and expressed in Eq. (2.55). The expected value of the total profit per unit time is given in Eq. (2.56):

$$\text{TPU}(y) = D\left(s - v + \frac{hy}{2x} + \frac{h_1 y}{2x}\right) + \left(\frac{D}{1-p}\right) \\ \times \left(v - \frac{K}{y} - C - C_1 - \frac{hy}{2x} - \frac{h_1 y}{2x}\right) - \frac{hy(1-p)}{2} \quad (2.55)$$

$$E[\text{TPU}(y)] = D\left(s - v + \frac{hy}{2x} + \frac{h_1 y}{2x}\right) + D\left(v - \frac{K}{y} - C - C_1 - \frac{hy}{2x} - \frac{h_1 y}{2x}\right) E\left(\frac{1}{1-p}\right) \\ - \frac{hy(1-E[p])}{2} \quad (2.56)$$

It can be easily shown that the $E[\text{TPU}(y)]$ is concave in y, and thus by minimizing the expected total profit per unit time, the optimal lot size is determined as:

$$y_1^* = \sqrt{\frac{2DKE[1/(1-p)]}{h(1-E[p]) + (D/x)(h+h_1)E[1/(1-p)] - (D/x)(h+h_1)}} \quad (2.57)$$

Let the optimal lot size be y_2 when the holding cost of the defective items and the good items is equal (i.e., $h = h_1$) and y_2 is given by:

$$y_2^* = \sqrt{\frac{2DKE[1/(1-p)]}{h[(1-E[p]) - 2(D/x)(1 - E[1/(1-p)])]}} \quad (2.58)$$

Regarding Eq. (2.58), the error appearing on Salameh and Jaber (2000) has been corrected on Cárdenas-Barrón (2001). A simple approach for determining the economic production quantity for an item with imperfect quality is also presented in Goyal and Cárdenas-Barrón (2002). Next, they have compared the optimal lot size derived in Maddah and Jaber (2008) with the one that they assigned different holding costs for the defective and good items. Since the cycle length T depends on the percentage rate of defective items, the cycle length is a random variable. Knowing $T = \frac{y(1-p)}{D}$ and $E[T] = \frac{y(1-E[p])}{D}$ (Wahab and Jaber 2010), the expected profit per cycle is:

$$E[\text{TP}(y)] = sy(1 - E[p]) + vyE[p]$$

$$- \left[K + Cy + C_1 y + h \left(\frac{E[p]y^2}{2x} + \frac{y^2 E\left[(1-p)^2\right]}{2D} \right) + h_1 \left(\frac{E[p]y^2}{2x} \right) \right] \quad (2.59)$$

As a renewal process, the expected profit per unit time will be:

$$E[\text{TPU}(y)] = \frac{D\{s(1 - E[p]) + vE[p] - C - C_1\}}{(1 - E[p])} - \frac{KD}{y(1 - E[p])}$$

$$- \frac{yD}{2(1 - E[p])} \left[h \left(\frac{E[p]}{x} + \frac{E\left[(1-p)^2\right]}{D} \right) + h_1 \left(\frac{E[p]}{x} \right) \right] \quad (2.60)$$

Because of the concavity of $E[\text{TPU}(y)]$, the optimum lot size is:

$$y_3^* = \sqrt{\frac{2DK}{hE\left[(1-p)^2\right] + (D/x)E[p](h + h_1)}} \quad (2.61)$$

By considering $h_1 = h$ in expression (2.61):

$$y_4^* = \sqrt{\frac{2DK}{h\left[E\left[(1-p)^2\right] + 2E[p]D/x\right]}} \quad (2.62)$$

which is the same as the one in Maddah and Jaber (2008). Jaber et al. (2008) considered learning effects in an EOQ model with imperfect items when the holding costs for the good and defective items are the same. Now different holding costs are assigned to the good and defective items. In this case, the total profit per unit time is the same as in Eq. (2.54). However, p is replaced with $p(n)$, which is the percentage of defective per shipment n. For example, $p(n)$ is expressed using a S-shaped logistic learning curve model as follows (Wahab and Jaber 2010):

$$p(n) = \frac{\alpha}{\gamma + e^{\beta n}} \quad (2.63)$$

Including the learning effects, the total profit per unit time is given as (Wahab and Jaber 2010):

$$\text{TPU}(y_n) = D \left(s - v + \frac{hy_n}{2x} + \frac{h_1 y_n}{2x} \right) + \left(\frac{D}{1 - p(n)} \right) \left(v - \frac{K}{y_n} - C - C_1 - \frac{hy_n}{2x} - \frac{h_1 y_n}{2x} \right)$$
$$- \frac{hy_n(1 - p(n))}{2}$$

$$(2.64)$$

In Eq. (2.64), n is not a decision variable. For a given n, $p(n)$ is a constant. Hence, the total profit per unit time is concave in y_n. By setting the first derivative equal to zero, the optimal y_n can be obtained as (Wahab and Jaber 2010):

2.3 EOQ Model with No Shortage

$$y_n^* = \sqrt{\frac{2DK[1/(1-p(n))]}{h(1-p(n)) + (D/x)(h+h_1)[1/(1-p(n))] - (D/x)(h+h_1)}} \quad (2.65)$$

Let the optimal lot size be z_n when the holding costs of defective and good items are equal (i.e., $h_1 = h$) and n is given by (Wahab and Jaber 2010):

$$z_n^* = \sqrt{\frac{2DK}{h(1-p(n))[(1-p(n)) + (2D/x)[p(n)/(1-p(n))]]}} \quad (2.66)$$

which is the same as the one in Jaber et al. (2008).

Example 2.4 (Without Learning Effects) In order to illustrate the behavior of the optimal lot sizes y_1, y_2, y_3, and y_4, let us consider an example with $D = 50{,}000$, $C = \$100$, $K = \$3000$, $s = \$200$, $v = \$50$, $x = 100$, $C_1 = \$0.5$, $h = \$20$ unit/year, and $h_1 = \$5$ unit/year. The percentage of defective item, p, is uniformly distributed, i.e., $p \sim U[a, b]$, where $E[p] = (a + b)/2$; $E[1/(1-p)] = [1/(a-b)]$. They considered $a = 0$ and b has been varied from 0.001 to 0.5 (Wahab and Jaber 2010). The optimal lot sizes for different values of b are given in Table 2.6 and y_1, y_2, y_3, and y_4 (Wahab and Jaber 2010). Optimal lot sizes y_2 and y_4 are computed by substituting $h_1 = h = \$20$ unit/year.

Example 2.5 (With Learning Effects) In this example, they considered the same parameters that they used in Example 2.1 except $p(n)$, as given in Eq. (2.63), and the values of $\alpha = 70.067$ and $\beta = 0.7932$. Using Eqs. (2.65) and (2.66), y_n^* and z_n^* are computed and values y_n^* and z_n^* are presented in Table 2.7. As the learning takes

Table 2.6 Comparison of order quantities (Wahab and Jaber 2010)

b	$f(p)$	$E[p]$	$E[1/(1-p)]$	$E[(1-p)^2]$	y_1^*	y_2^*	y_3^*	y_4^*
0.001	1000.00	0.0005	1.0005003	0.999000	3874.32	3873.95	3874.32	3873.95
0.01	100.00	0.005	1.0050336	0.990033	3886.34	3882.67	3886.31	3882.66
0.02	50.00	0.01	1.0101354	0.980133	3899.74	3892.34	3899.65	3892.27
0.03	33.33	0.015	1.0153069	0.970300	3913.19	3901.97	3912.99	3901.83
0.04	25.00	0.02	1.0205499	0.960533	3926.70	3911.58	3926.33	3911.32
0.05	20.00	0.025	1.0258659	0.950833	3940.26	3921.16	3939.68	3920.75
0.06	16.67	0.03	1.0312567	0.941200	3953.87	3930.70	3953.03	3930.11
0.07	14.29	0.035	1.0367242	0.931633	3967.52	3940.21	3966.37	3939.40
0.08	12.50	0.04	1.0422701	0.922133	3981.22	3949.68	3979.71	3948.63
0.09	11.11	0.045	1.0478964	0.912700	3994.98	3959.11	3993.04	3957.78
0.1	10.00	0.05	1.0536052	0.903333	4008.77	3968.50	4006.37	3966.85
0.2	5.00	0.1	1.1157178	0.813333	4149.06	4059.62	4138.72	4053.02
0.3	3.33	0.15	1.1889165	0.730000	4292.60	4143.91	4267.73	4129.32
0.4	2.50	0.2	1.2770641	0.653333	4437.47	4218.35	4390.69	4193.61
0.5	2.00	0.25	1.3862944	0.583333	4580.86	4279.33	4504.47	4243.91

Table 2.7 Comparison of order quantities with learning effects (Wahab and Jaber 2010)

n	P(n)	z_n	y_n
1	0.08524	4064.53	4105.47
2	0.08497	4063.82	4104.70
3	0.08436	4062.26	4103.02
4	0.08305	4058.92	4099.38
5	0.08030	4051.91	4091.74
6	0.07482	4038.16	4076.55
7	0.06502	4014.13	4049.45
8	0.05042	3979.72	4009.30
9	0.03369	3942.29	3963.66
10	0.01944	3911.99	3925.09

place in the system, the number of defective items is going to decrease. Therefore, both y_n^* and z_n^* decrease as the number of shipment increases.

2.3.4.2 Transfer of Learning

As presented before, Salameh and Jaber (2000) extended the traditional EOQ model by accounting for imperfect-quality items under 100% screening and poor-quality items. The behavior of inventory was presented in Fig. 2.2. In this section, the work of Salameh and Jaber (2000) under learning in inspection which is developed by Khan et al. (2010) will be introduced. They considered a learning curve fits well to the power form of learning suggested by Wright (1936) as below:

$$\pi_n = \pi_1 n^{-b} \tag{2.67}$$

where p_1 is the time to perform the first repetition, b is the learning exponent $0 < b < 1$, and n is the cumulative number of repetitions. Also according to the work of Khan et al. (2010), the forgetting curve takes the form:

$$\bar{\pi}_m = \bar{\pi}_1 m^f$$

where $\bar{\pi}_1 = \pi_1 m^{-(f+b)}$ and $\bar{\pi}_m = \pi_m$, where m is the equivalent number of items that could have been produced. The forgetting exponent in cycle i is determined by Jaber and Bonney (1996) as (Khan et al. 2010):

$$f_i = \frac{b(1-b)\log(u_i + y_i)}{\log(1 + V/\lambda_i)} \tag{2.68}$$

where V is the time for total forgetting to occur and is assumed to be an input parameter, u_i is the experience remembered in cycle i, and l_i is the time to inspect $(u_i + y_i)$ items without interruption. So (Khan et al. 2010):

2.3 EOQ Model with No Shortage

Table 2.8 Notations of a given problem

x_1	Initial inspection rate (units/unit time)
b	Learning exponent
f_i	Forgetting exponent in the ith cycle
V	Time for total forgetting to occur (time)
y_i	Batch size in the ith cycle (unit)
u_i	Experience of screening remembered from the previous i cycles
Z_i	Lost sales quantity in the ith cycle (unit)
B_i	Backorder quantity in the ith cycle (unit)
t_{si}	Time when the screening rate is equal to the demand in the ith cycle (time)
t_{Bi}	Time to inspect B_i and Dt_{Bi} units in the ith cycle (time)

$$\lambda_i = \frac{(u_i + y_i)^{1-b}}{x_1(1+b)}$$

$$u_i + 1 = (u_i + y_i)^{(f_i+b)/b} R_i^{-f_i/b} \quad (2.69)$$

with:

$$R_i = \left[x_1(1-b)(T-\tau_i) + (u_i+y_i)^{(1-b)}\right]^{1/(1-b)} \quad (2.70)$$

where R_i is the equivalent number of items that could have been screened if no interruption occurs of length $T - t_i$, and t_i is the screening time in cycle i. In case of breaks in screening, one should note that $0 < u_i < \sum_{j=1}^{i-1} y_j$ when $T - \tau_i < V$, for partial forgetting; $u_i = 0$ when $T - \tau_i \geq V$, for total forgetting; and $u_i = \sum_{j=1}^{i-1} y_j$, when V becomes infinite, for total transfer of learning will be used to determine the equivalent experience remembered at the start of each cycle, in case of partial and total transfer of learning (Khan et al. 2010).

In order to model the presented problem, some new notations which are specifically used are shown in Table 2.8.

Inspection is usually a manual task where an inspector tests incoming units for specific quality characteristics to determine if the units conform to the quality requirements. Time to inspect (or screen) each unit reduces as the number of inspected units increase as given in Eq. (2.67) and is represented as $x_n = x_1 n^b$, where $x_1 = 1/\pi_1$ and $x_n = 1/\pi_n$, whereas Salameh and Jaber (2000) assumed $x_1 = x_n > D$. Here, it is assumed that $x_1 < D$ for $y_S < y$. If $x_n < D$, then the demand met is x_n and the rate at which units lost or backordered is $D - x_n$ and D otherwise. The length of the stock-out period or the period over which backorders are accumulated is given as (Khan et al. 2010):

$$t_{si} = \int_{u_i}^{u_i+y_i} \frac{1}{x_1} n^{-b} dn - \int_{y_{si}+u_i}^{u_i+y_i} \frac{1}{x_1} n^{-b} dn = \frac{(y_{si}+u_i)^{1-b} - u_i^{1-b}}{(1-b)x_1} \quad (2.71)$$

where $y_{si} = \left[(1-b)x_1 t_s + u_i^{1-b}\right]^{1/(1-b)} - u_i$ and $y_{si} = (D/x_1)^{1/b} - u_i$. Now the time to screen y_i items in a cycle (Khan et al. 2010) is:

$$\tau_i = \frac{\left\{(y_i+u_i)^{1-b} - u_i^{1-b}\right\}}{(1-b)x_1} \quad (2.72)$$

where u_i is computed from Eq. (2.69). In case of no transfer of learning, that is, a worker does not retain any knowledge from earlier cycles ($u_i = 0$), it will be taken as (Khan et al. 2010):

$$y_s = y_{si} = (D/x_1)^{1/b} \quad (2.73)$$

$$t_s = t_{si} = \frac{D^{1-b/b}}{(1-b)x_1^{1/b}} \quad (2.74)$$

$$\tau = \tau_i = \frac{y^{1-b}}{(1-b)x_1} \quad (2.75)$$

Two cases (lost sales and backorders) will be considered now to deal with the shortages, in each of the three scenarios for the transfer of learning from one cycle to another. These models are a direct extension to the work of Salameh and Jaber (2000) extended by Khan et al. (2010).

In the case of lot sales, the demand that cannot be fulfilled due to slow screening will be taken as lost sale. The inventory level figure of this case is presented in Fig. 2.6 and the inventory level is shown in Eq. (2.76):

$$I_1(t) = \begin{cases} y_i - [(1-b)x_1 t]^{1/(1-b)} & 0 \leq t < t_{si} \\ y_i - y_{si} - D(t-t_{si}) & t_{si} \leq t < \tau_i \\ (1-p)y_i - y_{si} - D(t-t_{si}) & \tau_i < t \leq T_i \end{cases} \quad (2.76)$$

At time $t = T_i$, the inventory level is zero, i.e., $(1-p)y_i - y_{si} - D(T_i - t_{si}) = 0$, and the cycle time is given as (Khan et al. 2010):

$$T_i = \frac{(1-p)y_i}{D} - \frac{y_{si}}{D} + t_{si} \quad (2.77)$$

The holding costs for the three different behaviors of inventory shown in Fig. 2.9. are determined, respectively, from Eq. (2.76) as (Khan et al. 2010):

2.3 EOQ Model with No Shortage

Fig. 2.6 Learning in inspection with lost sales (Khan et al. 2010)

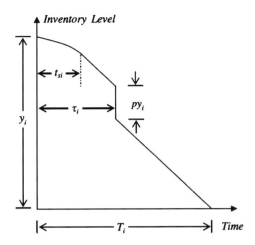

$$\text{HC}_{11} = h \int_0^{t_{si}} \left[y_i - [(1-b)x_1 t]^{1/(1-b)} \right] dt$$

$$= h y_i t_{si} - h \left(\frac{1-b}{2-b} \right) [(1-b)x_1]^{1/(1-b)} t_{si}^{2-b/1-b} \qquad (2.78)$$

$$\text{HC}_{21} = h \int_{t_s}^{\tau_i} [y_i - y_{si} - D(T_i - t_{si})] dt$$

$$= h(y_i - y_{si} + Dt_{si})(\tau_i - t_{si}) - \frac{hD}{2}(\tau_i^2 - t_{si}^2) \qquad (2.79)$$

$$\text{HC}_{31} = h \int_{\tau_i}^{T_i} [(1-p)y_i - y_{si} - D(T_i - t_{si})] dt$$

$$= h(y_i - y_{si} + Dt_{si})(T_i - \tau_i) - h p y_i (T_i - \tau_i) - \frac{hD}{2}(T_i^2 - \tau_i^2) \qquad (2.80)$$

Adding the costs for three different behaviors, they can get the total holding cost for the lost sales case as (Khan et al. 2010):

$$\begin{aligned}\text{HC}_\text{L} =\ & h(1-p)y_iT_i + hpy_i\tau_i - hT_i(y_{si} - Dt_{si}) + hy_{si}t_{si} \\ & - \frac{hD}{2}(T_i^2 - t_{si}^2) - h\left(\frac{1-b}{2-b}\right)[(1-b)x_1]^{1/(1-b)}t_{si}^{2-b/1-b}\end{aligned} \qquad (2.81)$$

After some simplifications and substitutions,

$$\begin{aligned}\text{HC}_\text{L} =\ & \frac{h}{2D}\left[y_i^2(1-p)^2 + 2y_iZ_i(1-p) + Z_i^2\right] - \frac{h}{2D}t_{si}^2 + hy_{si}t_{si} + \frac{hpy_i\left\{(y_i+u_i)^{1-b} - u_i^{1-b}\right\}}{x_1(1-b)} \\ & - h\left(\frac{1-b}{2-b}\right)[(1-b)x_1]^{1/(1-b)}t_{si}^{\frac{2-b}{1-b}}\end{aligned}$$

$$(2.82)$$

$$\text{Cost of the lost sales} = \widehat{\pi}(Dt_{si} - y_{si}) = \widehat{\pi}Z \qquad (2.83)$$

$$\text{Cost of inspection} = C_1\tau_i = \frac{C_1\left\{(y_i+u_i)^{1-b} - u_i^{1-b}\right\}}{(1-b)x_1} \qquad (2.84)$$

So the total profit per cycle is (Khan et al. 2010):

$$\begin{aligned}\text{TP}_\text{L} =\ & [s(1-p) + vp - C]y_i - K - \widehat{\pi}Z_i - \frac{h}{2D}\left[y_i^2(1-p)^2 + 2y_iZ_i(1-p) + Z_i^2\right] \\ & + \frac{hD}{2}t_{si}^2 - \frac{(hpy_i + C_1)\left\{(y_i+u_i)^{1-b} - u_i^{1-b}\right\}}{x_1(1-b)} - hy_{si}t_{si} + h\left(\frac{1-b}{2-b}\right)[(1-b)x_1]^{1/(1-b)}t_{si}^{2-b/1-b}\end{aligned}$$

$$(2.85)$$

And its expected value becomes:

$$\begin{aligned}E[\text{TP}_{i\text{L}}] =\ & \{s(1-E[p]) + vE[p] - C\}y_i - K - \widehat{\pi}Z_i \\ & - \frac{h}{2D}\left\{y_i^2 E\left[(1-p)^2\right] + 2y_iZ_i(1-E[p]) + Z_i^2\right\} + \frac{hD}{2}t_{si}^2 \\ & - \frac{(hy_iE[p] + C_1)\left\{(y_i+u_i)^{1-b} - u_i^{1-b}\right\}}{x_1(1-b)} - hy_{si}t_{si} + h\left(\frac{1-b}{2-b}\right)[(1-b)x_1]^{1/(1-b)}t_{si}^{2-b/1-b}\end{aligned}$$

$$(2.86)$$

The expected cycle time $E[T_{i\text{L}}]$ can be written as (Khan et al. 2010):

$$E[T_{i\text{L}}] = \frac{\{1-E[p]\}y_i}{D} - \frac{y_{si}}{D} + t_{si} \qquad (2.87)$$

Using renewal reward theorem:

2.3 EOQ Model with No Shortage

$$E[\text{TPU}_{iL}] = \frac{E[\text{TP}_{iL}]}{E[T_{iL}]} \tag{2.88}$$

According to the Khan et al. (2010), the experience gained in each cycle i will be taken as:

$$u_i = \sum_{j=1}^{i-1} y_j \tag{2.89}$$

This implies that the worker does not lose any knowledge in his break while he is not screening. In case of total forgetting, screening in each cycle starts with no prior knowledge, which means that the worker loses all the experience gained in the earlier cycles. In this case, the previous equations change to:

$$\tau = \frac{y^{1-b}}{(1-b)x_1} \tag{2.90}$$

$$\begin{aligned}\text{HC}_\text{L} &= \frac{h}{2D}\left[y^2(1-p)^2 + 2yZ(1-p) + Z^2\right] - \frac{hD}{2}t_s^2 + \frac{hpy^{2-b}}{x_1(1-b)} \\ &\quad + hy_s t_s - h\left(\frac{1-b}{2-b}\right)[(1-b)x_1]^{1/(1-b)} t_s^{2-b/1-b}\end{aligned} \tag{2.91}$$

In the expected profit per cycle and the expected annual profit of this case, Eqs. (2.90) and (2.91) should be replaced (Khan et al. 2010). But for the backorder case, the screening rate becomes equal to the demand rate at t_{si}. The dotted line shows that the backorder that piles up till t_{si} is fulfilled at the time $(t_{si} + t_{Bi})$. Inventory level diagram is presented in Fig. 2.7. So the maximum backorder level is:

$$B_i = Dt_{si} - y_{si} \tag{2.92}$$

Now, following its definition in the notations, y_{Bi} can be written as:

$$y_{Bi} = Dt_{Bi} + B_i + y_{si} = D(t_{Bi} + t_{si}) \tag{2.93}$$

The time t_{Bi} can be written as:

$$t_{Bi} = \int_0^{y_{Bi}} \frac{1}{x_1} n^{-b} dn - \int_0^{y_{si}} \frac{1}{x_1} n^{-b} dn = \frac{y_{B_i}^{1-b}}{x_1(1-b)} - \frac{y_{B_i}^{1-b}}{x_1(1-b)}$$

Substituting y_{Bi}:

Fig. 2.7 Learning in inspection with backorders (Khan et al. 2010)

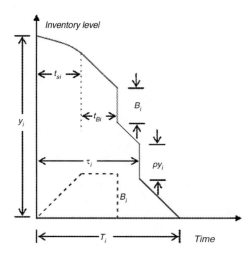

$$t_{Bi} = \frac{[D(t_{Bi} + t_{si})]^{1-b}}{x_1(1-b)} - t_{si} \text{ or } t_{Bi} + t_{si} = \frac{D^{1-b/b}}{x_1^{1/b}(1-b)^{1/b}} \quad (2.94)$$

This time can be taken as (Khan et al. 2010):

$$t_x = t_{Bi} + t_{si} = \frac{D^{1-b/b}}{x_1^{1/b}(1-b)^{1/b}} \quad (2.95)$$

Again, with the help of learning, the screening rate will become equal to or more than the demand rate, and there will not be any backorders after some cycles. The three scenarios of learning discussed in the lost sales case will be considered here to develop the expected annual profit of a buyer (Khan et al. 2010).

The inventory level in Fig. 2.7 can be represented as (Khan et al. 2010):

$$I_B(t) = \begin{cases} y_i - [(1-b)x_1 t]^{1/(1-b)} & 0 \le t \le t_{si} \\ y_i - y_{si} - D(t - t_{si}) & t_{si} \le t \le t_{si} + t_{Bi} \\ y_i - Dt & t_{si} + t_{Bi} \le t \le \tau_i \\ (1-p)y_i - Dt & \tau_i < t < T_i \end{cases} \quad (2.96)$$

At time $t = T_i$, the inventory level is zero, i.e., $(1-p)y_i - DT_i = 0$. The cycle time T_i is (Khan et al. 2010):

2.3 EOQ Model with No Shortage

$$T_i = \frac{(1-p)y_i}{D} \tag{2.97}$$

The holding costs for the four different time intervals are determined respectively from Eq. (2.96) as:

$$HC_{1Bi} = hyt_{si} - h\left(\frac{1-b}{2-b}\right)[(1-b)x_1]^{1/(1-b)}t_{si}^{2-b/1-b} \tag{2.98}$$

$$HC_{2Bi} = h(y_i - y_{si} + Dt_{si})t_{Bi} - \frac{hD}{2}\left[(t_{si} + t_{Bi})^2 - t_{si}^2\right] \tag{2.99}$$

$$HC_{3Bi} = h[\tau_i - (t_{si} + t_{Bi})]y_i - \frac{hD}{2}\left[\tau_i^2 - (t_{si} + t_{Bi})^2\right] \tag{2.100}$$

$$HC_{4Bi} = h(1-p)(T_i - \tau_i)y_i - \frac{hD}{2}\left(T_i^2 - \tau_i^2\right) \tag{2.101}$$

Adding Eqs. (2.98)–(2.101), the holding cost for the backorder case is:

$$HC_{Bi} = hy_i t_{si} - h\left(\frac{1-b}{2-b}\right)[(1-b)x_1]^{1/(1-b)}t_{si}^{2-b/1-b} + h(y_i - y_{si} + Dt_{si})t_{Bi} - \frac{hD}{2}\left[(t_{si} + t_{Bi})^2 - t_{si}^2\right]$$
$$+ h[\tau_i - (t_{si} + t_{Bi})]y_i - \frac{hD}{2}\left[\tau_i^2 - (t_{si} + t_{Bi})^2\right] + h(1-p)(T_i - \tau_i)y_i - \frac{hD}{2}\left(T_i^2 - \tau_i^2\right)$$

Using Eq. (2.97) and simplifying the above expression results in:

$$HC_{Bi} = h(1-p)y_i T_i + hpy_i \tau_i + ht_x B + hy_{si} t_{si} - \frac{hD}{2}\left(T_i^2 - t_{si}^2\right)$$
$$- h\left(\frac{1-b}{2-b}\right)[(1-b)x_1]^{1/(1-b)}t_{si}^{2-b/1-b} \tag{2.102}$$

Substituting t_i and T_i in terms of y_i from Eqs. (2.110) and (2.96) respectively, the above expression can be simplified as (Khan et al. 2010):

$$HC_{Bi} = \frac{h}{2D}\left[y_i^2(1-p)^2\right] - \frac{hD}{2}t_{si}^2 + \frac{hpy_i\left\{(y_i + u_i)^{1-b} - u_i^{1-b}\right\}}{x_1(1-b)}$$
$$+ hy_{si}t_{si} + ht_x B - h\left(\frac{1-b}{2-b}\right)[(1-b)x_1]^{1/(1-b)}t_{si}^{2-b/1-b} \tag{2.103}$$

It should be noted that the above holding cost reduces to the one in Salameh and Jaber (2000) once b, t_x, t_{si}, y_{si}, and u_i become zero (Khan et al. 2010).

Now the backorder cost in a cycle $= \frac{C_b t_s B_i}{2} + C_b(t_x - t_{si})B_i = C_b\left(t_x - \frac{t_{si}}{2}\right)B_i$

$$\text{TP}_{i\text{B}} = [s(1-p) + vp - C]y_i - K - C_\text{b}\left(t_x - \frac{t_{si}}{2}\right)B_i - \frac{(hpy_i + C_1)\left\{(y_i + u_i)^{1-b} - u_i^{1-b}\right\}}{x_1(1-b)}$$

$$- \frac{h}{2D}\left[y_i^2(1-p)^2\right] + \frac{hD}{2}t_{si}^2 - hy_{si}t_{si} - ht_x B + h\left(\frac{1-b}{2-b}\right)[(1-b)x_1]^{1/(1-b)}t_{si}^{2-b/1-b}$$

(2.104)

And the expected total profit per cycle is (Khan et al. 2010):

$$E[\text{TP}_{i\text{B}}] = \{s(1 - E[p]) + vE[p] - C\}y_i - K - C_\text{b}\left(t_x - \frac{t_{si}}{2}\right)B_i$$
$$- \frac{(hy_i E[p] + C_1)\left\{(y_i + u_i)^{1-b} - u_i^{1-b}\right\}}{x_1(1-b)} - \frac{h}{2D}\left\{y_i^2 E\left[(1-p)^2\right]\right\} + \frac{hD}{2}t_{si}^2 - hy_{si}t_{si}$$
$$- ht_x B + h\left(\frac{1-b}{2-b}\right)[(1-b)x_1]^{1/(1-b)}t_{si}^{2-b/1-b}$$

(2.105)

The expected cycle time $E[T_{i\text{B}}]$ can be written using Eq. (2.96) as:

$$E[T_{i\text{B}}] = \frac{(1 - E[p])y_i}{D}$$

So, the expected annual profit can be written as:

$$E[\text{TPU}_{i\text{B}}] = \frac{E[\text{TP}_{i\text{B}}]}{E[T_{i\text{B}}]} \quad (2.106)$$

In the case of backorders for total transfer of learning, the worker will retain all the experience gained in the earlier cycles. This experience will be calculated using Eq. (2.92). The holding cost, the expected profit per cycle, and the expected annual profit in Eqs. (2.103), (2.105), and (2.106), respectively, will be determined using this experience in each cycle (Khan et al. 2010).

In the case of backorders for total forgetting, the experience u_i in cycle i becomes zero, and the inspection time will be determined by Eq. (2.93). The holding cost in Eq. (2.103) will be written as (Khan et al. 2010):

$$\text{HC}_\text{B} = \frac{h}{2D}\left[y_i^2(1-p)^2\right] + \frac{hD}{2}t_{si}^2 + \frac{hpy^{2-b}}{x_1(1-b)} + hy_s t_s + ht_x B$$
$$- h\left(\frac{1-b}{2-b}\right)[(1-b)x_1]^{1/(1-b)}t_{si}^{2-b/1-b}$$

(2.107)

2.3 EOQ Model with No Shortage

The expected profit per cycle and the expected annual profit in Eqs. (2.105) and (2.106), respectively, will be determined using Eqs. (2.93) and (2.107) (Khan et al. 2010).

2.3.5 EOQ Models with Imperfect-Quality Items and Sampling

2.3.5.1 Inspection Shifts from Buyer to Supplier

In contrast to traditional EOQ models, which implicitly assume that all items are completely perfect, Salameh and Jaber (2000) have formulated the problem in situations where not all items are perfect. The imperfect items are separated from perfect ones by a full inspection and are used in another inventory situation. The implicit assumption of Salameh and Jaber's (2000) model, however, is that the supplier does not perform a full inspection; otherwise the received batches are expected to be completely perfect. In fact, the very presence of imperfect items in a batch depends on whether or not the supplier carries out a full inspection, which is why they outlined two different possible scenarios here (Rezaei and Salimi 2012):

Scenario 1. The supplier does not perform a full inspection, and, as a result, the batches received by the buyer contain some imperfect items. This implies that the buyer should conduct a full inspection.

Scenario 2. The supplier performs a full inspection, and, as a result, the batches received by the buyer contain no imperfect items. The first scenario was formulated and analyzed in Salameh and Jaber (2000), while the second scenario is the implicit assumption of traditional EOQ models. Based on these two scenarios, they examined the following research questions:

Assuming there is no relationship between the buyer's selling price, the purchasing price, and customer demand. According to Salameh and Jaber (2000) and Maddah and Jaber (2008), the buyer's expected profit per ordering cycle is as follows (Rezaei and Salimi 2012):

$TP(y) =$ Total sales of good-quality items $+$ Total sales of imperfect-quality items

$-$Ordering cost $-$ Purchasing cost $-$ Inspection cost $-$ Holding costs

or equivalently:

$$\text{TP}(y) = \overbrace{sy(1-p) + vyp}^{\text{Income}} - \underbrace{K}_{\text{Fixed cost}} - \overbrace{Cy}^{\text{Purchasing cost}} - \underbrace{h\left(\frac{[y(1-p)]^2}{2D} + \frac{py^2}{x}\right)}_{\text{Holding cost}} \quad (2.108)$$

Then the buyer's expected total profit per time unit is:

$$\text{ETPU}(y) = \frac{\left((s(1-E[p]) + vE[p] - C - C_1)D - KD/y - hy\left(E\left[(1-p)^2\right]/2 + E[p]D/x\right)\right)}{1 - E[p]} \quad (2.109)$$

Consequently, the optimal order quantity would be:

$$y^{\text{SMJ}*} = \left(\frac{2KD}{h\left(E\left[(1-p)^2\right] + 2E[p]D/x\right)}\right)^{1/2} \quad (2.110)$$

Given completely perfect batches ($p = 0$), the total profit per time unit is:

$$\text{TPU}_p = 0(y) = Ds - Dc' - \frac{KD}{y} - \frac{hy}{2} \quad (2.111)$$

And the optimal order quantity for this condition is the traditional EOQ:

$$y^{\text{trade}*} = \sqrt{\frac{2KD}{h}} \quad (2.112)$$

Now the maximum purchasing price for batches without imperfect items should be determined. First, determine the difference between the total profit per time unit when there are no imperfect items and the expected profit per time unit when there are $p\%$ imperfect items on average in each batch. If one considers c' as a variable here, then (Rezaei and Salimi 2012):

$$\text{TPU}(y^{\text{trade}*}, c') - \text{ETPU}(y^{\text{SMJ}*}) = D(s - c') - \frac{KD}{y^{\text{trade}*}} - \frac{hy^{\text{trade}*}}{2}$$
$$- \text{ETPU}(y^{\text{SMJ}*}) \quad (2.113)$$

The buyer accepts to pay more if and only if:

$$\text{TPU}(y^{\text{trade}*}, c') - \text{ETPU}(y^{\text{SMJ}*}) \geq 0 \quad (2.114)$$

which yields to:

2.3 EOQ Model with No Shortage

$$c' \leq s - \frac{K}{y^{\text{trade}*}} - \frac{hy^{\text{trade}*} + 2\text{ETPU}(y^{\text{SMJ}*})}{2D} \quad (2.115)$$

The right-hand side of this equation determines the maximum unit purchasing price (Mc) the buyer is willing to pay for batches without imperfect items.

To determine the optimal buyer's selling price and order quantity, they consider the expected total profit equation (Eq. 2.109) as an objective function, while both the buyer's selling price s and the order quantity y are decision variables. The expected total profit of the inventory problem should be maximized subject to the price–demand relationship function:

$$\max \bar{\pi}(y,s) = \frac{\left((s(1-E[p]) + vE[p] - C - C_1)D - KD/y - hy\left(E\left[(1-p)^2\right]/2 + E[p]D/x\right)\right)}{1 - E[p]}$$

s.t.
$D = f(s)$

$$\quad (2.116)$$

To obtain the optimal values of y, s, and D, the following Lagrangian function is used (Rezaei and Salimi 2012):

$$L = \bar{\pi}(y,s) - \lambda(D - f(s)) \quad (2.117)$$

Then the partial derivation of the Lagrangian function should be set with respect to y, s, D, and l to zero:

$$\frac{\partial L}{\partial D} = \frac{1}{1 - E[p]}\left(s(1-E[p]) + vE[p] - C - C_1 - \frac{K}{y} - \frac{hyE[p]}{x}\right) - \lambda = 0 \quad (2.118)$$

$$\frac{\partial L}{\partial s} = D + \lambda f'(s) = 0 \quad (2.119)$$

$$\frac{\partial L}{\partial y} = \frac{1}{1 - E[p]}\left(\frac{KD}{y^2} - h\left(E\left[(1-p)^2\right]/2 + E[p]D/x\right)\right) = 0 \quad (2.120)$$

$$\frac{\partial L}{\partial \lambda} = -D + f(s) = 0 \quad (2.121)$$

Then solving simultaneously, the equation system yields to (Rezaei and Salimi 2012):

$$y^{**} = \left(\frac{2KD}{h\left(E\left[(1-p)^2\right] + 2E[p]D/x\right)}\right)^{1/2} \qquad (2.122)$$

$$s^{**} = -\frac{D}{f'(s)} - \left(vE[p] - C - C_1 - \frac{K}{y} - \frac{hyE[p]}{x}\right)/(1 - E[p]) \qquad (2.123)$$

$$D^{**} = f(s) \qquad (2.124)$$

In Eq. (2.123), there is no specific price–demand relationship, which means it is a general formula designed to obtain the optimal value of buyer's selling price s. To determine the maximum purchasing price (Mc), Rezaei and Salimi (2012) first calculated the difference between the total profit when there are no imperfect items and the expected total profit when there are $p\%$ imperfect items on average in every batch:

$$\pi(s, y, c') - \bar{\pi}(y^{**}, s^{**}) = D\left(s - c' - \frac{k}{y}\right) - \frac{hy}{2} - \bar{\pi}(y^{**}, s^{**}) \qquad (2.125)$$

Here, the buyer agrees to pay more for each item in batches without imperfect items if and only if:

$$\pi(s, y, c') - \bar{\pi}(y^{**}, s^{**}) \geq 0 \qquad (2.126)$$

Consequently, one obtains (Rezaei and Salimi 2012):

$$c' \leq s - \frac{K}{y} - \frac{hy + 2\bar{\pi}(y^{**}, s^{**})}{2D} \qquad (2.127)$$

The right-hand side of this equation is the maximum purchasing price. To determine the highest value of c', one should find its maximum value (R). As R is a function of variables s and y, Rezaei and Salimi (2012) considered it a function that should be maximized subject to the price–demand relationship function:

$$\begin{aligned}\max R(s, y) &= s - \frac{K}{y} - \frac{hy + 2\bar{\pi}(y^{**}, s^{**})}{2D} \\ \text{s.t.} & \\ D &= f(s)\end{aligned} \qquad (2.128)$$

To obtain the optimal values of s, y, and D, they first made the following Lagrangian function:

2.3 EOQ Model with No Shortage

$$L = s - \frac{K}{y} - \frac{hy + 2\bar{\pi}(y^{**}, s^{**})}{2D} - \lambda(D - f(s)) \qquad (2.129)$$

The partial derivation of the Lagrangian function with respect to y, s, D, and l are:

$$\frac{\partial L}{\partial D} = \frac{hy + 2\bar{\pi}(y^{**}, s^{**})}{2D^2} - \lambda = 0 \qquad (2.130)$$

$$\frac{\partial L}{\partial s} = 1 + \lambda f'(s) = 0 \qquad (2.131)$$

$$\frac{\partial L}{\partial y} = \frac{K}{y^2} - \frac{h}{2D} = 0 \qquad (2.132)$$

$$\frac{\partial L}{\partial \lambda} = -D + f(s) = 0 \qquad (2.133)$$

Solving simultaneously equation system yields to (Rezaei and Salimi 2012):

$$s_p = \frac{(-f'(s)(2\bar{\pi}(y^{**}, s^{**}) + hy)/2)^{1/2} - f(0)}{f'(s)} \qquad (2.134)$$

$$y_p = \left(\frac{2KD}{h}\right)^{1/2} \qquad (2.135)$$

This is a general model to determine the values of s and y, which leads to determine the maximum value of c', Mc. Specifying a suitable price–demand relationship function, they found the optimal values of s and y. Putting the value of s_p and y_p on Eq. (2.127), the maximum value of c' can be determined. In order to make an optimal decision, Rezaei and Salimi (2012) proposed a decision rule as below which is presented in Fig. 2.8.

Example 2.6 Determining Mc assuming there is no relationship between the buyer's selling price, the purchasing price, and customer demand, Rezaei and Salimi (2012) adopted the same data as used in Salameh and Jaber (2000) and Maddah and Jaber (2008) as follows:

$$f(p) = \begin{cases} 25, 0 \le p \le 0.04 \\ 0, \text{otherwise} \end{cases} \Rightarrow E[p] = 0.02 \text{ and } E\left[(1-p)^2\right] = 0.96$$

$D = 50{,}000$ units/year, $C = \$25$/unit, $K = \$100$/cycle, $h = \$5$/unit/year, $x = 1$ unit/min, $C_I = \$0.5$/unit, $s = \$50$/unit, $v = 20$/unit, and the inventory operation operates on an 8 h/day, for 365 days a year. Using Eqs. (2.109)–(2.112) and (2.115), $y^* = 1434.48$, ETPU$^* = 1{,}212{,}274.30$, $y^{\text{trade}}* = 1414.21$, TPU$^{\text{trade}}* = 1{,}242{,}929$, and Mc $= 25.61$.

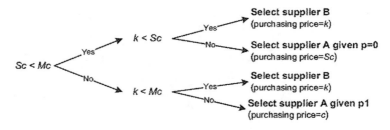

Fig. 2.8 Decision rule

Table 2.9 Average rate of imperfect items in each batch and the corresponding maximum purchasing price (Mc) (Rezaei and Salimi 2012)

$E[p]$	Mc	$E[p]$	Mc
0.01	25.56	0.14	26.40
0.02	25.61	0.15	26.48
0.03	25.67	0.16	26.56
0.04	25.73	0.17	26.64
0.05	25.79	0.18	26.72
0.06	25.85	0.19	26.80
0.07	25.92	0.20	26.89
0.08	25.98	0.21	26.98
0.09	26.05	0.22	27.07
0.10	26.12	0.23	27.16
0.11	26.19	0.24	27.25
0.12	26.26	0.25	27.35
0.13	26.33		

Table 2.9 shows the corresponding maximum purchasing price (Mc) for different average rates of imperfect items. As becomes clear, the higher the average rate of imperfect items $E[p]$, the higher the maximum purchasing price, Mc.

Example 2.7 To determining Mc assuming there is a relationship between the buyer's selling price, the purchasing price, and customer demand, Rezaei and Salimi (2012) considered the same data presented in Example 2.6. Commonly, two demand functions have been considered in literature: (1) the constant price–elasticity function and (2) the linear demand function (e.g., Rezaei and Davoodi in press). Here, they supposed a linear price–demand relationship function as $D = 100{,}000 - 1000s$. Using Eqs. (2.116) and (2.122)–(2.124), $s^{**} = 62.8477$; $y^{**} = 1238.39$; $D^{**} = 37{,}152.32$; and $y = 1{,}377{,}260.25$. Table 2.10 shows the optimal value of these variables (s^{**}, y^{**}, D^{**}, y, s_p, y_p, and Mc) for different values of $E[p]$.

2.3 EOQ Model with No Shortage

Table 2.10 The optimal value of s^{**}, y^{**}, D^{**}, $p^*(s, y)$, s_p, y_p, and Mc for different values of $E[p]$ (Rezaei and Salimi 2012)

$E[p]$	s^{**}	y^{**}	D^{**}	$p^*(s, y)$	s_p	y_p	Mc	$D \cdot Mc$
0.01	62.819	1229.16	37,180.96	1,379,381.21	62.819	1219.53	25.56	0.556
0.02	62.848	1238.39	37,152.32	1,377,260.25	62.848	1219.06	25.61	0.057
0.03	62.877	1247.66	37,123.08	1,375,096.78	62.877	1218.58	25.67	0.058
0.04	62.907	1256.98	37,093.23	1,372,889.51	62.906	1218.09	25.73	0.059
0.05	62.937	1266.33	37,062.75	1,370,637.11	62.937	1217.59	25.79	0.061
0.06	62.968	1275.72	37,031.61	1,368,338.19	62.968	1217.08	25.85	0.062
0.07	63.000	1285.14	36,999.80	1,365,991.29	63.000	1216.56	25.92	0.063
0.08	63.033	1294.59	36,967.28	1,363,594.91	63.032	1216.03	25.98	0.065
0.09	63.066	1304.06	36,934.05	1,361,147.48	63.065	1215.48	26.05	0.066
0.10	63.100	1313.54	36,900.07	1,358,647.34	63.099	1214.92	26.12	0.068
0.11	63.135	1323.04	36,865.32	1,356,092.80	63.134	1214.35	26.19	0.069
0.12	63.170	1332.54	36,829.77	1,353,482.06	63.169	1213.77	26.26	0.071
0.13	63.207	1342.05	36,793.40	1,350,813.26	63.205	1213.17	26.33	0.072
0.14	63.244	1351.55	36,756.17	1,348,084.47	63.242	1212.56	26.40	0.074
0.15	63.282	1361.03	36,718.06	1,345,293.63	63.280	1211.93	26.48	0.076
0.16	63.321	1370.50	36,679.03	1,342,438.63	63.319	1211.29	26.56	0.078
0.17	63.361	1379.94	36,639.06	1,339,517.24	63.359	1210.63	26.64	0.080
0.18	63.402	1389.34	36,598.10	1,336,527.14	63.400	1209.96	26.72	0.082
0.19	63.444	1398.71	36,556.12	1,333,465.88	63.442	1209.27	26.80	0.084
0.20	63.487	1408.02	36,513.09	1,330,330.93	63.485	1208.55	26.89	0.086
0.21	63.531	1417.28	36,468.96	1,327,119.59	63.529	1207.83	26.98	0.088
0.22	63.576	1426.47	36,423.68	1,323,829.08	63.574	1207.08	27.07	0.090
0.23	63.623	1435.58	36,377.22	1,320,456.45	63.620	1206.31	27.16	0.093
0.24	63.670	1444.61	36,329.53	1,316,998.61	63.668	1205.52	27.25	0.095
0.25	63.719	1453.55	36,280.56	1,313,452.31	63.717	1204.71	27.35	0.098

2.3.5.2 Sampling Inspection Plans

Rezaei (2016) considers a situation where a buyer wants to decide on economic order quantity of a product, where the received order contains a p percentage of imperfects, with a known probability density function, $f(p)$. In order to model the presented problem, some new notations which are specifically used are shown in Table 2.11.

Upon receiving the lot, the buyer draws a sample of size n from the lot, and based on the findings from the sample, one of the three following decisions is made:

- If p is above p_1, the whole lot will be rejected, and the supplier should send a same lot without any defective.
- If p is less than p_1 and above p_0, the lot is accepted. The buyer performs a full inspection. The imperfect items are separated from the perfect ones and used in another inventory situation.
- If p is below p_0, the lot is accepted and no inspection is conducted.

Table 2.11 Notations of a given problem

p_1	Maximum level of imperfect item (%)
p_0	Minimum level of imperfect item (%)
θ	Imperfect item quantity (unit)
α_0	Minimum level of imperfect item quantity (unit)
α_1	Maximum level of imperfect item quantity (unit)

Case I: $p > p_1$ Considering the lot size y and selling price s, the total profit is:

$$\text{TP}(y) = \underbrace{sy}_{\text{Revenue}} - \underbrace{Cy}_{\text{Purchasing cost}} - \underbrace{K}_{\text{Fixed cost}} - \underbrace{\frac{hy^2}{2D}}_{\text{Holding cost}} \tag{2.136}$$

Dividing TP(y) by the inventory cycle length $T = \frac{y}{D}$ gives the buyer's total profit per time unit, as follows:

$$\text{TPU}(y|p \geq p_1) = (s - C)D - \frac{KD}{y} - \frac{hy}{2} \tag{2.137}$$

So the optimal order quantity is:

$$y_R^* = \sqrt{\frac{2DK}{h}} \tag{2.138}$$

Case II: $p_0 < p < p_1$ This case has been proposed by Maddah and Jaber (2008). The total profit of this case is calculated as follows:

$$\text{TP}(y) = \underbrace{sy(1-p)}_{\substack{\text{Revenue of} \\ \text{perfect items}}} + \underbrace{vyp}_{\substack{\text{Revenue of} \\ \text{imperfect items}}} - \underbrace{K}_{\text{Fixed cost}} - \underbrace{Cy}_{\text{Purchasing cost}} - \underbrace{C_Iy}_{\text{Inspection cost}} - \underbrace{h\left(\frac{[y(1-p)]^2}{2D} + \frac{py^2}{x}\right)}_{\text{Holding cost}} \tag{2.139}$$

Dividing TP(y) by the expected inventory cycle length $E[T] = \frac{(1-E[p])y}{D}$, the buyer's expected total profit per time unit is:

2.3 EOQ Model with No Shortage

$E[\text{TPU}(y|p_0 \leq p \leq p_1)] =$

$$\frac{((s(1-E[p]) + vE[p] - C - C_1)D - KD/y - hy\left(E\left[(1-p)^2\right]/2 + E[p]D/x\right)}{1 - E[p]}$$

(2.140)

So, the optimum order size is:

$$y^*_{AI} = \left(\frac{2KD}{h\left(E\left[(1-p)^2\right] + 2E[p]D/x\right)}\right)^{\frac{1}{2}}$$

(2.141)

Case III: $p < p_0$ The total profit function of this case contains elements that are similar to the previous case, except that in this case, the buyer spends no cost on inspection, but has to spend for return costs. When a customer returns an item, the customer receives a new one. However, as no inspection is conducted in this case, there is still a chance (though very slim) that the new item delivered to the customer will also be imperfect. That is to say, from the order quantity y, py units are imperfect. This means that the buyer will initially receive py returned items. These customers are given new items. Again, from this py items, $p(py)$ are returned. If they considered the whole process including the next times, theoretically one obtains:

The total number of returned items $= y\left(p + p^2 + p^3 + \cdots\right)$ (2.142)

Inside the bracket is the sum of a geometric series, for which one obtains:

$$p + p^2 + p^3 + \cdots = \lim_{n \to \infty} \left(p + p^2 + p^3 + \cdots + p^n\right) = \frac{p}{1-p}$$

(2.143)

This means that

The total number of returned items $= y\dfrac{p}{1-p}$ (2.144)

The total costs of return $= C_I y \dfrac{p}{1-p}$ (2.145)

And finally the total profit becomes:

$$\text{TP}(y) = \overbrace{sy(1-p) + vyp}^{\text{Revenue of perfect and imperfect items}} - \underbrace{K}_{\text{Fixed cost}} -$$

$$\overbrace{Cy}^{\text{Purchasing cost}} - h\underbrace{\frac{y^2(1-p)}{2D}}_{\text{Holding cost}} - \overbrace{C_{\text{J}}y\frac{p}{1-p}}^{\text{Return cost}} \quad (2.146)$$

Dividing TP(y) by the expected inventory cycle length $E[T]$, the buyer's expected total profit per time unit is calculated as follows:

$$E[\text{TPU}(y|p \leq p_0)] = \frac{(s(1-E[p]) + vE[p])D - KD/y - CD - h\frac{y(1-E[p])}{2} - C_{\text{J}}DE\left[\frac{p}{1-p}\right]}{1 - E[p]}$$

(2.147)

So the optimum order quantity for this case would be:

$$y^*_{\text{AnI}} = \sqrt{\frac{2DK}{h(1 - E[p])}} \quad (2.148)$$

In order to plan a sampling, Rezaei (2016) suggested the following steps:

Step 1. Draw a random sample of n items from the lot, inspect the sample, and count θ.

Step 2. Based on the observed number θ, follow one of the following strategies:

(a) If $\theta > a_1$, reject the entire lot.
(b) If $a_0 < \theta < a_1$, accept the entire lot and conduct a 100% inspection.
(c) If $\theta < a_0$, accept the entire lot, and there is no need for inspection.

The expected total profit of the problem considering the sampling plan is:

$$\begin{aligned}\text{ETPU}(y|p \sim U(a,b)) =& \Pr(\theta \geq \alpha_1)\text{TPU}(y|p \geq p_1) \\ +\Pr(\alpha_0 \leq \theta \leq \alpha_1)&\text{ETPU}(y|p_0 \leq p \leq p_1) \\ +\Pr(\theta \leq \alpha_0)&\text{ETPU}(y|p \leq p_0)\end{aligned} \quad (2.149)$$

Before continuing, the sampling plan should be incorporated into the total profit functions of all cases.

Case IV: $p > p_1$ with Sampling Plan In this case, a new cost element (sampling costs) which is nC_I is added to the total profit function of the first case. In addition, if the buyer rejects a lot, the buyer receives a complete perfect lot, while, if the supplier later investigates and realizes that the lot has been wrongly rejected, or in fact $p \leq p_1$,

2.3 EOQ Model with No Shortage

the buyer is penalized for C_W. The chance of such an unjustified rejection by the buyer is (Rezaei 2016):

$$\psi = \Pr(p \leq p_1 | \theta \geq \alpha_1)$$

Finally, the total profit of the fourth case is:

$$\text{TPU}(y|p \geq P_1) = \overbrace{(s-c)D}^{\text{Revenue-purchasing cost}} - \overbrace{\frac{KD}{y}}^{\text{Fixed cost}} - \overbrace{\frac{C_1 nD}{y}}^{\text{Inspection cost}} - \overbrace{\frac{hy}{2}}^{\text{Holding cost}} - \overbrace{\frac{\psi C_W D}{y}}^{\text{Buyer's penalty cost}} \quad (2.150)$$

Case V: $p_0 < p < p_1$ with Sampling Plan In this case, since a 100% inspection is conducted and n is considered as part, the sampling cost is considered as part of the inspection cost. So the total profit of this case is similar to the second case.

Case VI: $p < p_0$ with Sampling Plan In this case, *inspection cost* should be added to the total profit of the third case:

$$E[\text{TPU}(y|p \leq p_0)]$$

$$= \frac{(s(1-E[p]) + vE[p])D - (K + C_1 n)D/y - CD - h\frac{y(1-E[p])}{2} - C_J DE\left[\frac{p}{1-p}\right]}{1 - E[p]} \quad (2.151)$$

Because of type I error and type II error and different parts of the probability density function $f(p)$ used for the three cases, the expected numbers of imperfect items for the three cases are slightly different. The expected numbers of imperfect items in a received lot, given the observed number of imperfect items in the sample is (1) greater than a_1, $E_1[p]$, (2) is between a_0 and a_1, and (3) is less than a_0, are calculated as follows:

$$\bar{E}_1[p] = \Pr(\theta \geq \alpha_1 \cap p \leq p_0) E_{p \leq p_0}[p] + \Pr(\theta \geq \alpha_1 \cap p_0 \leq p \leq p_1) E_{p_0 \leq p \leq p_1}[p]$$
$$+ \Pr(\theta \geq \alpha_1 \cap p \geq p_1) E_{p \geq p_1}[p] \quad (2.152)$$

$$\bar{E}_2[p] = \Pr(\alpha_0 \leq \theta \leq \alpha_1 \cap p \leq p_0) E_{p \leq p_0}[p] + \Pr(\alpha_0 \leq \theta \leq \alpha_1 \cap p_0 \leq p \leq p_1) E_{p_0 \leq p \leq p_1}[p]$$
$$+ \Pr(\alpha_0 \leq \theta \leq \alpha_1 \cap p \geq p_1) E_{p \geq p_1}[p] \quad (2.153)$$

$$\overline{E}_3[p] = \Pr(\theta \leq \alpha_0 \cap p \leq p_0)E_{p \leq p_0}[p] + \Pr(\theta \leq \alpha_0 \cap p_0 \leq p \leq p_1)E_{p_0 \leq p \leq p_1}[p]$$
$$+ \Pr(\theta \leq \alpha_0 \cap p \geq p_1)E_{p \geq p_1}[p]$$
(2.154)

$E_1[p]$, $E_2[p]$, and $E_3[p]$ are considered as the expected numbers of imperfect items in a lot when they respectively reject the lot, accept the lot and conduct the full inspection, and accept the lot and conduct no inspection. Considering Eqs. (2.140) and (2.150)–(2.154), Eq. (2.149) can now be rewritten as follows:

$$E[\text{TPU}(y|p \sim U(a,b))] = \Pr(\theta \geq \alpha_1)\left((s-C)D - \frac{KD}{y} - \frac{C_1 nD}{y} - \frac{hy}{2} - \frac{\psi RD}{y}\right)$$
$$+ \Pr(\alpha_0 \leq \theta \leq \alpha_1)\frac{((s(1-\overline{E}_2[p]) + v\overline{E}_2[p] - C - C_1)D - KD/y - hy\left(\overline{E}_2\left[(1-p)^2\right]/2 + \overline{E}_2[p]D/x\right)}{1 - \overline{E}_2[p]}$$
$$+ \Pr(\theta \leq \alpha_0)\frac{(s(1-\overline{E}_3[p]) + v\overline{E}_3[p])D - (K + C_1 n)D/y - CD - h\frac{y(1-\overline{E}_3[p])}{2} - C_J D\overline{E}_3\left[\frac{p}{1-p}\right]}{1 - \overline{E}_3[p]}$$
(2.155)

Using Eq. (2.155), the optimum order quantity considering the sampling plan becomes (Rezaei 2016):

$$y^*_{\text{Intg}} = \sqrt{\frac{2\left[\Pr(\theta \geq \alpha_1) + \Pr(\theta \leq \alpha_0)/(1-\overline{E}_3[p])\right](KD + C_1 nD)}{+ (\Pr(\alpha_0 \leq \theta \leq \alpha_1)/(1-\overline{E}_2[p]))KD + \Pr(\theta \geq \alpha_1)\psi RD}}$$
(2.156)

Example 2.8 To illustrate the proposed model, Rezaei (2016) considered $D = 50{,}000$ units/year, $C = \$25$/unit, $K = \$100$/cycle, $h = \$5$/unit/year, $x = 1$ unit/min, $C_I = \$0.5$/unit (the inventory operation operates on an 8 h/day, for 365 days a year), $s = \$50$/unit, $v = \$20$/unit, $C_J = \$15$/unit returned, $C_W = \$70$ and

$$f(p) = \begin{cases} 4, & 0 \leq p \leq 0.25 \\ 0, & \text{otherwise} \end{cases}.$$
(2.157)

It should note that for a uniform probability density function $P = U(a,b)$:

$$E[p] = \frac{a+b}{2}$$
(2.158)

2.3 EOQ Model with No Shortage

$$E\left[(1-p)^2\right] = \frac{1}{b-a} \int_a^b (1-p)^2 dp = \frac{a^2 + ab + b^2}{3} - a - b + 1 \quad (2.159)$$

$$E\left[\frac{p}{1-p}\right] = \frac{1}{b-a} \int_a^b \frac{p}{1-p} dp = \frac{-\ln(1-b) - b + \ln(1-a) + a}{b-a} \quad (2.160)$$

The buyer should first define a sampling strategy, identifying n, a_0, and a_1. As mentioned before, the buyer and supplier should agree on the maximum limit, while the minimum limit is identified by the buyer. Using the algorithm presented in the last section, and considering the data, they found $b = 0.063707$, which helps the buyer identify the minimum limit. They assumed that the buyer selects $b = 0.06$ for the minimum limit and, together with the supplier, chooses $b = 0.15$ for the maximum limit. Based on this information, the general probability density function can be further divided into three probability density functions, as follows:

$$f(p)_{\text{reject}} = \begin{cases} 10 & 0.15 \leq p \leq 0.25 \\ 0 & \text{otherwise} \end{cases} \quad (2.161)$$

$$f(p)_{\text{accept and inspection}} = \begin{cases} 11.1 & 0.06 \leq p \leq 0.15 \\ 0 & \text{otherwise} \end{cases} \quad (2.162)$$

$$f(p)_{\text{accept and no-inspection}} = \begin{cases} 16.6, & 0 \leq p \leq 0.06 \\ 0 & \text{otherwise} \end{cases} \quad (2.163)$$

The expected values of these three probability density functions are 0.2, 0.105, and 0.03, respectively. A random sample of 20 items ($n = 20$) is selected and inspected, based on the results of which the following decisions are made:

- If $\theta > 4$, the buyer rejects the entire lot and receives a replacement lot including no imperfect items.
- If $1 < \theta < 4$ (two or three imperfect items), the buyer accepts the entire lot, and then conducts a 100% inspection, after which the perfect items are sold to the customers ($50/unit), while the imperfect items are sold as a single batch at the end of inspection process in another inventory situation ($20/unit).
- If $\theta < 1$, the entire lot is accepted and the buyer does not inspect the lot.

The probability of observing θ imperfect items in a sample of size n can be calculated using a binomial distribution function as follows:

$$f(\theta, n, p) = \binom{n}{\theta} p^\theta (1-p)^{n-\theta} \quad (2.164)$$

Because the buyer makes decisions based on the sample, there is a chance of making wrong decision (the so-called type I and II errors), which is why the expected values, $E[p]$, have to be adjusted.

For instance, the first element of this matrix shows the probability of observing zero or one imperfect items and accepting the lot, meaning that the imperfect items of the sample follow the distribution function. In other words, it is likely (0.186) that they accepted the sample ($h > 1$), while the actual imperfect rate of the lot follows from the distribution function, and it is also likely (0.018) that the lot actually follows from the distribution function. The other elements of the matrix are interpreted in the same way. One has:

$$E_1[p] = 0.196; E_2[p] = 0.117; E_3[p] = 0.049 \qquad (2.165)$$

The optimal order quantity of the entire problem is calculated using Eq. (2.156), and is $y^*_{Intg} = 1485$, with an expected total profit of:

$$E\left[\text{TPU}\left(y|p\tilde{U}(a,b)\right)\right] = 1,216,570 \qquad (2.166)$$

The economic order quantity and expected total profit of these two models are shown below (Rezaei 2016):

Case II. EOQ $= 1541.04$, E[TPU] $= 1,178,298.12$ (Maddah and Jaber 2008)
Case III. EOQ $= 1511.86$, E[TPU] $= 1,077,530.75$

2.3.5.3 Instantaneous Replenishment Model with Sampling

Moussawi-Haidar et al. (2013) considered a periodic review EOQ-type inventory model with random supply and imperfect items. Upon receiving an order, an acceptance sampling plan is used to decide whether to accept the lot based on the number of defective items found in the sample or to reject the lot. It is assumed that a rejected lot is submitted to 100% screening. The acceptance sampling plan is characterized by two parameters, a sample of size n and an acceptance number a_n. The problem is to jointly determine the optimal lot size y and the optimal sampling plan, i.e., the sample size, n, and acceptance number a_n, that maximize the total expected profit subject to a constraint limiting the proportion of uninspected defective items passed to the customer.

In order to model the presented problem, some new notations which are specifically used are shown in Table 2.12.

In Moussawi-Haidar et al. (2013)'s model, the production of each lot varies randomly with probability density function $f(P)$. Then, the distribution of the number of lot defectives, N, in a lot of size y is given by the integral:

2.3 EOQ Model with No Shortage

Table 2.12 Notations of a given problem

a_n	Acceptance number/level
N	Number of lot defective (unit)
X	Number of sample defectives (unit)
P_a	Probability of accepting the lot
ICU	Average inventory cost per unit of time
QCU	Quality-related cost per unit time
$E[TRU(y, n, c)]$	Expected total revenue per unit time

$$g(N) = \int_0^1 b(N, y, P) f(P) dP$$

They showed that when P follows a beta distribution with parameters a and b, $g(N)$ follows the beta-binomial distribution, with parameters y, a, and b. The number of sample defectives, X, in a sample of size n is given by the conditional distribution $t(X|N)$, assumed to be a hypergeometric distribution. The joint distribution of N and X is obtained as:

$$\Pr(N, X) = t(X|N) g(N)$$

And the marginal distribution of X is beta binomial with parameters n, a, and b, which is given by:

$$h(X) = \sum_Y \Pr(N, X)$$

Upon screening a sample of size n, if the number of sample defective items X is less than the acceptance level a_n, the lot is accepted. So P_a becomes (Moussawi-Haidar et al. 2013):

$$P_a = P(X \le a_n) = \sum_{X=0}^{a_n} h(X)$$

Inventory-related costs. The inventory cost consists of the fixed ordering cost and the inventory holding cost. To compute the expected inventory cost, the dynamics of the inventory level are needed to consider as depicted in Fig. 2.9, for each of the two cases: (I) when the lot is accepted based on the sample defectives and (II) when the lot is rejected and subject to 100% screening (i.e., $X \cdot P_c$). The behavior of the inventory level for case I is illustrated by the solid line in Fig. 2.9, where T_1 represents the cycle time.

Screening the sample is completed at time $t_1 = n/x$, at which the inventory level drops by np, the average number of sample defectives withdrawn from the inventory at the end of the screening process. Thus, the inventory level at time t_1 drops to

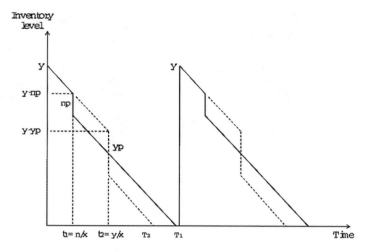

Fig. 2.9 Behavior of the inventory level over time, under two scenarios: $X < a_n$ and $X > a_n$ (Moussawi-Haidar et al. 2013)

$y - np$. The cycle time becomes $T_1 = (y - np)/D$. Referring to Fig. 2.9, the average inventory per cycle for this case, denoted by AI_I, can be written as follows:

$$AI_I = \left[\frac{n}{x}np + \frac{(y-np)^2}{2D}\right] \quad (2.167)$$

Using Eq. (2.167) and the renewal–reward theorem, Theorem 3.6.1 of Moussawi-Haidar et al. (2013), the average inventory per unit time for case I is:

$$AIU_I = AI_I/E(T_1) = \frac{\left[(n/x)np + \left((y-np)^2\right)/2D\right]}{E(T_1)} \quad (2.168)$$

In Case II which occurs with probability $1 - P_a$, the entire lot is subject to 100% screening. The inventory behavior is shown by the dotted line in Fig. 2.9, with T_2 being the cycle time and t_2 the screening time. Screening the lot is completed at time $t_2 = y/x$, at which the inventory level drops by yp, the average number of lot defectives. The cycle time is $T_2 = (y - yp)/D$. Referring to Fig. 2.9, the average inventory per cycle for case II, denoted by AI_{II}, can be written as:

$$AI_{II} = \left[\frac{n}{x}yp + \frac{(y-yp)^2}{2D}\right] \quad (2.169)$$

Using Eq. (2.169), in case II:

2.3 EOQ Model with No Shortage

$$\text{AIU}_{\text{II}} = \text{AI}_{\text{II}}/E(T_2) = \frac{\left[(y/x)yp + \left((y-yp)^2\right)/2D\right]}{E(T_2)} \quad (2.170)$$

And finally:

$$E[\text{ICU}] = P_a \times \text{AIU}_{\text{I}} + (1 - P_a) \times \text{AIU}_{\text{II}}$$

$$= P_a \left[\frac{(n/x)np + \left((y-np)^2\right)/2D}{E(T_1)}\right] + (1 - P_a)\left[\frac{(y/x)yp + \left((y-yp)^2\right)/2D}{E(T_2)}\right]$$

To compute the expected inventory ordering cost per unit time, they divided the ordering cost per cycle, K, by the expected cycle length, $E[T]$, which is obtained as the expectation over $E[T_1]$ and $E[T_2]$, as follows:

$$E[T] = P_a \times E[T_1] + (1 - P_a) \times E[T_2] = P_a \frac{(y-np)}{D} + (1 - P_a)\frac{y - yp}{D}$$
$$= \frac{y(1 - p + pP_a) - npP_a}{D} \quad (2.171)$$

Finally, $E[\text{ICU}]$, including the expected ordering, purchasing, and holding cost per unit time, is given as:

$$E[\text{ICU}(y, n, c)] = \frac{K + Cy}{E(T)} + h\left\{P_a\left[\frac{(n/x)np + \left((y-np)^2\right)/2D}{E(T_1)}\right]\right.$$
$$\left. +(1 - P_a)\left[\frac{(y/x)yp + \left((y-yp)^2\right)/2D}{E(T_2)}\right]\right\} \quad (2.172)$$

The quality-related cost consists of the cost of accepting the lot caused by the non-inspected defective items that are encountered in an accepted lot and cost of inspection. If the lot is accepted, only the sample is inspected. If the lot is rejected, the entire lot is inspected. To compute the quality-related cost, QCU can be expressed as:

$$E[\text{QCU}(y, n, a_n)] = \frac{C_d(y - n)pP_a + C_{\text{I}}nP_a + C_{\text{I}}y(1 - P_a)}{E(T)} \quad (2.173)$$

and

$$E[\text{TCU}(y,n,a_n)] = E[\text{ICU}(y,n,a_n)] + E[\text{QCU}(y,n,a_n)]$$

$$= \frac{K+Cy}{E(T)} + h\left\{ P_a\left[\frac{(n/x)np + \left((y-np)^2\right)/2D}{E(T_1)}\right]\right.$$

$$\left. + (1-P_a)\left[\frac{(y/x)yp + \left((y-yp)^2\right)/2D}{E(T_2)}\right]\right\} \quad (2.174)$$

$$+ \frac{C_d(y-n)pP_a + C_1nP_a + C_1y(1-P_a)}{E(T)}$$

Moussawi-Haidar et al. (2013) first expressed the expected revenue function. Then, they defined the integrated inventory-quality problem as an integer nonlinear program. So:

$$\text{ETRU}(y,n,a_n) = \frac{s(y-np)P_a + s(y-yp)(1-P_a)}{E(T)} \quad (2.175)$$

and

$$E[\text{TPU}(y,n,a_n)] = E[\text{TRU}(y,n,a_n)] - E[\text{TCU}(y,n,a_n)]$$

$$= \frac{[s(y-np)P_a + s(y-yp)(1-P_a)]D}{y(1-p+pP_a) - npP_a}$$

$$- h\left\{\frac{n^2pDP_a}{x(y-np)} - \frac{ypD(1-P_a)}{x(1-p)} - \frac{(y-np)P_a}{2} - \frac{y(1-p)(1-P_a)}{2}\right\} \quad (2.176)$$

$$- \frac{y(C+C_1 - C_1P_a + C_dpP_a)D + (K+C_1nP_a - C_dnpP_a)D}{y(1-p+pP_a) - npP_a}$$

Proposition 2.1 For a given acceptance level a_n, there exist a lot size y and a sample size n for which the Hessian matrix of the expected profit function in Eq. (2.176) is not negatively semidefinite (Moussawi-Haidar et al. 2013).

Proof See Moussawi-Haidar et al. (2013).

Lemma 2.1 The profit function $E[\text{TRU}(y, n, C)]$ given in Eq. (2.176) is concave in the order size y if the following condition holds: $C_d < K/(npP_a) + C_1/p$ (Moussawi-Haidar et al. 2013).

Proof For proof and more details, see Moussawi-Haidar et al. (2013).

Their objective was to maximize the expected profit per unit time in order to jointly determine the optimal y, n, and a_n using an integer nonlinear program with the following constraints:

2.3 EOQ Model with No Shortage

$$\max E[\text{TPU}(y, n, a_n)]$$

s.t.

$$P_a \frac{p(y-n)}{y} \leq \text{AOQL}$$

$\equiv$ quality of the outgoing material should be less than the AOQL

$$a_n < n < y$$

$$y, n, a_n \geq 0, \quad \text{Integers} \tag{2.177}$$

Moussawi-Haidar et al. (2013) developed the following numerical solution method:

Step 1. Fix a_n, and solve the nonlinear programming problem using a nonlinear program search method such as gradient search. Then, determine the optimal sample size $n(a_n)$ and lot size $y(a_n)$ and the corresponding profit $E[\text{TPU}(y(a_n), n(a_n))]$.

Step 2. Set $a_n = 1, 2, \ldots$, upper bound (e.g., upper bound = 30). For each value of a_n, repeat *Step* 1. They noted that after some value of c sets the probability of acceptance to 1, the optimal solution will remain the same. Hence, they may not need to reach the upper bound.

Step 3. Determine the optimal value of c by comparing the maximal expected profits $E[\text{TPU}(y, n, a_{n_1})]$, $E[\text{TPU}(y, n, a_{n_2})]$, ..., $E[\text{TPU}(y, n, a_{n_{\text{upper bound}}})]$. The c value associated with the highest of this list of profits is the optimal value of the acceptance level, a_n^*, with associated optimal sample size n^*, lot size y^*, and optimal profit $E[\text{TPU}(y^*, n^*, a_n^*)]$.

Example 2.9 Moussawi-Haidar et al. (2013) considered a situation with the following parameters: $D = 5000$ units/year, $K = \$100$/cycle, $h = \$5$/unit/year, $x = 175{,}200$ units/year, $C = \$50$/unit, $C_I = \$1$/unit, $C_d = \$25$ per unit defective, $s = \$70$/unit, and $p = 0.04$. The probability of acceptance, $P_a = P(X > n)$, is binomially distributed with parameters n and p and can be well approximated by a normal distribution when n is large or $np > 5$. In developing the numerical examples, they used the normal approximation of the binomial, with mean np and variance $np(1-p)$, so $P_a N(np, np(1-p))$.

Table 2.13 presents numerical results illustrating the effect of varying C_I, h, and C_d, one at a time, on the optimal solution, (y^*, n^*, a_n^*), and optimal profit, assuming that AOQL = 2.5%.

Table 2.13 Effect of varying C_I, h, and C_d, for AOQL = 2.5% (Moussavi-Haidar et al., 2013)

		a_n^*	n^*	y^*	Optimal profit
C_I	0.5	16	170.17	453.95	$89,832.2
	1	15	169.98	453.84	$88,880.56
	1.5	13	167.97	453.03	$87,928.5
	2	12	164.89	453.89	$86,976.61
	2.5	11	159.95	453.12	$86,024.56
h	5	12	165.09	452.54	$88,880.15
	10	9	116.12	320.30	$87,953.60
	15	8	95.702	261.72	$87,242.81
	20	8	84.33	226.81	$86,643.76
	25	8	75.88	202.97	$86,115.89
C_d	5	9	143.71	451.50	$91,417.40
	10	10	153.05	451.84	$90,783.30
	25	12	165.09	452.54	$88,880.15
	40	13	168.05	453.26	$86,976.70
	60	15	170.01	453.90	$84,438.77

2.3.6 Buy and Repair Options

In this section, a new model which is developed by Jaber et al. (2014) is presented. They developed an imperfect inventory system in which both repair and buy options for defective items are considered and compared.

In order to model the presented problem, some new notations which are specifically used are shown in Table 2.14.

2.3.6.1 Model I: Repair Case of Jaber et al. (2014)

According to Fig. 2.10, the maximum level of inventory or lot size is y and period length is T. The lot is subjected to a 100% inspection at a rate $X > D$, where ρy units (ρ is a percentage) of imperfect quality are withdrawn from inventory at the end of the screening period, t_I, and sent to a repair shop. Repaired items are returned after t_R units of time, which includes transportation and repair times where $t_I + t_R < T$ and $T = y/D$. Note that this model does not consider the case when $t_I + t_R > T$ as in our opinion it will logically be expensive favoring the buy option described in Case II as one has to account for backordering costs. They left this case as a mathematical exercise for interested readers. Further, assume that the repair process at the shop is always in control, which is not necessarily true. There are cases where the repair process may go out of control and restored through performing preventive maintenance (e.g., Jaber 2006; Liao and Sheu 2011, Liao 2012).

Figure 2.10 illustrates the behavior of inventory for the first case of the repair option. So, to repair the ρy items, the repair shop encounters the following costs: total cost to the repair shop $= K_R + 2K_s + \rho y \left(C_1 + 2C_T + h_2 t_R \right)$ where K_R is the repair setup cost, K_s is the transportation fixed cost (it is assumed that the transportation of

2.3 EOQ Model with No Shortage

Table 2.14 Notations of a given problem

$E[.]$	Expected value of a random variable
m	Markup percentage by the repair store
t_R	Transportation and repair time of imperfect products (time)
C_1	Material and labor cost to repair an item ($/unit)
K_R	Repay setup cost ($/setup)
h_2	Holding cost at the repair facility ($/unit/unit time)
C_E	Unit purchasing cost of an emergency order of a random variable ($/unit)
ρ	Fraction of defective items ($/unit/unit time)
t_R	Total transport time of defective units from the buyer to the repair shop and back to the buyer (time)
h_E	Holding cost of emergency order ($/unit/unit time)
C_R	Unit repair cost charged to the buyer ($/unit)

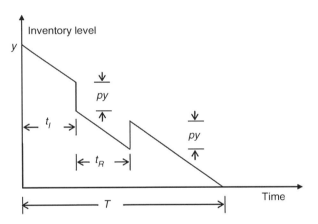

Fig. 2.10 Inventory level for the repair option for the defective items (Case I) (Jaber et al. 2014)

repairable items is charged to the repair shop), C_1 is the combined material and labor cost to repair an item, C_T is the unit transportation cost (from the inventory system to the repair shop and back to the system), and h_2 is the holding cost at the repair facility. Thus, each repaired item cost C_R, which could be agreed upon as a unit cost times a markup percentage, is:

$$C_R(y) = (1+m)\left(\frac{K_R + 2K_S}{\rho y} + C_1 + 2C_T + h_2 t_R\right) \quad (2.177)$$

where m is the markup percentage by the repair shop, $t_R = \rho y/R + t_T$, t_T is the total transport time of ρy units from the inventory system to the repair shop and back to the system, and R is the repair rate (where $R > D$), and these cost components are incurred by the repair shop. From this point onward, $C_R(y)$ will be used, which is paid by the firm for each repaired item. The holding cost per cycle is determined from Fig. 2.10 as:

$$HC = h\left(\frac{y^2(1-\rho)^2}{2D} + \rho\frac{y^2}{X}\right) + h_R\left(\rho\frac{y^2}{D} - \rho y\left(\frac{y}{X} + \frac{\rho y}{R} + t_T\right) - \rho^2\frac{y^2}{2D}\right) \quad (2.178)$$

where h_R is the holding cost of a repaired item. The cyclic total cost becomes:

$$TC(y) = \overbrace{K + Cy}^{\text{Fixed and purchasing cost}} + \overbrace{C_I y}^{\text{Inspection cost}} + \overbrace{C_R(y)\rho y}^{\text{Repair cost}}$$
$$+ \underbrace{h\left(\frac{y^2(1-\rho)^2}{2D} + \rho\frac{y^2}{X}\right) + h_R\left(\rho\frac{y^2}{D} - \rho y\left(\frac{y}{X} + \frac{\rho y}{R} + t_T\right) - \rho^2\frac{y^2}{2D}\right)}_{\text{Holding cost}} \quad (2.179)$$

Now, the total unit time profit is the total revenue per cycle less the total cost per cycle divided by the cycle time and is give as:

$$TPU(y) = sD - \frac{KD}{y} - CD - C_I D - C_R(y)\rho D - h\left(\frac{y(1-\rho)^2}{2} + \rho\frac{yD}{X}\right)$$
$$- h_R\left(\rho y - \rho D\left(\frac{y}{X} + \frac{\rho y}{R} + t_T\right) - \rho^2\frac{y}{2}\right)$$
$$= sD - \frac{KD}{y} - CD$$
$$- C_I D - (1+m)\frac{(K_R + 2K_S)D}{y} - (1+m)\left(C_1 + 2C_T + h_2\frac{\rho y}{R} + h_2 t_T\right)\rho D$$
$$- h\left(\frac{y(1-\rho)^2}{2} + \rho\frac{yD}{X}\right) - h_R\left(\rho y - \rho D\left(\frac{y}{X} + \frac{\rho y}{R} + t_T\right) - \rho^2\frac{y}{2}\right)$$
$$(2.180)$$

whose solution is given by setting the first derivative of Eq. (2.180) to zero and solving for y to get:

$$y^* = \sqrt{\frac{(K + (1+m)(K_R + 2K_S))D}{h\left(\frac{y(1-\rho)^2}{2} + \rho\frac{yD}{X}\right) + (1+m)h_2\frac{\rho^2 D}{R} + h_R\left(\rho - \rho\left(\frac{D}{X} + \frac{\rho D}{R}\right) - \frac{\rho^2}{2}\right)}} \quad (2.181)$$

The expected value of Eq. (2.180) is given as:

2.3 EOQ Model with No Shortage

$$E[TPU(y)] = sD - \frac{KD + (1+m)(K_R + 2K_S)D}{y} - CD - C_1D$$

$$-(1+m)\left(C_1 E[\rho] + 2C_T E[\rho] + h_2 \frac{E[\rho^2]y}{R} + h_2 t_T E[\rho]\right)D$$

$$-h\left(\frac{y(1 - 2E[\rho] + E[\rho^2])}{2} + E[\rho]\frac{yD}{X}\right)$$

$$-h_R\left(E[\rho]y - D\left(\frac{y}{X}E[\rho] + \frac{y}{R}E[\rho^2] + t_T E[\rho]\right) - E[\rho^2]\frac{y}{2}\right)$$

(2.182)

whose solution is given in a similar manner to Eq. (2.181) as:

$$y^* = \sqrt{\frac{2(K + (1+m)(K_R + 2K_S))D}{h(1 + E[\rho^2] + E[\rho](\frac{D}{X} - 1)) + 2(1+m)h_2 \frac{E[\rho^2]D}{R} + h_R(2E[\rho](1 - \frac{D}{X}) - E[\rho^2](\frac{2D}{R} + 1))}}$$

(2.183)

If the repaired items are received by time $y(1 - \rho) = D$ as shown in Fig. 2.9, then Eq. (2.181) is rewritten as (Jaber et al. 2014):

$$TPU(y) = sD - \frac{KD}{y} - CD - C_1D - C_R(y)\rho D - h\left(\frac{y(1-\rho)^2}{2} + \rho\frac{yD}{X}\right)$$

$$- h_R \rho^2 \frac{y}{2}$$

(2.184)

The expected value of Eq. (2.184) is given as:

$$E[TPU(y)] = sD - \frac{KD}{y} - CD - C_1D - (1+m)\frac{(K_R + 2K_S)D}{y}$$

$$-(1+m)\left((C_1 + 2C_T)E[\rho] + h_2 \frac{E[\rho^2]y}{R}\right)D$$

$$-h\left(\frac{y(1 - 2E[\rho] + E[\rho^2])}{2} + E[\rho]\frac{yD}{X}\right) - h_R E[\rho^2]\frac{y}{2}$$

(2.185)

whose solution is given in a similar manner to Eq. (2.181) as (Jaber et al. 2014):

$$y^* = \sqrt{\frac{2(K + (1+m)(K_R + 2K_S))D}{h((1 + E[\rho^2]) + 2E[\rho](\frac{D}{X} - 1)) + 2(1+m)\left(h_2 \frac{E[\rho^2]D}{R}\right) + h_R E[\rho^2]}}$$

(2.186)

2.3.6.2 Model II: Buy Case of Jaber et al. (2014)

In the buy scenario, like in Salameh and Jaber (2000), the imperfect-quality items are withdrawn from inventory by the end of the inspection/screening period and are sold (salvaged) as a single batch for v. The imperfect-quality items are substituted with an emergency order at a cost of C_E each, where $C_E < C < v$. The behavior of inventory for this model is the same as shown in Fig. 2.11 (Jaber et al. 2014).

The expected unit time profit is similar to Eq. (2.182), which is:

$$E[\text{TPU}(y)] = sD + vE[\rho]D - \frac{KD}{y} - CD - C_1D$$
$$-C_E E[\rho]D - h\left(\frac{y(1 - 2E[\rho] + E[\rho^2])}{2} + E[\rho]\frac{yD}{X}\right) - h_E E[\rho^2]\frac{y}{2} \quad (2.187)$$

Finally:

$$y^* = \sqrt{\frac{2KD}{h\left(1 - 2E[\rho] + E[\rho^2] + 2E[\rho]\frac{D}{X}\right) + h_E E[\rho^2]}} \quad (2.188)$$

The behaviors of Models I and II are illustrated numerically in the next section.

Example 2.10 In this section, the input parameters of the numerical examples are adopted from Salameh and Jaber (2000) and Jaber et al. (2014) as provided below in Table 2.15. Using data:

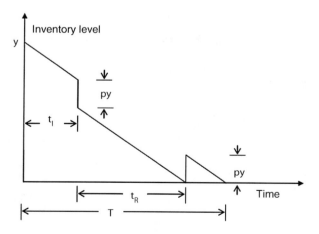

Fig. 2.11 Inventory level for the repair option for the defective items (Case II) (Jaber et al. 2014)

2.3 EOQ Model with No Shortage

Table 2.15 Data for the numerical analysis from Salameh and Jaber (2000) (Jaber et al. 2014)

Symbol	Value	Units	Symbol	Value	Units
D	50,000	Units/year	K_R	100	$
X	175,200	Units/year	K_S	200	$
s	50	$/unit	C_T	2	$/unit
K	100	$	C_I	5	$/unit
C	25	$/unit	h_2	4	$/unit/year
v	20	$/unit	R	50,000	Units/year
h	5	$/unit/year	t_T	2/220	Year
C_I	0.5	$/unit	h_R	6	$/unit/year
ρ	$U(0, 0.04)$		m	20%	
$f(\rho)$	$1/(0.04-0)$		C_E	40	$/unit
h_E	8	$/unit/year			

$$E[\rho] = \int_b^b \rho f(\rho) d\rho = \int_a^b \rho \frac{1}{b-a} d\rho = \frac{b+a}{2} = \frac{0.04+0}{2} = 0.02$$

$$E[\rho^2] = \int_a^b \rho^2 f(\rho) d\rho = \int_a^b \rho^2 \frac{1}{b-a} d\rho = \frac{a^2+ab+b^2}{3} = \frac{0+0+(0.04)^2}{3}$$

$$= 5.33E - 04.$$

The optimal policy for Model I Case I occurs when $y^* = 3732$ and $E[TPU(y^*)] = 1,195,455$, where $C_R = 18.89$. Solving for Case II when repaired items are received by time $y(1-p)/D$, the optimal policy occurs when $y = 3792$ and $E[TPU(y)] = 1,195,743$, where $C_R = 18.72$. It should be noted that C_R can be computed from Eq. (2.177) by substituting $\rho = E[\rho]$ (i.e., $E[C_R \rho y]/E[\rho y] = yE[C_R \rho]/yE[\rho] = E[C_R \rho]/[\rho]$). The optimal policy for Model II occurs when $y^* = 1434$ and $E[TPU(y^*)] = 1,198,026$ (Jaber et al. 2014).

2.3.7 Entropy EOQ

2.3.7.1 Model 1: Entropy EOQ Without Screening

In this section, the work of Jaber et al. (2009) in which entropy cost for EOQ model is considered will be presented. Jaber et al. (2009) analogized a production system with physical thermodynamic system. They expressed that a production system resembles a physical system operating within surroundings, including the market and the supply system. Similarly, a physical thermodynamic system is defined by its temperature, volume, pressure, and chemical composition. A production system could be described analogously by its characteristics, for example, the price (s)

Table 2.16 Notations of a given problem

s	The reduced selling price per unit ($/unit)
s_o	The higher selling price per unit ($/unit)
ρ	The Markov transition probability of production process (machine) shifts to an out-of-control state and the production process begins to produce
θ	The expected number of defective in a lot (unit)
h_1	The holding cost adjusted for the average additional cost of reworking defective items $h_1 > h$ ($/unit/unit time)
C^E	Entropy cost per unit ($/unit)
y_e^*	Optimal lost size under entropy cost (unit)
$TC_E(y)$	The total cost of EOQ model under entropy cost ($)

that the system ascribes to the commodity (or collection of commodities) that it produces. Reducing the price of the commodity below the market price may increase demand and produce a commodity flow (sales) from the system to its surroundings. This is similar to the flow of heat from a high temperature (source) to a low temperature (sink) in a thermodynamic system, where part of this heat is converted into useful work and some of the heat is lost from the system and wasted. So in the work of Jaber et al. (2009), a similar behavior between price in supply system and temperature in thermodynamic system presented and used a demand function $D = -W(s - s_o)$ to investigate the effects of price changes such as temperature changes.

Noting that when $s - s_o$ (analogues to difference in temperature in a thermal system), the direction of the commodity (heat) flow is from the system to its surrounding, i.e., from a high-temperature reservoir (low price) to a low-temperature reservoir (high price). The second law of thermodynamics applies to the spontaneous flow of heat from hot (low price) to cold (higher price). Heat flow ceases when the system and its surrounding have the same temperature (Jaber et al. 2009).

The new notations used in this section are presented in Table 2.16.

The EOQ model assumes constant commodity flow, an infinite planning horizon, no shortages (lost sales or backorders are not allowed), instantaneous replenishment of orders, zero lead time, no quantity discounts, no deterioration, and all units conform to quality and that the input cost parameters are constant over time. The EOQ model adopts the classical approach of minimizing the sum of the conflicting holding and procurement costs. The EOQ cost function is given as (Jaber et al. 2009):

$$\text{TC}(y) = \frac{KD}{y} + CD + h\frac{y}{2} = -\frac{KW(s - s_0)}{y} - CW(s - s_0) + h\frac{y}{2} \quad (2.189)$$

The optimal solution is obtained by setting the first derivative to zero to get (Jaber et al. 2009):

2.3 EOQ Model with No Shortage

$$y^* = \sqrt{\frac{2KD}{h}} = \sqrt{\frac{2KW(s-s_0)}{h}} \tag{2.190}$$

where the optimal cycle time is $T^* = y^*/D = y^*/W(s-s_0)$. Excluding the material cost $-CW(s-s_0)$, the optimal cost could be computed as (Jaber et al. 2009):

$$\text{TC}(y^*) = \sqrt{2hKW(s-s_0)} \tag{2.191}$$

Porteus (1986) denoted θ as the expected number of defectives in a lot of size y, given that the process is in control before beginning the lot. Then (Jaber et al. 2009):

$$\theta = \rho y + \beta d_{Q-1}, \quad \text{where } (\beta = 1 - \rho)$$

$$= \rho y \sum_{i=1}^{n-1} \beta^i - \rho \sum_{i=1}^{n-1} i\beta^i = \rho y \frac{(1-\beta^y)}{\rho} - \rho \frac{\beta[1+(y-1)\beta^y - Q\beta^{y-1}]}{\rho^2}$$

$$= y - \beta \frac{(1-\beta^y)}{\rho}$$

where θ could also be computed as $\theta = \sum_{i=1}^{y} \rho(y-i)(1-\rho)^{i-1}$. Porteus (1986) suggested that for very small ρ values, $\theta \approx \rho y^2/2$. This approximation has been adopted by several researchers (e.g., Chand 1989; Jaber 2006). The cost function (2.189) is then modified as follows (Jaber et al. 2009):

$$\begin{aligned}\text{TC}(y) &= \frac{KD}{y} + CD + h_1 \frac{y}{2} + C_R \rho D \frac{y}{2} \\ &= -\frac{KW(s-s_0)}{y} - CW(s-s_0) + h_1 \frac{y}{2} - C_R \rho W(s-s_0) \frac{y}{2}\end{aligned} \tag{2.192}$$

The holding cost for a single unit $h_1 = iC + i\rho C_R/2 = h + h\rho C_R/2C$, where $i = h/C$. Equation (2.192) can now be written as (Jaber et al. 2009):

$$\begin{aligned}\text{TC}(y) &= \frac{KD}{y} + CD + h\left(1 + \rho \frac{C_R}{2C}\right)\frac{y}{2} + C_R \rho D \frac{y}{2} \\ &= -\frac{KW(s-s_0)}{y} - CW(s-s_0) + h\left(1 + \rho \frac{C_R}{2C}\right)\frac{y}{2} - C_R \rho W(s-s_0)\frac{y}{2}\end{aligned} \tag{2.193}$$

The optimal solution is given by setting the first derivative of Eq. (2.193) to zero

$$y^{**} = \sqrt{\frac{2KD}{h(1+\rho(C_R/2C))+\rho C_R D}}$$

$$= \sqrt{\frac{2KW(s-s_0)}{h(1+\rho(C_R/2C))+\rho C_R W(s-s_0)}} \quad (2.194)$$

When $\rho = 0$ Eq. (2.194) reduces to Eq. (2.192). Substituting Eq. (2.194) in Eq. (2.193), and excluding the material cost, the optimal cost could be computed as:

$$\text{TC}(y^*) = \sqrt{\left[h\left(1+\rho\frac{C_R}{2C}\right)+\rho C_R W(s-s_0)\right] \times [2KW(s-s_0)]} \quad (2.195)$$

Considering Entropy Cost

Jaber et al. (2009) used the following equation as entropy cost per unit:

$$C^E = -\frac{s_0 s}{(s-s_0)} \quad (2.196)$$

Accounting for entropy cost is done by adding the entropy cost per unit time, C^E/T where $T = y/D$, to the other cost terms and substituting $D = E(s - s_0)$ in Eq. (2.193):

$$\text{TC}_E(y) = \frac{KD}{y} + CD + h\left(1+\rho\frac{C_R}{2C}\right)\frac{y}{2} + C_R\rho D\frac{y}{2} + \frac{s_0 s W}{y}$$

$$= -\frac{KW(s-s_0)}{y} - CW(s-s_0) + h\left(1+\rho\frac{C_R}{2C}\right)\frac{y}{2} - C_R\rho W(s-s_0)\frac{y}{2} + \frac{s_0 s W}{y} \quad (2.197)$$

The optimal solution is given by setting the first derivative of Eq. (2.197) to zero to get (Jaber et al. 2009):

$$y_e^{**} = \sqrt{\frac{2KD + 2s_0 s W}{h(1+\rho(C_R/2C))+\rho C_R D}}$$

$$= \sqrt{\frac{2KW(s_0-s) + 2s_0 s W}{h(1+\rho(C_R/2C))+\rho C_R W(s-s_0)}} \quad (2.198)$$

Substituting Eq. (2.198) in Eq. (2.197), and excluding the material cost, the optimal cost could be computed as (Jaber et al. 2009):

2.3 EOQ Model with No Shortage

$$\mathrm{TC_E}(y_e^*) = \sqrt{\left[h\left(1 + \rho\frac{C_R}{2C}\right) + \rho C_R W(s_0 - s)\right] \times [2KW(s - s_0) + 2s_0 sW]}$$

When $\rho = 0$, Eq. (2.198) reduces to (Jaber et al. 2009):

$$y_e^* = \sqrt{\frac{2KD + 2s_0 sW}{h}} = \sqrt{\frac{2KW(s_0 - s) + 2s_0 sW}{h}} \quad (2.199)$$

which is the entropic version of Eq. (2.190) whose cost function reduces from Eq. (2.197) to (Jaber et al. 2009):

$$\widehat{C}_E(y) = \frac{KD}{y} + CD + h\frac{y}{2} + \frac{s_0 sW}{y}$$

$$= -\frac{KW(s_0 - s)}{y} - CW(s_0 - s) + h\frac{y}{2} + \frac{s_0 sW}{y} \quad (2.200)$$

Similarly, substituting Eq. (2.200) in Eq. (2.199), and excluding the material cost, the optimal cost could be computed as (Jaber et al. 2009):

$$\mathrm{TC_E}(y_e^*) = \sqrt{h[2KW(s_0 - s) + 2s_0 sW]}$$

Example 2.11 Jaber et al. (2009) considered an inventory situation where the product price $s = 100$, the market equilibrium price $s_0 = 105$, and $W = 20$ represents the change in the flux for a change in the price of a commodity (e.g., units/day/$), corresponding to a commodity flow (demand) rate $D = W(s_o - s) = 20(105 - 100) = 100$. Also $h = 0.10$, $C = 60$, $C_R = 30$, $K = 100$, and $\rho = 0.0001$. In the numerical examples below, the term $CD = CW(ss_0)$ is excluded from the cost functions because it actually dominates the cost function ($CD = 60(100) = 6000$) although it does not affect the optimization.

In the case when there are no defects, $y = 447.21$ and $TC(y) = 44.72$. When the entropy cost is included in the cost function, the optimal entropic order quantity using Eq. (2.199) $y_e^* = 2097.62$, which minimizes, $\mathrm{TC_E}(y_e^*) = 209.76$, is obtained from Eq. (2.200). For a more detailed information about the model, readers can see Jaber et al. (2009).

2.3.7.2 Model 2: Entropy EOQ with Screening

Jaber et al. (2013) revisited their previous work (Jaber et al. 2009) by adding a screening topic. In this section, the work of Jaber et al. (2013) is presented. The behavior of inventory in this problem is depicted in Fig. 2.12 (Jaber et al. 2013). Upon replenishing the inventory instantaneously, the received lot is subjected to a 100% screening process. Items that do not confirm to quality are withdrawn from the inventory and sold at a discounted price as a single batch.

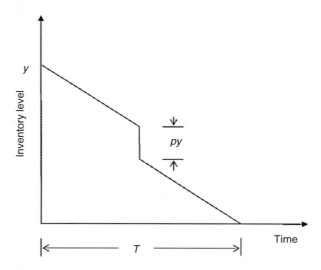

Fig. 2.12 Salameh and Jaber's model (Jaber et al. 2009)

Table 2.17 Notations of a given problem

s_0	Is the selling price of a good-quality item of a competitor brand in the market (\$/unit), where $s_0 > s$ (\$/unit)
g_0	Is the quality index of an item of a competitor brand (1/unit), where $0 < g_0 < 1$
g	Is the quality index of an item (1/unit), where $0 < g < 1$
$D(s, g)$	Demand rate (unit/year) as a function of s and g, where $s/g > s_0/g_0$
W	Elasticity (unit/year/\$) of the demand function $D(s, g)$
C_{inv}	Is the investment required to increase quality index g by 1% (\$/1%/year)

In order to model the presented problem, some new notations which are specifically used are shown in Table 2.17.

The total revenue per cycle is the sum of the revenue from selling good-quality items $Py(1 - \rho)$ and the revenue from salvaging imperfect-quality items, $vy\rho$, and is given as (Jaber et al. 2013):

$$\text{TR}(y) = sy(1 - \rho) + vy\rho$$

The total cost per cycle is (Jaber et al. 2013):

$$\text{TC}(y) = K + Cy + C_1 y + h\left(\frac{y(1 - \rho)}{2}T + \frac{\rho y^2}{x}\right)$$

Therefore, the profit function per unit of time is:

2.3 EOQ Model with No Shortage

$$\begin{aligned}\text{TPU}(y) &= \frac{\text{TR}(y) - \text{TC}(y)}{T} \\ &= \frac{D}{y(1-\rho)} \\ &\quad \times \left\{ sy(1-\rho) + vy\rho - K - Cy - C_1 y - h\left(\frac{y(1-\rho)}{2}T + \frac{\rho y^2}{x}\right) \right\}\end{aligned}$$

Then:

$$\text{TPU}(y) = D(s - v + hy/x) + D(v - hy/x - C - C_1 - K/y)\frac{1}{1-\rho} \\ - h\frac{y(1-\rho)}{2} \tag{2.201}$$

whose expected value is:

$$\text{ETPU}(y) = D(s - v + hy/x) + D(v - hy/x - C - C_1 - K/y)E\left[\frac{1}{1-\rho}\right] \\ - h\frac{y(1 - E[\rho])}{2} \tag{2.202}$$

The optimal solution is given as (Jaber et al. 2013):

$$y^* = \sqrt{\frac{2KDE[1/(1-\rho)]}{h(1 - E[\rho] - 2(d/x))(1 - E[1/(1-\rho)])}} \tag{2.203}$$

But in order to develop the thermodynamic version of the proposed model, Jaber et al. (2013) applied the per unit time profit function and the optimal lot size of Maddah and Jaber as below:

$$E[\text{TPU}(y)] = \frac{E[\text{TR}(y)] - E[\text{TC}(y)]}{E[T]} = \frac{E[\text{TP}(y)]}{E[T]}$$

$$= \frac{D\{s(1 - E[\rho]) + vE[\rho] - C - C_1\} - \left\{\frac{KD}{y} + hy\frac{E\left[(1-\rho)^2\right]}{2} + hy\frac{D}{x}E[\rho]\right\}}{1 - E[\rho]} \tag{2.204}$$

$$y^* = \sqrt{\frac{2KD}{hE\left[(1-\rho)^2\right] + 2hE[\rho]D/x}} \qquad (2.205)$$

Jaber et al. (2013) applied the laws of thermodynamics to inventory systems, where they postulated that commodity flow (demand) is similar to energy flow in thermal systems as presented in the previous model. They proposed commodity flow to be of the form:

$$D(t) = -W(s(t) - s_0(t)) \qquad (2.206)$$

where $s(t)$ and $s_0(t)$ are analogous, respectively, to the system temperature TH (high) and the surrounding temperature TL (low) and W (analogous to a thermal capacity) is the change in the flux for a change in the price $(s(t) - s_0(t))$ of a commodity and is measured in additional units per year per change in unit price, e.g., dollar (units/year/$). The difference between the temperature of a thermodynamic system and that of its surrounding (TH − TL > 0) creates a flow from the thermodynamic system to its surrounding (Jaber et al. 2013).

Jaber et al. (2013) assumed that the flow of commodity is one dimensional (price). The quality of the product was implicitly assumed to be the same as that of the competitor's brand in the market and was set at 1 (dimensionless measure between 0 and 1). They proposed a commodity flow function of the form:

$$D(t) = -W(V(t) - V_0(t)) = -W\left(\frac{s(t)}{g(t)} - \frac{s_0(t)}{g_0(t)}\right) \qquad (2.207)$$

where $0 < g(t) < 1$ and $0 < g_0(t) < 1$ are the goodness (quality) measures of the firm's product and that of its competitor brand, respectively, where $V(t)$ and $V_0(t)$ are the values as perceived by the customer. The smaller $V(t)$ is with respect to $V_0(t)$, the better it is for the customer as it is getting a higher-quality product for less. So, substituting $g(t) = 1$ and $g_0(t) = 1$ in Eq. (2.207) reduces it to Eq. (2.206), making it a special case of the latter. For commodity flow $\frac{s(t)}{g(t)} < \frac{s_0(t)}{g_0(t)}$ for every $t > 0$. The entropy generated per cycle of length T is determined from Eq. (2.207) as (Jaber et al. 2013):

$$\sigma(T) = \int_0^T \left(\frac{V(t)}{V_0(t)} + \frac{V_0(t)}{V(t)} - 2\right) dt \qquad (2.208)$$

Since the demand is considered as the constant rate, so assuming Eqs. (2.206)–(2.207) for constant situation, $V(t) = V$ and $V_0(t) = V_0$ for every $t > 0$, Eq. (2.206) reduces to: $D(t) = D(s, g) = -W(V - V_0) = -W(s/g - s_0/g_0)$ (Jaber et al. 2013).

The entropy cost per cycle is determined from Eqs. (2.207) and (2.208) as (Jaber et al. 2013):

2.3 EOQ Model with No Shortage

$$\text{En} = \text{En}(T) = \frac{\int_0^T -W(s/g - s_0/g_0)dt}{W\int_0^T \left(\frac{s/g}{s_0/g_0} + \frac{s_0/g_0}{s/g} - 2\right)dt} = -\frac{ss_0}{sg_0 - s_0g} \quad (2.209)$$

where $g_0 s < s_0 g \Rightarrow g_0/g < s_0/s$, $s < s_0 g/g_0$. Note that improving the quality of a product is usually associated with investment, and the function is $c_{inv}(g - g_0)/g_0$, where c_{inv} is the investment required to increase the quality by 1%. In this work, the salvage price is a decision variable as the imperfect-quality items may as well follow a commodity flow function like the one in Eq. (2.207) and is given as (Jaber et al. 2013):

$$y\rho = -WT\left(\frac{v}{g_s} - \frac{C}{g_0}\right) \quad (2.210)$$

$$y = -\frac{W}{\rho}T\left(\frac{v}{g_s} - \frac{C}{g_0}\right) \quad (2.211)$$

where $y\rho$ is the demand rate that occurs once in a cycle at y/x units of time and g_s is the quality of a salvaged item. The entropy cost for the flow of imperfect items is computed in a similar manner to Eq. (2.209) and is given as (Jaber et al. 2013):

$$\sigma_s(T) = WT\left(\frac{v/g_s}{C/g_0} + \frac{C/g_0}{v/g_s} - 2\right) \quad (2.212)$$

The entropy cost per cycle to manage the flow of imperfect-quality items is determined from Eqs. (2.210) and (2.212) in a similar manner to determining Eq. (2.209) as (Jaber et al. 2013):

$$\text{En}_s = \text{En}_s(T) = \frac{\rho y}{\sigma_s(T)} = -\frac{vC}{vg_0 - Cg_s} \quad \text{where } v < Cg_s/g_0 \quad (2.213)$$

In this section, they modified the model in Eq. (2.204) by replacing the constant D with $D(s,g)$ and subtracting the entropy cost per unit of time expressions and the investment cost from Eq. (2.204), which becomes (Jaber et al. 2013):

$$E[\text{TPU}_E(y)] = \frac{D(s,g)\left\{s(1-E[\rho]) + vE[\rho] - \frac{K}{y} - C - C_1\right\} - hy\left(\frac{E\left[(1-\rho)^2\right]}{2} + \frac{E[\rho]D(s,g)}{x}\right)}{1 - E[\rho]}$$
$$+ \frac{D(s,g)}{(1-E[\rho])y}\left[\frac{ss_0}{(sg_0 - s_0g)} + \frac{vC}{(vg_0 - Cg_s)}\right] - \frac{c_{inv}(g - g_0)}{g_0} \quad (2.214)$$

The optimal solution that maximizes is given as (Jaber et al. 2013):

$$y_E^* = \sqrt{\frac{2D(s,g)\left(K - \frac{ss_0}{(sg_0 - s_0 g)} - \frac{vC}{(vg_0 - Cg_s)}\right)}{hE\left[(1-\rho)^2\right] + 2h\frac{D(s,g)}{x}E[\rho]}} \qquad (2.215)$$

Equation (2.214) can be optimized for v and s solving a nonlinear programming model (NLPP) as (Jaber et al. 2013):

$$\begin{aligned} &\text{Maximize } E[\text{TPU}_E(s,v)], \\ &\text{s.t.} \\ &s > C, \\ &sg_0 < s_0 g, \\ &0 < v < Cg_s/g_0 \end{aligned} \qquad (2.216)$$

The proposed NLPP model can be solved by any suitable optimization package which can be used.

2.4 EOQ Model with Backordering

In this section, the imperfect EOQ model with backordering will be presented.

2.4.1 Imperfect Quality and Inspection

In this section, an EOQ model with imperfect items with backordering according to the work of Eroglu and Ozdemir (2007) is presented. In order to model the presented problem, some new notations which are specifically used are shown in Table 2.18.

Assume that y is replenished instantaneously at unit purchasing price of c and fixed cost of K per order similar to the previous models. The percentage of defectives in each lot is p, with a known probability density function, $f(p)$. A 100% screening policy with screening rate per unit time of x is applied. $(1 - \theta)\%$ of the defective items is imperfect and $\theta\%$ is scrap items. When screening process is finished, imperfect-quality items are sold as a single lot, and scrap items are subtracted from inventory with unit cost of C_d. The selling prices of good- and imperfect-quality items are s and v per unit, respectively, where $s < v$. Figure 2.13 presents the behavior of the inventory level. The rate of good-quality items which are screened during t_2 is $(1 - p)$. A part of these good-quality items meet the demand with a rate of D, and the remaining is used to eliminate backorders with a rate of $(1 - p)x - D = x(1 - p - D/x)$ (Eroglu and Ozdemir 2007).

The screening process finishes up at the end of time interval of t_3, and defective items of py are subtracted from inventory. Since the demand has been met from

2.4 EOQ Model with Backordering

Table 2.18 Notations of a given problem

t_1	Time to build up a backorder level of "w" units (time)
t_2	Time to eliminate the backorder level of w units (time)
t_3	Time to screen y units ordered per cycle (time)
B	The backorder level (unit)

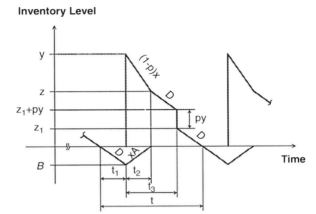

Fig. 2.13 Behavior of the inventory level over time (Eroglu and Ozdemir 2007)

perfect-quality items, the period length t is calculated by dividing the amount of perfect-quality items in a period to amount of demand in unit time:

$$t = \frac{(1-p)y}{D} \quad (2.217)$$

Since the percentage of defective items, p, is a random variable, the expected value of period length is given by:

$$E[t] = \frac{(1-E[p])y}{D} \quad (2.218)$$

Referring to Fig. 2.13, the findings are as follows. The time, t_1, needed to build up a backorder level of "B" units is:

$$t_1 = \frac{B}{D} \quad (2.219)$$

The time, t_2, needed to eliminate the backorder level of "B" units is:

$$t_2 = \frac{B}{x[1-p-D/x]} \quad (2.220)$$

$$t_2 = \frac{y-z}{(1-p)x} \tag{2.221}$$

Using Eqs. (2.229) and (2.221):

$$z = y - \frac{(1-p)B}{[1-p-D/x]} \tag{2.222}$$

According to Fig. 2.13:

$$t_3 = y/x \tag{2.223}$$

$$t_3 - t_2 = (z - z_1 - py)/D \tag{2.224}$$

And:

$$z_1 = [1 - p - D/x]y - B \tag{2.225}$$

Finally using Eqs. (2.217)–(2.225), the total cost function will be (Eroglu and Ozdemir 2007):

$$\begin{aligned}
TC &= \overbrace{(Cy+K)}^{\text{Purchasing cost}} + \overbrace{(C_1 y)}^{\text{Screening cost}} + \overbrace{(C_d \theta p y)}^{\text{Disposal cost}} + \overbrace{\left[\frac{C_b(t_1+t_2)B}{2}\right]}^{\text{Shortage cost}} \\
&+ \underbrace{\left\{ h \left[\frac{t_2(y+z)}{2} + \frac{(t_3-t_2)(z+z_1+py)}{2} + \frac{(t-t_1-t_3)z_1}{2} \right] \right\}}_{\text{Holding cost}} \\
&= (C + C_1 + C_d \theta p)y + K + \frac{h}{2} \times \left(\frac{2 - D/x}{x} + \frac{(1-p-D/x)^2}{D} \right) y^2 \\
&\quad - \frac{h(1-p)By}{D} + \frac{(h+C_b)(1-p)B^2}{2D(1-p-D/x)}
\end{aligned} \tag{2.226}$$

Moreover, the total revenue, TR, of both imperfect and good items is:

$$TR = s(1-p)y + v(1-\theta)py \tag{2.227}$$

Since cycle length is a variable, using the renewal reward theorem, the expected total profit per unit time is given as:

2.4 EOQ Model with Backordering

$$E[\text{TPU}] = \frac{E(\text{TR}) - E(\text{TC})}{E(t)}$$
$$= sD + \frac{vD(1-\theta)E(p)}{E_1} - \frac{D(C + C_1 + C_d\theta p)}{E_1} - \frac{KD}{yE_1} - \frac{hE_4 y}{2E_1} + hB - \frac{(h+C_b)E_2 B^2}{2yE_1}$$
(2.228)

where

$E_1 = 1 - E(p)$, $E_2 = E\left(\frac{1-p}{1-p-D/x}\right)$, $E_3 = E[(1 - p - D/x)^2]$, $E_4 = \frac{D(2-D/x)}{x} + E_3$.

Since the expected total profit is strictly concave, setting its partial derivatives with respect to B and y gives:

$$y^* = \sqrt{\frac{2KD}{h\left(E_4 - \frac{hE_1^2}{(h+C_b)E_2}\right)}} \quad (2.229)$$

$$B^* = \frac{hE_1 y^*}{(h + C_b)E_2} \quad (2.230)$$

But it should be noted that the derived optimal values are valid if both next conditions hold:

1. In order to avoid the backorders from the beginning of each cycle:

$$xE(1 - p - D/x) > 0 \text{ or } E(p) < 1 - D/x \quad (2.231)$$

and

$$x > D \quad (2.232)$$

2. t_3, must be at least equal or greater than $E(t_2)$. So:

$$E(t_2) \leq t_3$$

or

$$\frac{h}{h + C_b} \leq \frac{E(1 - p - D/x)E_2}{1 - E(p)} \quad (2.233)$$

If shortage cost is infinite, and scrap rate and unit scrap cost are zero, then the model with no shortages is attained. Thus, the following reduced forms of Eqs. (2.228)–(2.230) are achieved (Eroglu and Ozdemir 2007):

$$E[\text{TPU}] = sD + \frac{vDE(p)}{E_1} - \frac{D(C+C_1)}{E_1} - \frac{KD}{yE_1} - \frac{hE_4y}{2E_1} \qquad (2.234)$$

$$y^* = \sqrt{\frac{2KD}{hE_4}} \qquad (2.235)$$

$$B^* = 0 \qquad (2.236)$$

Suppose that defective's fraction, p, is zero. This yields t_3 and C_I, are zero and x, is infinite. So, $E_1 = E_2 = E_3 = E_4 = 1$ and Eqs. (2.228)–(2.230) are reduced to the following equations which are the same equations as those given by classical EOQ model with shortages:

$$\text{TPU} = (s-C)D - \frac{KD}{y} - \frac{hy}{2} + hB - \frac{(h+C_b)B^2}{2y} \qquad (2.237)$$

$$y^* = \sqrt{\frac{2KD(h+C_b)}{hC_b}} \qquad (2.238)$$

$$B^* = \frac{hy^*}{(h+C_b)} \qquad (2.239)$$

Example 2.12 Eroglu and Ozdemir (2007) presented a firm that orders a product as lots to meet outside demand. The defective fraction in each lot has a uniform distribution with the following probability density function:

$$f(p) = \begin{cases} 10, & 0 \le p \le 0.1 \\ 0, & \text{otherwise} \end{cases}$$

And the other model parameters are given as follows $D = 15{,}000$ units/year, $K = \$400$/cycle, $h = \$4$ unit/year, $C_b = 6$, $x = 60{,}000$ unit/year, $C_I = \$1$/unit, $C = \$35$/unit, $s = \$60$/unit, $v = \$25$/unit, $\theta = 0.2$, and $C_d = \$2$/unit. By using the above parameters, $E(p) = 0.05$, $E_1 = 0.95$, $E_2 = 1.357752$, $E_3 = 0.490833$, and $E_4 = 0.928333$, the optimum values of solution are calculated as $y^* = 2128.06$ units, $B^* = 595.59$ units, and $E[\text{TPU}]^* = \$341{,}116.89$. In addition, Eqs. (2.231)–(2.233) are valid herein.

2.4.2 Multiple Quality Characteristic Screening

Assume an imperfect single-item inventory system with constant demand rate and permitted backordering in which the lead time is negligible. Suppose that after the replenishment of items arrives from the supplier, the items have to go through n screening processes on each quality characteristic before delivering to customers. Denote by S_i the screening process for quality characteristic i and by x_i the corresponding screening rate of S_i, where $i = 1, \ldots, n$. Without loss of generality,

2.4 EOQ Model with Backordering

they supposed $x_1 \geq x_2 \geq \cdots \geq x_n$. The processes are performed in parallel and are completed when the last process is finished. Screening rate is greater than the demand rate, i.e., $x_i > D$ ($i = 1, \ldots, n$), and defective items exist in lot size y and are returned to the supplier when replenishment items arrive (Tai 2015).

In order to model the presented problem, some new notations which are specifically used are shown in Table 2.19.

Let p_i be the proportion of items in the batch that cannot pass S_i. Here they assumed that all p_i is independent of each other in the sense that whether an item can pass S_i does not depend on the result of other screening processes. For any $i < j$, the screening process S_i will finish before S_j. The first screening process S_1 starts at time 0 and finishes at time y/x_1. During this period, $p_1 y$ items cannot pass S_1, and they are removed from the inventory at time y/x_1. To save the transportation cost and administrative work, the items which cannot pass the screening process are removed from the inventory as one lot at the end of each screening process.

The next screening process S_2 finishes at time $y/x_2 \geq y/x_1$ since $x_2 \leq x_1$. The number of items which cannot pass S_2 is $p_2 y$. The items to be screened out are items which pass S_1 but not S_2. Hence, the number of items screened out by S_2 is $p_2(1 - p_1) y$.

By similar arguments, the items to be screened out by S_i are items which pass S_1, $\ldots$, $S_i - 1$ but not S_i. Because all screening processes are independent, the proportion of items in y that is screened out after S_i, which is denoted by ρ_i, is:

$$\rho_i = \begin{cases} p_1 & \text{for } i = 1 \\ \prod_{k=1}^{i-1} (1 - p_k) p_i & \text{for } i = 2, \ldots, n \end{cases}$$

The total number and portion of defective are $\sum_{i=1}^{n} \rho_i y$ and $\rho = \sum_{i=1}^{n} \rho_i$, respectively (Tai 2015).

The inventory level in a replenishment cycle is illustrated in Fig. 2.14. At the beginning of each cycle, all screening processes proceed simultaneously. The rate of items to complete all the screening processes depends only on the lowest screening

Table 2.19 Notations of a given problem

n	Number of quality characteristics of the product to be inspected
S_i	Screening process for quality characteristic i ($i = 1, \ldots, n$)
ρ_i	Random variable representing the proportion of items in y that screen out after S_i ($i = 1, \ldots, n$)
t_1	The length of time to make up backorders in a replenishment cycle (time)
t_2	The length of time of positive inventory level in a replenishment cycle (time)
t_3	The length of time to backorder shortage in a replenishment cycle (time)
t_4	The length of time of no backorder (time)
TR(y)	Cyclic total revenue ($)
TC (y, B)	Cyclic total cost ($)

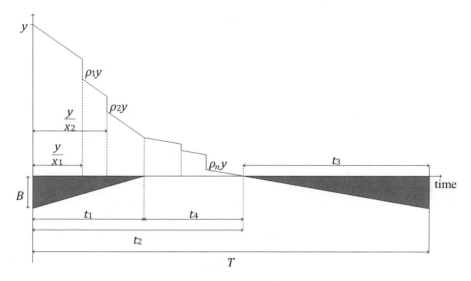

Fig. 2.14 The inventory level in a replenishment cycle inventory level (Tai 2015)

process rate x_n. The inventory level decreases at a rate of $(1 - \rho)x_n$, which is the rate of non-defective items that completed all the screening processes. After completing all the screening processes, the items are first shipped to satisfy the demands. The remaining items are then used to clear the outstanding backorders. Hence, the backordering quantity decreases at a rate of $(1 - \rho)x_n - D$ (see the left colored region in Fig. 2.14). Here Tai (2015) assumed that:

$$\text{prob}\,[(1 - \rho)x_n - D > 0] = 1 \quad (2.240)$$

The backorders are made up in time period t_1. Then, the inventory level decreases at a rate of D and reaches 0 at the end of time period t_2. Shortage is backordered in time period t_3 until the end of the replenishment cycle (see the right colored region in Fig. 2.14). The screening time for S_i is y/x_i. After the screening process S_i, the defective items ($\rho_i y$ units) are transferred to another inventory warehouse. They remarked that consolidating the defective items is a common policy in inventory management. The inventory level of defective items is illustrated in Fig. 2.15.

The number of items delivered to customers in a replenishment cycle is $(1 - \rho)\,y$. In which $(1 - \rho)y - B$ of them are used to satisfy the demands in the cycle. The remaining B items are used to make up the shortage in the last replenishment cycle. The same number of demands is backordered to the next cycle. So using Figs. 2.14 and 2.15 (Tai 2015):

2.4 EOQ Model with Backordering

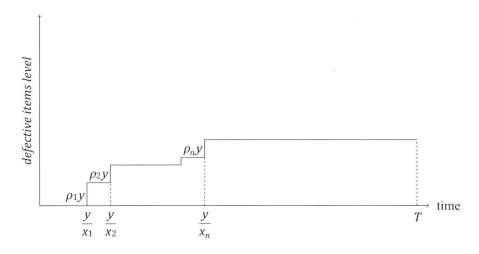

Fig. 2.15 The inventory level of defective items inventory level (Tai 2015)

$$T = \frac{(1-\rho)y}{D}$$

$$t_1 = \frac{B}{(1-\rho)x_n - D}$$

$$t_2 = T - t_1 = \frac{(1-\rho)y - B}{D}$$

$$t_3 = \frac{B}{D}$$

$$t_4 = t_2 - t_1 = \frac{(1-\rho)y - B}{D} - \frac{B}{(1-\rho)x_n - D}$$

TR(y) is the sum of the total sales of good-quality items and the amount received from the supplier for the return of the imperfect-quality items. One has (Tai 2015):

$$\text{TR}(y) = (1-\rho)ys + \rho yv$$

TC(y, B) consists of purchasing, fixed, screening, holding, and shortage costs. The purchasing, fixed, and screening costs are Cy, K, and $\sum_{i=1}^{n}(C_{1_i} \times y) = \left(\sum_{i=1}^{n} C_{1_i}\right)y$, respectively. The perfect items' holding cost is:

$$h\left[\frac{t_4^2 D}{2} + \frac{t_1^2(1-\rho)x_n}{2} + \sum_{i=1}^{n}\rho_i y \frac{y}{x_i} + t_1 t_4 D\right]$$

And after some simplifications, the holding cost becomes:

$$h\left[\frac{((1-\rho)y - B)^2}{2D} + \frac{B^2}{2((1-\rho)x_n - D)} + \sum_{i=1}^{n}\rho_i y \times \frac{y}{x_i}\right]$$

The defective items' holding cost is:

$$h_1\left[\sum_{i=1}^{n}\rho_i y\left(T - \frac{y}{x_i}\right)\right] = h_1\left[\sum_{i=1}^{n}\rho_i\left(\frac{(1-\rho)}{D} - \frac{1}{x_i}\right)\right]y^2$$

$$= h_1\left[\left(\frac{\rho(1-\rho)}{D}\right) - \sum_{i=1}^{n}\frac{\rho_i}{x_i}\right]y^2$$

The penalty cost for shortage is:

$$C_b\left(\frac{t_1 \times B}{2} + \frac{t_3 \times B}{2}\right) = C_b\left[\frac{1}{2D} + \frac{1}{2((1-\rho)x_n - D)}\right]B^2$$

$$= C_b\left[\frac{1-\rho}{2D((1-\rho) - D/x_n)}\right]B^2$$

For simplicity, let:
$P_1 = E\left[(1-\rho)^2\right]$, $P_2 = E[\rho(1-\rho)]$, $P_3 = \frac{1}{1-E[\rho]}$, $P_4 = \frac{E[\rho]}{1-E[\rho]}$, $R = \frac{1-E[\rho]}{E\left[\frac{1-\rho}{(1-\rho)-D/x_n}\right]}$. Let also.

$A_1 = \sum_{i=1}^{n}\frac{E[\rho_i]}{x_i}$ which depends on ρ_i, x_i ($i = 1, \ldots, n$) and $A_2 = \sum_{i=1}^{n}C_{I_i}$ which depends on C_{I_i} ($i = 1, \ldots, n$), since $R > 0$. Then the cyclic expected net profit is (Tai 2015):

$$E[\text{TP}(y,B)] = (1 - E[\rho])ys + E[\rho]y\upsilon$$
$$- \begin{bmatrix} Cy + A_2 y + h\left[\frac{P_1}{2D}y^2 - \frac{1-E[\rho]}{D}By + \frac{B^2}{2RP_3 D} + A_1 y^2\right] \\ +h_1\left[\left(\frac{P_2}{D} - A_1\right)\right]y^2 + \frac{C_b}{2RP_3 D}B^2 + K \end{bmatrix} \quad (2.241)$$

And the expected cycle length (Tai 2015) is:

2.4 EOQ Model with Backordering

$$E[T] = \frac{(1 - E[\rho])y}{D} \qquad (2.242)$$

Using renewal reward theorem:

$$E[\text{TPU}(y, B)] = \frac{E[\text{TP}(y, B)]}{E[T]} = \overbrace{sD + vDP_4 - (C + A_2)DP_3}^{\text{Independent of } B \ \& \ y} - \left\{ \frac{(h + C_b)B^2}{2Ry} - hB \right.$$
$$\left. + \left[\frac{hP_1}{2} + hA_1D + h_1(P_2 - A_1D)\right]P_3y + \frac{KDP_3}{y} \right\} \qquad (2.243)$$

The goal is to maximize the expected total profit using optimal decision variables y and B. According to Eq. (2.243), maximizing Eq. (2.243) can be reduced to minimizing $f(y, B)$ as below (Tai 2015):

$$f(y, B) = \frac{(h + C_b)B^2}{2Ry} - hB + \left[\frac{hP_1}{2} + hA_1D + h_1(P_2 - A_1D)\right]P_3y$$
$$+ \frac{KDP_3}{y} \qquad (2.244)$$

Differentiating $f(y, B)$ in Eq. (2.244) with respect to B and y, respectively, gives (Tai 2015):

$$\frac{\partial f}{\partial B} = \frac{(h + C_b)B}{Ry} - h \qquad (2.245)$$

$$\frac{\partial f}{\partial y} = -\frac{(h + C_b)B^2}{2Ry^2} + \left[\frac{hP_1}{2} + hA_1D + h_1(P_2 - A_1D)\right]P_3 - \frac{KDP_3}{2y^2} \qquad (2.246)$$

The second-order partial derivatives are (Tai 2015):

$$\frac{\partial^2 f}{\partial B^2} = \frac{(h + C_b)}{Ry}$$

$$\frac{\partial^2 f}{\partial y^2} = \frac{(h + C_b)B^2}{Ry^3} + \frac{2KDP_3}{y^3}$$

$$\frac{\partial^2 f}{\partial B \partial y} = -\frac{(h + C_b)B}{Ry^2}$$

$$\frac{\partial^2 f}{\partial B^2}\frac{\partial^2 f}{\partial y^2} - \left(\frac{\partial^2 f}{\partial B \partial y}\right)^2 = \frac{2(h + C_b)KDP_3}{Ry^4}$$

Since $R > 0$, one has $\frac{\partial^2 f}{\partial B^2} > 0$ and $\frac{\partial^2 f}{\partial B^2} \frac{\partial^2 f}{\partial y^2} - \left(\frac{\partial^2 f}{\partial y \partial B}\right)^2 > 0$. This implies that $f(y, B)$ is strictly convex for positive B and y. Hence, the unique global minimum for positive B and y can be obtained by solving $\frac{\partial f}{\partial B} = 0$ and $\frac{\partial f}{\partial y} = 0$, which gives (Tai 2015):

$$y^* = \sqrt{\frac{2KD}{hP_1 + 2hA_1D + 2h_1(P_2 - A_1D) - \frac{hR}{(h+C_b)P_3}}} \qquad (2.247)$$

So:

$$B^* = \frac{hR}{h + C_b} y^* \qquad (2.248)$$

When $n = 1$, the proposed model reduces to the one presented in Hsu and Hsu (2012), with:

$$y^* = \sqrt{\frac{2KD}{h\left[P_1 + 2A_1D - \frac{hR}{(h+C_b)P_3}\right]}} \qquad (2.249)$$

And:

$$B^* = \frac{hR}{h + C_b} y^* \qquad (2.250)$$

They remarked that the above expressions are simpler than those obtained in Hsu and Hsu (2012). As stated in Hsu and Hsu (2012), if the defective percentage p_1 follows uniform distribution with probability density function:

$$f_{p_1}(p_1) = \begin{cases} \frac{1}{\beta} & \text{for } 0 < p_1 < \beta \\ 0 & \text{otherwise} \end{cases}$$

Then $E[p_1] = \frac{\beta}{2}$, $E\left[(1 - p_1)^2\right] = 1 - \beta + \frac{\beta^2}{3}$, $E\left[\frac{1-p_1}{1-p_1-D/x_1}\right] = 1 + \frac{D}{\beta x_1} \ln\left(\frac{1-D/x_1}{1-\beta-D/x_1}\right)$.

In what follows, they gave an analysis which is absent in Hsu and Hsu (2012). Equations (2.249) and (2.250) can be reduced to (Tai 2015):

2.4 EOQ Model with Backordering

$$y^* = \sqrt{\frac{2KD}{h\left\{1 - \beta + \frac{\beta^2}{3} + \frac{D\beta}{x_1} - \frac{h(1-\beta/2)^2}{(h+C_b)}\left[1 + \frac{D\beta}{x_1}\ln\left(\frac{1-D/x_1}{1-\beta-D/x_1}\right)\right]^{-1}\right\}}} \quad (2.251)$$

and

$$B^* = \frac{h(1-\beta)}{2(h+C_b)}\left[1 + \frac{D}{\beta x_1}\ln\left(\frac{1-D/x_1}{1-\beta-D/x_1}\right)\right]^{-1} y^* \quad (2.252)$$

For the case when β is small, by the Taylor series expansion, the approximation can be used (Tai 2015):

$$\ln\left(\frac{1-D/x_1}{1-\beta-D/x_1}\right) \approx \frac{\beta}{1-\beta-D/x_1}$$

By neglecting the terms with β^2, one could further reduce Eqs. (2.211) and (2.212) to (Tai 2015):

$$y_1^* = \sqrt{\frac{2KD(h+C_b)}{h[h(1-\beta)D/x_1 + C_b(1-\beta-D\beta/x_1)]}} \quad (2.253)$$

and

$$B_1^* = \frac{h(2-\beta)(1-\beta-d/x_1)}{2(h+C_b)(1-\beta)} \cdot y_1^* \quad (2.254)$$

Finally, if all products are of good quality, which means the screening rate can be set as $x_1 \to \infty$, then $E[(1-\rho)^2] = 1$, $E[\rho] = 0$ and $E\left[\frac{1-\rho}{1-\rho-D/x_1}\right] = 1$; hence, y^* reduces to the classical EOQ with shortages (Tai 2015). When $b \to \infty$, i.e., no shortages are allowed, then (Tai 2015):

$$y^* \to \sqrt{\frac{2KD}{h(P_1 + 2A_1 D)}}$$

which is a generalization of the result obtained in Maddah and Jaber (2008).

Example 2.13 Tai (2015) presented a numerical example to demonstrate the use of the model. He applied the following parameters which are also used in Hsu and Hsu (2012) and Wee et al. (2007), $D = 50{,}000$ units/year, $K = 100$/cycle, $h = \$5$/unit/year, $C = \$25$/unit, $C_b = \$10$/unit/year, $s = \$50$/unit, and $v = \$20$/unit, and supposed that there are seven screening processes, S_i ($i = 1, \ldots, 7$). The probability density function of defective percentage, p_i ($i = 1, \ldots, 7$), is assumed as $p_i \sim U$

Table 2.20 Parameters for the screening processes S_i ($i = 1, \ldots, 7$) (Tai 2015)

	S_1	S_2	S_3	S_4	S_5	S_6	S_7
β_i	0.01	0.01	0.04	0.04	0.04	0.1	0.1
x_i	1	2	1	2	0.5	1	0.5
C_{I_i}	0.5	1	0.5	0.5	0.3	0.5	0.3

$[0, \beta]$. The values of β_i, x_i, and d_i for the screening process S_i ($i = 1, \ldots, 7$) are given in Table 2.20.

The optimal y^* and B^* from Eqs. (2.251) and (2.252) and the corresponding expected profit per unit time for each of the screening process are given in Table 2.21. They also gave the approximated y_1* and B_1* from Eqs. (2.247) and (2.248) to demonstrate the effectiveness of the approximations.

2.4.3 Rejection of Defective Supply Batches

Skouri et al. (2014) developed an imperfect EOQ model in which there is a fixed delivery schedule, where supply batches are delivered at equally spaced delivery intervals T. The batches may include defective and defective delivery occurrences that are independent of each other. Each supply batch is inspected upon arrival and, if found defective, the batch is rejected. So, the system operates under an "all or none" policy on supply. There are no emergency deliveries, and the total quantity of any rejected batch is routinely added to the batch quantity of the next planned delivery. In order to conform with the EOQ (with backorders) paradigm, the planned backorder level satisfies the inequality $J > 0$.

In order to model the expected cost, one needs to understand the system operation over time. Since defective batches are rejected, the length of inventory cycles is a random variable, depending on the respective history of consecutive previous defective deliveries and the probability of a defective batch p. Specifically, let X be a random variable representing the number of consecutive defective deliveries. Using this, the length of any inventory cycle can be directly expressed as $T' = (X + 1) T$. Since defective delivery occurrences are independent and X is a geometric random variable with parameter p and probability distribution function $P(X = x) = p^x (1 - p)$, $x \geq 0$. Therefore, under imperfect quality, there can be an infinite number of cycles of length T' (each occurring with a known probability). For any such cycle, they could directly determine the starting and ending inventory levels. Since the total quantity of successive defective deliveries becomes available with the first acceptable delivery (Assumption 7), the start-of-cycle inventory is $y - B$ (for any cycle length). The end-of-cycle inventory (representing backorders), however, depends on the cycle length and is $B + Xy$ (i.e., planned backorders the total quantity of the X successive defective deliveries, Xy) (Skouri et al. 2014).

Figure 2.16 depicts a possible situation with two different cycles, the first of length $T' = 2T$ (where a defective delivery at time T has been rejected) followed by a

2.4 EOQ Model with Backordering

Table 2.21 Numerical results for the screening processes S_i ($i = 1, \ldots, 7$) (Tai 2015)

	S_1	S_2	S_3	S_4	S_5	S_6	S_7
y^*	1624.85	1679.81	1638.40	1699.16	1534.16	1664.90	1542.35
B^*	384.34	477.24	379.32	474.22	209.17	368.63	194.28
$E[TPU(y^*, B^*)]$	1,217,432.76	1,192,509.48	1,213,159.67	1,213,382.37	1,222,940.58	1,204,203.81	1,214,227.78
y_1^*	1630.52	1682.95	1662.38	1712.65	1573.85	1732.15	1651.62
B_1^*	384.90	477.72	381.61	476.31	208.45	374.58	191.32
$E[TPU(y_1^*, B_1^*)]$	1,217,432.72	1,192,509.47	1,213,158.94	1,213,382.17	1,222,937.97	1,204,198.33	1,214,208.42

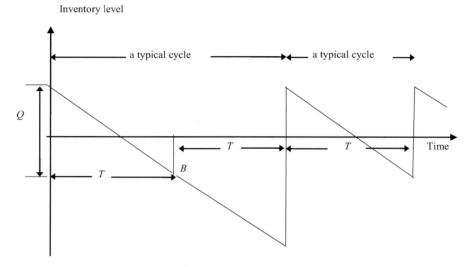

Fig. 2.16 Inventory realization with two consecutive cycles of length $2T$ and T (Skouri et al. 2014)

cycle of length $T' = T$. Observe that the batch quantity effectively rejected at delivery in time T has been made available at the delivery in time $2T$ (so both inventory cycles have the same start-of-cycle inventory). Also observe the end-of-cycle backorders for the first cycle which includes the quantity of the rejected delivery in time T (so backorders are $B + y$).

According to the above description, the problem has two decision variables (y, B). So they need to express system cost in terms of these variables. The cyclic inventory-related cost is:

$$\begin{aligned} \text{IC}(y, B, X) &= \frac{h(y-B)^2}{2D} + \frac{C_b B^2}{2D} + C_b BXT + \frac{C_b X^2 T^2 D}{2} \\ &= \frac{h(y-B)^2}{2D} + \frac{C_b B^2}{2D} + \frac{C_b BXy}{D} + \frac{C_b X^2 y^2}{2D} \end{aligned} \quad (2.255)$$

And the total cost in a cycle consists of setup and inventory-related costs and is (Skouri et al. 2014):

$$\begin{aligned} c(y, B, X) &= (X+1)K + \frac{h(y-B)^2}{2D} + \frac{C_b B^2}{2D} + C_b BXT + \frac{C_b X^2 T^2 D}{2} \\ &= (X+1)K + \frac{h(y-B)^2}{2D} + \frac{C_b B^2}{2D} + \frac{C_b BXy}{D} + \frac{C_b X^2 y^2}{2D} \end{aligned} \quad (2.256)$$

Since now X is a geometric random variable with parameter p, its expected value and second moment are respectively given by (Lefebvre 2008):

2.4 EOQ Model with Backordering

$$E(X) = \frac{p}{1-p}$$

$$E(X^2) = \text{Var}(X) + [E(X)]^2 = \frac{p}{(1-p)^2} + \frac{p^2}{(1-p)^2} = \frac{p(1+p)}{(1-p)^2}$$

Therefore, the expected value of Eq. (2.256) is given as:

$$C(y,B) = \frac{K}{1-p} + \frac{h(y-B)^2}{2D} + \frac{C_b B^2}{2D} + \frac{C_b B y p}{D(1-p)} + \frac{C_b y^2 p(1+p)}{2D(1-p)^2} \quad (2.257)$$

Now, recall that the length of any inventory cycle $T' = (X+1)T$, which is also a random variable. So, the expected length of each inventory cycle is:

$$E(T') = [E(X)+1]T = [E(T)+1]\frac{y}{D} = \frac{y}{(1-p)D} \quad (2.258)$$

Therefore, the following standard practice (e.g., Maddah and Jaber 2008; Nasr et al. 2013), expected cost per unit time, is finally obtained by invoking the renewal–reward theorem (Ross 1996, Theorem 3.6.1):

$$\text{TC}(y,B) = \frac{C(y,B)}{E(T')}$$
$$= K\frac{D}{y} + \frac{h(1-p)(y-B)^2}{2y} + \frac{C_b(1-p)B^2}{2y} + C_b p B$$
$$+ \frac{C_b p(1+p)y}{2(1-p)} \quad (2.259)$$

In order to solve the model, the first-order derivatives of cost function with respect to y and B are:

$$\frac{\partial \text{TC}(y,B)}{\partial y} = -\frac{KD}{y^2} + \frac{h(1-p)}{2} + \frac{C_b p(1+p)}{2(1-p)} - \frac{(h+C_b)(1-p)B^2}{2y^2} \quad (2.260)$$

$$\frac{\partial \text{TC}(y,B)}{\partial B} = p(h+C_b) - h + \frac{(1-p)(h+C_b)B}{y} \quad (2.261)$$

By equating relations (2.260) and (2.261) to zero (Skouri et al. 2014):

$$y^* = \sqrt{\frac{2KD(1-p)(h+C_b)}{C_b[h+p(h+C_b)]}} \quad (2.262)$$

$$B^* = \frac{[h - p(h + C_b)]y^*}{(1 - p)(h + C_b)} \geq 0 \qquad (2.263)$$

Since B^* should be nonnegative, Eqs. (2.261) and (2.263) indicate that two cases should be examined: (A) $h - p(h + C_b) \geq 0$ and (B) $h - p(h + C_b) < 0$ (denominator is always positive) (Skouri et al. 2014).

Case A: $p \leq \frac{h}{h + C_b}$. To determine the nature of optimal point, the second-order derivatives are derived (Skouri et al. 2014):

$$\frac{\partial^2 \text{TC}(y, B)}{\partial B^2} = \frac{(1 - p)(h + C_b)}{y} > 0 \qquad (2.264)$$

$$\frac{\partial^2 \text{TC}(y, B)}{\partial y^2} = \frac{2KD}{y^3} + \frac{(h + C_b)(1 - p)B^2}{y^3} > 0 \qquad (2.265)$$

$$\frac{\partial^2 \text{TC}(y, B)}{\partial y \partial B} = \frac{\partial^2 \text{TC}(y, B)}{\partial B \partial y} = -\frac{(h + C_b)(1 - p)B}{y^2} < 0 \qquad (2.266)$$

Since $\frac{\partial^2 \text{TC}(y,B)}{\partial B^2} > 0$, $\frac{\partial^2 \text{TC}(y,B)}{\partial y^2} > 0$, and $\left(\frac{\partial^2 \text{TC}(y,B)}{\partial B^2}\right)\left(\frac{\partial^2 \text{TC}(y,B)}{\partial y^2}\right) - \left(\frac{\partial^2 \text{TC}(y,B)}{\partial y \partial B}\right)^2 = \frac{2KD(h+C_b)(1-p)}{y^4} > 0$, so Eqs. (2.262) and (2.263) are optimal values and replacing them in Eq. (2.257) yields to (Skouri et al. 2014):

$$\text{TC}(y^*, B^*) = \sqrt{2KDC_b}\sqrt{\frac{h + p(h + C_b)}{(1 - p)(h + C_b)}} \qquad (2.267)$$

Case B: $p > \frac{h}{h + C_b}$. From relation (2.240), $\frac{\partial \text{TC}(y, B)}{\partial B}$ is positive for all $B > 0$ and therefore $\text{TC}(y, B)$ is increasing in B. So the minimum is located on the boundary of the constraint set $B = 0$. From the first-order condition for a minimum:

$$\frac{\partial \text{TC}(y, B = 0)}{\partial y} = -\frac{KD}{y^2} + \frac{h(1 - p)}{2} + \frac{C_b p(1 + p)}{2(1 - p)} = 0$$

They evaluated the respective order quantity:

$$y^* = \sqrt{\frac{2KD}{h}}\sqrt{\frac{1 - p}{(1 - p)^2 + \frac{cp(1+p)}{h}}} \qquad (2.268)$$

The point $(y^*, B^* = 0)$ is the unique global minimum and using Eq. (2.257), the respective expected cost per unit time is:

2.4 EOQ Model with Backordering

$$\text{TC}(y^*, B^* = 0) = \sqrt{2KD}\sqrt{\frac{h(1-p)^2 + C_b p(1+p)}{1-p}} \qquad (2.269)$$

Example 2.14 Skouri et al. (2014) presented an example using data $D = 4000$ per year, $K = \$250$ per delivery, $h = \$2$ per unit per year, $C_b = \$8$ per backordered per year, and $p = 0.10$. According to the parameter value since $p \leq h/(h + C_b)$, the solution is given by Case A. So, using Eqs. (2.262) and (2.263), $y^* = 866$, and $B^* = 96$, $\text{TC}(y^*, B^*) = 2309.4\$$.

2.4.4 Rework and Backordered Demand

Consider a production system allowing shortages which are backorders. A percentage of manufactured items are imperfect. The faction of imperfect items is a random variable and its distribution function is known. In order to have a quality control, during production, all items manufactured are inspected. Thus, during the production period, the imperfect items are identified and kept separately from perfect items. During production period, shortages do not occur. This means that production rate is greater than demand rate plus the product of percent of imperfect items by production rate; mathematically speaking this is $P - D - \gamma P > 0$ (Taleizadeh et al. 2016c).

Here, the imperfect items are reworkable. But repair of them in the production systems is not possible due to some restriction such as avoiding interruptions in the production program. On the other hand, the imperfect products have a significant value to the company; therefore, the rework of imperfect items is outsourced. It is assumed that after repair process, the products are as good as perfect ones. The repair cost and the holding cost of repaired products which is higher than the initial holding cost are taken into account. The total cost at repair shop is comprised of fixed and variable costs. The fixed cost is comprised of the repair setup cost and round trip fixed cost. The variable cost consists of the unit transportation cost, unit setup cost, and unit holding cost at the repair facility. Additionally, it is assumed that in the repair shop, the repair process is always in control and all imperfect products can be repaired. Also, repaired items are added to inventory in the same production cycle according to the following three cases (Taleizadeh et al. 2016c):

Case I. The repaired products are received when the inventory level is positive.
Case II. The repaired products are received when the inventory level is zero.
Case III. The repaired products are received when shortage quantity is equal to imperfect products' quantity.

The main objective of the proposed inventory model is to determine the optimal value for production lot size and backorder level in order to maximize the total profit (Taleizadeh et al. 2016c).

In order to identify imperfect products, during production period, all manufactured products are screened. The production rate (P) is greater than the demand rate (D) plus the product of percent of imperfect items (γ) by production rate (P). This means $P - D - \gamma P > 0$ or $(1 - \gamma) - \frac{D}{P} > 0$. All imperfect items are sent to the repair store. Repair duration is t_R and it contains the repair time at rate R and total transportation time t_T of imperfect products ($t_R = \gamma y/R + t_T$). The fixed cost at the repair shop is calculated with $K_R + 2K_s$ where K_R is the repair setup fixed cost and K_s is the transportation fixed cost. The variable cost per imperfect product is calculated with $C_1 + 2C_T + h_2 t_R$ where C_1 is the material and labor cost per unit, C_T is the transportation cost per unit, and h_2 is the unit holding cost in the repair store. So the cost at the repair shop per unit is given by $C_R = (K_R + 2K_s)/\gamma y + (C_1 + 2C_T + h_2 t_R)$. The repairer claims this cost with an m margin as repair charges. Thus, the total repair charge per unit is determined with:

$$(1+m)\left(\frac{K_R + 2K_S}{\gamma y} + C_1 + 2C_T + h_2 t_R\right)$$

$$(1+m)\left[\frac{K_R + 2K_S}{\gamma y} + C_1 + 2C_T + h_2\left(\frac{\gamma y}{R} + t_T\right)\right] \qquad (2.270)$$

Case I The behavior of inventory level is shown in Fig. 2.17. In this case, when the repaired items are returned to the company, the inventory level is positive. After receiving the repaired items, then the inventory level increases by repaired items' quantity. The shortages occur immediately after the inventory level reaches zero. The inventory cycle is divided into four sections. The production time starts at the beginning of t_1 ($t_1 = \frac{B}{P(1-\gamma)-D}$). During t_1 backorders and current demand are covered. At the end of the production time, the inventory level is I_{max}. Thus, the production period is equal to $t_1 + t_2$ where $t_2 = \frac{I_{max}}{P(1-\gamma)-D}$. During the production time, a lot size of y units is manufactured. Thus:

$$t_1 + t_2 = \frac{y}{P} = \frac{B + I_{max}}{P(1-\gamma) - D}$$

Consequently, I_{max} is equal to:

$$I_{max} = y\left[(1-\gamma) - \frac{D}{P}\right] - B \qquad (2.271)$$

At the end of the production period, imperfect items are sent to the repair store. The time t_3 represents the time in which the I_{max} is consumed, and it is given by $t_3 = \frac{I_{max} + \gamma y}{D}$. Repaired items after t_R period are returned to inventory as perfect items. In this case, t_R is always shorter than t_3. During t_4 ($t_4 = B/D$), the shortages occur and these are accumulated until B units. Inventory cycle duration (T) is equal to $t_1 + t_2 + t_3 + t_4$ and by substitution the t_i then T is obtained as:

2.4 EOQ Model with Backordering

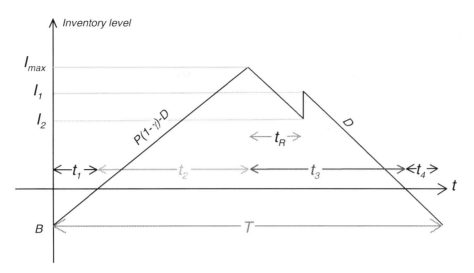

Fig. 2.17 Behavior of inventory level in Case I (Taleizadeh et al. 2016c)

$$T = t_1 + t_2 + t_3 + t_4 = \frac{y}{P} + \frac{I_{max} + \gamma y}{D} + \frac{B}{D}$$

$$T = \frac{y}{D} \tag{2.272}$$

Also, I_1 and I_2 are calculated as:

$$I_1 = (t_3 - t_R)D = I_{max} + \gamma y - t_R D$$

$$I_1 = y\left(1 - \frac{D}{P}\right) - B - \left(\frac{\gamma y}{R} + t_T\right)D \tag{2.273}$$

$$I_2 = I_{max} - t_R D = \frac{y}{P}(P(1 - \gamma) - D) - B - \left(\frac{\gamma y}{R} + t_T\right)D \tag{2.274}$$

The total sales revenue per time unit is:

$$\frac{sy}{T} = sD \tag{2.275}$$

The total fixed setup cost per time unit is:

$$\frac{K}{T} = \frac{KD}{y} \tag{2.276}$$

The total production and rework costs per time unit are, respectively:

$$\frac{Cy}{T} = CD \text{ and } C_R \frac{\gamma Q}{T} \tag{2.277}$$

The total inspection cost per time unit is:

$$\frac{C_1 y}{T} = C_1 D \tag{2.278}$$

The total holding cost (HC) per time unit is:

$$HC = \frac{h}{T}\left(\frac{I_{max}t_2}{2} + \frac{I_{max}I_{max}}{2D}\right) + \frac{h_R}{T}\left(\gamma y(t_3 - t_R) - \frac{\gamma^2 y^2}{2D}\right)$$

$$HC = \frac{h(1-\gamma)}{2}\left(Q\left[(1-\gamma) - \frac{D}{P}\right] - 2B\right) + \frac{h_R(1-\gamma)}{2y\left[(1-\gamma) - \frac{D}{P}\right]}B^2 \tag{2.279}$$

$$+\gamma h_R \left(y\left(1 - \frac{D}{P}\right) - B - \frac{\gamma y}{2} - D\left(\frac{\gamma y}{R} + t_T\right)\right)$$

The total backordering cost (BC) per time unit is:

$$BC = \frac{C_b}{T}\left(\frac{t_1 B}{2} + \frac{t_4 B}{2}\right)$$

$$BC = \frac{C_b D}{y}\left(\frac{B^2}{2(P(1-\gamma) - D)} + \frac{B^2}{2D}\right) = \frac{C_b(1-\gamma)}{2y\left[(1-\gamma) - \frac{D}{P}\right]}B^2 \tag{2.280}$$

Therefore, the total manufacturer's profit is determined by the total sales revenues minus the total costs incurred expressed as:

$$TP = sD - \left[CD + \frac{KD}{y} + C_1 D + \frac{\gamma y}{T}C_R + HC + BC\right] \tag{2.281}$$

Substituting and simplifying HC, BC, T, and C_R into Eq. (2.366), the total profit becomes:

$$TP = sD$$

$$- \begin{pmatrix} CD + \frac{KD}{y} + C_1 D + \gamma D(1+m)\left(\frac{K_R + 2K_S}{\gamma y} + C_1 + 2C_T + h_2\left(\frac{\gamma y}{R} + t_T\right)\right) \\ + \frac{h(1-\gamma)}{2}\left(y\left[(1-\gamma) - \frac{D}{P}\right] - 2B\right) \\ + \gamma h_R\left[y\left(1 - \frac{D}{P}\right) - B - \frac{\gamma y}{2} - D\left(\frac{\gamma y}{R} + t_T\right)\right] + \frac{(h+C_b)(1-\gamma)}{2y\left[(1-\gamma) - \frac{D}{P}\right]}B^2 \end{pmatrix}$$

$$\tag{2.282}$$

2.4 EOQ Model with Backordering

Obviously, the total profit is maximized if and only if the mathematical expression in brackets is minimized. It is important to remark that the mathematical expression in brackets is the total cost. Also the terms which are independent of decision variables (y, B) can be eliminated from the total cost and is represented by N (y, B). Hence, the total cost N(y, B) is given by:

$$N(y,B) = \frac{KD}{y} + \gamma D(1+m)\left(\frac{K_R + 2K_S}{\gamma y} + \frac{\gamma y h_2}{R}\right) + \frac{(h+C_b)(1-\gamma)}{2y\left[(1-\gamma) - \frac{D}{P}\right]} B^2$$
$$+ \frac{h_R(1-\gamma)}{2}\left(y\left[(1-\gamma) - \frac{D}{P}\right] - 2B\right) + \gamma h_R\left[y\left(1 - \frac{D}{P}\right) - B - \frac{\gamma Q}{2} - D\left(\frac{\gamma y}{R} + t_T\right)\right]$$
(2.283)

In order to obtain the optimal value for the decision variables, it is necessary to prove that N(y, B) is convex. In other words, it is sufficient to show that the Hessian matrix of N(y, B) is positive definite. It is worth to mention that if N(y, B) is minimized, then the total profit (TP) is maximized. The detailed optimization procedure is given below (Taleizadeh et al. 2016c):

$$\frac{\partial N(y,B)}{\partial y} = -\frac{D(K + (1+m)(K_R + 2K_S))}{y^2} + \frac{\gamma^2 D h_2(1+m)}{R}$$
$$+ \frac{h_R(1-\gamma)}{2}\left[(1-\gamma) - \frac{D}{P}\right] + \gamma h_R\left[\left(1 - \frac{D}{P}\right) - \frac{\gamma}{2} - \frac{\gamma D}{R}\right] - \frac{(h+C_b)(1-\gamma)}{2y^2\left[(1-\gamma) - \frac{D}{P}\right]} B^2$$
(2.284)

$$\frac{\partial N(y,B)}{\partial B} = -h(1-\gamma) - \gamma h_R + \frac{(h+C_b)(1-\gamma)}{y\left[(1-\gamma) - \frac{D}{P}\right]} B \quad (2.285)$$

$$\frac{\partial^2 N(y,B)}{\partial y^2} = \frac{2D(K + (1+m)(K_R + 2K_S))}{y^3} + \frac{(h+C_b)(1-\gamma)}{y^3\left[(1-\gamma) - \frac{D}{P}\right]} B^2 \quad (2.286)$$

$$\frac{\partial^2 N(y,B)}{\partial B \partial y} = \frac{-(h+C_b)(1-\gamma)}{y^2\left[(1-\gamma) - \frac{D}{P}\right]} B \quad (2.287)$$

$$\frac{\partial^2 N(y,B)}{\partial y \partial B} = \frac{-(h+C_b)(1-\gamma)}{y^2\left[(1-\gamma) - \frac{D}{P}\right]} B \quad (2.288)$$

$$\frac{\partial^2 N(y,B)}{\partial B^2} = \frac{(h+C_b)(1-\gamma)}{y\left[(1-\gamma) - \frac{D}{P}\right]} \quad (2.289)$$

Note that $(1-\gamma) - \frac{D}{P} > 0$ and the second derivatives of N(y, B) with respect to y and B are both positive, and the Hessian matrix determinant is positive (Taleizadeh et al. 2016c):

$$\frac{2D(h+C_b)(1-\gamma)(K+(1+m)(K_R+2K_S))}{y^4\left[(1-\gamma)-\frac{D}{P}\right]} > 0$$

Then the Hessian matrix is positive definite. Therefore, $N(y, B)$ is convex. In order to obtain the optimal value for the decision variables it is sufficient to set the first derivatives of $N(y, B)$ with respect to y and B equal to zero. Thus, the optimal values of lot size (y) and backorder level (B) are given by (Taleizadeh et al. 2016c):

$$y = \sqrt{\frac{2D(K+(1+m)(K_R+2K_S))\left[(1-\gamma)-\frac{D}{P}\right]+(h+C_b)(1-\gamma)B^2}{\left[(1-\gamma)-\frac{D}{P}\right]\left(\frac{2Dh_2\gamma^2(1+m)}{R}+h(1-\gamma)\left[(1-\gamma)-\frac{D}{P}\right]+2\gamma h_R\left[\left(1-\frac{D}{P}\right)-\frac{\gamma}{2}-\frac{\gamma D}{R}\right]\right)}} \quad (2.290)$$

$$B = \frac{y(h(1-\gamma)+\gamma h_R)\left[(1-\gamma)-\frac{D}{P}\right]}{(h+C_b)(1-\gamma)} \quad (2.291)$$

Substituting Eq. (2.291) into Eq. (2.290), the optimal value of y independent of B is obtained as shown below:

$$y = \sqrt{\frac{2D(K+(1+m)(K_R+2K_S))}{\left(\frac{2Dh_2\gamma^2(1+m)}{R}+h(1-\gamma)\left[(1-\gamma)-\frac{D}{P}\right]+2\gamma h_R\left[\left(1-\frac{D}{P}\right)-\frac{\gamma}{2}-\frac{\gamma D}{R}\right]\right)-\left(\frac{(h(1-\gamma)+\gamma h_R)^2}{(h+C_b)(1-\gamma)}\left[(1-\gamma)-\frac{D}{P}\right]\right)}} \quad (2.292)$$

Note that γ is a random variable, so the expected value of γ must be substituted in Eqs. (2.281), (2.291), and (2.292). Thus, the expected value of total profit, economic production quantity, and backorder level are (Taleizadeh et al. 2016c):

$$TP = sD - \begin{bmatrix} CD+\frac{KD}{y}+C_1D+E(\gamma)D(1+m)\left[\frac{K_R+2K_S}{\gamma y}+C_1+2C_T+h_2\left(\frac{E(\gamma)y}{R}+t_T\right)\right] \\ +\frac{h(1-E(\gamma))}{2}\left(y\left[(1-E(\gamma))-\frac{D}{P}\right]-2B\right)+\frac{(h+C_b)(1-E(\gamma))}{2y\left[(1-E(\gamma))-\frac{D}{P}\right]}B^2 \\ +E(\gamma)h_R\left[y\left(1-\frac{D}{P}\right)-B-\frac{E(\gamma)y}{2}-D\left(\frac{E(\gamma)y}{R}+t_T\right)\right] \end{bmatrix} \quad (2.293)$$

$$y = \sqrt{\frac{2D(K+(1+m)(K_R+2K_S))}{\left(\frac{2E(\gamma^2)Dh(1+m)}{R}+h(1-E(\gamma))\left[(1-E(\gamma))-\frac{D}{P}\right]+2E(\gamma)h_R\left[\left(1-\frac{D}{P}\right)-\frac{E(\gamma)}{2}-\frac{E(\gamma)D}{R}\right]\right) - \left(\frac{[(h(1-E(\gamma))+E(\gamma)h_1)]^2\left[(1-E(\gamma))-\frac{D}{P}\right]}{(h+C_b)(1-E(\gamma))}\right)}} \quad (2.294)$$

2.4 EOQ Model with Backordering

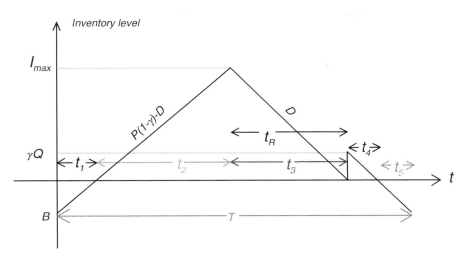

Fig. 2.18 Behavior of inventory level in Case II (Taleizadeh et al. 2016c)

$$B = \frac{y(h(1 - E(\gamma)) + E(\gamma)h_R)((1 - E(\gamma)) - \frac{D}{P})}{(h + C_b)(1 - E(\gamma))} \quad (2.295)$$

Case II Here, when the repaired items are returned to the company, the inventory level is zero. After receiving the repaired products, the inventory level increases as much as the repaired products' quantity, and then the inventory level is positive. Figure 2.18 illustrates the behavior of the inventory level for Case II. In this case, the inventory cycle is divided into five sections. The times t_1, t_2, and the maximum inventory level ($I_{\max}$) are the same as in Case I (Taleizadeh et al. 2016c). At the beginning of t_1 ($t_1 = \frac{B}{P(1-\gamma)-D}$) period, backordered shortage has its maximum value, and the production period is started. At the end of this period, the inventory level is zero. The production period is equal to $t_1 + t_2$ where $t_2 = \frac{I_{\max}}{P(1-\gamma)-D}$. Thus, $I_{\max}$ is derived as follows:

$$t_1 + t_2 = \frac{y}{P} = \frac{B + I_{\max}}{P(1 - \gamma) - D} \rightarrow I_{\max} = y\left[(1 - \gamma) - \frac{D}{P}\right] - B \quad (2.296)$$

At the end of the production period, the imperfect items are sent to the repair store. The t_3 ($t_3 = \frac{I_{\max}}{D}$) is a fraction of cycle period in which production is not occurring but the inventory level is positive; and the inventory is being consumed by demand. Repaired items after t_R period enter into inventory and t_4 period is given by $t_4 = \frac{\gamma y}{D}$. Obviously, in this case, t_R is equal to t_3. Also t_5 ($t_5 = \frac{B}{D}$) is a fraction of cycle period in which shortage is occurring. Cycle time duration (T) is equal to $t_1 + t_2 + t_3 + t_4 + t_5$, and by substitution t_i then T is obtained as $T = t_1 + t_2 + t_3 + t_4 + t_5 = \frac{y}{P} + \frac{I_{\max}}{D} + \frac{\gamma y}{D} + \frac{B}{D}$ (Taleizadeh et al. 2016c). Hence:

$$T = \frac{y}{D} \quad (2.297)$$

The total sales revenue per time unit is given by:

$$\frac{sy}{T} = sD \quad (2.298)$$

The fixed setup cost, the production cost, the inspection cost, and the outsourced rework cost are the same as Case I. The total holding cost (HC) per time unit is expressed as:

$$\text{HC} = \frac{h}{T}\left(\frac{t_2 I_{\max}}{2} + \frac{t_3 I_{\max}}{2}\right) + \frac{h_R}{T}\left(\frac{t_4 I_1}{2}\right)$$

Here I_1 is given by γQ:

$$\text{HC} = \frac{h(1-\gamma)}{2}(y[(1-\gamma) - D/P] - 2B) + \frac{h_R \gamma^2 y}{2}$$
$$+ \frac{h(1-\gamma)}{2y[(1-\gamma) - D/P]} B^2 \quad (2.299)$$

The total backordered cost (BC) per time unit is derived as:

$$\text{BC} = \frac{C_b}{T}\left(\frac{t_1 B}{2} + \frac{t_5 B}{2}\right)$$

$$\text{BC} = \frac{C_b D}{y}\left(\frac{B^2}{2(P(1-\gamma) - D)} + \frac{B^2}{2D}\right) = \frac{C_b(1-\gamma)}{2y\left[(1-\gamma) - \frac{D}{P}\right]} B^2 \quad (2.300)$$

So, the total manufacturer's profit is determined by the total sales revenues minus the total costs incurred:

$$\text{TP} = sD - \left[CD + \frac{KD}{y} + C_1 D + C_R \frac{\gamma y}{T} + \text{HC} + \text{BC}\right] \quad (2.301)$$

The mathematical expressions of HC, BC, T, and C_R are substituted into Eq. (2.301) and the total profit (TP) becomes:

$$\text{TP} = sD$$
$$- \left[CD + \frac{KD}{y} + C_1 D + D\gamma(1+m)\left[\frac{K_R + 2K_S}{\gamma y} + C_1 + 2C_T + h_2\left(\frac{\gamma y}{R} + t_T\right)\right] \right.$$
$$\left. + \frac{h(1-\gamma)\left(y\left[(1-\gamma) - \frac{D}{P}\right] - 2B\right)}{2} + \frac{h_R \gamma^2 y}{2} + \frac{(h+C_b)(1-\gamma)}{2y\left[(1-\gamma) - \frac{D}{P}\right]} B^2 \right]$$
$$(2.302)$$

2.4 EOQ Model with Backordering

It is easy to see that the total profit is maximized if and only if the mathematical expression in brackets is minimized. The constant terms can be eliminated from the total profit function; consequently $N(y, B)$ is expressed as:

$$N(B,y) = \frac{KD}{y} + D\gamma(1+m)\left[\frac{K_R + 2K_S}{\gamma y} + \frac{\gamma h_2 y}{R}\right] + \frac{h(1-\gamma)\left(y\left[(1-\gamma) - \frac{D}{P}\right] - 2B\right)}{2}$$
$$+ \frac{h_R \gamma^2 y}{2} + \frac{(h+C_b)(1-\gamma)}{2y\left[(1-\gamma) - \frac{D}{P}\right]} B^2$$
(2.303)

In order to obtain the optimal value for decision variables y and B, it is necessary to prove that $N(y, B)$ is convex, and it is sufficient to show that the Hessian matrix of $N(y, B)$ is positive definite. The detailed optimization procedure is given below:

$$\frac{\partial N(y,B)}{\partial y} = \frac{-KD}{y^2} - \frac{D(1+m)(K_R + 2K_S)}{y^2} + \frac{Dh_2\gamma^2(1+m)}{R} + \frac{h(1-\gamma)[(1-\gamma) - D/P]}{2} + \frac{h_R\gamma^2}{2}$$
$$- \frac{(h+C_b)(1-\gamma)}{2y^2[(1-\gamma) - D/P]} B^2$$
(2.304)

$$\frac{\partial N(y,B)}{\partial B} = -h(1-\gamma) + \frac{(h+C_b)(1-\gamma)}{y\left[(1-\gamma) - \frac{D}{P}\right]} B \quad (2.305)$$

$$\frac{\partial^2 N(y,B)}{\partial y^2} = \frac{2D[K + (1+m)(K_R + 2K_S)]}{y^3} + \frac{(h+C_b)(1-\gamma)}{y^3\left[(1-\gamma) - \frac{D}{P}\right]} B^2 \quad (2.306)$$

$$\frac{\partial^2 N(y,B)}{\partial B \partial y} = -\frac{(h+C_b)(1-\gamma)}{y^2\left[(1-\gamma) - \frac{D}{P}\right]} B \quad (2.307)$$

$$\frac{\partial^2 N(y,B)}{\partial y \partial B} = -\frac{(h+C_b)(1-\gamma)}{y^2\left[(1-\gamma) - \frac{D}{P}\right]} B \quad (2.308)$$

$$\frac{\partial^2 N(y,B)}{\partial B^2} = \frac{(h+C_b)(1-\gamma)}{y\left[(1-\gamma) - \frac{D}{P}\right]} \quad (2.309)$$

Note that $(1-\gamma) - \frac{D}{P} > 0$ and the second derivatives of $N(y, B)$ with respect to y and B are positive. Also the Hessian matrix determinant is positive:

$$\frac{2D(1-\gamma)(h+C_b)[K + (1+m)(K_R + 2K_S)]}{y^4\left[(1-\gamma) - \frac{D}{P}\right]} > 0$$

Since the Hessian matrix is positive definite, therefore the $N(y, B)$ is convex. Setting the first derivatives of $N(y, B)$ with respect to y and B equal to zero, then the optimal values for the decision variables y and B are derived as:

$$y = \sqrt{\frac{2D(K + (1+m)(K_R + 2K_S))\left[(1-\gamma) - \frac{D}{P}\right] + (h + C_b)(1-\gamma)B^2}{\left[(1-\gamma) - \frac{D}{P}\right]\left(\frac{2Dh_2\gamma^2(1+m)}{R} + h(1-\gamma)\left[(1-\gamma) - \frac{D}{P}\right] + h_1\gamma^2\right)}} \quad (2.310)$$

$$B = \frac{hy\left[(1-\gamma) - \frac{D}{P}\right]}{(h + C_b)} \quad (2.311)$$

And by substituting Eq. (2.310) into Eq. (2.311), the optimal value for y independent of B is obtained as:

$$y = \sqrt{\frac{2D(K + (1+m)(K_R + 2K_S))}{\left(\frac{2Dh_2\gamma^2(1+m)}{R} + h_1\gamma^2\right) + (1-\gamma)\left[(1-\gamma) - \frac{D}{P}\right]\left(\frac{hC_b}{(h+C_b)}\right)}} \quad (2.312)$$

It is important to remark that γ is a random variable, so in Eqs. (2.302), (2.310), and (2.311), the expected value of γ must be substituted. Therefore, the expected values of the total profit, economic production quantity, and optimal amount of backordered shortage are given by:

$$TP = sD - \begin{bmatrix} CD + \frac{KD}{y} + C_1 D + DE(\gamma)(1+m)\left[\frac{K_R + 2K_S}{E(\gamma)y} + C_1 + 2C_T + h_2\left(\frac{E(\gamma)y}{R} + t_T\right)\right] \\ + \frac{h(1 - E(\gamma))}{2}\left(y\left[(1 - E(\gamma)) - \frac{D}{P}\right] - 2B\right) + \frac{h_1 E(\gamma^2)y}{2} + \frac{(h + C_b)(1 - E(\gamma))}{2y\left[(1 - E(\gamma)) - \frac{D}{P}\right]}B^2 \end{bmatrix}$$

(2.313)

$$y = \sqrt{\frac{2D(K + (1+m)(K_R + 2K_S))}{\left(\frac{2DhE(\gamma^2)(1+m)}{R} + h_1 E(\gamma^2)\right) + (1 - E(\gamma))\left[(1 - E(\gamma)) - \frac{D}{P}\right]\left(\frac{hC_b}{(h+C_b)}\right)}} \quad (2.314)$$

$$B = \frac{hy\left[(1 - E(\gamma)) - \frac{D}{P}\right]}{(h + C_b)} \quad (2.315)$$

Case III In this case, when the repaired items return to the system, the inventory level is negative; in other words, shortage exists. Notice that in this case, the shortage quantity is exactly the same as the imperfect products' quantity. After receipt, the items that repaired the inventory levels are increased. The shortages are covered and the inventory level reaches zero and shortage time continues again. The cycle time is divided into five sections. At the beginning of t_1 period ($t_1 = \frac{B}{P(1-\gamma)-D}$), backordered shortage has its maximum value and production period starts. At the end of t_1, the

2.4 EOQ Model with Backordering

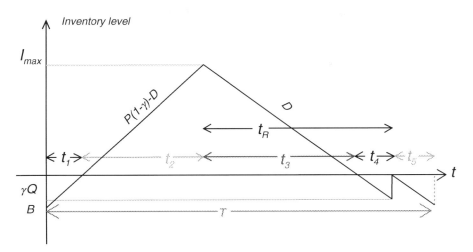

Fig. 2.19 Behavior of inventory level in Case III (Taleizadeh et al. 2016c)

inventory level is zero. Production time is equal to $t_1 + t_2$ where $t_2 = \frac{I_{max}}{P(1-\gamma)-D}$. The maximum inventory level (I_{max}) is calculated as:

$$t_1 + t_2 = \frac{y}{P} = \frac{B + I_{max}}{P(1-\gamma) - D} \rightarrow I_{max} = y\left[(1-\gamma) - \frac{D}{P}\right] - B \quad (2.316)$$

At the end of production time, the imperfect items are sent to the repair store. Here, t_3 ($t_3 = \frac{I_{max}}{D}$) is a fraction of cycle period in which the manufacturing system is not producing but the inventory level is positive. The repaired items after t_R period are returned to inventory, this time occurring also at the end of $t_4 = \gamma y/D$. In t_4 period, the shortage occurs. In this case, t_R is equal to $t_3 + t_4$. Also t_5 ($t_5 = \frac{B}{D}$) is a fraction of cycle time in which the manufacturing system does not produce items and shortage is occurring. Figure 2.19 shows the behavior of the inventory level in this case. Inventory cycle duration (T) is equal to $t_1 + t_2 + t_3 + t_4 + t_5$. The corresponding mathematical expression for each t_i ($i = 1, 2, 3, 4, 5$) is substituted. Thus, T is obtained as (Taleizadeh et al. 2016c):

$$T = t_1 + t_2 + t_3 + t_4 + t_5 = \frac{y}{P} + \frac{I_{max}}{D} + \frac{\gamma y}{D} + \frac{B}{D}$$

Hence:

$$T = \frac{y}{D} \quad (2.317)$$

The total sales revenue and the fixed setup cost, the production cost, the inspection cost, and the outsourced rework cost are the same as Case I and Case II.

Thus, the holding cost (HC) per time unit is:

$$HC = \frac{h}{T}\left(\frac{t_2 I_{\max}}{2} + \frac{t_3 I_{\max}}{2}\right)$$

$$HC = \frac{h(1-\gamma)}{2y\left[(1-\gamma) - \frac{D}{P}\right]}\left(y\left[(1-\gamma) - \frac{D}{P}\right] - B\right)^2$$

$$HC = \frac{h(1-\gamma)}{2}\left(y\left[(1-\gamma) - \frac{D}{P}\right] - 2B\right) + \frac{h(1-\gamma)}{2y\left[(1-\gamma) - \frac{D}{P}\right]}B^2 \quad (2.318)$$

The total backordered cost (BC) per time unit is:

$$BC = \frac{C_b}{T}\left(\frac{t_1 B}{2} + \frac{t_4 \gamma y}{2} + \frac{t_5 B}{2}\right)$$

$$BC = \frac{C_b D}{y}\left(\frac{B^2}{2(P(1-\gamma) - D)} + \frac{\gamma^2 y^2}{2D} + \frac{B^2}{2D}\right) = \frac{C_b(1-\gamma)}{2y\left[(1-\gamma) - \frac{D}{P}\right]}B^2 + \frac{C_b \gamma^2 y}{2}$$

$$(2.319)$$

Then the total manufacturer's profit is determined by the total sales revenues minus the total costs incurred:

$$TP = sD - \left[CD + \frac{KD}{y} + C_1 D + C_R \frac{\gamma y}{T} + HC + BC\right] \quad (2.320)$$

The mathematical expressions for HC, BC, T, and C_R are substituted into Eq. (2.320) and the total profit is rewritten as:

$$TP = sD$$

$$-\left[\begin{array}{l}CD + \frac{KD}{y} + C_1 D + D\gamma(1+m)\left[\frac{K_R + 2K_S}{\gamma y} + C_1 + 2C_T + h_2\left(\frac{\gamma y}{R} + t_T\right)\right] \\ + \frac{h(1-\gamma)}{2}\left(y\left[(1-\gamma) - \frac{D}{P}\right] - 2B\right) + \frac{(h+C_b)(1-\gamma)}{2y\left[(1-\gamma) - \frac{D}{P}\right]}B^2 + \frac{C_b \gamma^2 y}{2}\end{array}\right]$$

$$(2.321)$$

Notice that the total profit in Case II and Case III (Eqs. 2.313 and 2.321) differs on the holding cost of reworked items and the backorder cost. Thus, the concavity of the total profit function in Case III is proved. Therefore, setting the first derivatives of the total profit with respect to y and B equal to zero, then the optimal value for the lot size (y) and shortage level (B) are determined as (Taleizadeh et al. 2016a):

2.4 EOQ Model with Backordering

$$\frac{\partial TP}{\partial y} = 0 \rightarrow$$

$$-\left[\frac{-KD}{y^2} - \frac{D(1+m)(K_R+2K_S)}{y^2} + \frac{Dh_2\gamma^2(1+m)}{R} + \frac{h(1-\gamma)\left[(1-\gamma)-\frac{D}{P}\right]}{2} - \frac{(h+C_b)(1-\gamma)}{2y^2\left[(1-\gamma)-\frac{D}{P}\right]}B^2 + \frac{C_b\gamma^2}{2}\right] = 0$$

(2.322)

$$y = \sqrt{\frac{2D(K+(1+m)(K_R+2K_S))\left[(1-\gamma)-\frac{D}{P}\right] + (h+C_b)(1-\gamma)B^2}{\left[(1-\gamma)-\frac{D}{P}\right]\left(\frac{2Dh_2\gamma^2(1+m)}{R} + h(1-\gamma)\left[(1-\gamma)-\frac{D}{P}\right] + C_b\gamma^2\right)}}$$

(2.323)

$$\frac{\partial TP}{\partial B} = 0 \rightarrow$$

$$-h(1-\gamma) + \frac{(h+C_b)(1-\gamma)}{y\left[(1-\gamma)-\frac{D}{P}\right]}B = 0$$

(2.324)

$$B = \frac{hy\left[(1-\gamma)-\frac{D}{P}\right]}{(h+C_b)}$$

(2.325)

By substituting Eq. (2.325) into Eq. ((2.323), the optimal value of y independent of B is calculated as (Taleizadeh et al. 2016c):

$$y = \sqrt{\frac{2D(K+(1+m)(K_R+2K_s))}{\left(\frac{2Dh_2\gamma^2(1+m)}{R} + C_b\gamma^2\right) + (1-\gamma)\left[(1-\gamma)-\frac{D}{P}\right]\left(\frac{hC_b}{(h+C_b)}\right)}}$$

(2.326)

Again note that γ is a random variable, so in Eqs. (2.321), (2.325), and (2.326), the expected value of γ must be substituted. Thus, the expected values for the total profit, economic production quantity, and backorder level are given by (Taleizadeh et al. 2016c):

$$TP = sD - \left[\begin{array}{l} CD + \frac{KD}{y} + C_1D + DE(\gamma)(1+m)\left[\frac{K_R+2K_S}{E(\gamma)y} + C_1 + 2C_T + h_2\left(\frac{E(\gamma)y}{R} + t_T\right)\right] \\ + \frac{h(1-E(\gamma))}{2}\left(y\left[(1-E(\gamma))-\frac{D}{P}\right] - 2B\right) + \frac{C_bE(\gamma^2)y}{2} + \frac{(h+C_b)(1-E(\gamma))}{2y\left[(1-E(\gamma))-\frac{D}{P}\right]}B^2 \end{array}\right]$$

(2.327)

$$y = \sqrt{\frac{2D(K+(1+m)(K_R+2K_S))}{\left(\frac{2Dh_2E(\gamma^2)(1+m)}{R} + C_bE(\gamma^2)\right) + (1-E(\gamma))\left[(1-E(\gamma))-\frac{D}{P}\right]\left(\frac{hC_b}{(h+C_b)}\right)}}$$

(2.328)

$$B = \frac{hy\left[(1-E(\gamma))-\frac{D}{P}\right]}{(h+C_b)}$$

(2.329)

Table 2.22 Data for the numerical example (Salameh and Jaber 2000; Taleizadeh et al. 2016c)

Description	Symbol	Value
Percentage of defectives	γ	$U \sim (0, 0.04)$
Probability density function	$f(\gamma)$	$1/(0.04-0)$

Table 2.23 Data for the numerical example (Jaber et al. 2014; Taleizadeh et al. 2016c)

Description	Symbol	Value	Units
Repair setup cost	K_R	100	$/setup
Transportation fixed cost	K_S	200	$/trip
Unit transportation cost	C_T	2	$/unit
Unit material and labor cost	C_1	5	$/unit
Unit holding cost in repair shop	h_2	4	$/unit/year
Repair rate	R	50,000	Units/year
Total transport time	t_T	2/220	Year
Holding cost of repaired product	h_R	6	$/unit/year
Markup percentage	m	20%	

Table 2.24 Data for the numerical example (Hsu and Hsu, 2014; Taleizadeh et al. 2016c)

Description	Symbol	Value	Units
Selling price	s	300	$/unit
Demand rate	D	1000	Units/year
Production cost	C	100	$/unit
Production rate	P	3000	Units/year
Holding cost	h	5	$/unit/year
Backorder cost	C_b	10	$/unit/year
Inspection cost	C_I	0.5	$/unit

Table 2.25 The optimal value for each case (Taleizadeh et al. 2016c)

	TP	y	B	T
Case I	197,178.22	949.65	209.72	0.95
Case II	197,225.99	972.26	209.58	0.97
Case III	197,224.96	971.77	209.47	0.97

Example 2.15 This section presents a numerical example. The data for the numerical example is given in Tables 2.22, 2.23, and 2.24. Table 2.22 contains data from Salameh and Jaber (2000). Table 2.23 gives data from Jaber et al. (2014), and Table 2.24 presents data from Hsu and Hsu (2016).

The imperfect percentage follows a uniform distribution ($U \sim (0, 0.04)$). Thus, the expected value of the defective products (γ) is $E(\gamma) = \int_a^b \gamma f(\gamma) d\gamma = \int_a^b \gamma \frac{1}{b-a} d\gamma = \frac{b+a}{2} = \frac{0.04+0}{2} = 0.02$, and the $E(\gamma^2)$ and $(1 - E(\gamma^2))^2$ are given by $E(\gamma^2) = \text{var}(\gamma) + E^2(\gamma) = \frac{(b-a)^2}{12} + E^2(\gamma) = \frac{(0.04)^2}{12} + (0.02)^2 = 0.0005333333$.

Firstly, for each profit function, the following condition must be satisfied: $(1 - \gamma) - \frac{D}{P} = 0.65 > 0$.

Since the condition is satisfied, then the problem can be optimized. Table 2.25 shows the optimal results for the expected total profit and the decision variables and cycle duration (Taleizadeh et al. 2016c).

2.4 EOQ Model with Backordering

Fig. 2.20 Behavior of the inventory level over time (Konstantaras et al. 2012)

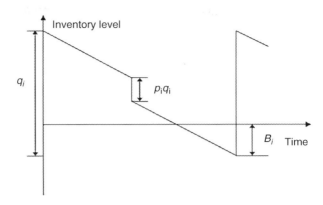

Table 2.26 Notations of a given problem

p_i	The fraction of defective items in shipment of size q_i (percent)
q_i	ith shipment size (unit)
n	Number of cycle
TR(y)	Total revenue per cycle ($)

According to Table 2.25, it is easy to see that the highest profit for this example is obtained in Case II. It is important to mention that the total profit in Case II and Case III differs only in the holding cost of repaired items and backorder cost. Here, it is easy to show that if the holding cost of repaired items is lower than the backordering cost, then the total profit in Case II is higher than the total profit in Case III and vice versa.

2.4.5 Learning in Inspection

Konstantaras et al. (2012) developed an EOQ imperfect system with learning in inspection. They assumed that 100% inspection of items is performed for each shipment, and the screening rate is faster than the demand rate. The defective items are sold at a discounted price; the fraction of defective items follows a learning curve that is either of an S-shape or of a power form learning curve.

To avoid shortages during the screening time, it is assumed that the number of good items in shipment i of order size q_i is at least equal to the demand during the screening time, i.e., $q_i(1 - p_i) \geq \frac{Dq_i}{x} \Rightarrow p_i \leq 1 - \frac{D}{x}$, where p_i is the fraction of defective items in shipment of size q_i, D is the demand rate, and x is screening rate in units per unit of time. The behavior of the inventory level in a given cycle i is shown in Fig. 2.20, which is the same as that in Wee et al. (2007).

In order to model the presented problem, some new notations which are specifically used are shown in Table 2.26.

The total revenue per cycle, TR(q_i), is the sum of revenues from selling good- and imperfect-quality items and is given as (Konstantaras et al. 2012):

$$\mathrm{TR}(q_i) = sq_i(1-p_i) + vq_ip_i \qquad (2.330)$$

where s is the unit price of a good-quality item and v is the unit price of an imperfect-quality item. The total cost per cycle, $\mathrm{TC}(q_i, B_i)$, is the sum of ordering cost in addition to penalty cost purchasing cost, screening cost, holding cost, and backordering cost and is given as (Konstantaras et al. 2012):

$$\mathrm{TC}(q_i, B_i) = \underbrace{K}_{\text{Fixed cost}} + \underbrace{Cq_i}_{\text{Purchasing cost}} + \underbrace{C_Iq_i}_{\text{Inspection cost}}$$

$$+ \underbrace{h\left\{\frac{(q_i - p_iq_i - B_i)^2}{2D} + \frac{p_iq_i^2}{x}\right\}}_{\text{Holding cost}} + \underbrace{\frac{C_bB_i^2}{2D}}_{\text{Shortage cost}} \qquad (2.331)$$

where K is the fixed cost, C is the unit purchasing cost, C_I is the unit screening cost, h is the holding cost per unit per unit of time, C_b is the backordering cost per unit per unit of time, and B_i is the maximum backordering level in cycle i. The profit per unit of time is obtained from Eqs. (2.330) and (2.331) as (Konstantaras et al. 2012):

$$\mathrm{TP}(q_i, B_i) = \frac{\mathrm{TR}(q_i) - \mathrm{TC}(q_i, B_i)}{T_i}$$

$$= sD + vD\frac{p_i}{1-p_i} - \frac{(C+C_I)D}{1-p_i} - \frac{KD}{q_i(1-p_i)} + hB_i - \frac{(h+C_b)B_i^2}{2q_i(1-p_i)} - \frac{hDp_iq_i}{(1-p_i)x} - \frac{h(1-p_i)q_i}{2}$$
$$(2.332)$$

where T_i is the length. Note that i in Eq. (2.332) is an input parameter, where p_i is a constant. The objective here is to maximize the profit per unit time given in Eq. (2.332). By taking the partial derivatives of $\mathrm{TP}(q_i, B_i)$ with respect to q_i and B_i and by setting the results to zero, one has (Konstantaras et al. 2012):

$$\frac{\partial \mathrm{TP}(q_i, B_i)}{\partial B_i} = h - \frac{(h+C_b)}{q_i(1-p_i)}B_i = 0 \text{ where } B_i = \frac{h(1-p_i)q_i}{h+C_b} \qquad (2.333)$$

$$\frac{\partial \mathrm{TP}(q_i, B_i)}{\partial q_i} = \frac{KD}{(1-p_i)q_i^2} + \frac{(h+C_b)}{2q_i^2(1-p_i)}B_i^2 - \frac{hDp_i}{(1-p_i)x} - \frac{h(1-p_i)}{2} = 0 \quad (2.334)$$

After some simple algebra, the unique solution is determined by substituting Eq. (2.333) in Eq. (2.334) and solving for q_i to get:

$$q_i^* = \sqrt{\frac{2KDx(h+C_b)}{h\left[x(1-p_i)^2C_b + 2Dp_i(h+C_b)\right]}} \qquad (2.335)$$

2.4 EOQ Model with Backordering

$$B_i^* = \sqrt{\frac{2KDhx(1-p_i)^2}{(h+C_b)\left[xC_b(1-p_i)^2 + 2Dp_i(h+C_b)\right]}} \quad (2.336)$$

The second partial derivatives of TP(q_i, B_i) are:

$$\frac{\partial^2 \text{TP}(q_i, B_i)}{\partial B_i^2} = -\frac{(h+C_b)}{q_i(1-p_i)} < 0 \quad (2.337)$$

$$\frac{\partial^2 \text{TP}(q_i, B_i)}{\partial q_i^2} = -\frac{2KD}{q_i^3(1-p_i)} - \frac{(h+C_b)}{q_i^3(1-p_i)} B_i^2 < 0 \quad (2.338)$$

$$\frac{\partial^2 \text{TP}(q_i, B_i)}{\partial q_i \partial B_i} = \frac{\partial^2 \text{TP}(q_i, B_i)}{\partial B_i \partial q_i} = \frac{(h+C_b)}{q_i^2(1-p_i)} B_i > 0 \quad (2.339)$$

And the following sufficient condition:

$$\left(\frac{\partial^2 \text{TP}(q_i, B_i)}{\partial B_i^2}\right) \cdot \left(\frac{\partial^2 \text{TP}(q_i, B_i)}{\partial q_i^2}\right) - \left(\frac{\partial^2 \text{TP}(q_i, B_i)}{\partial q_i \partial B_i}\right)^2 = \frac{2KD(h+C_b)}{q_i^4(1-p_i)^2} > 0$$

is satisfied and:

$$\frac{\partial^2 \text{TP}(q_i, B_i)}{\partial B_i^2} < 0$$

The behavior of inventory over several cycles is illustrated in Fig. 2.21. The depletion of inventory during the interval [t_i, s_i] of the ith replenishment cycle is due to the joint effect of the demand and the withdrawal of non-conforming items from

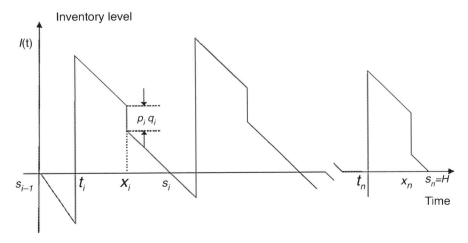

Fig. 2.21 Behavior of the inventory level over finite planning horizon (Konstantaras et al. 2012)

the system. Hence, the differential equation that describes the variation in the inventory level, $I(t)$, over time, t, is given as (Konstantaras et al. 2012):

$$\frac{dI(t)}{dt} = -D, \quad t_i \leq t < x_i \qquad (2.340)$$

with a boundary condition $I(x_i^-) - I(x_i^+) = p_i[I(t_i^-) + I(t_i^+)]$, $i = 1, 2, \ldots, n$. Since the number of non-conforming (defective) items in the ith replenishment cycle is removed from inventory by time x_i, then one obtains:

$$\frac{dI(t)}{dt} = -D, \quad t_i \leq t < x_i \qquad (2.341)$$

with a boundary condition $I(s_i) = 0$, $i = 1, 2, \ldots, n$. The solutions for Eqs. (2.340) and (2.341) are given, respectively, as:

$$I(t) = -D(t - s_i) + \frac{p_i D(s_i - s_{i-1})}{1 - p_i}, \quad t_i \leq t < x_i \qquad (2.342)$$

$$I(t) = D(s_i - t), \quad x_i \leq t < s_i \qquad (2.343)$$

The variation of the inventory level, $I(t)$, over the time interval $[s_i - 1, t_i]$ can be described by the following differential equation:

$$\frac{dI(t)}{dt} = -D, \quad s_{i-1} \leq t < t_i \qquad (2.344)$$

with a boundary condition $I(s_{i-1}) = 0$, $i = 1, 2, \ldots, n$ and whose solution is:

$$I(t) = -D(t - s_{i-1}), \quad s_{i-1} \leq t < t_i \qquad (2.345)$$

The order quantity for the ith shipment is then given from Eqs. (2.342) and (2.345) as:

$$q_i = |I(t_i^+)| + |I(t_i^-)| = \frac{D(s_i - s_{i-1})}{1 - p_i}$$

The total revenue for any policy with n replenishments is given by:

$$\mathrm{TR}(n, s_i, t_i) = \sum_{i=1}^{n} \left[sD(s_i - s_{i-1}) + v \frac{p_i D(s_i - s_{i-1})}{1 - p_i} \right] \qquad (2.346)$$

The total cost for n cycles is given from Eqs. (2.331) and (2.342)–(2.345) as:

2.4 EOQ Model with Backordering

$$TC(n, s_i, t_i) = nK + \sum_{i=1}^{n}\left[(C+C_1)\frac{D(s_i - s_{i-1})}{1-p_i}\right] + \sum_{i=1}^{n}\left[C_b \int_{s_{i-1}}^{t_i} D(t - s_{i-1})dt\right]$$

$$+ \sum_{i=1}^{n}\left[h \int_{t_i}^{x_i}\left(-Dt + Ds_i + \frac{p_i D(s_i - s_{i-1})}{1-p_i}\right)dt + h \int_{x_i}^{s_i} D(s_i - t)dt\right]$$

(2.347)

The total profit for any policy with n replenishments is given from Eqs. (2.346) and (2.347) as

$$TP(n, s_i, t_i) = TR(n, s_i, t_i) - TC(n, s_i, t_i)$$

$$= \sum_{i=1}^{n}\left[sD(s_i - s_{i-1}) + v\frac{p_i D(s_i - s_{i-1})}{1-p_i}\right] - nK - \sum_{i=1}^{n}\left[(C+C_1)\frac{D(s_i - s_{i-1})}{1-p_i}\right] - \sum_{i=1}^{n}\left[C_b \int_{s_{i-1}}^{t_i} D(t - s_{i-1})dt\right]$$

$$- \sum_{i=1}^{n}\left[h \int_{t_i}^{x_i}\left(-Dt + Ds_i + \frac{p_i D(s_i - s_{i-1})}{1-p_i}\right)dt + h \int_{x_i}^{s_i} D(s_i - t)dt\right]$$

(2.348)

Since each shipment of items undergoes a 100% screening, then the following relations should be in effect (Konstantaras et al. 2012):

$$x_i - t_i = \frac{|I(t_i^-)| + |I(t_i^+)|}{y_s} \Rightarrow x_i = t_i + \frac{D(s_i - s_{i-1})}{y_s(1-p_i)} \quad (2.349)$$

$$0 \leq s_{i-1} \leq t_i \leq s_i \quad (2.350)$$

The mathematical programming problem can be written as (Konstantaras et al. 2012):

$$(p)\begin{cases} \min FTP(n, s_1, t_i) \\ \text{s.t.} \\ x_i = t_i + \frac{D(s_i - s_{i-1})}{y_s(1-p_i)} \\ 0 \leq s_{i-1} \leq t_i \leq s_i \end{cases}$$

where

$$\text{FTP}(n, s_i, t_i) = -\text{TP}(n, s_i, t_i)$$

$$= nK + \sum_{i=1}^{n}\left[(C+C_1)\frac{D(s_i - s_{i-1})}{1-p_i}\right] + \sum_{i=1}^{n}\left[C_b \int_{s_{i-1}}^{t_i} D(t - s_{i-1})dt\right]$$

$$+ \sum_{i=1}^{n}\left[h\int_{t_i}^{x_i}\left(-Dt + Ds_i + \frac{p_i D(s_i - s_{i-1})}{1-p_i}\right)dt + h\int_{x_i}^{s_i} D(s_i - t)dt\right] \quad (2.351)$$

$$- \sum_{i=1}^{n}\left[sD(s_i - s_{i-1}) + v\frac{p_i D(s_i - s_{i-1})}{1-p_i}\right]$$

To develop a solution procedure for the problem (P), the initial value of n is fixed and the second constraint $0 \le s_{i-1} \le t_i \le s_i$ is ignored. Take the first-order derivatives of FTP(n, s_i, t_i) with respect to t_i and s_i and equate them to zero yields to (Konstantaras et al. 2012):

$$\frac{\partial \text{FTP}(n, s_i, t_i)}{\partial t_i} = h\left[\left(-Dx_i + Ds_i + \frac{p_i D(s_i - s_{i-1})}{1-p_i}\right)\frac{\partial x_i}{\partial t_i} - \left(-Dt_i + Ds_i + \frac{p_i D(s_i - s_{i-1})}{1-p_i}\right)\right]$$

$$+ h\left[-D(s_i - x_i)\frac{\partial x_i}{\partial t_i}\right] + C_b D(t_i - s_{i-1}) = 0$$

$$\Rightarrow \frac{\partial \text{FTP}(n, s_i, t_i)}{\partial t_i} = (h + C_b)(t_i - s_{i-1}) - h(s_i - s_{i-1}) = 0 \Rightarrow \frac{t_i - s_{i-1}}{s_i - s_{i-1}} = \frac{h}{h + C_b}$$

$$(2.352)$$

Next, Konstantaras et al. (2012) presented that for $t_1 P_0$ the constraints in Eq. (2.350) satisfy the necessary conditions obtained from Eq. (2.353).

Example 2.16 In order to illustrate the behavior of the optimal policy for the finite planning horizon model, which is more complex than the infinite one, an example with the following input parameters is considered: $D = 500$ units/year, $x = 17{,}520$ units/year, $K = \$300$, $h = \$1$/unit/year, $C_b = \$5$/unit/year, $C = \$25$/unit, $s = \$50$/unit, $C_1 = \$0.5$/unit, $v = \$20$/unit, and $H = 6$. In this example, the percentage of defectives per shipment n, p_n, is expressed using an S-shaped logistic learning curve model as $p_n = \frac{a}{g + e^{nm}}$ where $a = 70.067$, $g = 819.76$, and $m = 0.7932$ which are positive model parameters (Konstantaras et al. 2012).

The results in Table 2.27 show that the optimal replenishment policy occurs when $n' = 5$ where the maximum profit is TP(n, s_i, t_i) = 68,985.0. The table also shows the corresponding values of s_i, t_i, and q_i, $i = 1, 2, \ldots, 5$, for the optimal solution. The optimal solution was determined by programming the solution described in solution procedure in Mathematica 6.0.

The results in Table 2.28 show that as the learning parameter m increases, the total order quantity ($\sum_{i=1}^{n} q_i$) decreases and the profits increase. Learning in quality suggests ordering in smaller lots as the fraction of defective items decrease with

2.4 EOQ Model with Backordering

Table 2.27 Total profit for different values of n and the overall optimal replenishment policy (Konstantaras et al. 2012)

n	1	2	3	4	5	6	7	8	9	10
TP	64,110.1	67,589.0	68,552.7	68,891.0	68,985.0	68,966.1	68,896.3	68,809.6	68,720.4	68,626.3
$n = 5$										
t_i	0.1978	1.3851	2.5749	3.7705	4.9784	–	–	–	–	–
s_i	1.1868	2.3760	3.5698	4.7740	6	–	–	–	–	–
q_i	645.21	646.52	649.02	654.67	666.52	–	–	–	–	–

Table 2.28 The optimal t_i, s_i, q_i, n, and TP(n, s_i, t_i) values for increasing values of m under S-shaped learning curve (Konstantaras et al. 2012)

m	t_i	s_i	q_i	n	TP(n, s_i, t_i)
$m = 0.00$	0.20000	1.20000	656.01	5	68,949.5
	1.40000	2.40000	656.01		
	2.60000	3.60000	656.01		
	3.80000	4.80000	656.01		
	5.00000	6.00000	656.01		
$m = 0.10$	0.19996	1.19979	655.84	5	68,950.2
	1.39977	2.39967	655.89		
	2.59967	3.59966	655.95		
	3.79968	4.79976	656.00		
	4.99980	6.00000	656.08		
$m = 0.20$	0.19991	1.19944	655.57	5	68,951.3
	1.39939	2.39911	655.69		
	2.59910	3.59905	655.84		
	3.79910	4.79932	656.02		
	4.99943	6.0000	656.25		
$m = 0.30$	0.19982	1.19890	655.14	5	68,952.9
	1.39878	2.39819	655.36		
	2.59816	3.59800	655.64		
	3.79809	4.79852	656.03		
	4.99877	6.0000	656.55		
$m = 0.40$	0.19967	1.19804	654.46	5	68,955.4
	1.39782	2.39669	654.79		
	2.59661	3.59624	655.28		
	3.79639	4.79713	656.02		
	4.99760	6.00000	657.10		
$m = 0.50$	0.19945	1.19670	653.38	5	68,959.1
	1.39630	2.39429	653.86		
	2.59412	3.59332	654.65		
	3.79355	4.79472	655.94		
	4.99560	6.00000	658.06		
$m = 0.60$	0.19910	1.19462	651.68	5	68,964.6
	1.39393	2.39047	652.35		
	2.59015	3.58855	653.57		
	3.78891	4.79067	655.77		
	4.99222	6.00000	659.70		
$m = 0.70$	0.19856	1.19140	649.02	5	68,973.1
	1.39024	2.38447	649.93		
	2.58387	3.58090	651.76		
	3.78141	4.78397	655.38		
	4.98664	6.00000	662.44		
$m = 0.80$	0.19775	1.18647	644.92	5	68,986.0
	1.38459	2.37519	646.14		

(continued)

2.4 EOQ Model with Backordering

Table 2.28 (continued)

m	t_i	s_i	q_i	n	TP(n, s_i, t_i)
	2.57413	3.56883	648.81		
	3.76955	4.77316	654.62		
	4.97763	6.00000	666.86		

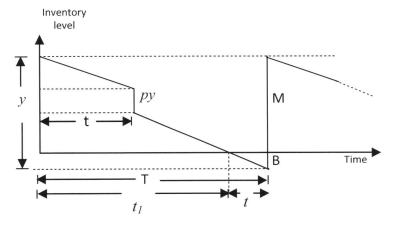

Fig. 2.22 Behavior of the inventory level over time (Rezaei 2005)

every shipment. This results in lower holding, screening, and backordering cost and higher revenue from selling good-quality items, thus increasing profits. When there is no learning, $m = 0$, the optimal policy is to order equal lots in each shipment (Konstantaras et al. 2012).

2.4.6 EOQ Model for Imperfect-Quality Items

Rezaei (2005) developed a simple EOQ model upon Salameh and Jaber's work. To avoid any possible confusion, the notations and assumptions utilized in Salameh and Jaber are employed in this article. He assumed that items, received or produced, are not of perfect quality and not necessarily defective; thus, they could be used in another production/inventory situation. Each lot received contains percentage defectives with a known probability density function. Good-quality items have a selling price for per unit, and defective items are sold as a single batch at a discounted price. A 100% screening process of the lot is conducted and shortage is permitted.

Based on the above assumptions, a mathematical model is developed that is closer to the real world because of exploitation quality and shortage simultaneously. The behavior of the inventory level is illustrated in Fig. 2.22. With the above assumptions, the total cost per cycle for the modified EOQ model with backorder for imperfect items is (Rezaei 2005):

$f(y, B)$ = Fixed cost of placing an order + Variable cost of lot size
+ Screening cost of lot size + Holding cost + Shortage cost.

$$f(y, B) = K + Cy + C_1 y + h\left(\frac{[y(1-p)] - B}{2} t_1 + \frac{py^2}{x}\right) + \frac{C_b t_2}{2} B \quad (2.353)$$

Also, the total revenue per cycle is $g(y, B)$ = total sales volume of good quality + total sales volume of imperfect-quality items:

$$g(y, B) = sy(1 - p) + vyp \quad (2.354)$$

The total profit per cycle is the total revenue per cycle minus the total cost per cycle, $\pi(y, B) = g(y, B) - f(y, B)$, and it is given as:

$$\pi(y,B) = \overbrace{sy(1-p) + vyp}^{\text{Revenue}} - \left\{ \overbrace{K}^{\substack{\text{Fixed}\\\text{cost}}} + \underbrace{Cy}_{\substack{\text{Purchasing}\\\text{cost}}} + \overbrace{C_1 y}^{\substack{\text{Inspection}\\\text{cost}}} + \underbrace{h\left(\frac{[y(1-p)] - B}{2} t_1 + \frac{py^2}{x}\right)}_{\text{Holding cost}} + \overbrace{\frac{C_b B}{2} t_2}^{\substack{\text{Shortage}\\\text{cost}}} \right\}$$

(2.355)

By dividing the total profit per cycle by the cycle length, $T = \frac{y(1-E[p])}{D}$, replacing t_1 by $\frac{y(1-E[p])-B}{D}$ and t_2 by B/D in (2.355), the total profit per unit time can be written as:

$$\pi_U(y, B) = D\left(s - v + \frac{hy}{x}\right) + D\left(v - \frac{hy}{x} - C - C_I - \frac{K}{y}\right)\left(\frac{1}{1-p}\right)$$
$$- \left[h\left(\frac{y(1-p)}{2} - B\right) + \frac{(C_b + h)}{2y(1-p)} B^2\right] \quad (2.356)$$

$$E[\pi_U(y, B)] = D\left(s - v + \frac{hy}{x}\right) + D\left(v - \frac{hy}{x} - C - C_I - \frac{K}{y}\right) E\left[\frac{1}{1-p}\right]$$
$$- \left[h\left(\frac{y(1-E[p])}{2} - B\right) + \frac{(C_b + h)}{2y(1-E[p])} B^2\right] \quad (2.357)$$

In order to solve this nonlinear programming problem, take the first partial derivatives of $E\pi_U(y, B)$ with respect to y, B, respectively. One obtains:

2.4 EOQ Model with Backordering

$$\frac{\partial E[\text{TPU}(y,B)]}{\partial y} = \frac{D}{x}\left(1 - E\left(\frac{1}{1-p}\right)\right) + \frac{DK}{y^2}E\left(\frac{1}{1-p}\right) - \frac{h(1-E[p])}{2}$$
$$+ \frac{(h+C_b)}{2y(1-E[p])}B^2 \qquad (2.358)$$

$$\frac{\partial E[\text{TPU}(y,B)]}{\partial B} = h - \frac{(h+C_b)}{y(1-E[p])}B$$

By setting Eqs. (2.356)–(2.358) equal to zero, one obtains:

$$y^* = \sqrt{\frac{2DKE\left(\frac{1}{1-p}\right)}{h\left[(1-E[p])\left(\frac{C_b}{C_b+h}\right) - \left(\frac{2D}{x}\right)\left(1 - E\left[\frac{1}{1-p}\right]\right)\right]}} \qquad (2.359)$$

$$B^* = \frac{y^*(1-E[p])h}{h+C_b} \qquad (2.360)$$

From Fig. 2.22 for maximum inventory level, one obtains:

$$M = y - B \qquad (2.361)$$

On the other hand, by replacing Eqs. (2.359) and (2.360) in Eq. (2.361), one obtains:

$$M^* = \frac{y^*(1-E[p])C_b}{h+C_b} \qquad (2.362)$$

In order to examine the second-order sufficient conditions (SOSC) for a maximum value, they first obtained the Hessian matrix H as follows (Rezaei 2005):

$$H = \begin{bmatrix} \dfrac{\partial^2 E[\pi_U(y,B)]}{\partial y^2} & \dfrac{\partial^2 E[\pi_U(y,B)]}{\partial y \partial B} \\ \dfrac{\partial^2 E[\pi_U(y,B)]}{\partial B \partial y} & \dfrac{\partial^2 E[\pi_U(y,B)]}{\partial B^2} \end{bmatrix} \qquad (2.363)$$

where

$$\frac{\partial^2 E[\pi_U(y,B)]}{\partial y^2} = \frac{-\left(2DKE\left[\frac{1}{1-p}\right](1-E[p]) + (h+C_b)B^2\right)}{(1-E[p])y^3} \qquad (2.364)$$

$$\frac{\partial^2 E[\pi_U(y,B)]}{\partial y \partial B} = \frac{(h+C_b)}{(1-E[p])y^2} B \qquad (2.365)$$

$$\frac{\partial^2 E[\pi_U(y,B)]}{\partial B \partial y} = \frac{(h+C_b)}{(1-E[p])y^2} B \qquad (2.366)$$

$$\frac{\partial^2 E[\pi_U(y,B)]}{\partial B^2} = \frac{-(h+C_b)}{(1-E[p])y} \qquad (2.367)$$

Then they proceeded by evaluating the principal minor determinants of H at point (y^*, B^*). The first principal minor determinant of H is (Rezaei 2005):

$$|H_{11}| = \frac{-\left(2DKE\left[\frac{1}{1-p}\right](1-E[p]) + B^{*2}(h+C_b)\right)}{(1-E[p])y^{*3}} < 0 \qquad (2.368)$$

$$|H_{22}| = \frac{2DKE\left[\frac{1}{1-p}\right](1-E[p])(h+C_b) + B^{*2}(h+C_b)^2}{(1-E[p])^2 y^{*4}} > 0 \qquad (2.369)$$

Therefore, the Hessian matrix H is negative definite at point (y^*, B^*) which implies that there exist unique values.

Example 2.17 Rezaei (2005) presented an example using $D = 50{,}000$ units/year, $K = 100$/cycle, $h = \$5$/unit/year, $x = 1$ unit/min (175,200 units/year), $C_I = \$0.5$/unit, $C = \$25$/unit, $s = \$50$/unit, and $v = C_b = \$20$/unit, and the percentage defective random variable, p, is uniformly distributed with its p.d.f.:

$$f(p) = \begin{cases} 25 & 0 \le p \le 0.04 \\ 0 & \text{otherwise} \end{cases}$$

So the maximum values of y and B that maximize Eq. (2.357) are $y^* = 1601.58$, $B^* = 313.9$, and $M^* = 1287.68$.

2.5 EOQ Model with Partial Backordering

2.5.1 EOQ Model of Imperfect-Quality Items

Roy et al. (2011) developed an imperfect inventory system under partial backordering. Similar to previous models, the ordering lot size is y. Among these, $(1-p)y$ is of perfect-quality and py of imperfect-quality products. The fraction p follows a probability density function. Generally, it follows a uniform distribution function. The inventory cycle starts with shortages and it continues up to time t_1. In the beginning of the cycle, shortages may occur due to lead time, the time gap

2.5 EOQ Model with Partial Backordering

Table 2.29 Notations of a given problem

$E(.)$	Expected value operator
$Q_s(t)$	The level of negative inventory at time t (unit)
$Q_l(t)$	The lost sale quantity at time t (unit)
$Q_i(t)$	On-hand inventory at time t (unit)

between placing and receiving of an order, and problems with labor staff, management systems, etc.

In order to model the presented problem, some new notations which are specifically used are shown in Table 2.29.

$\left(y - \frac{B}{1-p}\right)$ units of total units y are used for the period $[t_1, t_1 + t_2]$ and the remaining $\frac{B}{1-p}$ units are used for the period $[0, t_1]$. During the period $[t_1, t_1 + t_2]$, the total demand Dt_2 is adjusted with $[y(1-p) - B]$ units. The screening rate x per unit time is always greater than D. During the time span $[t_1, t_1 + t_2]$, $E(1-p)x \geq D$ must be held to avoid shortages. Also, to meet the total shortage B at time t_1, the screening rate x must be satisfied such that $E(1-p)x \geq B + D$. Hence, to avoid shortage within the screening time, $E(1-p)x \geq \text{Max}(B+D, D)$, i.e., $E(1-p)x \geq B + D$ must be satisfied. During stock-out period $[0, t_1]$, the demand $De^{-\delta(t_1-t)}$ at time t is met, and the rest of the demand, $D\left(1 - e^{-\delta(t_1-t)}\right)$, remains unsatisfied. Here $(t_1 - t)$ is the waiting time up to the replenishment at time t_1 and is a positive constant (Roy et al. 2011).

The differential equation of the level of negative inventory at any time t is (Roy et al. 2011):

$$\frac{dQ_s(t)}{dt} = -De^{-\delta(t_1-t)}, \quad 0 \leq t \leq t_1 \text{ with } Q_s(0) = 0 \qquad (2.370)$$

And its solution will be:

$$Q_s(t) = -\frac{D}{\delta}\left(e^{-\delta(t_1-t)} - e^{-\delta t_1}\right), \quad 0 \leq t \leq t_1$$

The lost sale quantity at time t is:
$Q_1(t) = D\left(1 - e^{-\delta(t_1-t)}\right), \quad 0 \leq t \leq t_1$
Here the maximum backorder level is:

$$B = -Q_s(t_1) = \frac{D}{\delta}\left(1 - e^{-\delta t_1}\right) \qquad (2.371)$$

The total backordering and lost sale costs are shown in Eqs. (2.372) and (2.373):

$$BC = C_b \int_0^{t_1} (-Q_s(t))dt$$

$$= \frac{C_b D}{\delta} \int_0^{t_1} \left(e^{-\delta(t_1-t)} - e^{-\delta t_1}\right) dt$$

$$= \frac{C_b D}{\delta} e^{-\delta t_1} \int_0^{t_1} (e^{\delta t} - 1) dt \qquad (2.372)$$

$$= \frac{C_b D}{\delta} e^{-\delta t_1} \left[\left(\frac{e^{\delta t}}{\delta} - t\right)\right]$$

$$= \frac{C_b D}{\delta^2} e^{-\delta t_1} (e^{\delta t} - \delta t_1 - 1)$$

$$= \frac{C_b D}{\delta^2} \left(1 - \delta t_1 e^{-\delta t_1} - e^{-\delta t_1}\right)$$

$$LSC = C_1 \int_0^{t_1} Q_1(t)dt = C_1 D \int_0^{t_1} \left(1 - e^{-\delta(t_1-t)}\right) dt$$

$$= C_1 D \left(t - \frac{e^{-\delta(t_1-t)}}{\delta}\right) = \frac{C_1 D}{\delta} \left(\delta t_1 - 1 + e^{-\delta t_1}\right) \qquad (2.373)$$

The differential equation of the inventory level at any time t is:

$$\frac{dQ_i(t)}{dt} = -D, \quad t_1 \le t \le t_1 + t_2 \text{ with } Q_i(t_1) = (1-p)y - B \qquad (2.374)$$

The solution of Eq. (2.374) is:

$$Q_i(t) = [(1-p)y - B] - D(t - t_1), \quad t_1 \le t \le t_1 + t_2 \qquad (2.375)$$

Considering $(1-p)y - B = Dt_2$, the total cycle length is (Roy et al. 2011):

$$T = t_1 + t_2 = \frac{(1-p)y - B}{D} + t_1 \qquad (2.376)$$

The inventory holding cost HC is:

$$h \left[\frac{((1-p)y - B)^2}{2D} + \frac{py^2}{x}\right] \qquad (2.377)$$

2.5 EOQ Model with Partial Backordering

$$\mathrm{TC}(y, t_1) = \overbrace{K}^{\text{Fixed cost}} + \overbrace{Cy}^{\text{Purchasing cost}} + \overbrace{C_1 y}^{\text{Screening cost}} + \overbrace{h \left[\frac{((1-p)y - B)^2}{2D} + \frac{p y^2}{x} \right]}^{\text{Holding cost}}$$

$$+ \underbrace{\frac{C_b D}{\delta^2} \left(1 - \delta t_1 e^{-\delta t_1} - e^{-\delta t_1} \right)}_{\text{Backordering cost}} + \underbrace{\frac{\hat{\pi} D}{\delta} \left(\delta t_1 - 1 + e^{-\delta t_1} \right)}_{\text{Lost sale cost}}$$

(2.378)

The expected total cost per cycle is (Roy et al. 2011):

$$E[\mathrm{TC}(y, t_1)] = K + Cy + C_1 y + h \left[\frac{E((1-p)y - B)^2}{2D} + \frac{E[p]y^2}{x} \right]$$
$$+ \frac{C_b D}{\delta^2} \left(1 - \delta t_1 e^{-\delta t_1} - e^{-\delta t_1} \right) + \frac{\hat{\pi} D}{\delta} \left(\delta t_1 - 1 + e^{-\delta t_1} \right)$$

(2.379)

The expected cyclic total revenue is:

$$E[\mathrm{TR}(y, t_1)] = s(1 - E(p))y + v E(p) y \tag{2.380}$$

And the expected cycle length is:

$$E(T) = \frac{(1 - E(p))y - B}{D} + t_1 \tag{2.381}$$

The expected average profit per cycle using renewal–reward theorem changes to:

$$E[\mathrm{TP}(y, t_1)] = \frac{E[\mathrm{TR}(y, t_1)] - E[\mathrm{TC}(y, t_1)]}{E(T)}$$

$$= \frac{1}{\frac{(1 - E(p))y - B}{D} + t_1} \bigg[s(1 - E(p))y + v E(p) y - K - C y$$

$$- C_1 y - h \left[\frac{E((1-p)y - B)^2}{2D} + \frac{E[p]y^2}{x} \right]$$

$$- \frac{C_b D}{\delta^2} \left(1 - \delta t_1 e^{-\delta t_1} - e^{-\delta t_1} \right) - \frac{\hat{\pi} D}{\delta} \left(\delta t_1 - 1 + e^{-\delta t_1} \right) \bigg]$$

$$= \frac{1}{\frac{ey - B}{D} + t_1} \bigg[\left(s - v + \frac{hB}{D} \right) e_1 y + (v - C - C_1) y - \frac{h e_1 y^2}{2D}$$

$$- \frac{1}{2D} \left\{ 2KD + hB^2 + \frac{2D^2}{\delta^2} \left((C_b - \hat{\pi}\delta)(1 - e^{-\delta t_1}) - \delta t_1 (C_b e^{-\delta t_1} - \hat{\pi}\delta) \right) \right\} \bigg]$$

(2.382)

where

$$e_1 = 1 - E(p) > 0$$

$$e_2 = E\left[(1-p)^2\right] + \frac{2E(p)D}{x} = e_1^2 + \text{var}(p) + \frac{2E(p)D}{x} > 0$$

because

$$E(1-p)^2 = E(1-m-p+m)^2$$

where

$$m = E(p) = E\left[(1-m)^2 - 2(1-m)(p-m) + (p-m)^2\right]$$
$$= (1-m)^2 - 2(1-m)(E(p) - m) + E\left[(p-m)^2\right]$$
$$= (1-E(p))^2 + \text{var}(p)$$
$$= e_1^2 + \text{var}(p)$$

In order to determine the optimal values of decision variable lot size y and shortage period t_1, which maximize the expected average profit, Roy et al. (2011) set $\frac{\partial E[TP]}{\partial y} = 0 = \frac{\partial E[TP]}{\partial t_1}$. Now $\frac{\partial E[TP]}{\partial y} = 0$ gives:

$$he_1e_2y^2 + 2he_2\gamma(t_1)y - [\alpha(t_1)\gamma(t_1) + e_1\beta(t_1)] = 0 \quad (2.383)$$

where

$$\alpha(t_1) = 2D\left[e_1\left(s - v + \frac{hB}{D}\right) + (v - C - C_1)\right]$$
$$\beta(t_1) = 2KD + hB^2 + \frac{2D^2}{\delta^2}\left((C_b - \widehat{\pi}\delta)(1 - e^{-\delta t_1}) - \delta t_1(C_b e^{-\delta t_1} - \widehat{\pi}\delta)\right)$$
$$\gamma(t_1) = Dt_1 - B$$

The equation can be written as:

$$Fy^2 + Gy - H = 0$$

The solution of the above equation is:

$$y^*(t_1) = \frac{1}{2F}\left[-G + \sqrt{G^2 + 4FH}\right] \quad (2.384)$$

where

2.5 EOQ Model with Partial Backordering

$$F = he_1e_2 > 0$$
$$G = 2he_2\gamma(t_1) > 0$$
$$H = \alpha(t_1)\gamma(t_1) + e_1\beta(t_1)$$

For a positive value of y, G must be positive. Now $\frac{\partial E[TP]}{\partial t_1} = 0$ gives:

$$y^{**}(t_1) = \frac{1}{2L}\left[M + \sqrt{M^2 + 4LN}\right] \quad (2.385)$$

where

$$L = e_1 \frac{\partial \alpha(t_1)}{\partial t_1} + he_2 \frac{\partial \gamma(t_1)}{\partial t_1} = Dh\left[2e_1^2 e^{-\delta t_1} + \left(1 - e^{-\delta t_1}\right)e_2\right] > 0$$

$$M = e_1 \frac{\partial \beta(t_1)}{\partial t_1} + \alpha(t_1)\frac{\partial \gamma(t_1)}{\partial t_1} - \gamma(t_1)\frac{\partial \alpha(t_1)}{\partial t_1}$$

$$= 2D^2 e_1\left[\frac{h}{g}\left(1 - e^{-\delta t_1}\right)^2 + \left(1 - e^{-\delta t_1}\right)\left(\frac{2h}{\delta}e^{-\delta t_1} + s - v + \frac{(v - s - C_I)}{e_1} + \widehat{\pi}\right) + t_1 e^{-\delta t_1}(C_b - h)\right]$$

$$N = \gamma(t_1)\frac{\partial \beta(t_1)}{\partial t_1} - \beta(t_1)\frac{\partial \gamma(t_1)}{\partial t_1}$$

$$= 2D^3\left[t_1^2 C_b e^{-\delta t_1} - \left(\frac{K}{D} - \frac{ht_1 e^{-\delta t_1}}{\delta}\right)\left(1 - e^{-\delta t_1}\right) - \left(\frac{C_b + he^{-\delta t_1}}{\delta^2}\right)\left(1 - e^{-\delta t_1}\right)^2 - \frac{h}{2\delta^2}\left(1 - e^{-\delta t_1}\right)^3\right]$$

From Eqs. (2.384) and (2.385), they got for feasibility of their model (Roy et al. 2011):

$$\frac{1}{2F}\left[-G + \sqrt{G^2 + 4FH}\right] = \frac{1}{2L}\left[M + \sqrt{M^2 + 4LN}\right] \quad (2.386)$$

Solving Eq. (2.386), and the intersecting point of the functions $y^*(t_1)$ and $y^{**}(t_1)$, they obtained the optimum value t_1^*, and hence, using either Eq. (2.384) or Eq. (2.385), they got the optimum value of the lot size y, i.e., y^*.

Example 2.18 Roy et al. (2011) supposed that the imperfect-quality items in each lot follow a uniform distribution with the following probability density function:

$$f(p) = \begin{cases} 10 & 0 \le p \le 0.1 \\ 0 & \text{otherwise} \end{cases}$$

K = \$400/lot, D = 16,000 units/year, h = \$4 unit/time, C_b = \$6 unit/time, x = 60,000 units/year, C_I = \$0.5/unit, C = \$40/unit, s = \$62/unit, $\widehat{\pi}$ = \$23/unit, and v = \$30/unit. Using the above parameters, they had e_1 = 0.95, Var(p) = 0.0025/3, e_2 = 0.93, and the optimum solutions y^* = 2082.47 units, t_1^* = 0.0264187 = 0.03, and $E[TP]^*$ = \$328,690 because the Hessian matrix H is negative definite.

Example 2.19 Roy et al. (2011) considered that all the parameters of Example 2.18 are the same except $\hat{\pi} = \$8$. Then, the optimal solution is $y^* = 2100.06$ units, $t_1^* = 0.03$ units, and $E[TP] = \$328,849$. Since at the derived solution values, the Hessian matrix H is negative definite, all are optimal solutions.

2.5.2 Screening

Wang et al. (2015) developed an imperfect inventory system with partial backordering. They assumed that the replenishment rate is infinite and all items are screened 100% with a known screening rate. At the end of the screening period, the imperfect-quality items are removed from the stock. Furthermore, to meet demand, similar to previous case $(1 - p)x - D > 0$. The screening and demand proceed simultaneously. The shortages are backordered constrained by the screening rate. During the backordering period, it is assumed that the existed backorders are filled before the new incoming orders are met.

In order to model the presented problem, some new notations which are specifically used are shown in Table 2.30.

Since the new demands cannot be filled immediately, it is also assumed that only a fraction of the new orders would wait during the backordering period, i.e., the demand rate is βD during the backordering period (see Fig. 2.23). The following condition $C + C_I < (1 - p)s + pv$ must be satisfied. The purchase and screening cost for each product is $C + C_I$. Since $(1 - p) \times 100\%$ good and $p \times 100\%$ imperfect items are sold, $(1 - p)s + pv$ is the average unit price. This constraint is to ensure the retailer has the benefit from selling products.

According to Fig. 2.23 during the shortage period t_1, a portion of demand with a rate of βD is backordered. At the beginning of screening interval $t_2 + t_3$, products start being screened upon being delivered to the retailer. The products are screened to remove the good ones from imperfect items. Since the delivered products contain p rate imperfect items, the rate of good-quality items is $1 - p$. Since the screening rate is x, the good items with a rate of $(1 - p)x$ are used to meet the demand and eliminate backorders during the period t_2. Since the demand rate during the backordering period t_2 is βD, during the period t_2, one can see that backorders are eliminated at a rate of $(1 - p)x - \beta D$, and the stock decreases at a rate of $-(1 - p)x$. During the period t_3 and t_4, all incoming demands are filled; the stock decreases at a rate of D. The screening process terminates at the end of the period t_3; the imperfect items are subtracted from the stock and these are sold at a discounted price (Wang et al. 2015).

Table 2.30 Notations of a given problem

t_2	Time to eliminate backorders (decision variable) (time)
t	$t_2 + \beta t_3$ (years) (decision variable) (time)
t_3	Time to screen after eliminating backorders (time)

2.5 EOQ Model with Partial Backordering

Fig. 2.23 Graphic representation of the inventory system (see online version for colors) (Wang et al. 2015)

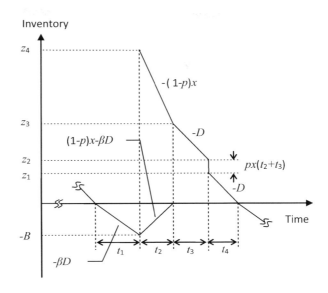

Wang et al. (2015) first used t_2 and t_3 to obtain the equations for related expressions. Referring to Fig. 2.23, the maximal backordering level is given by:

$$B = ((1-p)x - \beta D)t_2 \tag{2.387}$$

using,

$$B = \beta D t_1 = ((1-p)x - \beta D)t_2 \tag{2.388}$$

in Fig. 2.23 holds, the shortage interval is

$$t_1 = ((1-p)x - \beta D)t_2/\beta D$$

Since during $t_2 + t_3$ the order size is screened at a rate of x, one has (Wang et al. 2015):

$$z_4 = x(t_2 + t_3) \tag{2.389}$$

According to Fig. 2.23, we have:

$$z_3 = z_4 - (1-p)xt_2 = (pt_2 + t_3)x \tag{2.390}$$

$$z_2 = z_3 - Dt_3 = (pt_2 + t_3)x - Dt_3 \tag{2.391}$$

$$z_1 = z_2 - p(t_2 + t_3)x = ((1-p)x - \beta D)t_3 \tag{2.392}$$

$$t_4 = \frac{z_1}{D} = \frac{((1-p)x - \beta D)t_3}{D} \tag{2.393}$$

$$T = t_1 + t_2 + t_3 + t_4 = (1-p)x(t_2 + \beta t_3)/\beta D \tag{2.394}$$

The total inventory cost per cycle is (Wang et al. 2015):

$$\begin{aligned}\text{TC}(t_2, t_3) =& \overbrace{Cx(t_2+t_3)}^{\text{Purchasing cost}} + \overbrace{C_1 x(t_2+t_3)}^{\text{Screening cost}} + \overbrace{K}^{\text{Fixed cost}} + \overbrace{h\left[\frac{t_2(z_4+z_3)}{2} + \frac{t_3(z_3+z_2)}{2} + \frac{t_4 z_1}{2}\right]}^{\text{Holding cost}} \\ &+ \underbrace{\frac{C_b(t_2+t_1)B}{2}}_{\text{Backordering cost}} + \underbrace{\widehat{\pi}(1-\beta)D(t_2+t_1)}_{\text{Lost sale cost}} \\ =& Cx(t_2+t_3) + C_1 x(t_2+t_3) + K + h\frac{Dx[(1+p)t_2^2 + 2(1+p)t_2 t_3 + 2p_3^2 + x^2(1-p)^2 t_3^2]}{2D} \\ &+ C_b\left[\frac{(1-p)^2 x^2 t_2^2}{2\beta D} - \frac{(1-p)xt_2^2}{2}\right] + \widehat{\pi}\frac{(1-p)(1-\beta)xt_2}{\beta} \end{aligned} \tag{2.395}$$

Since it is not possible to derive the closed-form optimal solution of (t_2, t_3), Wang et al. (2015) developed an alternative variable transformation approach where $t_2 + \beta t_3$ is replaced by t in Eq. (2.395) and derived the optimal solution of (t_2, t):

$$\begin{aligned}\text{TC}(t_2, t = t_2 + \beta t_3) =& \frac{Cx(t-(1-\beta)t_2)}{\beta} + \frac{C_1 x(t-(1-\beta)t_2)}{\beta} + K \\ &+ h\left(\frac{(1-p)^2 x^2 (t-t_2)^2}{2\beta^2 D} + \frac{x(2p(t-t_2)^2 + 2(1+p)(t-t_2)t_2\beta + (1+p)\beta^2 t_2^2)}{2\beta^2}\right) \\ &+ C_b\left(\frac{(1-p)^2 x^2 t_2^2}{2\beta D} - \frac{(1-p)xt_2^2}{2}\right) + \widehat{\pi}\frac{(1-p)(1-\beta)xt_2}{\beta}\end{aligned} \tag{2.396}$$

The total revenue per cycle, TR, is comprised of sales revenue of good items and salvage value of imperfect items:

$$\begin{aligned}\text{TR}(t_2, t) &= s(1-p)x(t_2+t_3) + vpx(t_2+t_3) \\ &= \frac{(1-p)sx(t-(1-\beta)t_2)}{\beta} + \frac{pvx(t-(1-\beta)t_2)}{\beta}\end{aligned} \tag{2.397}$$

From Eq. (2.395), the ordering cycle can be rewritten as:

2.5 EOQ Model with Partial Backordering

$$T = t_1 + t_2 + t_3 + t_4 = \frac{(1-p)x(t_2 + \beta t_3)}{\beta} = \frac{(1-p)xt}{\beta D} \quad (2.398)$$

The expected total profit per unit time is given as:

$$E[\text{TPU}(t_2,t)] = \frac{E[\text{TR}] - E[\text{TC}]}{E[T]} = \left(sD - sD(1-\beta)\frac{t_2}{t}\right) + \left(\frac{vE_1D}{E_2} - \frac{vE_1D}{E_2}(1-\beta)\frac{t_2}{t}\right)$$
$$-\left(\frac{CD}{E_2} - \frac{CD}{E_2}(1-\beta)\frac{t_2}{t}\right) - \left(\frac{C_1D}{E_2} - \frac{C_1D}{E_2}(1-\beta)\frac{t_2}{t}\right) - \frac{KD\beta}{E_2xt}$$
$$-\frac{h(2E_1D + E_3x)}{2E_2\beta} t\left(1 - \frac{t_2}{t}\right)^2 - \frac{h(1+E_1)Dt_2}{E_2} + \frac{h(1+E_1)D(2-\beta)t_2^2}{2E_2t}$$
$$-\frac{C_b(E_3x - E_2D\beta)t_2}{2E_2t} - \widehat{\pi}D(1-\beta)\frac{t_2}{t}$$
$$(2.399)$$

where $E_1 = E[p]$, $E_2 = E[1-p]$, and $E_3 = E[(1-p)^2]$. Taking the expectation value for the constraint of $(1-p)x - D > 0$ yields $E_2x - D > 0$. After algebraic manipulation, Eq. (2.399) is written as (Wang et al. 2015):

$$E[\text{TPU}(t_2,t)] = \left(s + \frac{vE_1}{E_2} - \frac{C}{E_2} - \frac{C_1}{E_2}\right)D$$
$$- \left[\begin{array}{l} \frac{KD\beta}{E_2xt} + \frac{h(2E_1D + E_3x)}{2E_2\beta} t\left(1 - \frac{t_2}{t}\right)^2 + \frac{h(1+E_1)Dt_2}{E_2} \\ - \frac{h(1+E_1)D(2-\beta)t_2^2}{2E_2t} + \frac{C_b(E_3x - E_2D\beta)t_2}{2E_2} \frac{t_2}{t} \\ + \left(s + \frac{vE_1}{E_2} - \frac{C}{E_2} - \frac{C_1}{E_2} + \widehat{\pi}\right)D(1-\beta)\frac{t_2}{t} \end{array}\right]$$
$$(2.400)$$

Their objective was to maximize $E[\text{TPU}(t_2, t)]$ subject to $t > 0$, $t_2 \geq 0$, and $t \geq t_2$. According to Fig. 2.23, it can be seen $t_3 \geq 0$. Since $t = t_2 + \beta t_3$, one has $t_3 = \frac{t-t_2}{\beta} \geq 0$, i.e., $t \geq t_2$. If $t_3 < 0$, it shows t_4 in Eq. (2.393) and z_1 in Eq. (2.392) are less than zero. This implies that the model cannot fill all demand and backorders during an ordering cycle. It is obvious that (Wang et al. 2015):

$$\begin{cases} \text{Max } E[\text{TPU}(t_2, t_3)] \\ \text{s.t.} \\ t_2 + t_3 > 0 \\ t_2 \geq 0 \\ t_3 \geq 0 \end{cases} \text{ and } \begin{cases} \text{Max } E[\text{TPU}(t_2, t_3)] \\ \text{s.t.} \\ t > 0 \\ t_2 \geq 0 \\ t \geq t_2 \end{cases} \text{ are equivalent.}$$

Since $\left(s + \frac{vE_1}{E_2} - \frac{C}{E_2} - \frac{C_1}{E_2}\right)D$ is a constant with respect to the decision variables, t_2 and t, Wang et al. (2015) maximized $E[\text{TPU}(t_2, t)]$ by minimizing the expression in the square bracket in Eq. (2.400), which is defined as:

$$E[\text{TVCU}(t_2,t)] = \frac{KD\beta}{E_2xt} + \frac{h(2E_1D + E_3x)}{2E_2\beta}t\left(1 - \frac{t_2}{t}\right)^2 + \frac{h(1+E_1)Dt_2}{E_2} - \frac{h(1+E_1)D(2-\beta)t_2^2}{2E_2t}$$
$$+ \frac{C_b(E_3x - E_2D\beta)t_2}{2E_2} + \left(s + \frac{vE_1}{E_2} - \frac{C}{E_2} - \frac{C_1}{E_2} + \hat{\pi}\right)D(1-\beta)\frac{t_2}{t}$$

(2.401)

For further ease of notation, Wang et al. (2015) defined:

$$r_0 = sD(1-\beta) \tag{2.402}$$

$$r_1 = \frac{vE_1 D(1-\beta)}{E_2} \tag{2.403}$$

$$r_2 = \frac{CD(1-\beta)}{E_2} \tag{2.404}$$

$$r_3 = \frac{C_1 D(1-\beta)}{E_2} \tag{2.405}$$

$$r_4 = \frac{KD\beta}{E_2 x} \tag{2.406}$$

$$r_5 = \frac{h(2E_1 D + E_3 x)D\beta}{2E_2\beta} \tag{2.407}$$

$$r_6 = \frac{h(1+E_1)D}{E_2} \tag{2.408}$$

$$r_7 = \frac{h(1+E_1)D(2-\beta)}{2E_2} \tag{2.409}$$

$$r_8 = \frac{C_s(E_3 x - E_2 D\beta)}{2E_2} \tag{2.410}$$

$$r_9 = \hat{\pi} D(1-\beta) \tag{2.411}$$

Therefore, Eq. (2.401) is written as (Wang et al. 2015):

$$E[\text{TVCU}(t_2, t)] = (r_0 + r_1 - r_2 - r_3 + r_9)\frac{t_2}{t} + \frac{r_4}{t} + r_5 t\left(1 - \frac{t_2}{t}\right)^2 + r_6 t_2$$
$$- r_1 \frac{t_2^2}{t} + r_8 \frac{t_2^2}{t} \tag{2.412}$$

From Eq. (2.389), they saw if $t_2 = 0$, then $t_1 = 0$ (or $t_1 + t_2 = 0$) which implies backorders are not allowed. If the time t is given, Lemma 2.1 provides a criterion to

2.5 EOQ Model with Partial Backordering

decide whether a shortage period is greater than zero or not. Therefore, Lemma 1 (1) (a) and (2) (a) of Wang et al. (2015) show that if the time t is less than a specific value, the optimal policy is to fill all demand without backorders. Lemma 1 (1) (b) and (2) (b) of Wang et al. (2015) imply that if the time t is greater than a particular value, the optimal policy is to allow shortage period. Since $\beta > 0$ and the definition $t = t_2 + \beta t_3$, they saw $t_2 = t$ yields $t_3 = 0$, and t_4 in Eq. (2.394) is zero. Therefore, Lemma 1 (1) (c) of Wang et al. (2015) indicates that the remaining backorders are fulfilled at once when the order lot finishes the screening process. Note that if an optimal t is given, the conditions for the optimal t_2 in Lemma 1 of Wang et al. (2015) still hold.

Let t_2^+ and t^+ be solutions that satisfy the first-order condition of $E[\text{TVCU}(t_2, t)]$. Taking the first-order derivatives of $E[\text{TVCU}(t_2, t)]$ with respect to t_2 and t, one has:

$$\frac{\partial [\text{TVCU}(t_2, t)]}{\partial t_2} = \frac{(r_0 + r_1 - r_2 - r_3 + r_9) - (2r_5 - r_6)t + 2(r_5 - r_7 + r_8)t_2}{t} \tag{2.413}$$

$$\frac{\partial [\text{TVCU}(t_2, t)]}{\partial t} = \frac{r_4 - r_5 t^2 + (r_0 + r_1 - r_2 - r_3 + r_9)t_2 + 2(r_5 - r_7 + r_8)t_2^2}{t^2} \tag{2.414}$$

Letting $\frac{\partial [\text{TVCU}(t_2, t)]}{\partial t_2} = 0$ and $\frac{\partial [\text{TVCU}(t_2, t)]}{\partial t} = 0$, then

$$t_2(t) = \frac{-(r_0 + r_1 - r_2 - r_3 + r_9) + (2r_5 - r_6)t}{2(r_5 - r_7 + r_8)} \tag{2.415}$$

$$t(t_2) = \sqrt{\frac{r_4 + (r_0 + r_1 - r_2 - r_3 + r_9)t_2 + (r_5 - r_7 + r_8)t_2^2}{r_5}} \tag{2.416}$$

Substituting t_2 in Eq. (2.415) into Eq. (2.416), one can derive the solution of t as (Wang et al. 2015):

$$t^+ = \sqrt{\frac{4r_4(r_5 - r_7 + r_8) - (r_0 + r_1 - r_2 - r_3 + r_9)^2}{4r_5(r_6 - r_7 + r_8) - r_6^2}} \tag{2.417}$$

Substituting t^+ in Eq. (2.416) into Eq. (2.417), the solution of t_2 is given as (Wang et al. 2015):

$$t_2^+ = \frac{-(r_0 + r_1 - r_2 - r_3 + r_9) + (2r_5 - r_6)t^+}{2(r_5 - r_7 + r_8)} \tag{2.418}$$

If $t^+ \geq \frac{r_0 + r_1 - r_2 - r_3 + r_9}{2r_5 - r_6}$, it can be seen that t_2^+ in Eq. (2.418) is greater than or equal to zero.

Let the optimal t_2 and t be t_2^* and t^*, respectively. If shortage period is not allowed, substituting t_2 as zero into $E[\text{TVCU}(t_2, t)]$ of Eq. (2.412), one can derive the optimal t as (Wang et al. 2015):

$$t^* = \sqrt{\frac{r_4}{r_5}} = \beta\sqrt{\frac{2DK}{xh(2E_1 + E_3x)}} \tag{2.419}$$

The t^* in Eq. (2.419) can also be confirmed by substituting t_2 as zero into Eq. (2.416). The optimal expected variable cost per unit time without backorders is (Wang et al. 2015):

$$E\left[\text{TVCU}\left(0, \sqrt{\frac{r_4}{r_5}}\right)\right] = \frac{\sqrt{2DhK(2E_1D + E_3x)}}{E_2\sqrt{x}} \tag{2.420}$$

If t^* results in $t = 0$, they obtained that t^* in Eq. (2.419) must be less than $\frac{r_0+r_1-r_2-r_3+r_9}{2r_5-r_6}$. Thus, if $\sqrt{\frac{r_4}{r_5}} < \frac{r_0+r_1-r_2-r_3+r_9}{2r_5-r_6}$, then t_2^* is derived, i.e., the optimal solution is $\left(0, \sqrt{\frac{r_4}{r_5}}\right)$.

If $\beta = 1$ (i.e., complete backordering), it results in (Wang et al. 2015):

$$\sqrt{\frac{r_4}{r_5}} \geq \frac{r_0 + r_1 - r_2 - r_3 + r_9}{2r_5 - r_6} \quad (\because r_0 + r_1 - r_2 - r_3 + r_9 = 0)$$

and

$$2r_7 - 2r_8 - r_6 = \frac{-C_b(E_3x - E_2D)}{E_2} \leq 0 \, (E_3x - E_2D > 0)$$

If (t_2^{**}, t) is (t_2^{++}, t), substituting t^+ in Eq. (2.417) into Eq. (2.398), and taking the expectation value, the optimal cycle is:

$$T^* = \frac{E_2xt^+}{\beta D} \tag{2.421}$$

and the optimal ordering size is:

$$y^* = z_4^* = x(t_2^* + t_3^*) = x\frac{t^+ - (1-\beta)t_2^*}{\beta} \tag{2.422}$$

The optimal cycle length and the optimal ordering size without backorders are derived as (Wang et al. 2015):

2.5 EOQ Model with Partial Backordering

$$T^* = \frac{E_2 x t^+}{\beta D}\sqrt{\frac{r_4}{r_5}} = \frac{E_2 x}{\beta D}\beta\sqrt{\frac{2KD}{hx(E_3 x - 2E_1 D)}} = E_2\sqrt{\frac{2Kx}{hD(E_3 x - 2E_1 D)}} \quad (2.423)$$

$$y^* = \frac{x}{\beta}\sqrt{\frac{r_4}{r_5}} = \sqrt{\frac{2KDx}{h(E_3 x + 2E_1 D)}} = \sqrt{\frac{2KD}{h\left(E\left[(1-p)^2\right] + 2E[p]D/x\right)}} \quad (2.424)$$

Subsequently, to compare with the existed EOQ models, they adapted Theorem 1 of Wang et al. (2015) using the critical value of β as β^* (see, e.g., Montgomery et al. 1973; Rosenberg 1979; Park 1982; Pentico and Drake 2009) to determine the optimal policy. Let β^* be the solution of β derived from:

$$\sqrt{\frac{r_4}{r_5}} - \frac{r_0 + r_1 - r_2 - r_3 + r_9}{2r_5 - r_6} = 0$$

It gives (Wang et al. 2015):
$$\beta^* = 1 - \frac{\sqrt{2Kh}[E_3 x - (1 - E_1)D]}{\sqrt{Dx(E_3 x + 2E_1 D)}\left(E_2(s+\widehat{\pi}) + E_1 v - C - C_1\right) - \sqrt{2Kh}(1+E_1)D} \qquad \beta \geq \beta^* \quad \text{means}$$
backlogging unfilled demand would have more cost-efficiency than holding stock throughout the cycle. If $\beta < \beta^*$, the optimal policy is to fulfill all demand without backorders, i.e., (t_2^{**}, t) is $\left(0, \sqrt{\frac{r_4}{r_5}}\right)$. Although β^* provides a critical value to find the optimal policy, one also needs to check whether the resulting $E[\text{TVCU}(t_2^{**}, t)]$ is less than the cost of losing all sales (e.g., Zhang 2009). They have derived the cost of lost sales per unit time $(E_2(s+\widehat{\pi}) + E_1 v - C - C_1)\frac{D}{E_2}$ (Wang et al. 2015).

In order to solve the model, the following solution procedure is developed by Wang et al. (2015):

1. Calculate E_i and β^* from Eq. (2.427).
2. If $\beta < \beta^*$, let $(t_2^*, t^*) = \left(0, \beta\sqrt{\frac{2DK}{xh(2E_1 D + E_3 x)}}\right)$, where t^* is from Eq. (2.419).
   Calculate $E\left[\text{TVCU}\left(0, \beta\sqrt{\frac{2DK}{xh(2E_1 D + E_3 x)}}\right)\right] = \sqrt{\frac{2DhK(2E_1 D + E_3 x)}{\sqrt{x E_2}}}$ from Eq. (2.420).

   (a) If $\sqrt{\frac{2DhK(2E_1 D + E_3 x)}{\sqrt{x E_2}}} < (E_2(s+\widehat{\pi}) + E_1 v - C - C_1)\frac{D}{E_2}$, the optimal inventory policy is to meet all demand without backorders. The optimal solution is $\left(0, \beta\sqrt{\frac{2DK}{xh(2E_1 D + E_3 x)}}\right)$.

   (b) If $\sqrt{\frac{2DhK(2E_1 D + E_3 x)}{\sqrt{x E_2}}} \geq (E_2(s+\widehat{\pi}) + E_1 v - C - C_1)\frac{D}{E_2}$, the optimal policy is to lose all.

3. If $\beta \geq \beta^*$, calculate r_i from Eqs. (2.402) to (2.411) and (t_2^+, t^+) from Eqs. (2.418) and (2.419).

   (a) If $2r_7 - 2r_8 - r_6 > 0$ and $t^+ < \frac{r_0 + r_1 - r_2 - r_3 + r_9}{2r_7 - 2r_8 - r}$, let $(t_2^*, t^*) = (t_2^+, t^+)$.

(b) If $2r_7 - 2r_8 - r_6 > 0$ and $t^+ \geq \frac{r_0+r_1-r_2-r_3+r_9}{2r_7-2r_8-r}$, let $(t_2*, t^*) = (t_2^+, t^+)$.
(c) If $2r_7 - 2r_8 - r_6 \leq 0$, let $(t_2*, t^*) = (t_2^+, t^+)$.
(d) Substitute (t_2, t) by (t_2*, t^*) into Eq. (2.412) to derive $E[\text{TVCU}(t_2^*, t^*)]$.
(e) If $E[\text{TVCU}(t_2^*, t^*)] < (E_2(s + \hat{\pi}) + E_1 v - C - C_1)\frac{D}{E_2}$, the optimal policy is to meet the demand with partial backordering resulting in an optimal solution (t_2*, t^*).
(f) If $E[\text{TVCU}(t_2^*, t^*)] \geq (E_2(s + \hat{\pi}) + E_1 v - C - C_1)\frac{D}{E_2}$, the optimal policy is to lose all sales.

Example 2.20 Wang et al. (2015) presented an example using:

$$f(p) = \begin{cases} 25 & 0 \leq p \leq 0.04 \\ 0 & \text{otherwise} \end{cases}$$

$K = \$500$/lot, $D = 1000$ units/year, $h = \$10$ unit/year, $C_b = \$5$ unit/year, $x = 4000$ units, $\hat{\pi} = \$1$/unit, $C_I = \$1$/unit, $C = \$40$/unit, $s = \$50$/unit, and $v = \$10$/unit. Using the above parameters, $E_1 = E[p] = 0.02$, $E_2 = E[1-p] = 0.98$, and $E_3 = E[(1-p)^2] = 0.960533$ and $\beta^* = 0.72531$. If a given $\beta = 0.5 < \beta^* = 0.72531$, Step 2 should be applied to derive the optimal solution. Calculate $t^* = \sqrt{\frac{r_4}{r_5}} = \beta\sqrt{\frac{2DK}{xh(2E_1+E_3x)}} = 0.072222$ from Eq. (2.241). Since $\sqrt{\frac{2DhK(2E_1D+E_3x)}{\sqrt{x}E_2}} = 3178.92 < (E_2(s + \hat{\pi}) + E_1 v - C - C_1)\frac{D}{E_2} = 9367.35$, the optimal solution is $(t_2^*, t) = (0, 0.072222)$. But for a given $\beta = 0.9 \geq \beta^* = 0.72531$, Step 3 should be applied. So $r_0 = 5000$, $r_1 = 20.41$, $r_2 = 4081.63$, $r_4 = 114.80$, $r_5 = 22,007.56$, $r_6 = 10,408.16$, $r_7 = 5724.49$, $r_8 = 7551.36$, $r_9 = 100$ and $t^* = 0.10194, t_2^* = 0.05222$. Since $2r_7 - 2r_8 - r_6 = -14,061.9 \leq 0$, $(t_2*, t^*) = (t_2^* = 0.05222, t^* = 0.10194)$. Since $E[\text{TVCU}(t_2^*, t^*)] = \$2732.4 < (E_2(s + \hat{\pi}) + E_1 v - C - C_1)\frac{D}{E_2} = \$9367.35$, the optimal policy is to meet the demand with partial backordering. The optimal solution is $(t_2^*, t) = (0.05222, 0.10194)$ with total variable cost per year of \$2732.04. Using Eqs. (2.421) and (2.422), $T^* = \frac{E_2 x t^+}{\beta D} = 0.44401$, $y^* = z_4^* = x(t_2^* + t_3^*) = x\frac{t^+ - (1-\beta)t_2^*}{\beta} = 157.699$, $B = ((1-p)x - \beta D)t_2 = 157.699$.

2.5.3 Reparation of Imperfect Products

In this section, the work of Taleizadeh et al. (2016b) is presented. Consider the situation where there exists a purchaser which buys the products from a supplier that is located far away. Due to process failure or due to the mishandling of products during transportation, it is possible that the lot contains some imperfect products. By doing an inspection process, the buyer detects imperfect products, and these must be

2.5 EOQ Model with Partial Backordering

Table 2.31 Notations of a given problem

t_i	Inspection time of products (time/unit)
t_R	Transportation, repair, and return time of imperfect products (time/unit)
t_T	Total transportation time of imperfect products (time/unit)
x	Inspection rate (units/time unit)
K	Buyer's ordering cost ($/order)
K_s	Repair setup cost ($/setup)
h_2	Holding cost at the repair facility ($/unit/time unit)
C_1	Material and labor cost to repair a product ($/unit)
m	Markup percentage by the repair shop (%)
Decision variables	
T	Cycle time (time unit)
F	Percentage of duration in which inventory level is positive (%)
Q	Order quantity (units)

replaced by perfect products. Due to the fact that the lead time is high because the supplier is far away, the buyer cannot make an additional order to the same supplier with the purpose of substituting the imperfect products. In some situations, the imperfect products have a significant value and they must be repaired. Here, it is considered that the imperfect products are repaired. The demand is constant and known. Shortages are allowed and partially backordered. After the batch is received, firstly, the backordered demand is satisfied and then products are inspected in order to find imperfect products. The imperfect products can be repairable. It is assumed that after the repair process, the products are as good as new. So at the end of the screening period, the imperfect products are withdrawn and sent to a local repair shop in order to be repaired. After their reparation, they are added to inventory. There exists a repair cost and the holding cost of repaired products is higher than the initial holding cost. The total cost at the repair shop consists of fixed and variable cost. The fixed cost is comprised of the repair setup cost and round trip fixed cost to the repair shop. The variable cost consists of the unit transportation cost, unit material and labor cost, and unit holding cost at the repair facility. Additionally, it is assumed that in the repair shop, the repair process is always in control and all the imperfect products can be repaired.

In order to model the presented problem, some new notations which are specifically used are shown in Table 2.31.

According to the time when the repaired products are added to inventory, four cases are identified and studied. These are the following:

Case I. The repaired products are received when the inventory level is positive.
Case II. The repaired products are received when the inventory level is zero.
Case III. The repaired products are received when the shortage quantity is equal to imperfect products' quantity.
Case IV. The repaired products are received when shortage still remains.

The main goal of the inventory model is to obtain the optimal cycle time (T), the percentage of the cycle time (F), and the lot size (Q) in order to maximize the total profit.

In order to identify imperfect products, the whole lot is inspected at rate x where the inspection time is $t_i = I_{max}/x$. The buyer must send imperfect products to the repair shop in order to convert them into perfect products. The inspection rate is greater than the demand rate ($x > D$). Both demand and inspection rates are constant and known. The proportion of products which is imperfect (ρ) and its probability density function are given and known. At the end of the screening period (t_i), the imperfect products are withdrawn and sent to a local repair shop. Repaired products return after t_R units of time that includes repair and transportation times; here it is assumed that $t_i + t_R \leq T$ where T is the cycle duration. The repair process at the repair shop is in control. The fixed cost in the repair shop is determined with $K_R + 2K_S$ where K_R is the repair setup cost and K_S is the transportation fixed cost. The variable cost per imperfect product is calculated with $C_1 + 2C_T + h_2 t_R$ where C_1 is the material and labor cost to repair a product, C_T is the transportation cost per unit, and h_2 is the unit holding cost at the repair store. The time t_R consists of repair time at rate R and total transportation time t_T of imperfect products.

The unit holding cost of repaired products is h_1. Here, h_1 is greater than the initial unit holding cost (h). FT is the time in which inventory level is positive and $(1-F)T$ is the time in which the shortage occurs.

Case I At the beginning of cycle time (T), the maximum inventory level is $I_{max} = FTD$. The products are screened and the inspection time is $t_i = FTD/x$. During the screening process, a ρ percent of products is identified as imperfect products (ρFTD). At the end of inspection time, imperfect products are withdrawn from inventory and sent to the repair shop. After the repair time, the repaired products are added to inventory. Here, it is considered that when the repaired products are received, the inventory level is still positive. Therefore, the inventory level increases ρFTD units when the repaired products arrive. When the inventory level reaches zero, then the shortages start to occur; the shortage quantity is determined by $(1-F)TD$. The $\beta(1-F)TD$ is the shortage backordered quantity and the rest is lost sales quantity $(1-\beta)(1-F)TD$. Figure 2.24 illustrates the behavior of inventory for Case I. Repair duration is $t_R = \rho FTD/R + t_T$, and the total cost in the repair shop is $K_R + 2K_S + \rho FTD(C_1 + 2C_T + h_2 t_R)$, and m margin per unit is claimed as repair charge. The unit repair cost charged to the buyer is therefore (Taleizadeh et al. 2016b):

$$C_R(FTD) = (1+m)\left[\left(\frac{K_R + 2K_S}{\rho FTD}\right) + (C_1 + 2C_T + h_2 t_R)\right] \quad (2.425)$$

The order quantity of products per cycle is $Q = FTD + \beta(1-F)TD$. The total holding cost (HC) per time unit is given by (Taleizadeh et al. 2016b):

2.5 EOQ Model with Partial Backordering

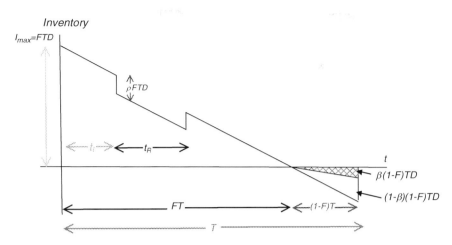

Fig. 2.24 Inventory level in Case I (Taleizadeh et al. 2016b)

$$HC = h\left[\frac{(1-\rho)^2 F^2 TD}{2} + \frac{\rho T(FD)^2}{x}\right]$$
$$+ h_R\left[\rho F^2 TD - \rho FD\left(\frac{FTD}{x} + \frac{\rho FTD}{R} + t_T\right) - \rho^2 \frac{F^2 TD}{2}\right] \quad (2.426)$$

where h is the holding cost of perfect products and h_R is the holding cost of repaired products, and the shortage cost (SC) per time unit is (Taleizadeh et al. 2016b):

$$SC = C_b \frac{\beta(1-F)^2 TD}{2} + g(1-\beta)(1-F)D \quad (2.427)$$

where C_b is the backordered cost and g is lost sales cost. β is the percentage of backordered demand. Obviously, the main goal of the buyer is to maximize his/her own profit per time unit (TP). The total profit (TP) per time unit is equal to the total revenue per time unit less the total cost per time unit. Thus, TP(T, F) is expressed as (Taleizadeh et al. 2016b):

$$TP(T,F) = s(FD + \beta(1-F)D) - \left[\frac{K}{T} + C(FD + \beta(1-F)D) + C_1 FD \right.$$
$$+ (\rho FD)(1+m)\left[\left(\frac{K_R + 2K_S}{\rho FTD}\right) + \left(C_1 + 2C_T + h_2\left(\frac{\rho FTD}{R} + t_T\right)\right)\right] + h\left(\frac{(1-\rho)^2 F^2 TD}{2} + \frac{\rho T(FD)^2}{x}\right)$$
$$\left. + h_R\left(\rho F^2 TD - \rho FD\left(\frac{FTD}{x} + \frac{\rho FTD}{R} + t_T\right) - \rho^2 \frac{F^2 TD}{2}\right) + C_b \frac{\beta(1-F)^2 TD}{2} + g(1-\beta)(1-F)D\right] \quad (2.428)$$

where s is the unit price of sales, K is the fixed ordered cost, C is the unit product cost, and C_I is the unit inspection cost. The optimal value for the percentage of duration of the cycle period (F) and the optimal value for length duration (T) are given below (Taleizadeh et al. 2016b):

$$F = \frac{C_b\beta T - (C_I + \rho(1+m)(C_1 + 2C_T + h_2 t_T) - h_R t_T \rho - \widehat{\pi}(1-\beta))}{2\left(\rho^2 D \frac{h(1+m)}{R} + \rho h\left[\frac{\rho}{2} + \frac{D}{x} - 1\right] - \rho h_R\left[\frac{\rho D}{R} + \frac{\rho}{2} + \frac{D}{x} - 1\right] + \left[\frac{h + C_b\beta}{2}\right]\right)T} \quad (2.429)$$

$$T = \sqrt{\frac{(K + (K_R + 2K_S)(1+m))\left(\rho^2 D \frac{h(1+m)}{R} + \rho h\left[\frac{\rho}{2} + \frac{D}{x} - 1\right] - \rho h_R\left[\frac{\rho D}{R} + \frac{\rho}{2} + \frac{D}{x} - 1\right] + \frac{h + C_b\beta}{2}\right)}{\frac{C_b\beta}{2}\left(\rho^2 D^2 \frac{h_2(1+m)}{R} + \rho Dh\left[\frac{\rho}{2} + \frac{D}{x} - 1\right] - \rho Dh_R\left[\frac{\rho D}{R} + \frac{\rho}{2} + \frac{D}{x} - 1\right] + \frac{Dh}{2}\right)}}$$

$$\qquad\qquad -\frac{D}{4}(C_I + \rho(1+m)(C_1 + 2C_T + h_2 t_T) - h_R t_T \rho - \widehat{\pi}(1-\beta))^2$$

(2.430)

where $\widehat{\pi}$ is given by $\widehat{\pi} = s + g - C$. Note that the expected value of ρ and ρ^2 must be substituted in Eqs. (2.429) and (2.430). The expected value of Eq. (2.428) is given as:

$$\text{ETP}(T,F) = s(FD + \beta(1-F)D) - \left[\frac{K}{T} + C(FD + \beta(1-F)D) + C_I FD\right.$$

$$+ (1+m)\left(\frac{K_R + 2K_S}{T} + E(\rho)FD(C_1 + 2C_T + h_2 t_T) + \frac{E(\rho^2)h_2 D^2 F^2 T}{R}\right)$$

$$+ \frac{E\left((1-\rho)^2\right)hF^2 TD}{2} + \frac{E(\rho)D^2 hTF^2}{x} + C_b\frac{\beta(1-F)^2 TD}{2} + g(1-\beta)(1-F)D$$

$$+ h_R\left(E(\rho)F^2 TD - \frac{E(\rho)D^2 F^2 T}{x} - \frac{E(\rho^2)D^2 F^2 T}{R} - E(\rho)FDt_T - \frac{E(\rho^2)DF^2 T}{2}\right)\right]$$

(2.431)

As a result, Eqs. (2.430) and (2.431) are rewritten as:

$$F = \frac{C_b\beta T - (C_I + E(\rho)(1+m)(C_1 + 2C_T + h_2 t_T) - E(\rho)h_1 t_T - \widehat{\pi}(1-\beta))}{2\left(E(\rho^2)D\frac{h(1+m)}{R} + h\left[\frac{E(\rho^2)}{2} + \frac{E(\rho)D}{x} - E(\rho)\right] - h_1\left[\frac{E(\rho^2)D}{R} + \frac{E(\rho^2)}{2} + \frac{E(\rho)D}{x} - E(\rho)\right] + \left[\frac{h + C_b\beta}{2}\right]\right)T}$$

(2.432)

2.5 EOQ Model with Partial Backordering

$$T = \sqrt{\dfrac{\begin{bmatrix}[K+(K_R+2K_S)(1+m)][\dfrac{(E(\rho^2)Dh_2(1+m)}{R}+h\left[\dfrac{E(\rho^2)}{2}+\dfrac{E(\rho)D}{x}-E(\rho)\right]\\ -h_R\left[\dfrac{E(\rho^2)D}{R}+\dfrac{E(\rho^2)}{2}+\dfrac{E(\rho)D}{x}-E(\rho)\right]+\dfrac{h+C_b\beta}{2}\\ -\dfrac{D}{4}(C_I+E(\rho)(1+m)(C_I+2C_T+h_2 t_T)-h_R t_T E(\rho)-\widehat{\pi}(1-\beta))^2\end{bmatrix}}{\dfrac{C_b\beta}{2}\left(E(\rho^2)D^2\dfrac{h_2(1+m)}{R}+Dh\left[\dfrac{E(\rho^2)}{2}+\dfrac{E(\rho)D}{x}-E(\rho)\right]-Dh_R\left[\dfrac{E(\rho^2)D}{R}+\dfrac{E(\rho^2)}{2}+\dfrac{E(\rho)D}{x}-E(\rho)\right]+\dfrac{Dh}{2}\right)}}$$

(2.433)

where $\widehat{\pi}$ is given by $\widehat{\pi}=s+g-C$. Appendix A of Taleizadeh et al. (2016b) shows that the denominator of Eq. (2.433) is positive. The numerator of Eq. (2.433) must be positive. Thus (Taleizadeh et al. 2016b):

$$w_1 = [K+(K_R+2K_S)(1+m)]\begin{bmatrix}\dfrac{(E(\rho^2)Dh_2(1+m)}{R}+h\left[\dfrac{E(\rho^2)}{2}+\dfrac{E(\rho)D}{x}-E(\rho)\right]\\ -h_R\left[\dfrac{E(\rho^2)D}{R}+\dfrac{E(\rho^2)}{2}+\dfrac{E(\rho)D}{x}-E(\rho)\right]+\dfrac{h+C_b\beta}{2}\end{bmatrix}$$
$$-\dfrac{D}{4}(C_I+E(\rho)(1+m)(C_I+2C_T+h_2 t_T)-h_R t_T E(\rho)-\widehat{\pi}(1-\beta))^2 > 0$$

If $w_1 > 0$, then T^* is equal to T in Eq. (2.433). If $w_1 \leq 0$, then T^* is equal zero. Also, if further condition $\dfrac{E(\rho^2)D}{R}+\dfrac{E(\rho^2)}{2}+\dfrac{E(\rho)D}{x} \leq E(\rho)$ is not satisfied, then T^* is equal to zero.

Case II In this case also, at beginning of the cycle, the inventory level is *FTD*. The products are screened at rate x; consequently, the inspection duration is equal to $t_i = FTD/x$. A ρ percent of products is found imperfect (ρFTD). The imperfect products are withdrawn and sent to a repair shop. The repair duration is $t_R = \rho FTD/R + t_T$. The repair charge per unit is determined as in Case I with the following expression (Taleizadeh et al. 2016b):

$$(1+m)\left[\left(\dfrac{K_R+2K_S}{\rho FTD}\right)+(C_I+2C_T+h_2 t_R)\right]$$

Additionally, this case assumes that the repaired products arrive to the buyer's store when the inventory level is exactly zero. Therefore, after adding the repaired products to inventory, then the inventory level reaches ρFTD units. At the end of the cycle, the shortage level is $(1-F)TD$. Figure 2.25 illustrates the behavior of inventory of Case II. The order quantity per cycle is $Q = FTD + \beta(1-F)TD$. The holding cost (HC) per time unit is calculated as:

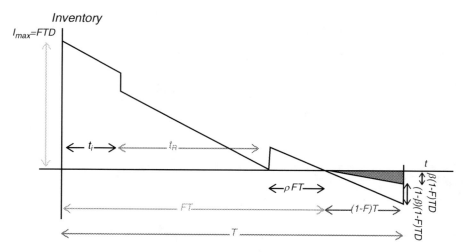

Fig. 2.25 Inventory level for Case II (Taleizadeh et al. 2016b)

$$\text{HC} = h\left(\frac{(1-\rho)^2 F^2 TD}{2} + \frac{\rho T(FD)^2}{x}\right) + h_R \frac{(\rho F)^2 DT}{2} \quad (2.434)$$

The shortage cost per time unit is given as in Case I:

$$\text{SC} = C_b \frac{\beta(1-F)^2 TD}{2} + g(1-\beta)(1-F)D \quad (2.435)$$

The total profit per time unit is equal to the total revenue per time unit less the total cost per time unit, and it is expressed below:

$$\text{TP}(T,F) = sD(F + \beta(1-F)) \\ - \begin{bmatrix} \frac{K}{T} + C(FD + \beta(1-F)D) + C_I FD + h\left[\frac{(1-\rho)^2 F^2 TD}{2} + \frac{\rho T(FD)^2}{x}\right] \\ + (\rho FD)(1+m)\left[\left(\frac{K_R + 2K_S}{\rho FTD}\right) + C_1 + 2C_T + h_2\left(\frac{\rho FTD}{R} + t_T\right)\right] \\ + h_R \frac{(\rho F)^2 DT}{2} + C_b \frac{\beta(1-F)^2 TD}{2} + g(1-\beta)(1-F)D \end{bmatrix} \\ \quad (2.436)$$

In Case II, the percentage of duration which inventory level is positive (F) is obtained as:

2.5 EOQ Model with Partial Backordering

$$F = \frac{C_b\beta T - (C_1 + \rho(1+m)[C_1 + 2C_T + h_2 t_T] - \widehat{\pi}(1-\beta))}{\left(\frac{2h_2\rho^2 D(1+m)}{R} + h(1-\rho)^2 + \frac{2\rho h D}{x} + h_R\rho^2 + C_b\beta\right)T} \quad (2.437)$$

The ngth duration (T) is:

$$T = \sqrt{\frac{(K+(1+m)(K_R+2K_S))\left(\frac{h\rho^2 D(1+m)}{R} + \frac{h(1-\rho)^2}{2} + \frac{\rho h D}{x} + h_R\frac{\rho^2}{2} + \frac{C_b\beta}{2}\right) - \frac{D}{4}(C_1 + \rho(1+m)[C_1 + 2C_T + h_2 t_T] - \widehat{\pi}(1-\beta))^2}{\frac{C_b\beta}{2}\left[\frac{h_2\rho^2 D^2(1+m)}{R} + \frac{hD(1-\rho)^2}{2} + \frac{\rho h D^2}{x} + h_R\frac{\rho^2 D}{2}\right]}}$$

(2.438)

where $\widehat{\pi}$ is given by $\widehat{\pi} = s + g - C$. The expected value of Eq. (2.438) is given as:

$$E(\mathrm{TP}(T,F)) = sD(F + \beta(1-F)) - \begin{bmatrix} \frac{K}{T} + C(FD + \beta(1-F)D) + C_1 FD \\ + h\left[\frac{E\left((1-\rho)^2\right)F^2 TD}{2} + \frac{E(\rho)D^2 F^2 T}{x}\right] \\ + (1+m)\left(\frac{K_R + 2K_S}{T} + E(\rho)FD(C_1 + 2C_T + h_2 t_T) + \frac{E(\rho^2)h_2 D^2 F^2 T}{R}\right) \\ + h_R\frac{E(\rho^2)DF^2 T}{2} + C_b\frac{\beta(1-F)^2 TD}{2} + g(1-\beta)(1-F)D \end{bmatrix}$$

(2.439)

Hence, Eqs. (2.438) and (2.439) are re-expressed as:

$$F = \frac{C_b\beta T - (C_1 + E(\rho)(1+m)[C_1 + 2C_T + h_2 t_T] - \widehat{\pi}(1-\beta))}{\left(\frac{2h_2 DE(\rho^2)(1+m)}{R} + hE\left((1-\rho)^2\right) + \frac{2E(\rho)hD}{x} + h_R E(\rho^2) + C_b\beta\right)T} \quad (2.440)$$

$$T = \sqrt{\frac{(K+(1+m)(K_R+2K_S))\left(\frac{hE(\rho^2)D(1+m)}{R} + \frac{hE\left((1-\rho)^2\right)}{2} + \frac{E(\rho)hD}{x} + h_R\frac{E(\rho^2)}{2} + \frac{C_b\beta}{2}\right) - \frac{D}{4}(C_1 + E(\rho)(1+m)[C_1 + 2C_T + h_2 t_T] - \widehat{\pi}(1-\beta))^2}{\frac{C_b\beta}{2}\left[\frac{h_2 E(\rho^2)D^2(1+m)}{R} + \frac{hDE\left((1-\rho)^2\right)}{2} + \frac{E(\rho)hD^2}{x} + h_R\frac{E(\rho^2)D}{2}\right]}}$$

(2.441)

It is easy to see that the denominator of Eq. (2.427) is always positive. The numerator of Eq. (2.427) must be greater than zero. Thus:

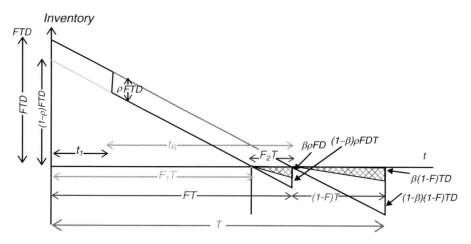

Fig. 2.26 Inventory level for Case III (Taleizadeh et al. 2016b)

$$w_2 = (K + (1+m)(K_R + 2K_S)) \left(\frac{h_2 E(\rho^2) D(1+m)}{R} + \frac{hE\left((1-\rho)^2\right)}{2} + \frac{E(\rho)hD}{x} + h_R \frac{E(\rho^2)}{2} + \frac{C_b \beta}{2} \right)$$
$$- \frac{D}{4}(C_1 + E(\rho)(1+m)[C_1 + 2C_T + h_2 t_T]) - \hat{\pi}(1-\beta))^2 > 0$$

If $w_2 > 0$, then T^* is equal to T in Eq. (2.441). Otherwise, if $w_2 \leq 0$, then T^* is equal zero.

Case III The behavior of inventory level for Case III is illustrated in Fig. 2.26. F represents the fraction of cycle time in which the inventory level (without considering the imperfect products) is positive. It is equal to $F_1 + F_2$. These are given by $F_1 = (1 - \rho)F$ and $F_2 = \rho F$. F_i ($i = 1, 2$) is within the interval [0, 1]. The buyer sends the imperfect products to the repair shop. The repair cost (Taleizadeh et al. 2016b) per unit is given by $(1+m)\left[\left(\frac{(K_R + 2K_S)}{\rho FTD}\right) + C_1 + 2C_T + h_2\left(\frac{\rho FTD}{R} + t_T\right)\right]$

While imperfect products are being repaired, the inventory system faces shortages. Then, the repaired products arrive when the shortage level is exactly the same as the repaired products' quantity. Now, consequently, the inventory level is equal to zero.

In this case, the order quantity per cycle is $Q = F_1 TD + \beta F_2 TD + \beta(1 - F)TD = F_1 TD + \beta(1 - F_1)TD$. The holding cost (HC) is equal to:

$$\text{HC} = h\left(\frac{(1-\rho)^2 F^2 TD}{2} + \frac{\rho F^2 TD^2}{x} \right) \tag{2.442}$$

and the shortage cost (SC) is:

2.5 EOQ Model with Partial Backordering

$$SC = C_b \frac{\beta \rho^2 F^2 TD}{2} + C_b \frac{\beta(1-F)^2 TD}{2} + g(1-\beta)F_2 D + g(1-\beta)(1-F)D$$

$$SC = C_b \frac{\beta \rho^2 F^2 TD}{2} + C_b \frac{\beta(1-F)^2 TD}{2} + g(1-\beta)(1-F_1)D \qquad (2.443)$$

Thus, the total profit per time unit is defined as:

$$TP(T,F) = s(F_1 D + \beta(1-F_1)D)$$

$$- \begin{bmatrix} \frac{K}{T} + CD(F_1 + \beta(1-F_1)) + h\left(\frac{(1-\rho)^2 F^2 TD}{2} + \frac{\rho F^2 TD^2}{x}\right) \\ +\rho FD(1+m)\left[\left(\frac{(K_R + 2K_S)}{\rho FTD}\right) + C_1 + 2C_T + h_2\left(\frac{\rho FTD}{R} + t_T\right)\right] \\ +C_1 FD + \frac{C_b \beta TD}{2}\left(\rho^2 F^2 + (1-F)^2\right) + g(1-\beta)(1-F_1)D \end{bmatrix} \qquad (2.444)$$

The optimal values for decision variables, the length duration of cycle time (T) and the percentage of cycle time in which the inventory is positive (F), are obtained as:

$$T = \sqrt{\frac{(K + (1+m)(K_R + 2K_S))\left(\frac{h\rho^2 D(1+m)}{R} + \frac{h(1-\rho)^2}{2} + \frac{\rho hD}{x} + \frac{C_b \beta \rho^2}{2} + \frac{C_b \beta}{2}\right)}{\frac{C_b \beta}{2}\left(\frac{h_2 \rho^2 D^2(1+m)}{R} + \frac{h(1-\rho)^2 D}{2} + \frac{\rho hD^2}{x} + \frac{C_b \beta \rho^2 D}{2}\right)}} \qquad (2.445)$$

$$F = \frac{C_b \beta T - (C_1 + \rho(1+m)(C_1 + 2C_T + h_2 t_T) - \widehat{\pi}(1-\rho)(1-\beta))}{\left(\frac{2h_2 \rho^2 D(1+m)}{R} + h(1-\rho)^2 + \frac{2\rho hD}{x} + C_b \beta \rho^2 + C_b \beta\right)T} \qquad (2.446)$$

The expected value of Eq. (2.443) is given as:

$$ETP(T,F) = s(F_1 D + \beta(1-F_1)D) - \begin{bmatrix} \frac{K}{T} + CD(F_1 + \beta(1-F_1)) + h\left(\frac{E\left((1-\rho)^2\right)F^2 TD}{2} + \frac{E(\rho)F^2 TD^2}{x}\right) \\ +(1+m)\left(\frac{K_R + 2K_S}{T} + E(\rho)FD(C_1 + 2C_T + h_2 t_T) + \frac{E(\rho^2)h_2 D^2 F^2 T}{R}\right) \\ +C_1 FD + \frac{C_b \beta TD}{2}\left(E(\rho^2)F^2 + (1-F)^2\right) + g(1-\beta)(1-F_1)D \end{bmatrix} \qquad (2.447)$$

Thus, Eqs. (2.445) and (2.446) are rewritten as:

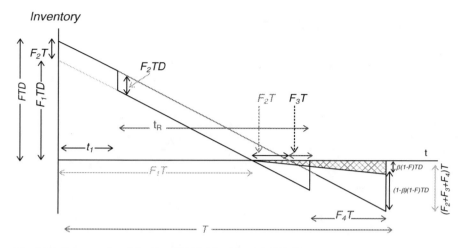

Fig. 2.27 Inventory level for Case IV (Taleizadeh et al. 2016b)

$$T = \sqrt{\frac{(K+(1+m)(K_R+2K_S))\left(\frac{E(\rho^2)h_2D(1+m)}{R}+\frac{hE\left((1-\rho)^2\right)}{2}+\frac{E(\rho)hD}{x}+\frac{C_b\beta E(\rho^2)}{2}+\frac{C_b\beta}{2}\right)}{\frac{C_b\beta}{2}\left(\frac{E(\rho^2)h_2D^2(1+m)}{R}+\frac{E((1-\rho)^2)hD}{2}+\frac{E(\rho)hD^2}{x}+\frac{E(\rho^2)C_b\beta D}{2}\right)} - \frac{D}{4}(C_I + E(\rho)(1+m)(C_1+2C_T+h_2t_T) - \hat{\pi}(1-E(\rho))(1-\beta))^2}$$

(2.448)

$$F = \frac{C_b\beta T - (C_I + E(\rho)(1+m)(C_1+2C_T+h_2t_T) - \hat{\pi}(1-E(\rho))(1-\beta))}{\left(\frac{2h_2E(\rho^2)D(1+m)}{R} + hE\left((1-\rho)^2\right) + \frac{2E(\rho)hD}{x} + C_b\beta E(\rho^2) + C_b\beta\right)T}$$

(2.449)

The denominator of Eq. (2.449) is positive. On the other hand, its numerator could be positive, zero, or negative:

$$w_3 = (K+(1+m)(K_R+2K_S))\left(\frac{E(\rho^2)h_2D(1+m)}{R}+\frac{hE\left((1-\rho)^2\right)}{2}+\frac{E(\rho)hD}{x}+\frac{C_b\beta E(\rho^2)}{2}+\frac{C_b\beta}{2}\right)$$
$$-\frac{D}{4}(C_I + E(\rho)(1+m)(C_1+2C_T+h_2t_T) - \hat{\pi}(1-E(\rho))(1-\beta))^2$$

If $w_3 > 0$, then T^* is equal T in Eq. (2.449). Otherwise, if $w_3 \leq 0$, then T^* is equal zero.

Case IV The inventory pattern for Case IV is shown in Fig. 2.27. In Case IV, the duration length of cycle time is divided into four parts. Each part is represented by F_i and it is in the interval [0, 1] where $i = 1, 2, 3, 4$. Again, F denotes the fraction of

2.5 EOQ Model with Partial Backordering

length duration of the cycle time in which the inventory level is positive (Taleizadeh et al. 2016b).

Here, F is equal to $F_1 + F_2$ and $(1 - F)$ is equal to $F_3 + F_4$. The inventory level at the beginning of the cycle is FTD. The products are inspected by rate x with duration of t_i ($t_i = FTD/x$). At the end of inspection duration, the imperfect products ($\rho FTD = F_2 TD$) are sent to the buyer's store. Repair duration is $t_R = \frac{\rho(F_1+F_2)TD}{R} + t_T = \frac{\rho FTD}{R} + t_T$ and repair cost per unit is $(1 + m)\left[\left(\frac{K_R + 2K_S}{\rho FTD}\right) + (C_1 + 2C_T + h_2 t_R)\right]$.

Notice that when repaired products arrive at the buyer's store, the shortage level is $(F_2 + F_3)TD$. After adding them, the inventory level is still negative. It means that the shortage remains until the end of the cycle; at this time, the shortages reach to $(F_2 + F_3 + F_4)TD$ units. In this case, the order quantity per cycle is $Q = (F_1 + F_2)TD + \beta(F_3 + F_4)TD = FTD + \beta(1 - F)TD$. The holding cost (HC) of products is computed as:

$$\text{HC} = h\left[\frac{(1-\rho)^2 F^2 TD}{2} + \frac{\rho(F)^2 TD^2}{x}\right] \tag{2.450}$$

and the shortage cost (SC) is determined as:

$$\text{SC} = C_b \frac{(F_2 + F_3 + F_4)(F_3 + F_4)\beta TD}{2} + g(1 - \beta)(F_3 + F_4)D \tag{2.451}$$

The total profit is calculated as follows: the total revenue per time unit less the total cost per time unit. It is illustrated as:

$$\text{TP}(T, F) = sD(F + \beta(1 - F))$$

$$- \begin{bmatrix} \frac{K}{T} + CD(F + \beta(1-F)) + h\left(\frac{(1-\rho)^2 F^2 TD}{2} + \frac{\rho(F)^2 TD^2}{x}\right) \\ + C_1(F_1 + F_2)D \\ + \rho DF(1 + m)\left[\left(\frac{K_R + 2K_S}{\rho FTD}\right) + C_1 + 2C_T + h_2\left(\frac{\rho FTD}{R} + t_T\right)\right] \\ + C_b \frac{(F_2 + F_3 + F_4)(F_3 + F_4)\beta TD}{2} + g(1-\beta)(1-F)D \end{bmatrix}$$

$$\tag{2.452}$$

It is clear that $F + \beta(1 - F) = 1 - 1 + F + \beta(1 - F) = 1 - (1 - \beta)(1 - F)$ and $F_2 + F_3 + F_4 = 1 - F_1 = 1 - (1 - \rho)F$. Thus, the profit function is simplified as:

$$TP(T,F) = (s-C)D - \hat{\pi}D(1-\beta) - \begin{bmatrix} \frac{K}{T} + C_1DF + \rho DF(1+m)\left[\left(\frac{K_R+2K_S}{\rho FTD}\right) + C_1 + 2C_T + h_2\left(\frac{\rho FTD}{R} + t_T\right)\right] \\ +h\left(\frac{(1-\rho)^2F^2TD}{2} + \frac{\rho F^2TD^2}{x}\right) \\ +(1-(1-\rho)F)(1-F)\frac{C_b\beta TD}{2} - \hat{\pi}D(1-\beta)F \end{bmatrix}$$

(2.453)

The optimal values for F and T are given below:

$$F = \frac{C_b\beta(2-\rho)T + 2\hat{\pi}(1-\beta) - 2C_1 - 2\rho(1+m)(C_1+2C_T+h_2t_T)}{4\left(\frac{(1+m)h_2\rho^2 D}{R} + \frac{(1-\rho)^2 h}{2} + \frac{\rho h D}{x} + \frac{C_b\beta(1-\rho)}{2}\right)T}$$

(2.454)

$$T = \sqrt{\frac{4(K+(1+m)(K_R+2K_S))\left(\frac{(1+m)h_2\rho^2 D}{R} + \frac{(1-\rho)^2 h}{2} + \frac{\rho h D}{x} + \frac{C_b\beta(1-\rho)}{2}\right)}{2C_b\beta\left(\frac{(1+m)h_2\rho^2 D^2}{R} + \frac{(1-\rho)^2 hD}{2} + \frac{\rho hD^2}{x} - \frac{C_b\beta D\rho^2}{8}\right)}}$$

(2.455)

The expected value of Eq. (2.453) is given as:

$$ETP(T,F) = (s-C)D - \hat{\pi}D(1-\beta)$$
$$- \begin{bmatrix} \frac{K}{T} + C_1DF + (1+m)\left(\frac{K_R+2K_S}{T} + E(\rho)DF(C_1+2C_T+h_2t_T) + \frac{E(\rho^2)h_2D^2F^2T}{R}\right) \\ +h\left(\frac{E\left((1-\rho)^2\right)F^2TD}{2} + \frac{E(\rho)F^2TD^2}{x}\right) + (1-(1-E(\rho))F)(1-F)\frac{C_b\beta TD}{2} - \hat{\pi}D(1-\beta)F \end{bmatrix}$$

(2.456)

Consequently, Eqs. (2.454) and (2.455) are re-expressed as:

$$F = \frac{C_b\beta(2-E(\rho))T + 2\hat{\pi}(1-\beta) - 2C_1 - 2E(\rho)(1+m)(C_1+2C_T+h_2t_T)}{4\left(\frac{(1+m)E(\rho^2)Dh_2}{R} + \frac{hE\left((1-\rho)^2\right)}{2} + \frac{E(\rho)hD}{x} + \frac{C_b\beta(1-E(\rho))}{2}\right)T}$$

(2.457)

$$T = \sqrt{\frac{4(K+(1+m)(K_R+2K_S))\left(\frac{(1+m)E(\rho^2)h_2D}{R} + \frac{E\left((1-\rho)^2\right)h}{2} + \frac{E(\rho)hD}{x} + \frac{C_b\beta(1-E(\rho))}{2}\right)}{2C_b\beta\left(\frac{(1+m)E(\rho^2)h'D^2}{R} + \frac{E((1-\rho)^2)hD}{2} + \frac{E(\rho)hD^2}{x} - \frac{E(\rho^2)C_b\beta D}{8}\right)}}$$

(2.458)

Notice that Eq. (2.458) is ever positive if its numerator becomes positive. Thus:

2.5 EOQ Model with Partial Backordering

Table 2.32 Data for the numerical example (Salameh and Jaber 2000; Taleizadeh et al. 2016b)

Description	Symbol	Value	Units
Demand rate	D	50,000	Units/year
Screening rate	x	175,200	Units/year
Selling price	s	50	$/unit
Order cost	K	100	$/order
Unit purchase cost	C	25	$/unit
Holding cost	h	5	$/unit/year
Unit inspection cost	C_I	0.5	$/unit
Percentage of defectives	ρ	$U \sim (0, 0.04)$	
Probability density function	$f(\rho)$	1/(0.04–0)	

Table 2.33 Data for the numerical example (Salameh and Jaber 2014; Taleizadeh et al. 2016b)

Description	Symbol	Value	Units
Repair setup cost	K_R	100	$/setup
Transportation fixed cost	K_S	200	$/trip
Unit transportation cost	C_T	2	$/unit
Unit material and labor cost	C_1	5	$/unit
Unit holding cost in repair shop	h_2	4	$/unit/year
Repaired rate	R	50,000	Units/year
Total transport time	t_T	2/220	Year
Holding cost of repaired product	h_R	6	$/unit/year
Markup percentage	m	20%	Percent

Table 2.34 Additional data for the numerical example (Taleizadeh et al. 2016b)

Description	Symbol	Value	Units
Backorder cost	C_b	20	$/unit/year
Lost sales cost	g	0.5	$/unit/year
Percentage of backordered demand	β	97%	Percent

$$w_4 = 4(K + (1+m)(K_R + 2K_S)) \left(\frac{(1+m)E(\rho^2)h_2 D}{R} + \frac{hE\left((1-\rho)^2\right)}{2} + \frac{E(\rho)hD}{x} + \frac{C_b \beta(1 - E(\rho))}{2} \right)$$
$$- D(C_1 - \widehat{\pi}(1-\beta) + E(\rho)(1+m)(C_1 + 2C_T + h_2 t_T))^2$$

must be positive. If $w_4 > 0$, then the optimal cycle time T^* is determined with Eq. (2.458). Otherwise, if $w_4 \leq 0$, then the cycle time T^* is equal to zero. Additionally, the following condition must be satisfied: $\frac{E\left((1-\rho)^2\right)h}{2} + \frac{E(\rho)hD}{x} - \frac{C_b \beta E(\rho^2)}{2} > 0$.

Example 2.21 This section presents a numerical example. The data for the numerical example is shown in Tables 2.32, 2.33, and 2.34. Table 2.32 summarizes the data from Salameh and Jaber (2000), Table 2.33 contains data from Jaber et al.

(2014), and Table 2.34 presents new data for the numerical example (Taleizadeh et al. 2016b).

The percentage of imperfect items follows a uniform distribution ($U \sim (0, 0.04)$). Thus, the expected value of the defective products (ρ) is:

$$E(\rho) = \int_a^b \rho f(\rho) d\rho = \int_a^b \rho \frac{1}{b-a} d\rho = \frac{b+a}{2} = \frac{0.04+0}{2} = 0.02$$

The $E(\rho^2)$ and $(1 - E(\rho^2))^2$ are given by:

$$E(\rho^2) = \text{var}(\rho) + E^2(\rho) = \frac{(b-a)^2}{12} + E^2(\rho) = \frac{(0.04)^2}{12} + (0.02)^2 = 0.0005333333$$
$$(1 - E(\rho))^2 = 1 - 2E(\rho) + E(\rho^2) = 1 - 2(0.02) + 0.0005333333 = 0.9605333355$$

Firstly, for each profit function, the w_i must be greater than zero.
$w_1 = 8519.089532 > 0$, $w_2 = 8464.869573 > 0$, $w_3 = 8482.853463 > 0$, $w_4 = 8327.949572531 > 0$.
Additionally, for Cases I and IV, the following conditions must be satisfied:

Case I: condition 1: $\frac{E(\rho^2)D}{R} + \frac{E(\rho^2)}{2} + \frac{E(\rho)D}{x} - E(\rho) = -0.0135 \leq 0$.
Case IV: condition 4: $\frac{E((1-\rho)^2)h}{2} + \frac{E(\rho)hD}{x} - \frac{\pi\beta E(\rho^2)}{2} = 2.42 > 0$.

Note that if all the conditions are satisfied, then the problem can be optimized. Table 2.35 shows the optimal results for the decision variables and the expected total profit (Taleizadeh et al. 2016b).

According to Table 2.35, the optimal policy in Case I happens when the cycle time is $T = 0.083598$ year, the percent of cycle time duration when the inventory level is positive is $F = 81.83131\%$, the order quantity is $Q = 4157.11965$ units, and the expected total profit is 1,197,016.99277. The optimal policy in Case II occurs when $T = 0.084679$ year, $F = 82.28241\%$, and $Q = 4211.44917$ units and the expected total profit is 1,197,197.0042. The optimal policy for Case III is when $T = 0.084706$ year, $F = 81.51335\%$, and $Q = 4209.78066$ units and expected total profit is 1,196,560.33681. The optimal policy in Case IV is when $T = 0.084019$ year, $F = 82.83673\%$, and $Q = 4179.32228$ units and the expected total profit is 1,197,083.77763. With these results, it is easy to see that the highest expected profit is obtained in Case II and the lowest profit in Case III.

It is important to remark that on the one hand the holding cost of repaired products in Case II is less than Case I. On the other hand, the shortage cost in Case II is less than Cases III and IV, and the shortage cost in Case III is more than other cases (Taleizadeh et al. 2016b).

2.5 EOQ Model with Partial Backordering

Table 2.35 Optimal results for each case (Taleizadeh et al. 2016b)

	TP	T	F (%)	F_1	F_2	Q
Case I	1,197,016.99277	0.083598053	81.83131	–	–	4157.119653
Case II	1,197,197.00424	0.084679076	82.28241	–	–	4211.449176
Case III	1,196,560.33681	0.084706825	81.51335	79.88309%	1.630267%	4209.780665
Case V	1,197,083.77763	0.084019058	82.83673%	81.17999%	1.65673%	4179.322286

2.5.4 Replacement of Imperfect Products

Taleizadeh et al. (2018b) developed their previous work (Taleizadeh et al. 2016b) by considering replacement instead of reparation. In order to model the presented problem, some new notations which are specifically used are shown in Table 2.36, and other parameters are the same as those used in models of the previous section as presented in Table 2.31.

They have considered three cases including the following:

Case I. The reordered items are received when the inventory level is zero.
Case II. The reordered items are received when the backordered quantity is equal to the imperfect items' quantity.
Case III. The reordered items are received when shortage still remained.

All these cases are presented in Figs. 2.28, 2.29, and 2.30, respectively.
For the first case, they derived the total profit function as:

Table 2.36 Notations of a given problem

c_E	Unit purchasing cost of an emergency order (\$/unit)
h_E	Holding cost of emergency purchased unit (\$/unit/time unit)
t_i	Inspection time of products (time unit)

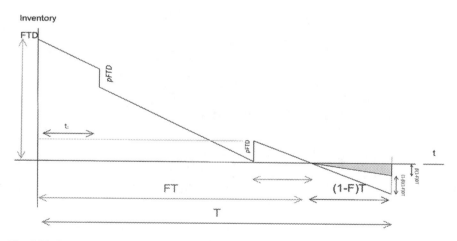

Fig. 2.28 Inventory level for Case I (Taleizadeh et al. 2018b)

2.5 EOQ Model with Partial Backordering

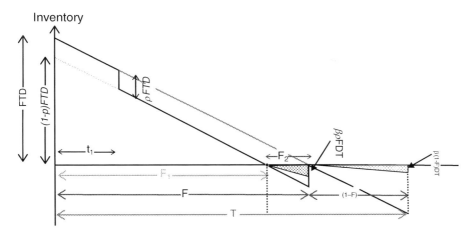

Fig. 2.29 Inventory level for Case II (Taleizadeh et al. 2018b)

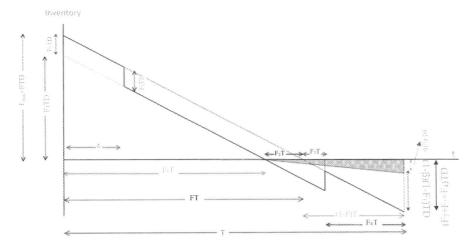

Fig. 2.30 Inventory level for Case III (Taleizadeh et al. 2018b)

$$\mathrm{TP}(T,F) = \overbrace{s(FD + \beta(1-F)D) + v\rho FD}^{\text{Revenue}} - [\overbrace{\frac{K}{T}}^{\text{Fixed cost}} + \overbrace{C(FD + \beta(1-F)D) + C_1 FD}^{\text{Purchasing cost}} + \overbrace{C_E \rho FD}^{\text{Emergency ordering cost}}$$

$$+ h\underbrace{\left[\frac{(1-\rho)^2 F^2 TD}{2} + \frac{\rho T(FD)^2}{x}\right]}_{\text{Holding cost}} + \underbrace{h_E \frac{(\rho F)^2 DT}{2}}_{\substack{\text{Holding cost for}\\ \text{emergency order}}} + \underbrace{C_b \frac{\beta(1-F)^2 TD}{2}}_{\text{Backordering cost}} + \underbrace{g(1-\beta)(1-F)D}_{\text{Lost sale cost}}]$$

(2.459)

And after some simplifications and algebra, they derived similar to the previous case presented in Sect. 2.5.3:

$$T = \sqrt{\frac{K\left(\frac{h(1-E(\rho))^2}{2} + \frac{E(\rho)hD}{x} + \frac{h_E E(\rho^2)}{2} + \frac{C_b\beta}{2}\right) - \frac{D}{4}(C_I + (C_E - v)E(\rho) - (s+g-C)(1-\beta))^2}{\frac{C_b\beta}{2}\left(\frac{hD(1-E(\rho))^2}{2} + \frac{E(\rho)hD^2}{x} + \frac{h_E E(\rho^2)D}{2} + \frac{C_b\beta D}{2}\right)}} \quad (2.460)$$

$$F(T) = \frac{C_b\beta - C_I - (E_E - v)E(\rho) + (s+g-C)(1-\beta)}{\left(h(1-E(\rho))^2 + \frac{2E(\rho)hD}{x} + h_E E(\rho^2) + C_b\beta\right)T} \quad (2.461)$$

Take $w_1 = K\left(\frac{h(1-E(\rho))^2}{2} + \frac{E(\rho)hD}{x} + \frac{h_E E(\rho^2)}{2} + \frac{C_b\beta}{2}\right) - \frac{D}{4}(C_I + (C_E - v)E(\rho) - (s+g-C)(1-\beta))^2$. If $w_1 > 0$, T^* is equal T in Eq. (2.460). On the other hand, if $w_1 \leq 0$, T^* is equal to zero, meaning the optimal value of shortage is infinite and no inventory level exists.

Also for the second case, the total profit is defined as:

$$TP(T,F) = \overbrace{s(F_1 D + \beta(1-F_1)D) + v\rho FD}^{\text{Revenue}}$$

$$- \left[\underbrace{\frac{K}{T}}_{\text{Fixed cost}} + \underbrace{C(FD + \beta(1-F)D)}_{\text{Purchasing cost}} + \underbrace{C_I FD}_{\text{Inspection cost}} + \underbrace{C_E \rho FD}_{\text{Emergency order cost}} + \overbrace{h\left(\frac{(1-\rho)^2 F^2 TD}{2} + \frac{\rho F^2 TD^2}{x}\right)}^{\text{Holding cost}}\right.$$

$$\left. + \underbrace{C_b \frac{\beta\rho^2 F^2 TD}{2} + C_b \frac{\beta(1-F)^2 TD}{2}}_{\text{Backordering cost}} + \underbrace{g(1-\beta)(1-F_1)D}_{\text{Lost sale cost}}\right] \quad (2.462)$$

And after some simplifications and algebra:

$$T = \sqrt{\frac{K\left(\frac{h(1-E(\rho))^2}{2} + \frac{E(\rho)hD}{x} + \frac{C_b\beta E(\rho^2)}{2} + \frac{C_b\beta}{2}\right) - \frac{D}{4}(C_I + (C_E - v)E(\rho) - (s+g-C)(1-\beta))^2}{\frac{C_b\beta}{2}\left(\frac{h(1-E(\rho))^2 D}{2} + \frac{E(\rho)hD^2}{x} + \frac{C_b\beta E(\rho^2)D}{2}\right)}} \quad (2.463)$$

and

$$F(T) = \frac{C_b\beta T - (C_I + (C_E - v)E(\rho) - (s+g-C)D(1-\beta))}{\left(h(1-E(\rho))^2 + \frac{2E(\rho)D}{x} + C_b\beta E(\rho^2) + C_b\beta\right)T} \quad (2.464)$$

2.5 EOQ Model with Partial Backordering

Considering $w_2 = K\left(\frac{h(1-E(\rho))^2}{2} + \frac{E(\rho)hD}{x} + \frac{C_b\beta E(\rho^2)}{2} + \frac{C_b\beta}{2}\right) - \frac{D}{4}(C_1 + (C_E - c_s)E(\rho) - (s+g-C)(1-\beta))^2$, if $w_2 > 0$, then T^* is equal T in Eq. (2.463). If $w_2 \leq 0$, then T^* is equal zero, meaning the optimal value of shortage is infinite, and no inventory cycle exists.

And for the third case, similarly, the total profit is defined as:

$$TP(T,F) = \overbrace{sD(F+\beta(1-F)) + v\rho FD}^{\text{Revenue}}$$

$$-\left[\underbrace{\frac{K}{T}}_{\substack{\text{Fixed}\\\text{cost}}} + \underbrace{CD(F+\beta(1-F))}_{\text{Purchasing cost}} + \underbrace{C_1FD}_{\substack{\text{Inspection}\\\text{cost}}} + \underbrace{C_E\rho FD}_{\text{Emergency order cost}} + \overbrace{h\left(\frac{(1-\rho)^2 F^2 TD}{2} + \frac{\rho(F)^2 TD^2}{x}\right)}^{\text{Holding cost}}\right.$$

$$\left. + \underbrace{C_b\frac{(1-F_1)(1-F)\beta TD}{2}}_{\text{Backordering cost}} + \underbrace{g(1-\beta)(1-F)D}_{\text{Lost sale cost}}\right]$$

(2.465)

After some algebra:

$$T = \sqrt{\frac{4K\left(\frac{(1-E(\rho))h}{2} + \frac{E(\rho)hD}{x} + \frac{C_b\beta(1-E(\rho))}{2}\right) - D(d + (C_E - v)E(\rho) - (s+g-C)(1-\beta))^2}{4\frac{C_b\beta}{2}\left(\frac{(1-E(\rho))^2 hD}{2} + \frac{E(\rho)hD^2}{x} - \frac{C_b\beta D(E(\rho))^2}{2}\right)}}$$

(2.466)

$$F(T) = \frac{C_b\beta(2-E(\rho))T + 2(C_E - v)(1-\beta) - 2C_1 - 2(s+g-C)E(\rho)}{4\left(\frac{(1-E(\rho))h}{2} + \frac{E(\rho)hD}{x} + \frac{C_b\beta(1-E(\rho))}{2}\right)T}$$

(2.467)

Considering $w_3 = K\left(\frac{(1-E(\rho))^2 h}{2} + \frac{E(\rho)hD}{x} + \frac{C_b\beta(1-E(\rho))}{2}\right) - \frac{D}{4}(C_1 + (C_E - v)E(\rho) - (s+g-C)(1-\beta))^2$, if $w_3 > 0$, then T^* is given by Eq. (2.466). If $w_3 \leq 0$, then T^* is equal zero, meaning the optimal value of shortage is infinite and no inventory cycle exists. Also additional condition of Case III is $\frac{(1-\rho)^2 h}{2} + \frac{\rho hD}{x} > \frac{C_b\beta\rho^2}{2}$, so if this condition is violated, this means the backordered cost is very huge, and in this situation, shortage is meaningless and economic order quantity is (Jaber et al. 2014):

$$Q^* = \sqrt{\frac{2KD}{h\left(1 - 2E(\rho) + E(\rho^2) + 2E(\rho)\frac{D}{x}\right) + h_E E(\rho^2)}}$$

(2.468)

Table 2.37 Data taken from Salameh and Jaber (2000)

Symbol	Value	Units	Symbol	Value	Units
D	50,000	Units/year	v	20	$/unit
x	175,200	Units/year	h	5	$/unit/year
s	50	$/unit	C_I	0.5	$/unit
K	100	$/order	ρ	U ~ (0, 0.04)	
C	25	$/unit	$f(\rho)$	1/(0.04–0)	

Table 2.38 Date extracted from Jaber et al. (2014)

Symbol	Value	Units
c_E	40	$/unit
h_E	8	$/unit/year

Table 2.39 Additional data (Taleizadeh et al. 2018b)

Description	Symbol	Value	Units
Backordered cost per unit	C_b	20	$/unit
Lost sales cost per unit	g	0.5	$/unit
Percentage of backordered demand	β	97%	

Example 2.22 This section illustrates the use of the three cases with a numerical example. The optimal values of the decision variables and the lot size are calculated. Table 2.37 provides the data taken from Salameh and Jaber (2000). Table 2.38 presents data taken from Jaber et al. (2014). Table 2.39 shows additional data for the numerical example.

According to Table 2.37, the distribution of percentage of imperfect items is assumed that follows a uniform distribution [U ~ (0, 0.04)]. Thus, the expected value of the defective items (ρ) is:

$$E(\rho) = \int_a^b \rho f(\rho) d\rho = \int_a^b \rho \frac{1}{b-a} d\rho = \frac{b+a}{2} = \frac{0.04+0}{2} = 0.02$$

Moreover, $E(\rho^2)$ is:

$$E(\rho^2) = \text{var}(\rho) + E^2(\rho) = \frac{(b-a)^2}{12} + E^2(\rho) = \frac{(0.04)^2}{12} + (0.02)^2 = 0.0005333333$$
$$(1 - E(\rho))^2 = 1 - 2E(\rho) + E(\rho^2) = 1 - 2(0.02) + 0.000533 = 0.9605333355$$

Firstly, the significance of each profit functions is verified:
$w_1 = 985.3880 > 0$, $w_2 = 931.1284 > 0$, $w_3 = 965.7747 > 0$.

Table 2.40 Optimal value for the decision variables and the economic order quantity

	TP	T	F (%)	Q
Case I	1,200,732.887	0.0289	60.70	1428.138
Case II	1,200,277.629	0.0281	57.88	1385.718
Case III	1,200,667.453	0.0286	60.70	1414.757

For Case III, the following constraint $\frac{(1-E(\rho))^2 h}{2} + \frac{E(\rho)hD}{x} - \frac{\pi\beta E(\rho^2)}{2} = 2.4247 > 0$ must be satisfied. Therefore, all of the conditions are satisfied. Hence, the optimal value for the decision variables and the economic order quantity for each case are given in Table 2.40.

References

Alamri, A. A., Harris, I., & Syntetos, A. A. (2016). Efficient inventory control for imperfect quality items. *European Journal of Operational Research, 254*(1), 92–104.

Bakker, M., Riezebos, J., & Teunter, R. H. (2012). Review of inventory systems with deterioration since 2001. *European Journal of Operational Research, 221*(2), 275–284.

Cárdenas-Barrón, L.E., (2001). The economic production quantity (EPQ) with shortage derived algebraically. *International Journal of Production Economics, 70*(3), 289–292

Chan, W. M., Ibrahim, R. N., & Lochert, P. B. (2003). A new EPQ model: Integrating lower pricing, rework and reject situations. *Production Planning and Control, 14*(7), 588–595.

Chiu, Y.-S.P., Wang, S.-S., Ting, C.-K., Chuang, H.-J., Lien, Y.-L., (2008). Optimal run time for EMQ model with backordering, failure-in- rework and breakdown happening in stock-piling time, *WSEAS Transactions on Information Science and Applications, 5*(4), pp. 475–486

Dave, U. (1986). A probabilistic scheduling period inventory model for deteriorating items with lead times. *Zeitschrift für Operations Research, 30*(5), 229–237.

Diabat, A., Taleizadeh, A. A., & Lashgari, M. (2017). A Lot Sizing Model with partial down-stream delayed payment, partial up-stream advance payment, and partial backordering for deteriorating items. *Journal of Manufacturing Systems, 45*, 322–342.

Dohi, T., & Osaki, S. (1995). Optimal inventory policies under product obsolescent circumstance. *Computers and Mathematics with Applications, 29*, 23–30.

Elmaghraby, W., & Keskinocak, P. (2003). Dynamic pricing in the presence of inventory considerations: Research overview, current practices, and future directions. *Management Science, 49*(10), 1287–1309.

Eroglu, A., & Ozdemir, G. (2007). An economic order quantity model with defective items and shortages. *International Journal of Production Economics, 106*(2), 544–549.

Ferguson, M. E., & Ketzenberg, M. E. (2005). *Sharing information to manage perishables* (2nd ed.). Georgia Institute of Technology.

Goyal, S.K., & Cárdenas-Barrón, L.E. (2002). Note on: Economic production quantity model for items with imperfect quality - A practical approach, *International Journal of Production Economics, 77*(1), 85–87.

Harris, F. W. (1913). What quantity to make at once. In *The library of factory management. Operation and costs* (The factory management series) (Vol. 5, pp. 47–52). Chicago, IL: A.W. Shaw.

Hasanpour, J., Sharafi, E., & Taleizadeh, A. A. (2019). A lot sizing model for imperfect and deteriorating product with destructive testing and inspection errors. *International Journal of Systems Sciences: Operations and Logistic.* https://doi.org/10.1080/23302674.2019.1648702.

Hou, K. L., Lin, L. C., & Lin, T. Y. (2015). Optimal lot sizing with maintenance actions and imperfect production processes. *International Journal of Systems Science, 46*(15), 2749–2755.

Hsu L.F., Hsu, J.T., (2016). Economic production quantity (EPQ) models under an imperfect production process with shortages backordered, *International Journal of Systems Science 47* (4): 852–867

Huang, C. K. (2004). An optimal policy for a single-vendor single-buyer integrated production-inventory problem with process unreliability consideration. *International Journal of Production Economics, 91*(1), 91–98.

Jaber, M.Y., Bonney, M. (1996). Production breaks and the learning curve: The forgetting phenomenon, *Applied Mathematical Modelling, 20*(2), 162–169.

Jaber, M., Goyal, S., & Imran, M. (2008). Economic production quantity model for items with imperfect quality subject to learning effects. *International Journal of Production Economics, 115*(1), 143–150.

Jaber, M. Y., Bonney, M., & Moualek, I. (2009). An economic order quantity model for an imperfect production process with entropy cost. *International Journal of Production Economics, 118*(1), 26–33.

Jaber, M. Y., Zanoni, S., & Zavanella, L. E. (2013). An entropic economic order quantity (EnEOQ) for items with imperfect quality. *Applied Mathematical Modelling, 37*(6), 3982–3992.

Jaber, M. Y., Zanoni, S., & Zavanella, L. E. (2014). Economic order quantity models for imperfect items with buy and repair options. *International Journal of Production Economics, 155*, 126–131.

Joglekar, P., & Lee, P. (1993). An exact formulation of inventory costs and optimal lot size in face of sudden obsolescence. *Operations Research Letters, 14*, 283–290.

Jaber, M., Nuwayhid, R. Y., & Rosen, M. A. (2004). Price-driven economic order systems from a thermodynamic point of view. *International Journal of Production Research, 42*(24), 5167–5184.

Jaber, M. Y. (2006). Learning and forgetting models and their applications. In A. B. Badiru (Ed.), *Handbook of industrial & systems engineering* (pp. 30.1–30.27). Boca Raton, FL: CRC Press.

Jaggi, C. K., & Mittal, M. (2011). Economic order quantity model for deteriorating items with imperfect quality. *Revista Investigación Operacional, 32*(2), 107–113.

Jaggi, C. K., Goel, S. K., & Mittal, M. (2011). Economic order quantity model for deteriorating items with imperfect quality and permissible delay on payment. *International Journal of Industrial Engineering Computations, 2*, 237–248.

Keshavarz, R., Makui, A., Tavakkoli-Moghaddam, & Taleizadeh, A. A. (2019). Optimization of imperfect economic manufacturing models with a power demand rate dependent production rate. *Sadhana - Academy Proceedings in Engineering Sciences, 44*(9), 206.

Khan, M., Jaber, M. Y., Guiffrida, A. L., & Zolfaghari, S. (2011). A review of the extensions of a modified EOQ model for imperfect quality items. *International Journal of Production Economics, 132*, 1–12.

Khan, M., Jaber, M. Y., & Wahab, M. I. M. (2010). Economic order quantity model for items with imperfect quality with learning in inspection. *International Journal of Production Economics, 124*(1), 87–96.

Khan, M., Jaber, M. Y., & Bonney, M. (2011). An economic order quantity (EOQ) for items with imperfect quality and inspection errors. *International Journal of Production Economics, 133*(1), 113–118.

Konstantaras, I., Skouri, K., & Jaber, M. Y. (2012). Inventory models for imperfect quality items with shortages and learning in inspection. *Applied Mathematical Modelling, 36*(11), 5334–5343.

Kalantary, S. S., & Taleizadeh, A. A. (2018). Mathematical modelling for determining the replenishment policy for deteriorating items in an EPQ model with multiple shipments. *International Journal of Systems Science: Operations and Logistics, 7*, 164–171.

Leung, S. C. H., & Ng, W.-L. (2007). A stochastic programming model for production planning of perishable products with postponement. *Production Planning & Control, 18*(3), 190–202.

References

Lashgary, M., Taleizadeh, A. A., & Sana, S. S. (2016). An inventory control problem for deteriorating items with back-ordering and financial considerations under two levels of trade credit linked to order quantity. *Journal of Industrial and Management Optimization, 12*(3), 1091–1119.

Lashgary, M., Taleizadeh, A. A., & Sadjadi, S. J. (2018). Ordering policies for non-instantaneous deteriorating items under hybrid partial prepayment, partial delay payment and partial backordering. *Journal of Operational Research Society, 69*(8), 1167–1196.

Maddah, B., & Jaber, M. (2008). Economic order quantity for items with imperfect quality. Revisited. *International Journal of Production Economics, 112*(2), 808–815.

Metters, R. (1997). Quantifying the bullwhip effect in supply chains. *Journal of the Operations Management, 15*(2), 89–100.

Moussawi-Haidar, L., Salameh, M., & Nasr, W. (2013). An instantaneous replenishment model under the effect of a sampling policy for defective items. *Applied Mathematical Modelling, 37*(3), 719–727.

Mohammadi, B., Taleizadeh, A. A., Noorossana, R., & Samimi, H. (2015). Optimizing integrated manufacturing and products inspection policy for deteriorating manufacturing system with imperfect inspection. *Journal of Manufacturing Systems, 37*, 299–315.

Moussawi-Haidar, L., Salameh, M., & Nasr, W. (2014). Effect of deterioration on the instantaneous replenishment model with imperfect quality items. *Applied Mathematical Modelling, 38*(24), 5956–5966.

Nobil, A. H., Kazemi, A., & Taleizadeh, A. A. (2019). Single-machine lot scheduling problem for deteriorating items with negative exponential deterioration rate. *RAIRO Operation Research, 53*(4), 1297–1307.

Paknejad, J., Nasri, F., & Affisco, J. F. (2005). Quality improvement in an inventory model with finite-range stochastic lead times. *Journal of Applied Mathematics and Decision Sciences, 3*, 177–189.

Pal, B., Sana, S. S., & Chaudhuri, K. S. (2013). A mathematical model on EPQ for stochastic demand in an imperfect production system. *Journal of Manufacturing Systems, 32*(1), 260–270.

Papachristos, S., Konstantaras, I. (2006). Economic ordering quantity models for items with imperfect quality, International *Journal of Production Economics, 100*(1), 148–154.

Porteus, E. L. (1986). Optimal lot sizing, process quality improvement and setup cost reduction. *Operations Research, 34*(1), 137–144.

Rezaei, J. (2016). Economic order quantity and sampling inspection plans for imperfect items. *Computers & Industrial Engineering, 96*, 1–7.

Rezaei, J., & Salimi, N. (2012). Economic order quantity and purchasing price for items with imperfect quality when inspection shifts from buyer to supplier. *International Journal of Production Economics, 137*(1), 11–18.

Rezaei, J. (2005). Economic order quantity model with backorder for imperfect quality items. In *Proceedings of the 2005 IEEE International Engineering Management Conference* (Vol. 2, pp. 466–470). Piscataway, NJ: IEEE.

Roy, A., Sana, S. S., & Chaudhuri, K. (2015). Optimal pricing of competing retailers under uncertain demand a two-layer supply chain model. *Annals of Operations Research, 260*, 481–500.

Roy, M. D., Sana, S. S., & Chaudhuri, K. (2011). An economic order quantity model of imperfect quality items with partial backlogging. *International Journal of Systems Science, 42*(8), 1409–1419.

Salameh, M. K., & Jaber, M. Y. (2000a). Economic production quantity model for items with imperfect quality. *International Journal of Production Economics, 64*(1), 59–64.

Sana, S. S. (2012). An economic order quantity model for nonconforming quality products. *Service Science, 4*(4), 331–348.

Salameh, M. K., & Jaber, M. Y. (2000). Economic production quantity model for items with imperfect quality. *International Journal of Production Economics, 64*(1), 59–64.

Shah, B. J., Shah, N. H., & Shah, Y. K. (2005). EOQ model for time-dependent deterioration rate with a temporary price discount. *Asia Pacific Journal of Operational Research, 22*(4), 479–485.

Skouri, K., Konstantaras, I., Lagodimos, A. G., & Papachristos, S. (2014). An EOQ model with backorders and rejection of defective supply batches. *International Journal of Production Economics, 155*, 148–154.

Song, J. S., & Zipkin, P. (1996). Managing inventory with the prospect of obsolescence. *Operations Research, 44*, 215–222.

Tai, A. H. (2015). An EOQ model for imperfect quality items with multiple quality characteristic screening and shortage backordering. *European Journal of Industrial Engineering, 9*(2), 261–276.

Taleizadeh, A. A., & Zamani-Dehkordi, N. (2017a). Optimizing setup cost in (R,T) inventory system model with imperfect production process, quality improvement and partial backordering. *Journal of Remanufacturing, 7*, 199–215.

Taleizadeh, A. A., Akram, R., Lashgari, M., & Heydari, J. (2016a). Imperfect economic production quantity model with upstream trade credit periods linked to raw material order quantity and downstream trade credit periods. *Applied Mathematical Modelling, 40*, 8777–8793.

Taleizadeh, A. A., Khanbaglo, M. P. S., & Cárdenas-Barrón, L. E. (2016b). An EOQ inventory model with partial backordering and reparation of imperfect products. *International Journal of Production Economics, 182*, 418–434.

Taleizadeh, A. A., Pourrezaie Khaligh, P., & Moon, I. (2018a). Hybrid NSGA-II for an imperfect production system considering product quality and returns under two warranty policies. *Applied Soft Computing, 75*, 333–348.

Taleizadeh, A. A., Noori-Daryan, M., & Tavakkoli-Moghadam, R. (2015). Pricing and ordering decisions in a supply chain with imperfect quality items and inspection under a buyback of defective items. *International Journal of Production Research, 53*(15), 4553–4582.

Taleizadeh, A. A., & Zamani-Dehkordi, N. (2017b). Stochastic lot sizing model with partial backordering and imperfect production processes. *International Journal of Inventory Research, 4*(1), 75–96.

Taleizadeh, A. A., Khanbaglo, M. P. S., & Cárdenas-Barrón, L. E. (2016c). Outsourcing rework of imperfect items in the EPQ inventory model with backordered demand. *IEEE Transactions on Systems, Man, and Cybernetics: Systems, 49*(12), 2688–2699.

Taleizadeh, A. A., Perak Sari-Khanbeglo, M., & Cárdenas-Barrón, L. E. (2018b). Replenishment of imperfect items in an EOQ inventory model with partial backordering. *RAIRO-Operation Research, 54*(2), 413–434.

Taleizadeh, A. A. (2014). An economic order quantity model for deteriorating item in a purchasing system with multiple prepayments. *Applied Mathematical Modeling, 38*, 5357–5366.

Taleizadeh, A. A., & Nematollahi, M. R. (2014). An inventory control problem for deteriorating items with backordering and financial engineering considerations. *Applied Mathematical Modeling, 38*, 93–109.

Taleizadeh, A. A., & Rasouli-Baghban, A. (2015). Pricing and inventory decisions for deteriorating product under shipment consolidation. *International Journal of Advanced Logistics, 4*(4), 89–99.

Taleizadeh, A. A., & Rasouli-Baghban, A. (2018). Pricing and lot sizing of a decaying item under group dispatching with time-dependent demand and decay rates. *Scientia Iranica, 25*(3E), 1656–1670.

Taleizadeh, A. A., Wee, H. M., & Jolai, F. (2013a). Revisiting fuzzy rough economic order quantity model for deteriorating items considering quantity discount and prepayment. *Mathematical and Computer Modeling, 57*(5-6), 1466–1479.

Taleizadeh, A. A., Mohammadi, B., Cárdenas-Barron, L. E., & Samimi, H. (2013b). An EOQ model for perishable product with special sale and shortage. *International Journal of Production Economics, 145*(1), 318–338.

Taleizadeh, A. A., Nouri-Dariyan, M., & Cárdenas-Barrón, L. E. (2015). Joint optimization of price, replenishment frequency, replenishment cycle and production rate in vendor managed inventory system with deteriorating items. *International Journal of Production Economics, 159,* 285–295.

Taleizadeh, A. A., Satariyan, F., & Jamili, A. (2016). Optimal multi discount selling prices schedule for deteriorating product. *Scientia Iranica, 22*(6), 2595–2603.

Taleizadeh, A. A., Pourmohammadzia, N., & Konstantaras, I. (2019). Partial linked to order delayed payment and life time effects on decaying items ordering. *Operational Research.*

Tat, R., Taleizadeh, A. A., & Esmaeili, M. (2015). Developing EOQ model with non-instantaneous deteriorating items in vendor-managed inventory (VMI) system. *International Journal of Systems Science, 46*(7), 1257–1268.

Tavakkoli, S., & Taleizadeh, A. A. (2017). An EOQ model for decaying item with full advanced payment and conditional discount. *Annals of Operations Research, 259,* 415–436.

Wang, W. T., Wee, H. M., Cheng, Y. L., Wen, C. L., & Cárdenas-Barrón, L. E. (2015). EOQ model for imperfect quality items with partial backorders and screening constraint. *European Journal of Industrial Engineering, 9*(6), 744–773.

Wahab, M. I. M., & Jaber, M. Y. (2010). Economic order quantity model for items with imperfect quality, different holding costs, and learning effects: A note. *International Journal of Production Economics, 58*(1), 186–190.

Wee, H. M. (1993). Economic production lot size model for deteriorating items with partial backordering. *Computers & Industrial Engineering, 24*(3), 449–458.

Wee, H. M., Yu, J., & Chen, M. C. (2007). Optimal inventory model for items with imperfect quality and shortage backordering. *Omega, 35*(1), 7–11.

Wee, H. M. (1993). Economic production lot size model for deteriorating items with partial backordering. *Computers & Industrial Engineering, 24*(3), 449–458.

Chapter 3
Scrap

3.1 Introduction

As stated in Chap. 2, the economic order quantity (EOQ) model was first introduced in 1913. Seeking to minimize the total cost, the model generated a balance between holding and ordering costs and determined the optimal order size. Later, the EPQ model considered items produced by machines inside a manufacturing system with a limited production rate, rather than items purchased from outside the factory. Despite their age, both models are still widely used in major industries. Their conditions and assumptions, however, rarely pertain to current real-world environments. To make the models more applicable, different assumptions have been proposed in recent years, including random machine breakdowns, generation of imperfect and scrap items, and discrete shipment orders. The assumption of discrete shipments using multiple batches can make the EPQ model more applicable to real-world problems. The EPQ inventory models assume that all the items are manufactured with high quality and defective items are not produced. However, in fact, defective items appear in the most of manufacturing systems; in this sense, researchers have been developing EPQ inventory models for defective production systems. In these production systems, defective items are of two types: scrapped items and reworkable items.

Chung (1997) investigated bounds for production lot sizing with machine breakdown conditions. Rosenblatt and Lee (1986) proposed an EPQ model that deals with imperfect quality. They assumed that at some random point in time, the process might shift from an in-control to an out-of-control state. Chiu et al. (2007) investigated an EPQ model with scrap, rework, and stochastic machine breakdowns to determine the optimal run time and production quantity. Sarkar et al. (2014) developed an EPQ model with a random defective rate, rework process, and backorders for a single-stage production system. Chiu et al. (2010) presented a robust production–inventory model with an imperfect rework process for imperfect products. Machine breakdown can occur in that model, and failure time is considered a random variable. Chiu et al. (2011a) developed the EPQ model with discrete shipment and generation of imperfect-quality items. They assumed that a portion of imperfect items are scrapped and that the rest can be repaired or salvaged. At the end of the repair run, repaired items are delivered to customers in discrete batches. In that research, the objective function was to determine the economic production quantity that would minimize the total cost. Chiu et al. (2011b) investigated an EPQ model with discrete shipment and generation of imperfect and scrap items. During a production and repair run, the initial delivery of batches was considered to satisfy demand, and the rest were discretely delivered to the customer at the end of the machine repair time. Chiu et al. (2012) provided an EPQ model with random machine breakdowns and the generation of imperfect and scrap items. Akbarzadeh et al. (2015, 2016) developed EPQ models with scrapped items using vendor-managed inventory systems.

The main aim of this chapter is providing a comprehensive framework of analyzing the EPQ models considering the scrap. Hence, some important EPQ inventory models in which scrapped items are categorized in two sections including EPQ models with shortage and without shortage are presented. The shortage sections are divided to two subsections which are partial and full backordering shortage. Also, the models based on the delivery policies including discreet and continuous deliveries are categorized. Therefore, two key features, EPQ models and scrapped items, are the main goals that are investigated in this chapter.

In this chapter, a comprehensive review of some major EPQ models in which scrapped items is considered is presented. First, a brief introduction and literature review of mentioned problem is provided. Then, a framework of problem and mathematical model is presented, and a numerical example for each problem is solved. According to above description, three categorizes, no shortage, partial backordering, full backordering, are provided as shown in Fig. 3.1. All categories are investigated in the next sections.

All categorized problems in Fig. 3.1 are investigated separately. The common notations of EPQ model considering scrap are shown in Table 3.1. To integrate the report, these notations for all models are used. The main decision variables of this field on inventory are Q and B, but in some studies, other decision variables are considered too.

3.2 No Shortage

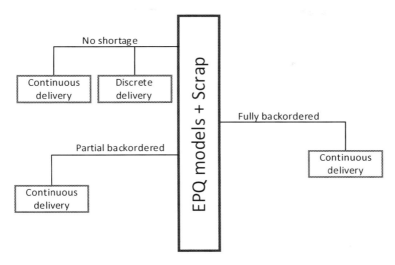

Fig. 3.1 Categories of EPQ model considering scrap

3.2 No Shortage

3.2.1 *Continuous Delivery*

In this section, the EPQ models with scrap and continuous delivery policy without shortage are presented.

3.2.1.1 Considering Steady Production Rate

Chiu et al. (2003) studied the effect of the steady production rate of scrap items on the economic production quantity (EPQ) model. Many research efforts on the EPQ model assumed that the manufacturing facility functions perfectly during a production run. But in most practical settings, defective items may be generated, by an imperfect production process due to process deterioration or other factors.

Chiu et al. (2003) extended the work of Hayek and Salameh (2001) and studied the effect of the steady production rate of the scrap items on the classical finite production model. The imperfect production process may generate x percent of imperfect-quality items. In this problem, it is assumed that the imperfect-quality items are all scrap items and have a defective production rate d'. Since shortage is not allowed, backordering all demands must be satisfied at all times. Hence, the replenishment policy is to restart a new production run (cycle) whenever on-hand inventories run out.

The production rate P is constant and is much larger than demand rate D. The defective production rate d of imperfect-quality items could be expressed as the product of the production rate times the percentage of defective items produced.

Table 3.1 Notations

Q	Production lot size per cycle (unit)
P	Production rate per unit time (units per unit time)
T	Length of production inventory cycle (time)
ts	Setup time to produce item (time)
H, I, I_1, H_1	Maximum level of on-hand inventory in units (unit)
$I(t)$	On-hand inventory of perfect-quality items at time t (unit)
$f(x)$	Probability density function of x
X	The percentage of the imperfect-quality items produced, x may be a random variable for some models with known probability of density function
D	Demand rate (units per unit time)
H	Holding cost per unit per unit time (\$/unit/unit time)
B	Size of the backorders (unit)
C_b	Backorder cost per item per unit time (\$/unit/unit time)
$\hat{\pi}$	Unit lost sale cost (\$/unit)
K	Setup cost (\$/setup)
K_S	Fixed delivery cost per shipment (\$/shipment)
d	Production rate of scrap items (units per unit time)
C	Production cost per item (\$/unit)
s	The selling price per unit for good-quality items (\$/unit)
v	The selling price per unit for defective items (\$/unit)
C_d	Disposal cost for each scrap item (\$/unit)
C_T	Unit delivery cost (\$/unit)
C_I	Inspection cost per item (\$/unit)
C_M	Machine repair cost per breakdown (\$/breakdown)
h	Holding cost per unit per unit of time (\$/unit/unit time)
h_1	Holding cost of defective items per unit per unit of time (\$/unit/unit time)
g, t_r	Time needed to repair and restore the machine after breakdown
TCU	Total inventory costs per unit time (\$)
TC	Total inventory costs per cycle (\$)
$E(.)$	Denotes the expected value operator

Therefore, the defective production rate d of imperfect-quality items can be written as (Chiu et al. 2003):

$$d = P \cdot x, \qquad (3.1)$$

The production rate of perfect-quality items must always be greater than or equal to the sum of the demand rate and the rate at which defective items are produced. Therefore,

3.2 No Shortage

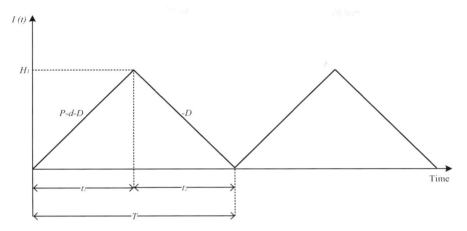

Fig. 3.2 On-hand inventory of non-defective items (Chiu et al. 2003)

$$P - d - D \geq 0,$$
$$\left(1 - \frac{D}{P}\right) \geq x \geq 0, \tag{3.2}$$

For the following derivation, the solution procedures are those used by Hayek and Salameh (2001), referring to Fig. 3.2 (Chiu et al. 2003):

$$T = t_1 + t_2 \text{ and } T = \frac{Q(1-x)}{D} \tag{3.3}$$

The production uptime t_1 needed to build up H_1 units of perfect-quality items is (Chiu et al. 2003):

$$t_1 = \frac{Q}{P} \tag{3.4}$$

and

$$H_1 = (P - d - D)t_1 = Q\left(1 - x - \frac{D}{P}\right), \tag{3.5}$$

The production downtime t_2 needed to consume the maximum on-hand inventory H_1 is (Chiu et al. 2003):

$$t_2 = \frac{H_1}{D} = Q\left(\frac{1}{D} - \frac{x}{D} - \frac{1}{P}\right), \tag{3.6}$$

The imperfect-quality items which build up during production uptime t_1 as shown in Fig. 3.3 are:

Fig. 3.3 On-hand inventory of defective items (Chiu et al. 2003)

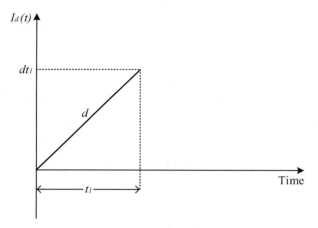

$$dt_1 = xQ, \qquad (3.7)$$

In real-life situations, the percentage of imperfect-quality items may be a random variable, with a known probability density function. For instance, if x follows the uniformly distributed over the range $[X_u, X_l]$, the probability density function $f(x)$ is (Chiu et al. 2003):

$$f(x) = \begin{cases} \dfrac{1}{X_u - X_l}; X_l < x < X_u \\ 0 \end{cases} \qquad (3.8)$$

Then, for an inventory system which does not allow backorder, to prevent shortages from occurring, one must restart a new production cycle when on-hand inventory is equal to zero. So the cyclic inventory cost is (Chiu et al. 2003):

$$\text{TC}(Q) = \overbrace{CQ}^{\text{Production cost}} + \overbrace{C_d(xQ)}^{\text{Disposal cost}} + \overbrace{K}^{\text{Setup cost}} + \overbrace{h\frac{H_1}{2}T + h\frac{xQ}{2}t_1}^{\text{Holding cost}} \qquad (3.9)$$

Because the percentage of imperfect-quality items is a random variable, the expected annual cost is $E[\text{TCU}(Q)] = E[\text{TC}(Q)]/E[T]$; from Eq. (3.3) through Eq. (3.9), one obtains (Chiu et al. 2003):

$$E[\text{TCU}(Q)] = D\left[\frac{C}{1 - E(x)} + C_d \frac{E(x)}{1 - E(x)}\right] + \frac{KD}{Q} \frac{1}{1 - E(x)} + \frac{hQ}{2}\left(1 - \frac{D}{P}\right)\frac{1}{1 - E(x)}$$
$$- hQ\left(1 - \frac{D}{P}\right)\frac{E(x)}{1 - E(x)} + \frac{hQ}{2}\frac{E(x^2)}{1 - E(x)}$$
$$(3.10)$$

3.2 No Shortage

Let $E_0 = \frac{1}{1-E(x)}$; $E_1 = \frac{E(x)}{1-E(x)}$; $E_2 = \frac{E(x^2)}{1-E(x)}$;

$$E(\text{TCU}(Q)) = D[CE_0 + C_d E_1] + \frac{KD}{Q}E_0 + \frac{hQ}{2}\left(1 - \frac{D}{P}\right)E_0 \\ - hQ\left(1 - \frac{D}{P}\right)E_1 + \frac{hQ}{2}E_2 \qquad (3.11)$$

Differentiating $E[\text{TCU}(Q)]$ with respect to Q twice, as shown below, one can easily find that $E[\text{TCU}(Q)]$ is convex. The optimal production quantity Q is obtained by setting the first derivative equal to zero (Chiu et al. 2003):

$$\frac{dE(\text{TCU}(Q))}{dQ} = \frac{-KD}{Q^2}E_0 + \frac{h}{2}\left(1 - \frac{D}{P}\right)E_0 - h\left(1 - \frac{D}{P}\right)E_1 + \frac{h}{2}E_2 \qquad (3.12)$$

$$\frac{d^2 E(\text{TCU}(Q))}{d^2 Q} = \frac{2KD}{Q^3}E_0 > 0 \qquad (3.13)$$

From $\frac{dE[\text{TCU}(Q)]}{dQ} = 0$, one obtains (Chiu et al. 2003):

$$Q^* = \sqrt{\frac{2KD}{h\left(1 - \frac{D}{P}\right) - 2h\left(1 - \frac{D}{P}\right)E(x) + hE(x^2)}} \qquad (3.14)$$

Suppose that no defective items are produced, then $x = 0$, and the same equation as that of the classical finite production rate model will be derived (Nahmias and Cheng 2005):

$$Q^* = \sqrt{\frac{2KD}{h\left(1 - \frac{D}{P}\right)}} \qquad (3.15)$$

Example 3.1 Chiu et al. (2003) presented a manufactured product which has experienced a relatively flat demand of 4000 units per year. This item is produced at a rate of 10,000 units per year. The accounting department has estimated that it costs $450 to initiate a production run, each unit costs the company $2 to manufacture, the cost of holding is $0.6 per item per year, and the disposal cost is $0.3 for each scrap item. The production rate of defective items is uniformly distributed over the interval [0, 0.1]. Thus, $P = 10{,}000$ units per year, $D = 4000$ units per year, $x = $ uniformly distributed over the interval [0, 0.1], $K = \$450$ for each production run, $C = \$2$ per item, $C_s = \$0.3$ for each scrap item, and $h = \$0.6$ per item per unit time.

The optimal production lot size, computed from Eq. (3.14), is $Q^* = 3323$ units (Chiu et al. 2003).

The expected annual inventory cost, from Eq. (3.10), is $E[TCU(Q)] = \$9625$ per year (Chiu et al. 2003).

If no defective items are produced, that is, when $x = 0$, the model is changed to the classical finite production rate model. From Eq. (3.15) it is obtained that $Q^* = 3162$, and annual inventory costs, $E[TCU(Q)] = \$9138$ per year comparing these values, it could be noticed that in the case of scrap items which are produced in EPQ model, the optimal lot size must be determined using the new equations (Chiu et al. 2003).

3.2.1.2 Multi-product Multi-machine

Nobil et al. (2016) considered a multi-product problem with nonidentical machines. This manufacturing system consists of various machine types with different production capacities, production costs, setup times, production rates, and failure rates.

In real-world problems, there are different options to purchase machines, considering different factors like production rate, floor space limitation for production, budget constraint, and so forth. As a result, production managers cope with decisions about minimizing machine utilization expenditure and inventory costs simultaneously. Moreover, when factory produces more than one item, factory may need to buy more than one machine. This section investigates this problem.

Nobil et al. (2016) considered a multi-machine production system producing m different items. The problem is an EPQ problem with unrelated parallel machine in which utilized production machines are considered. It is an extension of the single-machine multi-product EPQ problem with defective items, in which the determination of machine's number and items allocation is considered simultaneously. Each machine has a particular failure rate based on the characteristics of product. The proposed inventory model minimizes total cost of the inventory system, including utilization, setup, production, holding, and disposal costs.

The following notations presented in Table 3.2 are used for machines $i = 1, 2, \ldots, n$ and items $j = 1, 2, \ldots, n$ to model the problem.

The inventory problem under study is shown in Fig. 3.4 (Nobil et al. 2016). It is evident from Fig. 3.4 that the maximum on-hand inventory of the jth item produced by ith machine is determined by Eq. (3.16):

$$I_j = \left((1 - x_{ij})P_{ij} - D_j\right)\frac{Q_j}{P_{ij}} \tag{3.16}$$

In Fig. 3.4, the cycle length of jth item produced by ith machine consists of two periods: the uptime production period denoted by Tp_{ij} and downtime period Td_{ij}. The lengths of these periods are calculated as follows:

3.2 No Shortage

Table 3.2 Notations of given problem (Nobil et al. 2016)

N	Number of machines
M	Number of items
K_{ij}	Setup cost of ith machine to produce jth item (\$/setup)
α_{ij}	Binary parameter, $\alpha_{ij} = 1$ if $(1 - x_{ij})P_{ij} \geq D_j$; otherwise, $\alpha_{ij} = 0$
f_i	The fixed cost of utilizing for ith machine (\$/use)
r_i	Required space of ith machine (square meter)
C_{ij}	Production cost of jth item per unit on machine i (\$/unit)
N_i	Number of cycles per unit time for ith machine
Tp_{ij}	Uptime of the jth item produced by ith machine (time)
Td_{ij}	Downtime of the jth item produced by ith machine (time)
BC	Maximum available budget (\$)
F_M	Maximum available space (square meter)
CP	Total production cost of all items (\$)
CU	Total utilization cost of all machines (\$)
CH	Total holding cost of all items (\$)
CK	Total setup cost of all items (\$)
CD	Total disposal cost of all items (\$)
y_i	$y_i =$ if machine i is utilized; otherwise, $y_i = 0$ decision variables
z_{ij}	$z_{ij} = 1$ if jth item produced by ith machine; otherwise, $z_{ij} = 0$ decision variables

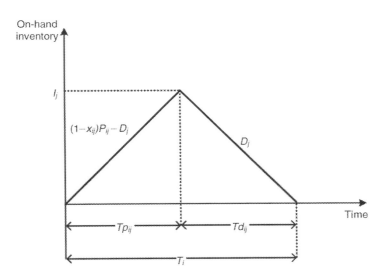

Fig. 3.4 The cycle length of the on-hand inventory of jth item produced by ith machine (Nobil et al. 2016)

$$\text{Tp}_{ij} = \frac{I_j}{(1-x_{ij})P_{ij} - D_j} = \frac{Q_j}{P_{ij}} \tag{3.17}$$

$$\text{Td}_{ij} = \frac{I_j}{D_j} = \frac{Q_j((1-x_{ij})P_{ij} - D_j)}{P_{ij}D_j} \tag{3.18}$$

Consequently, the length of a cycle for ith machine is:

$$T_i = \text{Tp}_{ij} + \text{Td}_{ij} = \frac{(1-x_{ij})Q_j}{D_j} \tag{3.19}$$

Hence,

$$Q_j = \frac{D_j T_i}{(1-x_{ij})} \tag{3.20}$$

Besides, all items are manufactured on each machine with a limited capacity and a common cycle length of $T_{i1} = T_{i2} = \ldots = T_{im}$. The total inventory system cost consists of the following costs, utilization, setup, production, holding, and disposal, as shown in Eq. (3.21):

$$TC = CU + CK + CP + CH + CD \tag{3.21}$$

In what follows, all the components of Eq. (3.20) are derived. The total utilization cost is calculated by Eq. (3.22):

$$CU = \sum_{i=1}^{n} y_i f_i \tag{3.22}$$

As the setup cost of ith machine to produce jth item in a cycle is K_{ij} and there are N_i cycles in a specific period of time (i.e., a year), then the total setup cost is given by:

$$CK = \sum_{i=1}^{n} \sum_{j=1}^{m} N_i z_{ij} K_{ij} \tag{3.23}$$

In addition, based on the joint production policy, $N_i = 1/T_i$. Hence,

$$CK = \sum_{i=1}^{n} \sum_{j=1}^{m} \frac{z_{ij} K_{ij}}{T_i} \tag{3.24}$$

3.2 No Shortage

The total production cost of the inventory system based on the production cost of the jth item per unit per cycle on ith machine (C_{ij}) is computed using Eq. (3.25) as:

$$\text{CP} = \sum_{i=1}^{n}\sum_{j=1}^{m} N_i z_{ij} C_{ij} Q_j = \sum_{i=1}^{n}\sum_{j=1}^{m} \frac{x_{ij} C_{ij}}{T_i}\left(\frac{D_j T_i}{(1-x_{ij})}\right)$$

$$= \sum_{i=1}^{n}\sum_{j=1}^{m} \frac{z_{ij} C_{ij} D_j}{(1-d_{ij})} \quad (3.25)$$

Based on Fig. 3.4, the total holding cost of the inventory system under study is determined as:

$$\text{CH} = \sum_{i=1}^{n}\sum_{j=1}^{m} \frac{z_{ij} h_j}{T_i}\left[\frac{I_j}{2}(T_i)\right] \quad (3.26)$$

Inserting I_j from Eq. (3.16) results in:

$$\text{CH} = \sum_{i=1}^{n}\sum_{j=1}^{m} \frac{z_{ij} h_j}{T_i}\left[\frac{T_i}{2}\left((1-x_{ij})P_{ij} - D_j\right)\frac{Q_j}{P_{ij}}\right]$$

$$= \sum_{i=1}^{n}\sum_{j=1}^{m} \frac{z_{ij} h_j D_j}{2(1-x_{ij})}\left(1 - x_{ij} - \frac{D_j}{P_{ij}}\right)T_i \quad (3.27)$$

The total disposal cost of the inventory system based on the disposal cost of the jth item per unit per cycle (C_{dj}) is calculated by Eq. (3.28) as:

$$\text{CD} = \sum_{i=1}^{n}\sum_{j=1}^{m} N_i z_{ij} C_{d_j} x_{ij} Q_j$$

$$= \sum_{i=1}^{n}\sum_{j=1}^{m} \frac{x_{ij} C_{d_j} z_{ij}}{T_i}\left(\frac{D_j T_i}{(1-x_{ij})}\right) = \sum_{i=1}^{n}\sum_{j=1}^{m} \frac{x_{ij} C_{d_j} z_{ij} D_j}{(1-x_{ij})} \quad (3.28)$$

Finally, the total cost is:

$$\text{TC} = \text{CU} + \text{CK} + \text{CP} + \text{CH} + \text{CD} = \sum_{i=1}^{n} y_i f_i + \sum_{i=1}^{n}\sum_{j=1}^{m} z_{ij}$$

$$\times \left[\frac{K_{ij}}{T_i} + \frac{h_j D_j}{2(1-x_{ij})}\left(1 - x_{ij} - \frac{D_j}{P_{ij}}\right)T_i + \frac{C_{ij} D_j + C_{d_j} x_{ij} D_j}{(1-x_{ij})}\right] \quad (3.29)$$

Because of characteristics of the proposed problem, a constrained problem is modeled, as shown below:

$$\sum_{i=1}^{n} \alpha_{ij} z_{ij} = 1 \tag{3.30}$$

$$\begin{cases} \alpha_{ij} = 1 & (1-x_{ij})P_{ij} - D_j \geq 0 \\ \alpha_{ij} = 0 & \text{Otherwise} \end{cases} \tag{3.31}$$

$$z_{ij} \leq y_i \quad i = 1, 2, \ldots, n; j = 1, 2, \ldots, m. \tag{3.32}$$

$$\sum_{i=1}^{n} f_i y_i \leq \text{BC} \tag{3.33}$$

$$\sum_{i=1}^{n} r_i y_i \leq F_M \tag{3.34}$$

$$\sum_{j=1}^{m} z_{ij}\left(\text{Tp}_{ij} + \text{ts}_{ij}\right) \leq T_i; \quad i = 1, 2, \ldots, n \tag{3.35}$$

Inserting Tp_{ij} from Eq. (3.17) results in:

$$\left(\frac{\sum_{j=1}^{m} z_{ij}\text{ts}_{ij}}{1 - \sum_{j=1}^{m} \frac{zD_j}{(1-x_{ij})P_{ij}}} \right) \leq T_i \quad i = 1, 2, \ldots, n \tag{3.36}$$

Equation (3.30) shows the constraint that every item must only be allocated to a machine, where α_{ij} permits jth item to be assigned to machine i such that Eq. (3.32) shows the constraint that every item can be produced by a machine if and only if the machine is utilized. In the proposed inventory model, the capital required for machines utilization must be smaller than or equal to its maximum available budget (see Eq. 3.33). The space required for all the machines must be smaller than or equal to its maximum available space (see Eq. 3.34). According to Eq. (3.35), the sum of the production and setup times for all items produced by ith machine cannot be greater than the common cycle length of ith machine, T_i.

In short, the mathematical formulation of the MINLP problem that minimizes the total inventory system cost under constraints is (Nobil et al. 2016):

$$\text{Min TC} = \sum_{i=1}^{n} y_i f_i + \sum_{i=1}^{n}$$
$$\times \sum_{j=1}^{m} z_{ij} \left[\frac{K_{ij}}{T_i} + \frac{h_j D_j}{2(1-x_{ij})} \left(1 - x_{ij} - \frac{D_j}{P_{ij}}\right) T_i + \frac{C_{ij}D_j + C_{d_j}x_{ij}D_j}{(1-x_{ij})} \right] \tag{3.37}$$

s.t.

3.2 No Shortage

$$\sum_{i=1}^{n} \alpha_{ij} z_{ij} = 1, \quad j = 1, 2, \ldots, m \tag{3.38}$$

$$x_{ij} \leq y_i, \quad i = 1, 2, \ldots, n; j = 1, 2, \ldots, m. \tag{3.39}$$

$$\sum_{i=1}^{n} f_i y_i \leq \text{BC} \tag{3.40}$$

$$\sum_{i=1}^{n} r_i y_i \leq F_M \tag{3.41}$$

$$\left(\frac{\sum_{j=1}^{m} z_{ij} \text{ts}_{ij}}{1 - \sum_{j=1}^{m} \frac{z D_j}{(1 - x_{ij}) P_{ij}}} \right) \leq T_i \quad i = 1, 2, \ldots, n \tag{3.42}$$

$$T_i > 0 \quad i = 1, 2, \ldots, n \tag{3.43}$$

$$y_i, x_{ij} \in \{0, 1\} \quad i = 1, 2, \ldots, n; j = 1, 2, \ldots, m \tag{3.44}$$

Example 3.2 Nobil et al. (2016) presented several examples solved using three different solution procedures and an optimization software. Input parameters of these problems are chosen randomly from Table 3.3, and results of three solution approaches are represented in Table 3.4. Each cell of Table 3.5 is an average of ten solutions of each problem with different sizes.

It is worth mentioning that in the proposed HGA, by combining derivatives method and GA optimal T_i using derivative based on the randomly generated y_i and x_{ij} are obtained. The conventional GA is utilized to evaluate efficiency of proposed HGA in large-scale problems. With regard to obtained results represented in Table 3.4, the proposed HGA finds the optimal solution acceptable for small-sized problems. Whereas the solution of the proposed HGA for medium-sized problems in all iterations is the same, it can be concluded that these solutions are near optimal solutions. Moreover, the proposed HGA has significant efficiency in comparison with conventional GA, because it finds better solutions with less computational effort. Finally, to evaluate the quality of HGA solutions for small-sized problems, the ten problems were solved using DICOPT, a GAMS solver for MINLP models. To do so, it was considered the case that there are no limitations for machines to produce the items. According to Table 3.5, it can be said this solver obtains a poor feasible solution in comparison with proposed HGA (Nobil et al. 2016).

Table 3.3 Input parameters of MINLP problem (Nobil et al. 2016)

$P_{ij} \sim U(15{,}000, 25{,}000)$; $x_{ij} \sim U(0.001, 0.007)$; $K_{ij} \sim U(100, 300)$; $C_{ij} \sim U(40, 70)$; $S_{ij} \sim U(0.03, 0.08)$; $r_i \sim U(5, 20)$; $BC \sim U(500{,}000, 800{,}000)$; $F_M \sim U(4000, 90{,}000)$; $f_i \sim U(120{,}000, 220{,}000)$; $D_j \sim U(1000, 3000)$; $h_j \sim U(10, 20)$; $d_j \sim U(20, 40)$

3.2.2 Discrete Delivery

In this section, EPQ models with scrap items and discrete delivery policy are considered where shortage is not permitted.

3.2.2.1 Multi-delivery

Chiu et al. (2011) employed a mathematical modeling and algebraic approach to derive the optimal manufacturing batch size and number of shipment for a vendor–buyer integrated economic production quantity (EPQ) model with scrap. Chiu et al. (2011) assume there is an x portion of defective items produced randomly at a production rate d during regular production time. All produced items are screened, and inspection cost per item is included in the unit production cost C. All non-conforming items are assumed to be scrap and will be discarded at the end of production. Under regular supply (not allowing shortages), the constant production rate P must be larger than the sum of demand rate D and production rate of scrap items d. That is, $(P - d - D) > 0$. The production rate of scrap items d can be expressed as $d = Px$.

A multi-delivery policy is considered in this study, and it is also assumed that the finished items can only be delivered to customers if the whole lot is quality assured at the end of production process. Fixed-quantity n installments of finished batch are delivered to customers at a fixed interval of time during the production downtime t_2 (see Fig. 3.5).

Some notations which are specifically used to model this problem are shown in Table 3.6.

$TC(Q, n)$, the total production–inventory–delivery costs per cycle consists of (1) setup cost, (2) variable production costs, (3) variable scrap disposal costs, (4) fixed delivery cost, (5) variable delivery costs, (6) variable holding costs at the supplier side for all items produced (defective and perfect-quality items) in t_1 and all items waiting to be delivered in t_2, and (7) holding cost for finished goods stocked at customer's end. Therefore, $TC(Q, n)$ is (Chiu et al. 2011):

3.2 No Shortage

Table 3.4 Comparison of algorithm for two measures (Nobil et al. 2016)

	Size $n \times m$	Proposed HGA Total cost	CPU time (s)	GA Total cost	CPU time (s)	Enumeration Total cost	CPU time (s)
1	2 × 5 (S)	721,778.12	34.78	722,412.07	52.13	721,778.12	9.48
2	3 × 5 (S)	812,287.30	35.51	813,145.36	67.51	812,287.30	28.25
3	4 × 6 (S)	1,916,984.20	41.23	1,920,841.59	99.12	1,916,984.20	56.12
4	4 × 8 (S)	2,189,670.23	49.22	2,193,949.74	130.97	2,189,670.23	99
5	5 × 10 (S)	2,931,095.37	62.80	3,075,214.55	201.26	2,931,095.37	250.98
6	6 × 15 (M)	3,854,269.34	90.63	4,097,709.70	270.45	Complex	More than 500
7	7 × 20 (M)	4,649,783.06	147.97	4,973,744.20	369.60	Complex	
8	10 × 20 (L)	5,072,974.86	220.82	5,404,672.06	506.25	Complex	
9	15 × 30 (L)	7,454,053.68	602.63	8,001,922.28	826.34	Complex	
10	20 × 40 (L)	9,370,322.75	1020.63	10,198,622.85	1573.59	Complex	

Table 3.5 Comparison of HGA and GAMS software (Nobil et al. 2016)

	Size $n \times m$	GAMS software Total cost	CPU time (s)	HGA Total cost	CPU time (s)
1	2×5	427,493.95	0.313	418,114.23	31.78
2	3×5	503,550.26	0.706	472,946.93	33.25
3	5×10	2,971,438.38	0.986	2,541,907	60.98

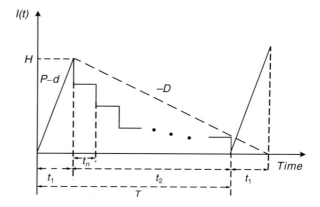

Fig. 3.5 On-hand inventory of perfect-quality items in the proposed EPQ model with scrap and a multiple shipment policy (Chiu et al. 2011)

Table 3.6 Notation of given problem (Chiu et al. 2011)

n	Number of fixed-quantity installments of the finished batch to be delivered to customers, a decision variable to be determined for each cycle
t_1	The production uptime (time)
t_2	Time required for delivering all finished products (time)
t_n	A fixed interval of time between each installment of finished products delivered during production downtime t_2 (time)
h_2	Holding cost for finished goods stocked at customer's end per unit per unit of time ($/unit/unit of time)

$$TC(Q,n) = K + CQ + C_d(xQ) + nK_S + C_T[Q(1-x)]$$
$$+ h\left[\frac{H + dt_1}{2}(t_1) + \left(\frac{n-1}{2n}\right)Ht_2\right] + \frac{h_2}{2}\left[\frac{H}{n}t_2 + T(H - Dt_2)\right]$$
(3.45)

Figure 3.6 shows supplier's inventory holding during delivery time t_2. The variable holding costs for finished products kept by the supplier in delivery time t_2 are (Chiu et al. 2011):

1. When $n = 1$, total holding cost in delivery time $= 0$.
2. When $n = 2$, total holding costs in delivery time become (see Fig. 3.7):

3.2 No Shortage

Fig. 3.6 On-hand inventory of the finished items kept by supplier during t_2 (Chiu et al. 2011)

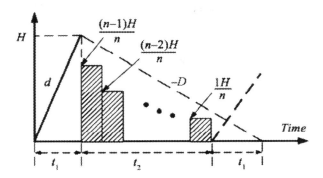

$$h\left(\frac{H}{2} \times \frac{t_2}{2}\right) = h\left(\frac{1}{2^2}\right) H t_2 \qquad (3.46)$$

3. When $n = 3$, total holding costs in delivery time are:

$$h\left(\frac{2H}{3} \times \frac{t_2}{3} + \frac{1H}{3} \times \frac{t_2}{3}\right) = h\left(\frac{2+1}{3^2}\right) H t_2 \qquad (3.47)$$

4. When $n = 4$, total holding costs in delivery time become:

$$h\left(\frac{3H}{4} \times \frac{t_2}{4} + \frac{2H}{4} \times \frac{t_2}{4} + \frac{1H}{4} \times \frac{t_2}{4}\right) = h\left(\frac{3+2+1}{4^2}\right) H t_2 \qquad (3.48)$$

Therefore, the following general term for total holding costs during t_2 can be obtained (as shown in Eq. (3.45)):

$$h\left(\frac{1}{n^2}\right)\left(\sum_{i=1}^{n-1} i\right) H t_2 = h\left(\frac{1}{n^2}\right)\left(\frac{n(n-1)}{2}\right) H t_2 = h\left(\frac{n-1}{2n}\right) H t_2 \qquad (3.49)$$

Taking randomness of scrap rate into consideration and employing the expected values of it, and with further derivations, the long-run average costs per unit time for the proposed EPQ model can be derived as follows (refer to a similar derivation procedure in Chiu et al. 2009):

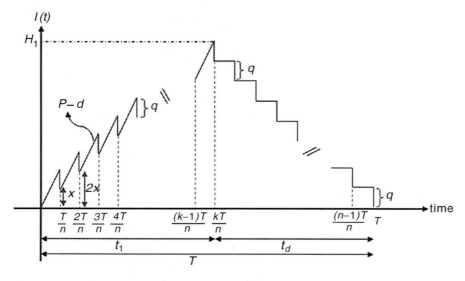

Fig. 3.7 The vendor's on-hand inventory of perfect-quality items when machine breakdown does not occur (Taleizadeh et al. 2017)

$$E[TCU(Q,n)] = \frac{E[TC(Q,n)]}{E(T)}$$
$$= \left[\frac{CD}{1-E(x)} + C_d \frac{E(x)}{1-E(x)}\right] + \frac{(K+nK_S)D}{Q}\frac{1}{1-E(x)} + C_T D$$
$$+ \frac{hQD}{2P} \times \frac{1}{1-E(x)} + \left(\frac{n-1}{n}\right)\left[\frac{hQ(1-E(x))}{2} - \frac{hQD}{2P}\right]$$
$$+ \frac{h_2 Q}{2}\left[\left(\frac{1}{n}\right)(1-E(x)) + \left(\frac{n-1}{n}\right)\frac{D}{P}\right]$$

(3.50)

This study employs algebraic approach to derive the optimal production–shipment policies, instead of using differential calculus on $E[TCU(Q, n)]$ with the need of proving its optimality (Grubbström and Erdem 1999; Chiu 2008; Chiu et al. 2010). In Eq. (3.50), both Q and n are decision variables; by rearranging terms in Eq. (3.50) as the constants Q^{-1}, Q, nQ^{-1}, and Qn^{-1}, one has (Chiu et al. 2011):

$$E[TCU(Q,n)] = \beta_1 + \beta_2(Q) + \beta_3(Q^{-1}) + \beta_4(nQ^{-1}) + \beta_5(Qn^{-1}) \quad (3.51)$$

where $\beta_1, \beta_2, \beta_3, \beta_4,$ and β_5 denote the following (Chiu et al. 2011):

3.2 No Shortage

$$\beta_1 = \left[\frac{CD}{1-E(x)} + C_d\frac{E(x)}{1-E(x)}\right] + C_TD \qquad (3.52)$$

$$\beta_2 = \left\{\frac{hD}{2P(1-E(x))} + \frac{h(1-E(x))}{2} - \frac{(h-h_2)D}{2P}\right\} \qquad (3.53)$$

$$\beta_3 = \frac{KD}{1-E(x)} \qquad (3.54)$$

$$\beta_4 = \frac{K_SD}{1-E(x)} \qquad (3.55)$$

$$\beta_5 = \left\{\left[\frac{D}{2P} - \frac{(1-E(x))}{2}(h-h_2)\right]\right\} \qquad (3.56)$$

With further rearrangements, Eq. (3.51) becomes (Chiu et al. 2011):

$$E[\text{TCU}(Q,n)] = \beta_1 + Q^{-1}[\beta_2Q^2 + \beta_3] + (Qn^{-1})\left[\beta_4(nQ^{-1})^2 + \beta_5\right] \qquad (3.57)$$

$$E[\text{TCU}(Q,n)] = \beta_1 + Q^{-1}\left[\left(\sqrt{\beta_2}Q\right)^2 + \left(\sqrt{\beta_3}\right)^2 - 2\left(\left(\sqrt{\beta_2}Q\right)\left(\sqrt{\beta_3}\right)\right)\right] + (Qn^{-1})$$

$$\left[\left(\sqrt{\beta_4}(nQ^{-1})\right)^2 + \left(\sqrt{\beta_5}\right)^2 - 2\sqrt{\beta_4}\sqrt{\beta_5}(nQ^{-1})\right]$$

$$+Q^{-1}\left[2\left(\sqrt{\beta_2}Q\right)\sqrt{\beta_3}\right) + (n^{-1}Q)\left[2\sqrt{\beta_4}\sqrt{\beta_5}(nQ^{-1})\right] \qquad (3.58)$$

$$E(\text{TCU}(Q,n)) = \beta_1 + Q^{-1}\left[\left(\sqrt{\beta_2}Q\right) - \left(\sqrt{\beta_3}\right)\right]^2$$

$$+(Qn^{-1})\left[\left(\sqrt{\beta_4}(nQ^{-1})\right) - \sqrt{\beta_5}\right]^2 + 2\left(\sqrt{\beta_2\beta_3} + \sqrt{\beta_4\beta_5}\right) \qquad (3.59)$$

It is noted that if the following square terms (Eqs. 3.60 and 3.61) are equal to zero, then Eq. (3.59) will be minimized (Chiu et al. 2011):

$$Q^{-1}\left[\left(\sqrt{\beta_2}Q - \left(\sqrt{\beta_3}\right)\right]^2 = 0 \qquad (3.60)$$

$$(Qn^{-1})\left[\left(\sqrt{\beta_4}(nQ^{-1})\right) - \sqrt{\beta_5}\right]^2 = 0 \qquad (3.61)$$

Or

$$Q^* = \sqrt{\frac{\beta_3}{\beta_2}} \tag{3.62}$$

And

$$n^* = \sqrt{\frac{\beta_5}{\beta_4}} Q^* \tag{3.63}$$

Substituting Eqs. (3.53) and (3.54) in Eq. (3.62), the optimal replenishment lot size Q^* can be obtained (Chiu et al. 2011):

$$Q^* = \sqrt{\frac{2KD}{\frac{hD}{P} + h(1 - E(x))^2 - \frac{D}{P}(h - h_2)(1 - E(x))}} \tag{3.64}$$

Substituting Eqs. (3.55), (3.56), and (3.64) in Eq. (3.63), the optimal number of shipments is (Chiu et al. 2011):

$$n^* = \sqrt{\frac{K(h - h_2)(1 - E(x))\left[\frac{D}{P} - (1 - E(x))\right]}{K_S\left[\frac{hD}{P} + h(1 - E(x))^2 - \frac{D}{P}(h - h_2)(1 - E(x))\right]}} \tag{3.65}$$

One notes that Eq. (3.65) is identical to what was obtained by using the conventional differential calculus method on $E[TCU(Q, n)]$ (Chiu et al. 2009). Further, from Eq. (3.51) the optimal cost function $E[TCU(Q^*, n^*)]$ is (Chiu et al. 2011):

$$E(\text{TCU}(Q^*, n^*)) = \beta_1 + 2\sqrt{\beta_2\beta_3} + 2\sqrt{\beta_4\beta_5} \tag{3.66}$$

Example 3.3 Chiu et al. (2011) considered a product with $P = 60{,}000$ and $D = 3400$ units per year. Random scrap rate follows a uniform distribution over the interval [0, 0.3]. In addition, the following values of related variables are considered: $C = \$100$ per item; $C_d = \$20$, per scrap item; $h = \$20$ per item per year; $h_2 = \$80$ per item kept at the customer's end per unit time; $K = \$20{,}000$ per production run; $K_S = \$4350$ per shipment, and $C_T = \$0.1$ per item delivered.

From Eq. (3.65), one obtains the optimal number of delivery $n^* = 3$. By plugging n back into Eq. (3.51) and resolving the algebraic solution, one finds the optimal production batch size $Q^* = 2652$. Calculating Eq. (3.66) one obtains the long-run average cost $E \times [\text{TCU}(Q^*, n)] = \$512{,}047$ (Chiu et al. 2011).

It is noted that n^* should practically be an integer number, but Eq. (3.65) gives a real number. In order to obtain the optimal integer value of n, one should compute the $E[TCU(Q, n)]$ for both integers that are adjacent to real number n^*, respectively (for instance, in this example, Eq. (3.65) gives $n = 3.1733$, so both $n = 3$ and $n = 4$

3.2 No Shortage

must be plugged in $E[TCU(Q, n)]$), and select the one with minimum cost as our optimal n^* (Chiu et al. 2011).

3.2.2.2 Random Machine Breakdown

Taleizadeh et al. (2017) developed an integrated inventory model to determine the optimal lot size and production uptime while considering stochastic machine breakdown and multiple shipments for a single buyer and single vendor. The proposed model considers a manufacturing system that generates imperfect products such that x percent of manufactured products are defective and that portion is generated randomly at a production rate of $d = P \cdot x$. The defective products are not repairable, and at the end of the production uptime, they will be discarded. The constant production rate of items, P, is greater than the annual demand rate D. They assumed that stochastic machine breakdown can occur, and the number of breakdowns in a year is a random variable, λ, that follows the Poisson probability distribution function. When a breakdown occurs, the production system follows the NR policy. In other words, immediately after a breakdown, the repair will be done, and production will not be started until all of the on-hand inventory is depleted. They assumed the machine repair time to be constant, and to prevent shortages, they considered safety stock. In this manufacturing system, both batch quantity and the distance between two shipments are identical. Also, the cost of transportation is paid by the buyer.

Some notations which are specifically used to model this problem are shown in Table 3.7.

t_1 is the production uptime, and t denotes the time before a breakdown occurs. They investigated two cases, $t < t_1$ and, $t \geq t_1$ separately, because machine breakdowns can occur randomly during the production uptime.

The First Case—$t \geq t_1$

In this case, a machine breakdown does not occur during the production uptime. For this case, the on-hand inventories of perfect and defective vendor items are shown in Figs. 3.10 and 3.11, respectively. Moreover, Fig. 3.7 represents the buyer's inventory level when a machine breakdown does not occur.

Vendor's Cost

According to Fig. 3.7, the setup, variable production, and disposal costs are SCV $(1) = K$, PC $(1) = CPt_1$, and dCV $(1) = C_S P t_1 x$, respectively, because x percent of all produced items are scrapped. According to Fig. 3.8, the holding cost for defective items is:

Table 3.7 Notation of given problem (Taleizadeh et al. 2017)

H_1	Maximum level of on-hand inventory when machine breakdown does not occur (unit)
H_2	Maximum level of on-hand inventory when machine breakdown occurs (unit)
K_1	Fixed ordering cost of buyer ($/order)
h_2	Holding cost of buyer ($/unit/unit time)
T	Cycle length when breakdown does not occur (time)
T'	Cycle length when breakdown occurs (time)
T_u	Cycle length for integrated case (time)
t	Production time before a random breakdown occurs (time)
t_d	Time required to deplete all available perfect-quality items when machine breakdown does not occur (time)
t_d'	Time required to deplete all available perfect-quality items when machine breakdown occurs (time)
t_1	Production uptime when a breakdown does not occur (time)
TC(t, Q)	Total inventory costs per cycle when machine breakdown occurs ($)
TC(t_1, Q)	Total inventory costs per cycle when machine breakdown does not occur ($)
TCU(t_1, Q)	Total inventory costs per unit time for integrated case ($)

$$\text{HdCV}(1) = \frac{h_1 d(t_1)^2}{2} \tag{3.67}$$

The holding cost of the safety stock during each cycle is:

$$\text{HSCV}(1) = hDt_r T \tag{3.68}$$

To calculate the holding cost of the perfect items, the calculation of number of perfect products per shipment is needed. Thus, the number of perfect items is calculated and added. To determine the holding cost of perfect items, it first requires determining the average inventory of perfect items in the production uptime which is as follows:

$$\begin{aligned} \text{AIPU} &= \frac{(Q+x)T}{2n} + \frac{(Q+x+2x)T}{2n} + \cdots + \frac{(Q+(2k-1)x)T}{2n} \\ &= \frac{kQT}{2n} + \frac{xT}{2n}(1+3+5+\cdots+(2k-1)) \\ &= \frac{kQT}{2n} + \frac{xT}{2n}\left(\frac{k(1+2k-1)}{2}\right) = \frac{kQT}{2n} + \frac{k^2 xT}{2n} \end{aligned} \tag{3.69}$$

The average inventory in the production downtime is:

3.2 No Shortage

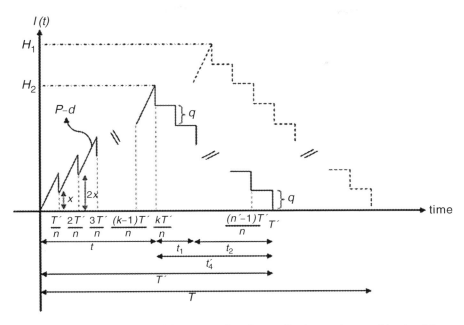

Fig. 3.10 The vendor's on-hand inventory of perfect-quality items when machine breakdown occurs (Taleizadeh et al. 2017)

Fig. 3.11 The vendor's on-hand inventory of defective items when machine breakdown occurs (Taleizadeh et al. 2017)

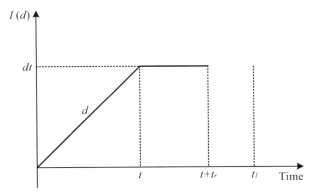

$$\text{AIPD} = \frac{kxT}{n} + \frac{(kx-Q)T}{n} + \frac{(kx-2Q)T}{n} \cdots + \frac{QT}{n} = \frac{(n-k)}{2n}(Q+kx)T \quad (3.70)$$

According to Eqs. (3.69) and (3.70), the average inventory of perfect items is (Taleizadeh et al. 2017):

Fig. 3.8 The vendor's on-hand inventory of defective items when machine breakdown does not occur (Taleizadeh et al. 2017)

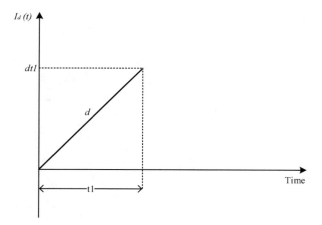

$$\begin{aligned} \text{AIPI} &= \frac{kQT}{2n} + \frac{k^2xT}{2n} + (n-k)\left[\frac{QT}{2n} + \frac{kxT}{2n}\right] \\ &= \left(\frac{QT}{2n} + \frac{kxT}{2n}\right)(k+n-k) = \frac{QT}{2} + \frac{kxT}{2} \end{aligned} \quad (3.71)$$

Figure 3.7 shows that:

$$x = \frac{(P-d)T}{n} - Q \quad (3.72)$$

$$T = \frac{nQ}{D} \quad (3.73)$$

Using Eqs. (3.72) and (3.73) in Eq. (3.71) gives:

$$\begin{aligned} \text{HCPI} &= \frac{Q\left(\frac{nQ}{D}\right)}{2} + \frac{k\left(\frac{nQ}{D}\right)}{2}\left(\frac{(P-d)T}{n} - Q\right) \\ &= \frac{nQ^2}{2D} + \frac{nkQ}{2D}\left(\frac{(P-d)nQ}{nD} - Q\right) \\ &= \frac{nQ^2}{2D} + \frac{nkQ^2}{2D}\left(\frac{(P-d)}{D} - 1\right) \\ &= \frac{nQ^2}{2D}\left[1 + k\left(\frac{(P-d)}{D} - 1\right)\right] \end{aligned} \quad (3.74)$$

According to Figs. 3.7 and 3.8,

3.2 No Shortage

$$T = \frac{(P-d)t_1}{D} \Rightarrow \frac{T}{t_1} = \frac{(P-d)}{D} = \frac{n}{k} \Rightarrow n = \frac{k(P-d)}{D} \quad (3.75)$$

Replacing n in Eq. (3.74) with $\frac{k(P-d)}{D}$, as shown in Eq. (3.75), gives the average inventory as: $\frac{nQ^2}{2D}[1+n-k]$.

Thus, the holding cost of perfect items is:

$$\text{HCPI} = h\left[\frac{nQ^2}{2D}(1+n-k)\right] \quad (3.76)$$

Therefore, the vendor's total cost is (Taleizadeh et al. 2017):

$$\text{TCV}(t_1, Q) = K + CPt_1 + C_S Pt_1 x + \frac{h_1 d(t_1)^2}{2} + hDt_r T + \frac{hnq^2}{2D}$$
$$\times (1+n-k) \quad (3.77)$$

Buyer's Cost

According to Figs. 3.7 and 3.9, the transportation and fixed ordering costs are SCB $(1) = nC_t$ and OCB $(1) = K_1$, respectively. Moreover, the holding cost is:

$$\text{HCB}(1) = h_2\left(\frac{QT}{2n}n\right) = h_2 \frac{QT}{2} = h_2 \frac{nQ^2}{2D} \quad (3.78)$$

Therefore, the buyers total cost is (Taleizadeh et al. 2017):

$$\text{TCB}(t_1, Q) = K_1 + nC_T + h_2 \frac{nQ^2}{2D} \quad (3.79)$$

From Eqs. (3.77) and (3.79), the total cost is (Taleizadeh et al. 2017):

$$\text{TC}(t_1, Q) = K + K_1 + nC_T + CPt_1 + C_S Pt_1 x + \frac{h_1 dt_1^2}{2}$$
$$+ hDt_r T + \frac{hnQ^2}{2D}(1+n-k) + h_2 \frac{nQ^2}{2D} \quad (3.80)$$

According to Figs. 3.7 and 3.9,

$$T = \frac{(P-d)t_1}{D} \quad (3.81)$$

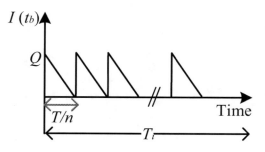

Fig. 3.9 The buyer's inventory level when machine breakdown does not occur (Taleizadeh et al. 2017)

$$k = \frac{Dt_1}{Q} \tag{3.82}$$

$$n = \frac{(P-d)t_1}{Q} \tag{3.83}$$

As the defective rate x, is a random variable with a known probability density function, its expected value can be used. Using all related parameters from Eqs. (3.80) to (3.83), the expected production–inventory cost per cycle, $E[TC(t_1, Q)]$, is (Taleizadeh et al. 2017):

$$\begin{aligned} E[TC(t_1, Q)] &= K + K_1 + CPt_1 + C_S Pt_1 E(x) + \frac{C_T(P-d)t_1}{q} + \frac{h_1 PE(x)t_1^2}{2} \\ &+ hDg\left(\frac{(P-d)t_1}{Q}\right) + h_2 \frac{nQ^2}{2D} \times \frac{(P-d)t_1}{Q} \\ &+ \frac{hQ^2}{2D}\left(\frac{(P-d)t_1}{Q}\right)\left(1 + \frac{(P-d)t_1}{Q} - \right)\frac{Dt_1}{Q} \end{aligned} \tag{3.84}$$

This can be simplified to:

$$\begin{aligned} E[TC(t_1, Q)] &= Pt_1 \left\{ C + C_S E(x) + \frac{C_T(1-E(x))}{Q} + hg(1-E(x)) + \frac{hQ(1-E(x))}{2D} \right. \\ &\left. + \frac{h_2 Q(1-E(x))}{2D} \right\} + \left\{ \frac{h_1 PE(x)}{2} + \frac{hP^2(1-E(x))^2}{2D} - \frac{hP(1-E(x))}{2} \right\} t_1^2 + \{K + K_1\} \end{aligned} \tag{3.85}$$

where $t_r = g$ is the fixed repair time.

3.2 No Shortage

The Second Case—$t < t_1$

In this case, a machine failure occurs during the production uptime, and the NR policy is assumed. When a machine breakdown occurs, the machine will immediately be repaired, and production will only restart when the inventory level reaches zero. The vendor's on-hand inventories of perfect-quality and defective items are shown in Figs. 3.10 and 3.11, respectively. Moreover, Fig. 3.12 depicts the buyer's inventory level in case of a breakdown.

Vendor's Cost

According to Fig. 3.10, the setup, variable production, and disposal costs are SCV (2) $= K$, PC (2) $= CPt_1$, and dCV (1) $= C_s Ptx$, respectively. According to Fig. 3.11, the holding cost of defective items is:

$$\text{HdCV}(2) = \frac{h_1 dt_1^2}{2} + h_1 t dt_r \tag{3.86}$$

And the holding cost of safety stock during each cycle is:

$$\text{HSCV}(2) = hDt_r T' \tag{3.87}$$

Similar to holding cost of safety stock during each cycle is:

$$\text{HCPI} = h \left[\frac{n'Q^2}{2D} (1 + n' - k') \right] \tag{3.88}$$

Moreover, the machine repair cost is assumed to be M, Therefore, the vendor's total cost is:

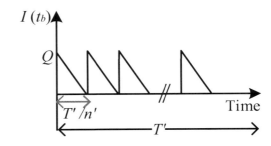

Fig. 3.12 The buyer's inventory level when machine breakdown occurs (Taleizadeh et al. 2017)

$$\mathrm{TCV}(t,Q) = K + C_M + CPt + C_S Ptx + \frac{h_1 dt^2}{2} + hDt_r T' + \frac{hn'Q^2}{2D}$$
$$\times (1 + n' - k') + h_1 t dt_r \qquad (3.89)$$

Buyer's Cost

According to Figs. 3.10 and 3.12, the transportation and fixed ordering costs are $SCB\,(2) = nC_T$ and $OCB\,(2) = K_1$, respectively. Moreover, the holding cost is:

$$\mathrm{HCB}(2) = h_2 \left(\frac{QT'}{2n'} n' \right) = h_2 \frac{QT'}{2} = h_2 \frac{n'Q^2}{2D} \qquad (3.90)$$

Therefore, the buyers total cost is:

$$\mathrm{TCB}(t,Q) = K_1 + n'C_T + h_2 \frac{n'Q^2}{2D} \qquad (3.91)$$

From Eqs. (3.89) and (3.90), the total cost is:

$$\mathrm{TC}(t,Q) = K + K_1 + C_M + CPt + C_S Ptx + \frac{h_1 dt^2}{2} + h_1 t dt_r$$
$$+ hDt_r T' + \frac{hn'Q^2}{2D}(1 + n' - k') + h_2 \frac{n'Q^2}{2D} + n'C_T \qquad (3.92)$$

According to Figs. 3.10 and 3.12,

$$T' = \frac{(P-d)t}{D} \qquad (3.93)$$

$$k' = \frac{Dt}{Q} \qquad (3.94)$$

$$n' = \frac{(P-d)t}{Q} \qquad (3.95)$$

$$T' = n'\frac{T}{n} \Rightarrow \frac{T'}{n'} = \frac{T}{n} \qquad (3.96)$$

As the defective rate, x, is a random variable with a known probability density function, this expected value can be used. Using all related parameters from Eqs. (3.93) to (3.96), the expected production–inventory cost per cycle, $E[\mathrm{TC}(t,Q)]$, is (Taleizadeh et al. 2017):

3.2 No Shortage

$$E[TC(t,Q)] = K + K_1 + C_M + Pt^2 \left\{ \frac{h_1 E(x)}{2} + \frac{hP(1-E(x))^2}{2D} - \frac{h(1-E(x))}{2} \right\}$$
$$+ \left\{ C + C_S E(x) + h_1 g E(x) + \left(hg + \frac{C_T}{PQ} + \frac{hQ}{2D} + \frac{h_2 Q}{2D} \right) (1-E(x)) \right\} Pt \quad (3.97)$$

where $t_r = g$ is the fixed repair time.

Now, Eqs. (3.85) and (3.97) can be rewritten as (Taleizadeh et al. 2017):

$$E[TC(t_1,Q)] = (K+K_1) + S_1 t_1 + S_2 t_1^2 \quad (3.98)$$

$$E[TC(t,Q)] = (K + K_1 + M) + (S_1 + h_1 g P E(x))t + S_2 t^2 \quad (3.99)$$

where

$$S_1 = CP + C_S PE(x) + \frac{C_T P(1-E(x))}{Q} + hgP(1-E(x))$$
$$+ \frac{hQP(1-E(x))}{2D} + \frac{h_2 Q(1-E(x))}{2D} \quad (3.100)$$

$$S_2 = \frac{h_1 PE(x)}{2} + \frac{hP^2(1-E(x))^2}{2D} - \frac{hP(1-E(x))}{2} \quad (3.101)$$

Integrating the EPQ Models with and Without Breakdowns

Because the defective rate and the number of breakdowns are random variables, the cycle length of the proposed model is not constant. Thus, the expected production–inventory cost per unit time, $E[TCU(t_1, Q)]$, can be obtained as (Taleizadeh et al. 2017):

$$E[TCU(t_1,Q)] = \frac{\left\{ \int_0^{t_1} E[TC(t,Q)] f(t) dt + \int_t^{\infty} E[TC(t_1,Q)] f(t) dt \right\}}{E(T_U)} \quad (3.102)$$

and

$$E(T_U) = \int_0^{t_1} E(T') f(t) dt + \int_t^{\infty} E(T) f(t) dt \quad (3.103)$$

The authors assume that the number of breakdowns per unit time is a random variable that follows a Poisson probability distribution function. Therefore, the time between two breakdowns follows the exponential probability distribution function with parameter λ. According to Eq. (3.103) (Taleizadeh et al. 2017),

$$E(T_U) = \int_0^{t_1} \frac{P(1-E(x))t}{D} f(t)dt + \int_{t_1}^{\infty} \frac{P(1-E(x))t_1}{D} f(t)dt \qquad (3.104)$$

$$\text{Given that } \int_0^{t_1} f(t)dt = F(t_1) = 1 - e^{-\lambda t_1} \qquad (3.105)$$

Then,

$$\int_0^{t_1} t \cdot f(t)dt = -t_1 e^{-\lambda t_1} - \frac{1}{\lambda} e^{-\lambda t_1} + \frac{1}{\lambda} \qquad (3.106)$$

Substituting Eqs. (3.105) and (3.106) into Eq. (3.104) gives (Taleizadeh et al. 2017)

$$\begin{aligned} E(T_U) &= \frac{P(1-E(x))}{D} \left\{ \int_0^{t_1} tf(t)dt + \int_{t_1}^{\infty} t_1 f(t)dt \right\} \\ &= \frac{P(1-E(x))}{D} \left\{ -t_1 e^{-\lambda t_1} - \frac{1}{\lambda} e^{-\lambda t_1} + \frac{1}{\lambda} + t_1 e^{-\lambda t_1} \right\} \\ &= \frac{P(1-E(x))}{\lambda D} \left(1 - e^{-\lambda t_1}\right) \end{aligned} \qquad (3.107)$$

Using $E[\text{TCU}(t_1, Q)]$, $E[\text{TCU}(t, Q)]$ and $E[T(U)]$ (shown in Eqs. (3.85), (3.97), and (3.107), respectively), the expected production–inventory cost per unit time becomes (see Appendix A of Taleizadeh et al. (2017) for detailed calculation):

$$\begin{aligned} E[\text{TCU}(t_1, Q)] = & \frac{(K+K_1)}{P(1-E(x))(1-e^{-\beta t_1})} + \frac{S_1 D}{P(1-E(x))} + \frac{h_1 g P E(x) D}{P(1-E(x))} \\ & + \frac{2S_2 D}{\lambda P(1-E(x))} + \frac{\lambda C_M D}{P(1-E(x))} - \frac{h_1 g P E(x) \lambda D t_1 e^{-\lambda t_1}}{P(1-E(x))(1-e^{-\lambda t_1})} \\ & - \frac{2S_2 D t_1 e^{-\lambda t_1}}{P(1-E(x))(1-e^{-\lambda t_1})} \end{aligned} \qquad (3.108)$$

To prove the convexity of $E[\text{TCU}(t_1, Q)]$, Taleizadeh et al. (2017) used the Hessian matrix to obtain the following derivatives:

3.2 No Shortage

$$\frac{\partial E[\text{TCU}(t_1, Q)]}{\partial t_1} = \frac{-De^{-\lambda t_1}}{(1-e^{-\lambda t_1})^2}\left[\frac{(K+K_1)\lambda^2}{P(1-E(x))} + \left(1-e^{-\lambda t_1}-\lambda t_1\right)\right.$$
$$\left.\times \left\{\frac{h_1 g P E(x)\lambda + 2S_2}{P(1-E(x))}\right\}\right] \quad (3.109)$$

$$\frac{\partial^2 E[\text{TCU}(t_1, Q)]}{\partial t_1^2} = \frac{-\lambda e^{-\lambda t_1} D}{P(1-E(x))}\left[(K+K_1)\lambda^2 \frac{(1-e^{-\lambda t_1})}{(1-e^{-\lambda t_1})^3}\right.$$
$$h_1 g P E(x) B + 2S_2 \left\{\frac{2(1-e^{-\lambda t_1}) - \beta t_1(1+e^{-\lambda t_1})}{(1-e^{-\lambda t_1})^3}\right\} \quad (3.110)$$

$$\frac{\partial^2 E[\text{TCU}(t_1, Q)]}{\partial t_1 \partial Q} = \frac{\partial^2 E[\text{TCU}(t_1, Q)]}{\partial Q \partial t_1} = 0 \quad (3.111)$$

$$\frac{\partial E[\text{TCU}(t_1, Q)]}{\partial Q} = \frac{D}{P(1-E(x))}\left\{-\frac{C_T P(1-E(x))}{Q^2} + \frac{hP(1-E(x))}{2D} + \frac{h_2 P(1-E(x))}{2D}\right\}$$
$$= -\frac{C_T D}{Q^2} + \frac{h}{2} + \frac{h_2}{2}$$
$$(3.112)$$

$$\frac{\partial^2 E[\text{TCU}(t_1, Q)]}{\partial Q^2} = \frac{2C_T D}{Q^3} \quad (3.113)$$

After substituting Eqs. (3.109), (3.110), (3.112), and (3.113) into the following Hessian matrix and making some simplifications:

$$\begin{bmatrix} t_1 & Q \end{bmatrix} \begin{pmatrix} \dfrac{\partial^2 E[\text{TCU}(t_1, Q)]}{\partial t_1^2} & \dfrac{\partial^2 E[\text{TCU}(t_1, Q)]}{\partial Q \partial t_1} \\ \dfrac{\partial^2 E[\text{TCU}(t_1, Q)]}{\partial t_1 \partial Q} & \dfrac{\partial^2 E[\text{TCU}(t_1, Q)]}{\partial Q^2} \end{pmatrix} \begin{bmatrix} t_1 \\ Q \end{bmatrix}$$
$$= \frac{\lambda e^{-\lambda t_1} D}{P(1-E(x))(1-e^{-\lambda t_1})^3}\left\{(K+K_1)\lambda^2 \left(1+e^{-\lambda t_1}\right)\right.$$
$$\left.+(h_1 g P E(x)\lambda + 2S_2)\left[2(1-e^{-\lambda t_1}) - \lambda t_1(1+e^{-\lambda t_1})\right]\right\} \quad (3.114)$$

From Eq. (3.114), because $\frac{\lambda e^{-\lambda t_1} D}{P(1-E(x))(1-e^{-\lambda t_1})^3}$ is greater than zero, $E[\text{TCU}(t_1, Q)]$ will be convex if and only if (Taleizadeh et al. 2017):

$$(K+K_1)\lambda^2 \left(1+e^{-\lambda t_1}\right) + (h_1 g P E(x)\lambda + 2S_2)$$
$$\times \left[2(1-e^{-\lambda t_1}) - \lambda t_1(1+e^{-\lambda t_1})\right]$$
$$> 0 \quad (3.115)$$

which can be simplified to:

$$t_1 \leq \frac{(K+K_1)\lambda}{h_1 gPE(x)\lambda + 2S_2} + \frac{2(1-e^{-\lambda t_1})}{\lambda(1+e^{-\lambda t_1})} \tag{3.116}$$

Hence, $E[TCU(t_1, q)]$ is convex if and only if (Taleizadeh et al. 2017):

$$0 \leq t_1 \leq \frac{(K+K_1)\lambda}{h_1 gPE(x)\lambda + 2S_2} + \frac{2(1-e^{-\lambda t_1})}{\lambda(1+e^{-\lambda t_1})} = w(t_1) \tag{3.117}$$

To determine the optimal values of t_1 and q, the first derivatives of $E[TCU(t_1, Q)]$ with respect to t_1 and Q should be made equal to zero, which gives (Taleizadeh et al. 2017):

$$\frac{\partial E[TCU(t_1, Q)]}{\partial Q} = 0 \Rightarrow -\frac{C_T D}{Q^2} + \frac{h}{2} + \frac{h_2}{2} = 0 \Rightarrow Q^* = \sqrt{\frac{2C_T D}{h+h_2}} \tag{3.118}$$

$$\frac{\partial E[TCU(t_1, Q)]}{\partial t_1} = \frac{-De^{-\lambda t_1}}{(1-e^{-\lambda t_1})^2 P(1-E(x))} \left[(K+K_1)\lambda^2 \right.$$
$$\left. + (1-e^{-\lambda t_1} - \lambda t_1)\{h_1 gPE(x)\lambda + 2S_2\} \right] = 0 \tag{3.119}$$

Since $\frac{-De^{-\lambda t_1}}{(1-e^{-\lambda t_1})^2 P(1-E(x))}$ is greater than zero, Eq. (3.119) can be rewritten as (Taleizadeh et al. 2017):

$$-(K+K_1)\lambda^2 + \left(e^{-\lambda t_1 -1} + \lambda t_1\right)\{h_1 gPE(x)\lambda + 2S_2\} = 0$$
$$\Rightarrow e^{-\lambda t_1 -1} + \lambda t_1 = \frac{(K+K_1)\lambda^2 + h_1 gPE(x)\lambda + 2S_2}{h_1 gPE(x)\lambda + 2S_2} = \frac{(K+K_1)\lambda^2}{h_1 gPE(x)\lambda + 2S_2} + 1$$
$$\tag{3.120}$$

Assuming $y = \frac{(K+K_1)\lambda^2}{h_1 gPE(x)\lambda + 2S_2}$

$$e^{-\lambda t_1 -1} + \lambda t_1 = \lambda^2 y + 1 \tag{3.121}$$

To find the optimal run time, it can be supposed that:

$$t_{1L}^* = \sqrt{2y} = \sqrt{\frac{2(K+K_1)}{h_1 gPE(x)\lambda + h_1 PE(x) + \frac{hP(1-E(x))^2}{D} - hP(1-E(x))}} \tag{3.122}$$

$$t_{1U}^* = \frac{\lambda y + \sqrt{\lambda^2 y^2 + \lambda y}}{2} \tag{3.123}$$

3.2 No Shortage

Theorem 3.1 The optimal run time must follow the relation $t_{1L}^* < t_1^* < t_{1U}^*$.

Proof It is proved in Appendix B of Taleizadeh et al. (2017).

Theorem 3.2 The total cost function TCU(t_1, Q) is convex.

Proof According to Eq. (3.117), $E[\text{TCU}(t_1, Q)]$ is convex if and only if $0 \leq t_1 \leq \frac{(K+K_1)\lambda}{h_1 g P E(x)\lambda + 2S_2} + \frac{2(1-e^{-\lambda t_1})}{\lambda(1+e^{-\lambda t_1})} = w(t_1)$ because both λ and t are positive, and $1 \leq (1 + e^{-\lambda t_1}) \leq 2$. Thus (Taleizadeh et al. 2017),

$$w(t_1) > \frac{(K+K_1)\lambda}{h_1 g P E(x)\lambda + 2S_2} + \frac{(1-e^{-\lambda t_1})}{\lambda} \tag{3.124}$$

if $v = \frac{(K+K_1)}{h_1 g P E(x)\lambda + 2S_2}$, Eq. (3.124) becomes (Taleizadeh et al. 2017)

$$w(t_1) = v\lambda + \frac{(1-e^{-\lambda t_1})}{\lambda} \tag{3.125}$$

Also, given that $\frac{\partial E[\text{TCU}(t_1, Q)]}{\partial t_1} = 0$,

$$e^{-\lambda t_1} + \lambda t_1 = \frac{(K+K_1)\lambda^2}{h_1 g P E(x)\lambda + 2S_2} + 1 \tag{3.126}$$

Or $\lambda t_1 = v\lambda^2 + (1 - e^{-\lambda t_1}) \Rightarrow t_1 = v\lambda + \frac{(1-e^{-\lambda t_1})}{\lambda}$ (3.127)

Combining Eqs. (3.126) and (3.127) gives (Taleizadeh et al. 2017):

$$w(t_1) > v\lambda + \frac{(1-e^{-\lambda t_1})}{\lambda} = t_1 \tag{3.128}$$

Thus, the total cost function TCU(t_1, Q) is convex.

Example 3.4 Taleizadeh et al. (2017) presented an example to illustrate the applicability of our proposed model. Assume that the production and demand rates are $P = 10{,}000$ and $D = 4000$ units per year, respectively. x percent of the items produced during the production time could be defective following a uniform probability distribution function over the interval [0, 0.2]. Machine breakdowns might occur during the production uptime. The number of machine failures follows a Poisson distribution function with $\beta = 0.5$. The other parameters are as follows: $K = \$450$ per production run, $K_1 = \$150$ per shipment, $h = \$0.6$ per item per unit time, $h_1 = \$0.8$ per defective item per unit time, $h_2 = \$0.9$ per item per unit time, $C = \$2$ per item, $C_s = \$0.3$ per scrapped item, $C_T = \$0.3$ per item, $C_M = \$500$ per each breakdown, and $t_r = 0.018$ per year (Taleizadeh et al. 2017).

From Eq. (3.118), the optimal batch size for each delivery is $Q^* = 730.2967$. According to Eqs. (3.108) and (3.123), $E[\text{TCU}(t_{1L}^*, Q^*)] = \$11{,}657.06$ and $t_{1L}^* = 0.3985$. Also, from Eqs. (3.108) and (3.123), $t_{1U}^* = 0.4188$ and $E[\text{TCU}(t_{1U}^*, Q^*)] = \$11{,}656.51$.

Given that $[t_{1L}^* = 0.3985, t_{1U}^* = 0.4188] \in [0, (t_{1U}^*) = 0.4569]$, obviously the expected total cost function, $E[\text{TCU}(t_1, Q)]$, is convex, meeting the required condition. Using Newton's method and the upper and lower bounds (t_{1L}^*, t_{1U}^*) as two initial points, the optimal production uptime will be equal to $t_1^* = 0.4122$. The optimal expected total cost is $E[\text{TCU}(t_1^*, Q^*)] = \$11{,}656.35$.

3.3 Fully Backordered

3.3.1 Continuous Delivery

In this section, EPQ models with scrap which considered continuous delivery policy with shortage (Fully backordered) are presented.

3.3.1.1 Proposing an Arithmetic–Geometric Mean Inequality Method

Shyu et al. (2014) proposed an arithmetic–geometric mean inequality method to simplify the algebraic method of completing perfect square established by Huang (2006) to find the optimal solution under which the expected annual cost minimized.

First consider that the shortage is not permitted. From Chiu (2006), the expected annual cost can be expressed as:

$$\begin{aligned} E[\text{TCU}(Q)] = D&\left[\frac{C}{1-E(x)} + C_S \frac{E(x)}{1-E(x)}\right] + \frac{KD}{Q}\frac{1}{1-E(x)} + \frac{hQ}{2}\left(1-\frac{D}{P}\right)\frac{1}{1-E(x)} \\ &- hQ\left(1-\frac{D}{P}\right)\frac{E(x)}{1-E(x)} + \frac{hQ}{2}\frac{E(x^2)}{1-E(x)} \\ &= E + \frac{F}{Q} + GQ, \end{aligned}$$

(3.129)

where the constants:

$$E = D\left[\frac{F}{1-E(x)} + C_S \frac{E(x)}{1-E(x)}\right] \qquad (3.130)$$

3.3 Fully Backordered

$$F = \frac{K\lambda}{1 - E(x)} \tag{3.131}$$

and

$$G = \left(\frac{1}{1 - E(x)}\right)\left\{\frac{h}{2}\left(\left(1 - \frac{D}{P}\right) + E(x^2)\right) - h\left(1 - \frac{D}{P}\right)E(x)\right\} \tag{3.132}$$

By using the arithmetic–geometric mean inequality, it can easily be obtained that:

$$E(\text{TCU}(Q)) = E + \frac{F}{Q} + GQ \geq E + 2\sqrt{FG}. \tag{3.133}$$

when the equality

$$\frac{F}{Q} = GQ, \tag{3.134}$$

Holds, $E[\text{TCU}(Q)]$, has a minimum. Therefore,

$$Q^* = \sqrt{\frac{F}{G}} = \sqrt{\frac{2KD}{h\left(1 - \frac{D}{P}\right) - 2h\left(1 - \frac{D}{P}\right)E(x) + hE(x^2)}}. \tag{3.135}$$

Hence, the minimum value of $E[\text{TCU}(Q)]$ is as follows (Hsu and Hsu 2016):

$$\begin{aligned} E[\text{TCU}(Q^*)] &= E + 2\sqrt{FG} \\ &= D\left[\frac{F}{1 - E(x)} + C_S \frac{E(x)}{1 - E(x)}\right] + 2KD\frac{1}{1 - E(x)} \\ &\quad \times \sqrt{\frac{\frac{h}{2}\left(\left(1 - \frac{D}{P}\right) + E(x^2)\right) - h\left(1 - \frac{D}{P}\right)E(x)}{KD}} \end{aligned} \tag{3.136}$$

Now under backordering, the problem will be remodeled.
Some notations which are specifically used to model this problem are shown in Table 3.8.
The expected annual cost can be expressed as:

Table 3.8 Notation of given problem (Hsu and Hsu 2016)

t_1	Production uptime (time)
t_2	Production downtime (time)
t_3	Time shortage permitted (time)
t_4	Time needed to satisfy all the backorders by the next production (time)

$$E[TCU(Q,B)] = D\left[\frac{C}{1-E(x)} + C_S \frac{E(x)}{1-E(x)}\right] + \frac{KD}{Q}\frac{1}{1-E(x)}$$
$$+ \frac{h}{2}\left[\left(1-\frac{D}{P}\right)Q - 2B\right]\frac{1}{1-E(x)} + \frac{(C_b+h)}{1-E(x)}E\left(\frac{1-x}{1-x-\frac{D}{P}}\right)\frac{B^2}{2Q}$$
$$+ h\left[B - \left(1-\frac{D}{P}\right)Q\right]\frac{E(x)}{1-E(x)} + \frac{hQ}{2}\frac{E(x^2)}{1-E(x)}$$
$$= E + \frac{F}{Q} + GQ + \frac{IB^2}{Q} + JB$$
$$= E + \frac{F}{Q} + \frac{I}{Q}\left[B + \frac{JQ}{2F}\right]^2 + \left[G - \frac{J^2}{4I}\right]Q$$

(3.137)

where the constants E, F, and G are the same as before case and the constants I and J are given as:

$$I = \frac{(C_b+h)}{2(1-E(x))}E\left(\frac{1-x}{1-x-\frac{D}{P}}\right) \text{ and } J = -h$$

Equation (3.138) implies that when Q is given, B can be set as $B = -\frac{G}{2F}Q$ to get the minimum value of $E[TCU(Q,B)]$ as follows:

$$E\left[TCU(Q,B(Q))\right] = E + \frac{F}{Q} + \left[G - \frac{J^2}{4I}\right]Q \quad (3.138)$$

By using the arithmetic–geometric mean inequality, one easily obtains that:

$$E\left[TCU(Q,B(Q))\right] = A + \frac{C}{Q} + \left[D - \frac{G^2}{4F}\right]Q \geq E + 2\sqrt{F\left[D - \frac{J^2}{4I}\right]} \quad (3.139)$$

when the equality

$$\frac{F}{Q} = \left[G - \frac{J^2}{4I}\right]Q \quad (3.140)$$

Holds $E[TCU(Q, B(Q))]$, has a minimum. Therefore, using Eq. (3.140), Q* is:

3.3 Fully Backordered

$$Q^* = \sqrt{\frac{F}{G - \frac{J^2}{4I}}}$$

$$= \sqrt{\frac{2KD}{h\left(1 - \frac{D}{P}\right) - \frac{h^2}{C_b + h} \frac{\{1-E(x)\}^2}{E\left(\frac{1-x}{1-x\frac{D}{P}}\right)} - 2h\left(1 - \frac{D}{P}\right)E(x) + hE(x^2)}} \quad (3.141)$$

and the optimal allowable backorder level is:

$$B^* = -\frac{J}{2I}Q^* \quad (3.142)$$

Therefore, the minimum value of the expected annual cost is:

$$E[\text{TCU}(Q^*, B^*)] = E + 2\sqrt{F\left[G - \frac{J^2}{4I}\right]} = D\left[\frac{F}{1 - E(x)} + C_s \frac{E(x)}{1 - E(x)}\right]$$

$$+ \sqrt{\frac{2hKD}{1 - E(x)} \left[\frac{\left(1 - \frac{D}{P}\right) + E(x^2)}{1 - E(x)} - 2\frac{\left(1 - \frac{D}{P}\right)E(x)}{1 - E(x)} - \frac{h}{C_b + h} \frac{1 - E(x)}{E\left[\frac{1 - x}{1 - x - \frac{D}{P}}\right]} B\right]}$$

$$(3.143)$$

3.3.1.2 Random Defective Rate

Hsu and Hsu (2016) developed economic production quantity (EPQ) models to determine the optimal production lot size and backorder quantity for a manufacturer under an imperfect production process. They considered EPQ models with shortages backordered. It is assumed that all customers are willing to wait for a new supply when there is a shortage. The production process is imperfect, and the fraction of defective items at the time of production is x. Once an item is produced, it is inspected immediately. The inspection time is negligible in comparison to the time taken to produce each item.

Some notations which are specifically used to model this problem are shown in Table 3.9.

They consider three cases regarding the time taken to sell the defective items (Hsu and Hsu 2016):

Case I. The defective items may be sold to a secondary market (when $v_d > 0$) or scrapped (i.e., $v_d = 0$) at the time identified and are not counted into inventory (see Fig. 3.13).
Case II. The defective items are kept in stock and sold at the end of the production period within each cycle (see Fig. 3.14).
Case III. The defective items are sold at the end of the production cycle or at the beginning of the next production run (see Fig. 3.15).

They considered the above three cases because different industries will dispose of the defective items at different timings; for example, pharmaceutical companies will scrap the defective products and not count them into inventory, and the furniture industry will sell the defective items at a discounted price as a secondary market. Moreover, the timing of disposing of the defective items may be different for different companies (which should be negotiated between the company and its secondary market customers):

$$t_1 = \frac{B}{P(1-x) - D}, \tag{3.144}$$

$$t_2 = \frac{Q}{P} - \frac{B}{P(1-x) - D}, \tag{3.145}$$

$$t_3 = \frac{(P(1-x) - D)Q}{PD} - \frac{B}{D}, \tag{3.146}$$

$$t_4 = \frac{B}{D}, \tag{3.147}$$

$$(t_1 + t_2) = \frac{Q}{P} \tag{3.148}$$

$$T = (t_1 + t_2 + t_3 + t_4) = \frac{Q(1-x)}{D} \tag{3.149}$$

Since the defective rate x is constant,

$$\text{Production cost} = C \frac{D}{(1-x)}, \tag{3.150}$$

$$\text{Inspection cost} = C_1 \frac{D}{(1-x)}, \tag{3.151}$$

$$\text{Setup cost} = K \frac{D}{Q(1-x)}, \tag{3.152}$$

3.3 Fully Backordered

Table 3.9 Notation of given problem (Hsu and Hsu 2016)

TC_C	The total cost per cycle ($/cycle)
$TC_j(Q, b)$	The total annual cost for case j (j = I, II, III) ($/year)
$TP_j(Q, b)$	The total annual profit for case j (j = I, II, III) ($/year)

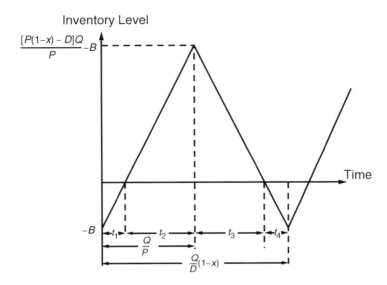

Fig. 3.13 Behavior of the inventory level over time for Case I (Hsu and Hsu 2016)

$$\text{Holding cost} = \frac{1}{2}h\left((P(1-x)-D)\frac{Q}{P} - B\right)\frac{(t_2+t_3)}{T}$$

$$= \frac{1}{2}h\left(\frac{(Q(1-x-\frac{D}{P})-B)^2}{Q(1-x-\frac{D}{P})}\right), \quad (3.153)$$

$$\text{Shortage cost} = \frac{1}{2}C_b B \frac{(t_1+t_4)}{T} = \frac{1}{2}C_b \frac{B^2}{Q(1-x-\frac{D}{P})} \quad (3.154)$$

$$TC_1(Q,B) = T_P + T_i + T_K + T_h + T_b$$
$$= \frac{CD}{(1-x)} + \frac{C_1 D}{(1-x)} + \frac{KD}{Q(1-x)} + \frac{1}{2}h\left(\frac{(Q(1-x-\frac{D}{P})-B)^2}{Q(1-x-\frac{D}{P})}\right) + \frac{C_b B^2}{2Q(1-x-\frac{D}{P})}$$
$$(3.155)$$

Note that since the total annual production and the inspection (screening) costs are independent of Q and B, the relevant costs to determining the optimal Q and B include only the setup, holding, and backordering costs. However, since both the

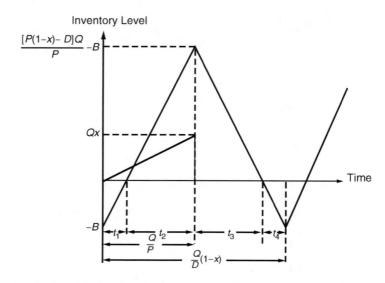

Fig. 3.14 Behavior of the inventory level over time for Case II (Hsu and Hsu 2016)

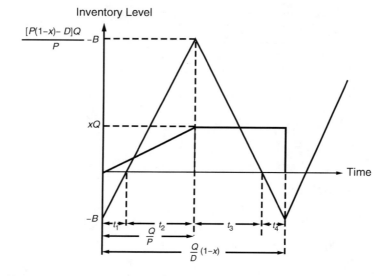

Fig. 3.15 Behavior of the inventory level over time for Case III (Hsu and Hsu 2016)

3.3 Fully Backordered

total annual production and inspection costs are functions of the defective rate x, these costs are included in the total annual cost for the purpose of sensitivity analyses. To obtain the total revenue per cycle, note that for each production lot of size Q, $Q(1-x)$ units are non-defective with a selling price of s per unit, and $Q \cdot x$ units are defective, which can be sold at v per unit. Dividing the total revenue per cycle by the cycle time $T = \frac{Q(1-x)}{D}$, the total annual revenue is equal to $sD + v\frac{Dx}{(1-x)}$. That is, if the defective rate is x, then in order to produce D units of good-quality items, a total of $\frac{D}{(1-x)}$ units will be produced, among which $\frac{Dx}{(1-x)}$ units are defective. The total annual profit is the total annual revenue less the total annual cost and is given as follows (Hsu and Hsu 2016):

$$\text{TP}_1(Q,B) = sD + v\frac{xD}{(1-x)} - C\frac{D}{(1-x)} - C_1\frac{D}{(1-x)} - K\frac{D}{Q(1-x)}$$
$$-\frac{1}{2}h\left(\frac{(Q(1-x-\frac{D}{P})-B)^2}{Q(1-x-\frac{D}{P})}\right) - \frac{1}{2}C_b\frac{B^2}{Q(1-x-\frac{D}{P})} \tag{3.156}$$

Since the revenue is also independent of the production lot size and the backorder quantity, maximizing the total annual profit is equivalent to minimizing the total annual cost. By taking the second derivative of $\text{TC}_1(Q,B)$ with respect to Q and B, one obtains:

$$\frac{\partial^2 \text{TC}_1(Q,B)}{\partial Q^2} = \frac{2KD}{Q^3(1-x)} + \frac{(h+C_b)B^2}{Q^3(1-x-\frac{D}{P})} \tag{3.157}$$

$$\frac{\partial^2 \text{TC}_1(Q,B)}{\partial B^2} = \frac{(h+C_b)}{Q(1-x-\frac{D}{P})} \tag{3.158}$$

$$\frac{\partial^2 \text{TC}_1(Q,B)}{\partial Q \partial B} = \frac{\partial^2 \text{TC}_1(Q,B)}{\partial B \partial Q} = \frac{-(h+C_b)}{Q^2(1-x-\frac{D}{P})}B \tag{3.159}$$

and

$$\left(\frac{\partial^2 \text{TC}_1(Q,B)}{\partial Q^2}\right)\left(\frac{\partial^2 \text{TC}_1(Q,B)}{\partial B^2}\right) - \left(\frac{\partial^2 \text{TC}_1(Q,B)}{\partial Q \partial B}\right)^2$$
$$= \frac{2BD(h+C_b)}{Q^4(1-x)(1-x-\frac{D}{P})} \tag{3.160}$$

Since the production process is imperfect with a defective rate of x, the effective annual production rate $P' = P(1-x)$. Note that to make sure that the production process has enough capacity to satisfy the customers' demand, the assumption that P

$(1-x) > D$ (i.e., $x < 1 - \frac{D}{P}$) should be held. If $x < 1 - \frac{D}{P}$, then $\frac{\partial^2 TC_1(Q,B)}{\partial Q^2} > 0$, $\frac{\partial^2 TC_1(Q,B))}{\partial B^2} > 0$ and which implies that the total annual cost is a convex function of Q and B (see, e.g., Ghorpade and Limaye 2010, Proposition 3.71 on page 137). Another way to prove the convexity of $TC_1(Q, B)$ is to use the Hessian matrix equations (see, e.g., Rardin and Rardin 1998) and verify the existence of the following equation:

$$[Q \ B] \begin{pmatrix} \frac{\partial^2 TC_1(Q,B)}{\partial Q^2} & \frac{\partial^2 TC_1(Q,B)}{\partial Q \partial B} \\ \frac{\partial^2 TC_1(Q,B)}{\partial Q \partial B} & \frac{\partial^2 TC_1(Q,B)}{\partial B^2} \end{pmatrix} \begin{bmatrix} Q \\ B \end{bmatrix} > 0, Q, B \neq 0.$$

Solving the elements of the Hessian matrix, one obtains:

$$[Q \ B] \begin{pmatrix} \frac{\partial^2 TC_1(Q,B)}{\partial Q^2} & \frac{\partial^2 TC_1(Q,B)}{\partial Q \partial B} \\ \frac{\partial^2 TC_1(Q,B)}{\partial Q \partial B} & \frac{\partial^2 TC_1(Q,B)}{\partial B^2} \end{pmatrix} \begin{bmatrix} Q \\ B \end{bmatrix}$$

$$= [Q \ B] \begin{bmatrix} \frac{2KD}{Q^3(1-x)} + \frac{(h+C_b)B^2}{Q^3\left(1-x-\frac{D}{P}\right)} & \frac{-(h+C_b)}{Q^2\left(1-x-\frac{D}{P}\right)}B \\ \frac{-(h+C_b)}{Q^2\left(1-x-\frac{D}{P}\right)}B & \frac{(h+C_b)}{Q\left(1-x-\frac{D}{P}\right)} \end{bmatrix} \begin{bmatrix} Q \\ B \end{bmatrix}$$

$$= \begin{bmatrix} \frac{2KD}{Q^2(1-x)} & 0 \end{bmatrix} \begin{bmatrix} Q \\ B \end{bmatrix} = \frac{2KD}{Q(1-x)} > 0 \text{ for } Q, B \neq 0$$

Hence, $TC_1(Q, B)$ is a strictly convex function for all Q and B different from zero. By taking the first derivative of $TC_1(Q, B)$ with respect to Q and B and setting the results to zero, the optimal production lot size Q^*_{CI} and the maximum backorder quantity B^*_{CI} for Case I are given as follows (Hsu and Hsu 2016):

$$Q^*_{CI} = \sqrt{\frac{2DK}{h} \frac{(h+C_b)}{C_b(1-x)\left(1-x-\frac{D}{P}\right)}} \qquad (3.161)$$

$$B^*_{CI} = Q^*_{CI}\left(1 - x - \frac{D}{P}\right) \frac{h}{(h+C_b)} \qquad (3.162)$$

3.3 Fully Backordered

Figure 3.16 depicts the behavior of the inventory over time for Case II. Note that the only difference between Cases I and II is that in Case II, the defective items are held until the end of the production period, so one obtains (Hsu and Hsu 2016):

$$\text{TC}_{\text{II}}(Q,B) = \text{TC}_{\text{I}}(Q,B) + \frac{1}{2}hQx\frac{(t_1+t_2)}{T} = \text{TC}_{\text{I}}(Q,B) + \frac{1}{2}h\frac{QxD}{P(1-x)}$$

$$= \frac{KD}{Q(1-x)} + \frac{1}{2}h\left(\frac{\left(Q\left(1-x-\frac{D}{P}\right)-B\right)^2}{Q\left(1-x-\frac{D}{P}\right)}\right) + \frac{C_b B^2}{2Q\left(1-x-\frac{D}{P}\right)} + \frac{hQxD}{2P(1-x)}$$

(3.163)

and the optimal solution for Case II is given as:

$$Q^*_{\text{CII}} = \sqrt{\frac{2DK}{h} \frac{(h+C_b)}{\left\{C_b(1-x)\left(1-x-\frac{D}{P}\right) + (h+C_b)\frac{D}{P}\right\}}} \qquad (3.164)$$

$$B^*_{\text{CII}} = Q^*_{\text{CII}}\left(1-x-\frac{D}{P}\right)\frac{h}{(h+C_b)} \qquad (3.165)$$

Figure 3.17 depicts the behavior of the inventory over time for Case III. Note that the only difference between Cases II and III is that in Case III, the defective items are held until the beginning of the next production run, so one obtains:

$$\text{TC}_{\text{III}}(Q,B) = \text{TC}_{\text{II}}(Q,B) + hQx\frac{(t_3+t_4)}{T} = \text{TC}_{\text{II}}(Q,B) + hQx\left(1 - \frac{D}{P(1-x)}\right)$$

$$= \frac{KD}{Q(1-x)} + \frac{1}{2}h\left(\frac{\left(Q\left(1-x-\frac{D}{P}\right)-B\right)^2}{Q\left(1-x-\frac{D}{P}\right)}\right)$$

$$+ \frac{1}{2}C_b\frac{B^2}{Q\left(1-x-\frac{D}{P}\right)} + \frac{1}{2}h\frac{QxD}{P(1-x)} + hQx\left(1 - \frac{D}{P(1-x)}\right)$$

(3.166)

and the optimal solution for Case III is given as (Hsu and Hsu 2016):

$$Q^*_{\text{CIII}} = \sqrt{\frac{2DK}{h} \frac{(h+C_b)}{\left\{C_b(1-x)\left(1-x-\frac{D}{P}\right) + (h+C_b)\left(2x(1-x)-x\frac{D}{P}\right)\right\}}} \qquad (3.167)$$

$$B^*_{\text{CIII}} = Q^*_{\text{CIII}}\left(1-x-\frac{D}{P}\right)\frac{h}{(h+C_b)} \qquad (3.168)$$

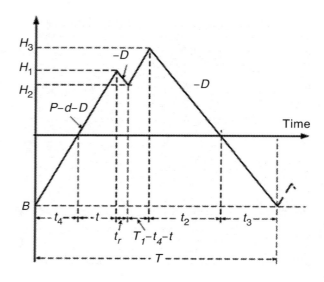

Fig. 3.16 On-hand inventory of perfect-quality items in EPQ model with scrap and breakdown occurring in inventory-stacking period (Chiu et al. 2008)

If x is a random variable with a probability density function $f(x)$, then Hsu and Hsu (2016) derived production, inspection, setup, holding, shortage, and total cyclic costs as presented in Eqs. (3.169)–(3.174):

$$CQ, \quad (3.169)$$

$$C_1 Q, \quad (3.170)$$

$$K, \quad (3.171)$$

$$\frac{1}{2} h \left(Q \left(1 - x - \frac{D}{P} \right) - B \right)(t_2 + t_3) = \frac{1}{2} h \left(Q \left(1 - x - \frac{D}{P} \right) - B \right) \times \left(\frac{Q}{D}(1-x) - \frac{B(1-x)}{D\left(1 - x - \frac{D}{P}\right)} \right) \quad (3.172)$$

$$\frac{1}{2} C_b B (t_1 + t_4) = \frac{1}{2} C_b \frac{B^2 (1-x)}{Q \left(1 - x - \frac{D}{P} \right)} \quad (3.173)$$

$$\mathrm{TC_C} = CQ + C_1 Q + K + \frac{1}{2} h \left(Q \left(1 - x - \frac{D}{P} \right) - B \right) \\ \times \left(\frac{Q}{D}(1-x) - \frac{B(1-x)}{D\left(1 - x - \frac{D}{P}\right)} \right) + \frac{C_b (1-x)}{2Q \left(1 - x - \frac{D}{P} \right)} B^2 \quad (3.174)$$

From Eq. (3.149), the expected cycle length would be:

3.3 Fully Backordered

Fig. 3.17 On-hand inventory of scrap items in EPQ model with scrap and breakdown occurring in inventory-stacking period (Chiu et al. 2008)

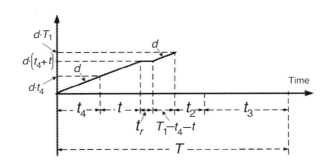

$$E(T)\frac{Q}{D}(1-E(x)) \tag{3.175}$$

Using the renewal reward theorem, the expected total annual cost would be (Hsu and Hsu 2016):

$$E(TC_1(Q,B)) = \frac{E(TC_C)}{E(T)} = \frac{CD}{1-E(x)} + C_1\frac{D}{1-E(x)} + \frac{KD}{Q(1-E(x))}$$
$$+\frac{hE(1-x)^2}{2(1-E(x))}Q + \frac{h}{2}\left(-\frac{DQ}{P}-2B\right) + \frac{(h+C_b)}{1-E(x)}A_1\frac{B^2}{2Q_b} \tag{3.176}$$

with $A_1 = E\left(\frac{1-x}{1-x-\frac{D}{P}}\right)$, and

$$E(TP_1(Q,B)) = sD + v\frac{DE(x)}{(1-E(x))} - E(TC_1(Q,B)) \tag{3.177}$$

By taking the second derivative of $E[TC_1(Q,B)]$ with respect to Q and B, one obtains:

$$\frac{\partial^2 E(TC_1(Q,B))}{\partial Q^2} = \frac{2KD}{Q^3(1-E(x))} + \frac{(h+C_b)B^2A_1}{Q^3(1-E(x))} \tag{3.178}$$

$$\frac{\partial^2 E(TC_1(Q,B))}{\partial B^2} = \frac{(h+C_b)A_1}{Q(1-E(x))} \tag{3.179}$$

$$\frac{\partial^2 E(TC_1(Q,B))}{\partial Q \partial B} = \frac{-B(h+C_b)A_1}{Q^2(1-E(x))} \tag{3.180}$$

And

$$\left(\frac{\partial^2 E(\mathrm{TC}_1(Q,B))}{\partial Q^2}\right)\left(\frac{\partial^2 E(\mathrm{TC}_1(Q,B))}{\partial B^2}\right) - \left(\frac{\partial^2 E(\mathrm{TC}_1(Q,B))}{\partial Q \partial B}\right)^2$$
$$= \frac{2KD(h+C_\mathrm{b})A_1}{Q^4(1-E(x))} \tag{3.181}$$

From Eqs. (3.178)–(3.181), one can see that if $A_1 > 0$ and $E(x) < 1$, then $\frac{\partial^2 E(\mathrm{TC}_1(Q,B))}{\partial Q^2} > 0$, $\frac{\partial^2 E(\mathrm{TC}_1(Q,B))}{\partial B^2} > 0$ and $\left(\frac{\partial^2 E(\mathrm{TC}_1(Q,B))}{\partial Q^2}\right)\left(\frac{\partial^2 E(\mathrm{TC}_1(Q,B))}{\partial B^2}\right) - \left(\frac{\partial^2 E(\mathrm{TC}_1(Q,B))}{\partial Q \partial B}\right)^2 > 0$, which implies that the expected total annual cost is a convex function of Q and B and that exist unique values of Q and B that minimize Eq. (3.176) and maximize Eq. (3.177). If x is uniformly distributed between 0 and b, then we have $A_1 = 1 + \frac{D}{bP}$ in $\left(\frac{1-\frac{D}{P}}{1-\frac{D}{P}-b}\right)$ (see the derivation in Hsu and Hsu 2016). For A_1 to be real, b (the maximum possible value of x) should be less than $1 - \frac{D}{P}$; otherwise, $\ln\left(1 - \frac{D}{P} - b\right)$ is undefined. If x follows a beta distribution with shape parameters $p = q = 2$, then b (the maximum possible value of x) should also be less than $1 - \frac{D}{P}$. Consider the following Hessian matrix (Hsu and Hsu 2016):

$$[Q \ B] \begin{pmatrix} \frac{\partial^2 E(\mathrm{TC}_1(Q,B))}{\partial Q^2} & \frac{\partial^2 E(\mathrm{TC}_1(Q,B))}{\partial Q \partial B} \\ \frac{\partial^2 E(\mathrm{TC}_1(Q,B))}{\partial Q \partial B} & \frac{\partial^2 E(\mathrm{TC}_1(Q,B))}{\partial B^2} \end{pmatrix} \begin{bmatrix} Q \\ B \end{bmatrix}$$

$$= [Q \ B] \begin{bmatrix} \frac{2KD}{Q^3(1-E(x))} + \frac{(h+C_\mathrm{b})B^2 A_1}{Q^3(1-E(x))} & \frac{-B(h+C_\mathrm{b})A_1}{Q^2(1-E(x))} \\ \frac{-B(h+C_\mathrm{b})A_1}{Q^2(1-E(x))} & \frac{(h+C_\mathrm{b})A_1}{Q(1-E(x))} \end{bmatrix} \begin{bmatrix} Q \\ B \end{bmatrix}$$

$$= \left[\frac{2KD}{Q^2(1-E(\gamma))} \ 0\right] \begin{bmatrix} Q \\ B \end{bmatrix} = \frac{2KD}{Q(1-E(x))} > 0 \text{ for } Q, B \neq 0$$

Since it is assumed that the maximum possible value of x is less than $1 - \frac{D}{P}$, $E(\mathrm{TC}_1(Q,B))$ is a strictly convex function for all nonzero Q and B. By taking the first derivative of $E[\mathrm{TP}_1(Q,B)]$ or $E[\mathrm{TC}_1(Q,B)]$ with respect to Q and B setting the results to zero, the optimal production lot size Q^*_{rI} and the maximum backorder quantity B^*_{rI} for Case I are given as follows (Hsu and Hsu 2016):

$$Q^*_{\mathrm{rI}} = \sqrt{\frac{2KD}{h\left(E(1-x)^2 - \frac{D}{P}(1-E(x)) - \frac{h}{(h+C_\mathrm{b})A_1}\{1-E(x)\}^2\right)}} \tag{3.182}$$

3.3 Fully Backordered

$$B_{rI}^* = Q_{rI}^*(1 - E(x))\frac{h}{(h + C_b)A_1} \qquad (3.183)$$

Note that if the defective rate x is constant, Eqs. (3.182) and (3.183) reduce to Eqs. (3.167) and (3.168), respectively. The expected cost per year and the optimal solution for Case II are given as (Hsu and Hsu 2016):

$$E[TC_{II}(Q,B)] = E[TC_{I}(Q,B)] + \frac{1}{2}hQE(x)\frac{D}{P(1-E(x))} \qquad (3.184)$$

$$Q_{rII}^* = \sqrt{\frac{2KD}{h\left(E(1-x)^2 - \frac{D}{P}(1-2E(x)) - \frac{h}{(h+C_b)A_1}\{1-E(x)\}^2\right)}} \qquad (3.185)$$

$$B_{rII}^* = Q_{rII}^*(1 - E(x))\frac{h}{(h + C_b)A_1} \qquad (3.186)$$

If γ is constant, Eqs. (3.185) and (3.186) reduce to Eqs. (3.164) and (3.165), respectively. For Case III, one obtains (Hsu and Hsu 2016):

$$E[TC_{III}(Q,B)] = E[TC_{II}(Q,B)] + hQ\left(\frac{E(x) - E(x^2) - E(x)\frac{D}{P}}{(1 - E(x))}\right) \qquad (3.187)$$

$$Q_{rIII}^* = \sqrt{\frac{2KD}{h\left(E(1-x)^2 - \frac{D}{P}2(E(x) - E(x^2)) - \frac{h}{(h+C_b)A_1}\{1-E(x)\}^2\right)}} \qquad (3.188)$$

$$B_{rIII}^* = Q_{rIII}^*(1 - E(x))\frac{h}{(h + C_b)A_1} \qquad (3.189)$$

If x is constant, Eqs. (3.188) and (3.189) reduce to Eqs. (3.167) and (3.168), respectively (Hsu and Hsu 2016).

Example 3.5 Hsu and Hsu (2016) presented an example in which a designer window treatments manufacturer in Taiwan is a subsidiary of an interior decorating company. The manufacturer supplies all the window curtains needed by the interior decorating company. Since the window curtains are tailored designs, customers are willing to wait when a shortage occurs. Those window curtains produced which are not of perfect quality are sold to night market retailers at a discounted price. It should be noted that there are many famous night markets in Taiwan where people can buy cheaper products. During the production downtimes, the manufacturer will use the capacity to produce other products such as designer cushions, designer slipcovers, or designer bedding sets. Suppose that the manufacturer has the following parameters: $P = 3000$ units/year, $D = 1000$ units/year, $K = \$400$/lot, $C = \$100$/unit, $C_I = \$2$/unit, $h = \$5$/unit/year, $C_b = \$10$/unit/year, $s = \$300$/unit, and $v = \$80$/unit.

First, it is assumed that the defective rate γ has a constant value and obtains the optimal solution of the three cases given in Table 3.10. Table 3.11 shows the optimal solutions when x is uniformly distributed between 0 and b (Hsu and Hsu 2016). If the defective rate x follows a uniform distribution with the probability density function (Hsu and Hsu 2016),

$$f(x) = \begin{cases} \frac{1}{b}, & 0 \le x \le b \\ 0, & \text{otherwise} \end{cases}$$

then

$$E[x] = \int_0^b \frac{x}{b} dx = \frac{b}{2},$$

$$E[x^2] = \int_0^b x^2 f(x) dx = \int_0^b \frac{x^2}{b} dx = \frac{b^2}{3},$$

$$E\left[(1-x)^2\right] = \int_0^b (1-x)^2 f(x) dx = \int_0^b \frac{(1-x)^2}{b} dx = 1 - b + \frac{b^2}{3},$$

$$A_1 = 1 + \frac{D}{bP} \ln\left[\frac{1 - \frac{D}{P}}{1 - \frac{D}{P} - b}\right].$$

3.3.1.3 Random Breakdown

Chiu et al. (2008) are concerned with determination of optimal lot size for an economic manufacturing quantity model with backordering, scrap, and breakdown occurring in inventory-stocking period. Also, they investigated the optimal manufacturing lot size for EMQ model with scrap, backlogging, and random breakdown occurring in inventory-stacking period.

Some notations which are specifically used to model this problem are shown in Table 3.12.

Let t denote production time before a breakdown taking place in the inventory-stacking period, and let the constant machine repair time $t_r = g$. From Fig. 3.16, one can obtain the following (Chiu et al. 2008):

3.3 Fully Backordered

$$H_1 = (P - d - D)t \tag{3.190}$$

$$H_2 = H_1 - t_r D = H_1 - gD \tag{3.191}$$

$$T_1 = \frac{Q}{P} \tag{3.192}$$

$$H_3 = H_2 + (P - d - D)(T_1 - t_4 - t) \tag{3.193}$$

$$T = T_1 + t_2 + t_3 + t_r, \tag{3.194}$$

$$t_2 = \frac{H_3}{D}, \tag{3.195}$$

$$t_3 = \frac{B}{D} \tag{3.196}$$

$$t_4 = \frac{B}{P - d - D} \tag{3.197}$$

where $d = P \cdot x$.

Total scrap items produced during production uptime T_1 are (see Fig. 3.17):

$$d \cdot T_1 = Q \cdot x \tag{3.198}$$

Total production–inventory cost per cycle is (Chiu et al. 2008):

$$\begin{aligned}
\text{TC}(T_1, B) &= K + C \cdot (P \cdot T_1) + C_S \cdot (P \cdot x \cdot T_1) + C_M \\
&+ h\left[\frac{H_1}{2}(t) + \frac{H_1 + H_2}{2}(t_r) + \frac{H_2 + H_3}{2}(T_1 - t_4 - t) + \frac{H_3}{2}(t_2)\right] \\
&+ h\left[\frac{d(t_4 + t)}{2}(t_4 + t) + (t_4 + t)t_r + \frac{(t_4 + t) + dT_1}{2}(T_1 - t_4 - t)\right] \\
&+ C_b\left[\frac{B}{2}(t_4) + \frac{B}{2}(t_3)\right]
\end{aligned} \tag{3.199}$$

Substituting all related parameters from Eqs. (3.190) to (3.198) in Eq. (3.199), one obtains:

$$\begin{aligned}
\text{TC}(T_1, B) &= K + C \cdot (P \cdot T_1) + C_S \cdot (P \cdot x \cdot T_1) + C_M - h\frac{P}{D}T_1 B(1-x) \\
&+ \frac{C_b(1-x)}{2D\left(1 - x - \frac{D}{P}\right)}B^2 + \frac{hg}{\left(1 - x - \frac{D}{P}\right)}(B + gD) - hPgT_1(1-x) + hPgt \\
&+ C_b\left[\frac{B}{2}(t_4) + \frac{B}{2}(t_3)\right]
\end{aligned} \tag{3.200}$$

Table 3.10 The optimal solutions when x is constant (Hsu and Hsu 2016)

x	0.01	0.02	0.03	0.04	0.05	0.1	0.15	0.2	0.25
1	607.6	615.39	623.4	631.61	640.06	685.99	739.25	801.78	876.36
2	133	132.65	132.3	131.94	131.57	129.58	127.32	124.72	121.72
3	196,447.82	196,224.51	195,996.6	195,763.96	195,526.43	194,259.79	192,844.5	191,252.78	189,449.51
4	605.27	610.59	615.96	621.37	626.82	654.65	683.21	712.07	740.66
5	132.49	131.62	130.72	129.8	128.85	123.66	117.66	110.77	102.87
6	196,442.71	196,214.08	195,980.63	195,742.21	195,498.65	194,197.76	192,740.07	191,095.64	189,226.5
7	596.37	592.98	589.82	586.88	584.15	573.38	566.95	564.43	565.69
8	130.54	127.82	125.17	122.59	120.08	108.31	97.64	87.8	78.57
9	196,422.79	196,174.38	195,921.3	195,663.4	195,400.52	194,005.3	192,457.57	190,728.31	188,781.05

1, Q^*_{cI}; 2, B^*_{cI}; 3, $TP_{II}(Q^*_{cI}, B^*_{cI})$; 4, Q^*_{cII}; 5, B^*_{cII}; 6, $TP_{II}(Q^*_{cII}, B^*_{cII})$; 7, Q^*_{cIII}; 8, B^*_{cIII}; 9, $TP_{III}(Q^*_{cIII}, B^*_{cIII})$

3.3 Fully Backordered

Table 3.11 The optimal solutions when x is uniformly distributed between 0 and b (Hsu and Hsu 2016)

b	0.02	0.04	0.06	0.08	0.1	0.2	0.3	0.4	0.5
1	0.01	0.02	0.03	0.04	0.05	0.1	0.15	0.2	0.25
2	607.57	615.28	623.13	631.12	639.25	682	727.93	776	823.58
3	132.99	132.61	132.21	131.77	131.3	128.32	123.93	117.34	107
4	196,447.76	196,224.26	195,996.04	195,762.93	195,524.78	194,252.19	192,824.69	191,211.35	189,371.51
5	605.25	610.48	615.7	620.9	626.07	651.18	674.24	693.83	707.97
6	132.48	131.58	130.63	129.64	128.59	122.52	114.79	104.91	91.98
7	196,442.65	196,213.84	195,980.07	195,741.19	195,497.03	194,190.51	192,721.74	191,058.72	189,160
8	596.39	593.06	589.98	587.16	584.58	574.96	570.29	570.05	573.84
9	130.54	127.82	125.18	122.59	120.07	108.18	97.09	86.2	74.56
10	196,422.83	196,174.55	195,921.68	195,664.07	195,401.57	194,009.57	192,467.3	190,745.78	188,807.84

1, $E[\gamma]$; 2, Q^*_{rI}; 3, B^*_{rI}; 4, $E[TP_I(Q^*_{rI}, B^*_{rI})]$; 5, Q^*_{rII}; 6, B^*_{rII}; 7, $E[TP_{II}(Q^*_{rII}, B^*_{rII})]$; 8, Q^*_{rIII}; 9, B^*_{rIII}; 10, $E[TP_{III}(Q^*_{rIII}, B^*_{rIII})]$

Table 3.12 Notation of given problem (Chiu et al. 2008)

T_1	The optimal production time to be determined (time)
t	Production time before a random breakdown occurs (time)
t_r	Time required for repairing and restoring the machine (time)
t_2	Time required for depleting all available perfect-quality on-hand items (time)
t_3	Shortage permitted time (time)
t_4	Time required for filling backorder quantity (time)
H_1	Level of on-hand inventory when machine breakdown occurs (unit)
H_2	Level of on-hand inventory when machine is repaired and restored (unit)
H_3	The maximum level of on-hand inventory for each production cycle (unit)
TCU(T_1, B)	Total production–inventory costs per unit time
TC(T_1, B)	Total production–inventory costs per cycle

The production cycle length is not constant due to the assumption of random scrap rate, and a uniformly distributed breakdown is assumed to occur in the inventory-stacking time. Thus, to take randomness of scrap and breakdown into account, one can use the renewal reward theorem in inventory cost analysis to cope with variable cycle length and use integration of TC(T_1, B) to deal with the random breakdown happening in inventory-stacking time. The expected total production–inventory costs per unit time can be calculated as follows (Chiu et al. 2008):

$$E[\text{TCU}(T_1, B)] = \frac{E\left(\int_0^{T_1-t_4} \text{TC}(T_1, B) \cdot f(t) dt\right)}{E(T)}$$

$$= \frac{E\left(\int_0^{T_1-t_4} \text{TC}(T_1, B) \cdot (1/t_4) dt\right)}{T_1 P(1-E(x))/D} \quad (3.201)$$

Substituting Eqs. (3.190) through (3.200) in Eq. (3.201), one obtains (Chiu et al. 2008):

$$E[\text{TCU}(T_1, B)] = D\left[\frac{C}{1-E(x)} + C_S \frac{E(x)}{1-E(x)}\right] + \frac{(K+C_M)D}{PT_1} \frac{1}{1-E(x)}$$

$$+ \frac{h}{2}\left[\left(1-\frac{D}{P}\right)PT_1 - 2B\right]\frac{1}{1-E(x)} + \frac{hPT_1}{2}\frac{E(x^2)}{1-E(x)}$$

$$+ \frac{B^2}{2PT_1}\frac{(C_b+h)}{1-E(x)} E\left(\frac{1-x}{1-x-\frac{D}{P}}\right) + h\left[B - \left(1-\frac{D}{P}\right)PT_1\right]\frac{E(x)}{1-E(x)}$$

$$+ \frac{hgD}{2PT_1}(B+gD)E\left(\frac{1}{\left(1-x-\frac{D}{P}\right)}\right)\frac{1}{1-E(x)} + hgD\frac{E(x)}{1-E(x)} - \frac{hgD}{2}\frac{1}{1-E(x)}$$

$$\quad (3.202)$$

3.3 Fully Backordered

Let $E_0 = \frac{1}{1-E(x)}$; $E_1 = \frac{E(x)}{1-E(x)}$; $E_2 = \frac{E(x^2)}{1-E(x)}$; $E_3 = \frac{1}{1-E(x)}E\left(\frac{1-x}{1-x-\frac{D}{P}}\right)$; $E_4 = \frac{1}{1-E(x)}E\left(\frac{1}{1-x-\frac{D}{P}}\right)$

Then Eq. (3.202) becomes (Chiu et al. 2008):

$$E[TCU(T_1, B)] = D[CE_0 + C_S E_1] + \frac{(K+C_M)D}{PT_1}E_0 + \frac{h}{2}\left[\left(1-\frac{D}{P}\right)PT_1 - 2B\right]E_0$$
$$+ \frac{hPT_1}{2}E_2 + \frac{B^2}{2PT_1}(C_b + h)E_3 + h\left[B - \left(1-\frac{D}{P}\right)PT_1\right]E_1$$
$$+ \frac{hgD}{2PT_1}(B+gD)E_4 + hgDE_1 - \frac{hgD}{2}E_0$$

(3.203)

The optimal inventory operating policy can be obtained by minimizing the expected cost function. For the proof of convexity of $E[TC(T_1, B)]$, one can utilize the Hessian matrix equation in Rardin and Rardin (1998) and verify the existence of the following:

$$[T_1 \quad B]\begin{pmatrix} \frac{\partial^2 E[TCU(T_1,B)]}{\partial T_1^2} & \frac{\partial^2 E[TCU(T_1,B)]}{\partial T_1 \partial B} \\ \frac{\partial^2 E[TCU(T_1,B)]}{\partial T_1 \partial B} & \frac{\partial^2 E[TCU(T_1,B)]}{\partial B^2} \end{pmatrix}\begin{bmatrix} T_1 \\ B \end{bmatrix} > 0 \quad (3.204)$$

$E[TC(T_1, B)]$ is strictly convex only if Eq. (3.204) is satisfied, for all T_1 and B different from zero. By computing all the elements of the Hessian matrix equation, one obtains (Chiu et al. 2008):

$$[T_1 \quad B]\begin{pmatrix} \frac{\partial^2 E[TCU(T_1,B)]}{\partial T_1^2} & \frac{\partial^2 E[TCU(T_1,B)]}{\partial T_1 \partial B} \\ \frac{\partial^2 E[TCU(T_1,B)]}{\partial T_1 \partial B} & \frac{\partial^2 E[TCU(T_1,B)]}{\partial B^2} \end{pmatrix}\begin{bmatrix} T_1 \\ B \end{bmatrix}$$
$$= \frac{2D(K+M)}{PT_1}E_0 + \frac{hg^2D^2}{PT_1}E_4 > 0$$

(3.205)

Equation (3.205) is positive, because all parameters are positive. Hence, $E[TC(T_1, B)]$ is a strictly convex function. It follows that for optimal production uptime T_1 and maximal backorder level B, one can differentiate $E[TC(T_1, B)]$ with respect to T_1 and with respect to B and solve linear systems of Eqs. (3.206) and (3.207) by setting these partial derivatives equal to zero:

$$\frac{\partial E[\text{TCU}(T_1,B)]}{\partial T_1} = -\frac{(K+C_M)D}{PT_1^2}E_0 + \frac{h}{2}\left[P\left(1-\frac{D}{P}\right)\right]E_0 + \frac{hP}{2}E_2$$
$$-(C_b+h)E_3\frac{B^2}{2PT_1^2} - hP\left(1-\frac{D}{P}\right)E_1 - \frac{hgDB}{2PT_1^2}E_4 - \frac{hg^2D^2}{2PT_1^2}E_4 \quad (3.206)$$

$$\frac{\partial E[\text{TCU}(T_1,B)]}{\partial B} = -hE_0 + \frac{B}{PT_1}(C_b+h)E_3 + hE_1 + \frac{hgD}{2PT_1}E_4 \quad (3.207)$$

$$T_1^* = \frac{1}{P}\sqrt{\frac{2D(K+C_M)E_0 + hD^2g^2E_4\left(1-\frac{hE_4}{4(C_b+h)E_3}\right)}{h\left(1-\frac{D}{P}\right)E_0 - \frac{h^2}{(C_b+h)E_3} - 2h\left(1-\frac{D}{P}\right)E_1 + hE_2}} \quad (3.208)$$

$$B^* = \frac{h}{(C_b+h)E_3}\left(PT_1^* - \frac{gD}{2}E_4\right) \quad (3.209)$$

Plugging E_0, E_2, E_3, and E_4 in Eqs. (3.208) and (3.209), the optimal production run time and optimal backordering quantity become (Chiu et al. 2008):

$$T_1^* = \frac{1}{P}\left[\frac{\left(2D(K+C_M) + \frac{hDg^2}{1-E(x)}\cdot E(1/(1-x-D/P))\right)\left(1 - \frac{h\cdot E(1/(1-x-D/P))}{4(C_b+h)\cdot E[(1-x)/(1-x-D/P)]}\right)}{h\left(1-\frac{D}{P}\right) - 2h\left(1-\frac{D}{P}\right)E(x) + hE(x^2) - \frac{h^2(1-E(x))^2}{(C_b+h)\cdot E[(1-x)/(1-x-D/P)]}}\right]^{\frac{1}{2}} \quad (3.210)$$

$$B^* = \frac{h\cdot P(1-E(x))\cdot T_1^* - \frac{Dgh}{2}E\left(\frac{1}{1-x-\frac{D}{P}}\right)}{(C_b+h)\cdot E\left(\frac{1-x}{1-x-\frac{D}{P}}\right)} \quad (3.211)$$

From Eq. (3.193) to Eqs. (3.208) and (3.209), one can obtain the optimal lot size Q^* and optimal backorder level B^* as follows (Chiu et al. 2008):

$$Q^* = \sqrt{\frac{2D(K+C_M)E_0 + hD^2g^2E_4\left(1-\frac{hE_4}{4(C_b+h)E_3}\right)}{h\left(1-\frac{D}{P}\right)E_0 - \frac{h^2}{(C_b+h)E_3} - 2h\left(1-\frac{D}{P}\right)E_1 + hE_2}} \quad (3.212)$$

$$B^* = \frac{h}{(C_b+h)E_3}\left(Q^* - \frac{gD}{2}E_4\right) \quad (3.213)$$

Plugging E_0, E_2, E_3, and E_4 in Eqs. (3.212) and (3.213), the optimal production lot size and optimal backordering quantity become (Chiu et al. 2008):

3.3 Fully Backordered

$$Q^* = \left[\frac{\left(2D(K+C_M) + \frac{hDg^2}{1-E(x)} \cdot E(1/(1-x-D/P))\right)\left(1 - \frac{h \cdot E(1/(1-x-D/P))}{4(C_b+h) \cdot E[(1-x)/(1-x-D/P)]}\right)}{h\left(1-\frac{D}{P}\right) - 2h\left(1-\frac{D}{P}\right)E(x) + hE(x^2) - \frac{h^2(1-E(x))^2}{(C_b+h) \cdot E[(1-x)/(1-x-D/P)]}} \right]^{\frac{1}{2}} \quad (3.214)$$

$$B^* = \frac{h \cdot P(1-E(x)) \cdot Q^* - \frac{Dgh}{2} E\left(\frac{1}{1-x-\frac{D}{P}}\right)}{(C_b + h)E\left(\frac{1-x}{1-x-\frac{D}{P}}\right)} \quad (3.215)$$

Chiu et al. (2008) supposed that machine breakdown factor is not an issue to be considered, then the cost and time for repairing failure machine $M = 0$ and $g = 0$, Eqs. (3.214) and (3.215) become the same equations as were given by:

$$Q^* = \sqrt{\frac{2KD}{h\left(1-\frac{D}{P}\right) - \frac{h^2}{C_b+h} \frac{\{1-E(x)\}^2}{E\left(\frac{1-x}{1-x-\frac{D}{P}}\right)} - 2h\left(1-\frac{D}{P}\right)E(x) + hE(x^2)}} \quad (3.216)$$

$$B^* = \frac{h}{(C_b+h)} \frac{1-E(x)}{E\left(\frac{1-x}{1-x-\frac{D}{P}}\right)} Q^* \quad (3.217)$$

Further, suppose that regular production process produces no defective items, i.e., $x = 0$, then Eqs. (3.216) and (3.217) become the same equations as were presented by the classic EPQ model with backordering permitted (Hillier and Lieberman 1995; Silver et al. 1998):

$$Q^* = \sqrt{\frac{2KD}{h\left(1-\frac{D}{P}\right)}} \sqrt{\frac{C_b+h}{C_b}} \quad (3.218)$$

$$B^* = \left[\frac{h}{C_b+h}\left(1-\frac{D}{P}\right)Q^*\right] \quad (3.219)$$

Numerical 3.6 Chiu et al. (2008) assumed annual production rate of a manufactured item is 18,000 units and demand of this item is 3000 units per year. The percentage of random scrap items produced x and follows a uniform distribution over the range [0, 0.15). Other parameters used are as follows (Chiu et al. 2008): $K = \$240$ for each production run, $C_s = \$1.00$ disposal cost for each scrap item, $C = \$2.00$ per item, $C_M = \$500$ repair cost for each breakdown, $h = \$0.6$ per item per unit time, $C_b = \$0.8$ per item backordered per unit time, and $g = 0.018$ years, time needed to repair and restore the machine.

Applying Eqs. (3.214), (3.215), and (3.202), one can obtain the optimal production time $T_1^* = 0.2345$ years, the optimal lot size $Q^* = 4221$, the backorder $B^* = 1368$, and $E[TCU(Q^*, B^*)] = \$7851.30$ (Chiu et al. 2008).

Numerical 3.7 Chiu et al. (2008) assumed another manufactured item can be produced at a rate of 44,000; its annual demand is 11,000 units. A random percentage of scrap items produced x follows a uniform distribution over the interval [0, 0.20]. Other parameters used are as follows (Chiu et al. 2008): $K = \$350$ for each production run, $C_s = \$0.80$ disposal cost for each scrap item, $C = \$2.40$ per item, $C_M M = \$600$ repair cost for each breakdown, $h = \$1.00$ per item per unit time, $C_b = \$1.40$ per item backordered per unit time, and $g = 0.036$ years, time needed to repair and restore the machine.

Applying Eqs. (3.210) and (3.211), one obtains the optimal $T_1^* = 0.1711$ years or 8.90 weeks, and backorder level $B^* = 2008$. From Eq. (3.203), the long-run average costs $E[\text{TCU}(T_1^*, B^*)] = \$28,324$. The optimal production run time T_1^* can be used to determine a multi-item production schedule. From Eq. (3.212), one also obtains optimal lot size $Q^* = 7526$ (Chiu et al. 2008).

3.3.1.4 Integrated Procurement–Production–Inventory Model

Nobil et al. (2018) derived an integrated procurement–production–inventory system for a single product and its raw materials without/with shortage. This system considers that the manufacturing process fabricates both perfect and defective finished products. The defective products are considered as scrapped items. The products are fabricated at P rate where $x\%$ of these are not useful, so non-defective items are produced with $(1 - x)P$ rate. In fact, the inventory level of finished product increases with $(1 - x)P - D$ rate, where D is demand rate of items. The finished product needs n type of raw materials to produce it; these are provided from outside suppliers. Thus, producer has to consider cost of ordering and purchasing raw materials in the inventory system costs in addition to cost of producing the item. Here, producer defines the quantity of raw material j that needs to be ordered for some producing periods (M_j) and store in his/her stock for beginning of production. Their model is developed without/with shortage for final finished product. For the shortage case is supposed that items in each cycle can have shortage up to B units (Nobil et al. 2018).

Some notations which are specifically used to model this problem are shown in Table 3.13.

First consider that the shortage is not permitted. In this case, the on-hand inventory graph of the raw material j and the finished product for the proposed problem without shortage are shown in Fig. 3.18.

From Fig. 3.18, the following equations are expressed (Nobil et al. 2018):

$$t_P = \frac{Q}{P} \tag{3.220}$$

$$I = [(1 - x)P - D]\frac{Q}{P} \tag{3.221}$$

3.3 Fully Backordered

$$t_d = \frac{I}{D} = \left(\frac{(1-x)P - D}{DP}\right)Q, \quad (3.222)$$

$$T = t_P + t_d = \frac{Q}{P} + \left(\frac{(1-x)P - D}{DP}\right)Q = \frac{(1-x)Q}{D}, \quad (3.223)$$

Then,

$$T_j = M_j T = \frac{(1-x)M_j Q}{D} \quad (3.224)$$

The total cost consists of the following costs: production cost, the disposal cost of scrapped items, the setup cost for manufacturing the finished product, holding cost of finished product, the ordering cost of raw materials, the purchasing cost of raw materials, and the holding cost for raw materials. Thus, these costs are obtained as follows (Nobil et al. 2018):

The production cost per cycle and the number of cycles per unit time are equal to CQ and $\frac{1}{T}$, respectively. Therefore, the production cost per unit time is calculated by:

$$\text{Production cost} = \frac{CQ}{T} \quad (3.225)$$

$$= \frac{DCQ}{Q(1-x)} = \frac{DC}{(1-x)} \quad (3.226)$$

$$\text{Disposal cost} = \frac{C_S x Q}{T} = \frac{DC_S x Q}{Q(1-x)} = \frac{DC_S x}{(1-x)} \quad (3.227)$$

$$\text{Set up cost} = \frac{K}{T} = \frac{DK}{Q(1-x)} = \frac{DK}{(1-x)}\left(\frac{1}{Q}\right) \quad (3.228)$$

$$\text{Holding cost for finished product} = \frac{H}{2T}[I \times T] \quad (3.229)$$

Substituting I and T from Eqs. (3.221) and (3.223) into Eq. (3.229), hence:

$$\text{Holding cost for finished product} = \frac{H}{2T}$$
$$\times \left[((1-x)P - D)\frac{Q}{P} \times \frac{Q(1-x)}{D}\right] \quad (3.230)$$

Thus, using Eq. (3.223) in above relation, thus (Nobil et al. 2018):

Table 3.13 Notation of given problem (Nobil et al. 2018)

N	Number of raw materials
a_j	Amount of raw material j required to produce one finished product (amount of raw material j/unit of finished product)
I	Maximum on-hand inventory of finished product (time)
O_j	Ordering cost of raw material j per order (time)
C_j^{Raw}	Purchasing cost of raw material j per unit ($/unit)
h_j	Holding cost of raw material j per item per unit time ($/unit/unit time)
T	Cycle length of the finished product (time)
T_j	Cycle length of the raw material j (time)
M_j	Number of cycles for raw material j

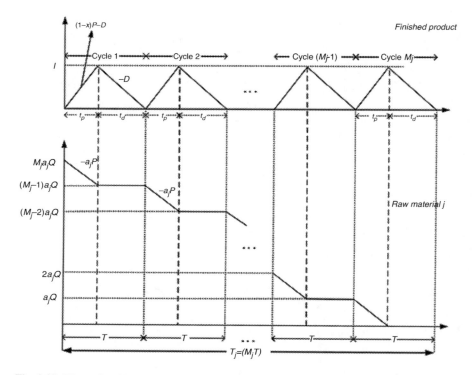

Fig. 3.18 The on-hand inventory graph for the problem without shortage (Nobil et al. 2018)

$$\text{Holding cost for finished product} = \frac{H[(1-x)P - D]}{2P} Q$$

$$= \frac{H}{2}\left(1 - x - \frac{D}{P}\right) Q \quad (3.231)$$

3.3 Fully Backordered

The ordering cost of the raw material j for M_j cycles and the number of cycles per unit time are equal to O_j and $\frac{1}{T_f}$ respectively. Thus, total ordering cost per unit time is obtained by:

$$\text{Ordering cost} = \sum_{j=1}^{n} \frac{O_j}{T_j} = \sum_{j=1}^{n} \frac{DO_j}{(1-x)} \left(\frac{1}{M_j Q}\right) \quad (3.232)$$

The purchasing cost of the raw material j for M_j cycles is equal to $M_j R_j \alpha_j Q$ that α_j is proportion of per unit of finished product. Thus, total purchasing cost per unit time is computed by:

$$\text{Raw material purchasing cost} = \sum_{j=1}^{n} \frac{M_j C_j^{\text{Raw}} \alpha_j Q}{T_j} = \sum_{j=1}^{n} \frac{C_j^{\text{Raw}} \alpha_j D}{(1-x)} \quad (3.233)$$

From Fig. 3.18, the area of the raw material in this figure is equal to:

$$\left\{ \frac{\alpha_j Q t_P}{2} + \frac{3\alpha_j Q t_P}{2} + \frac{5\alpha_j Q t_P}{2} + \cdots + \frac{(2M_j - 3)\alpha_j Q t_P}{2} + \frac{(2M_j - 1)\alpha_j Q t_P}{2} \right\}$$
$$+ \left\{ \alpha_j Q t_d + 2\alpha_j Q t_d + 3\alpha_j Q t_d + \cdots + (M_j - 2)\alpha_j Q t_d + (M_j - 2)\alpha_j Q t_d \right\}$$
$$(3.234)$$

Based on Appendix A of Nobil et al. (2018), the area of the raw material j can easily be determined as follows:

$$\text{Area of raw material } j = \left[\frac{\alpha_j}{2P} + \frac{\alpha_j}{2} \left(\frac{(1-x)P - D}{DP} \right) Q^2 M_j^2 \right] - \frac{\alpha_j}{2}$$
$$\times \left(\frac{(1-x)P - D}{DP} \right) Q^2 M_j \quad (3.235)$$

Thus, total holding cost per unit time for all the raw materials is obtained as follows:

$$\text{Holding cost of raw materials} = \sum_{j=1}^{n} \left\{ \frac{h_j}{T_j} \left(\left[\frac{\alpha_j}{2P} + \frac{\alpha_j}{2} \left(\frac{(1-x)P - D}{DP} \right) Q^2 M_j^2 \right] \right. \right.$$
$$\left. \left. - \frac{\alpha_j}{2} \left(\frac{(1-x)P - D}{DP} \right) Q^2 M_j \right) \right\}$$
$$= \sum_{j=1}^{n} \left\{ \frac{Dh_j}{(1-x)} \left[\frac{\alpha_j}{2P} + \frac{\alpha_j}{2} \left(\frac{(1-x)P - D}{DP} \right) QM_j \right] \right.$$
$$\left. - \frac{Dh_j \alpha_j}{2(1-x)} \left(\frac{(1-x)P - D}{DP} \right) Q \right\}$$
$$(3.236)$$

Now, based on Eqs. (3.226)–(3.228), (3.231)–(3.233), and (3.236), the total cost for the IPPI model without shortage, denoted by TC, is written as follows:

$$\text{TC} = \sum_{j=1}^{n} \left\{ \frac{D\left(C_j^{\text{Raw}} \alpha_j + C_S x + C\right)}{(1-x)} + \frac{DK}{(1-x)}\left(\frac{1}{Q}\right) + \frac{DO_j}{(1-x)}\left(\frac{1}{M_j Q}\right) \right.$$
$$+ \left[\frac{Dh_j}{(1-x)} \left[\frac{\alpha_j}{2P} + \frac{\alpha_j}{2}\left(\frac{(1-x)P - D}{DP}\right) \right] \right] QM_j$$
$$\left. + \left[\frac{h}{2}\left(1 - x - \frac{D}{P}\right) - \frac{Dh_j \alpha_j}{2(1-x)}\left(\frac{(1-x)P - D}{DP}\right) \right] Q \right\}$$

(3.237)

Therefore, the optimization problem is formulated as follows:

$$\text{TC} = \sum_{j=1}^{n} \Delta_1^j + \Delta_2 \left(\frac{1}{Q}\right) + \sum_{j=1}^{n} \Delta_3^j \left(\frac{1}{M_j Q}\right) + \sum_{j=1}^{n} \Delta_4^j (M_j Q) + \sum_{j=1}^{n} \Delta_5^j Q$$
$$\text{s.t.} Q > 0$$
$$M_j \geq 1 \ \& \ \text{integer}; j = 1, 2, \ldots, n$$

(3.238)

$$\Delta_1^j = \frac{D\left(C_j^{\text{Raw}} \alpha_j + C_S x + C\right)}{(1-x)} > 0; j = 1, 2, \ldots, n \qquad (3.239)$$

$$\Delta_2 = \frac{DK}{(1-x)} > 0 \qquad (3.240)$$

$$\Delta_3^j = \frac{DO_j}{(1-x)} > 0; j = 1, 2, \ldots, n \qquad (3.241)$$

$$\Delta_4^j = \frac{Dh_j}{(1-x)} \left[\frac{\alpha_j}{2P} + \frac{\alpha_j}{2}\left(\frac{(1-x)P - D}{DP}\right) \right] > 0; j = 1, 2, \ldots, n \qquad (3.242)$$

$$\Delta_5^j = \frac{h}{2}\left(1 - x - \frac{D}{P}\right) - \frac{Dh_j}{2(1-x)}\left(\frac{(1-x)P - D}{DP}\right); j = 1, 2, \ldots, n \qquad (3.243)$$

Now consider that shortage is permitted and is fully backordering. The on-hand inventory graph for the raw material j and the finished product for the model with shortage are shown in Fig. 3.19.

From Fig. 3.19, the following equations are deduced:

$$t_1 = \frac{B}{(1-x)P - D} \qquad (3.244)$$

3.3 Fully Backordered

$$I = [(1-x)P - D]\frac{Q}{P} - B \tag{3.245}$$

$$t_2 = \frac{I}{(1-x)P - D} = \frac{Q}{P} - \frac{B}{(1-x)P - D}, \tag{3.246}$$

$$t_P = t_1 + t_2 = \frac{Q}{P}, \tag{3.247}$$

$$t_3 = \frac{I}{D} = \left(\frac{(1-x)P - D}{DP}\right)Q - \frac{B}{D} \tag{3.248}$$

$$t_4 = \frac{B}{D} \tag{3.249}$$

$$t_d = t_3 + t_4 = \left(\frac{(1-x)P - D}{DP}\right)Q \tag{3.250}$$

$$T = t_P + t_d = \frac{Q}{P} + \left(\frac{(1-x)P - D}{DP}\right)Q = \frac{(1-x)Q}{D} \tag{3.251}$$

So,

$$T_j = M_j T = \frac{(1-x)M_j Q}{D} \tag{3.252}$$

The total cost is comprised of the following costs: production cost, the disposal cost of scrapped items, the setup cost for producing the finished product, the holding cost for finished product, the backorder cost, the ordering cost of raw materials, the purchasing cost of raw materials, and holding cost of raw materials. Thus, these costs are obtained as follows (Nobil et al. 2018):

The production cost per time unit for the finished product is calculated by:

$$\text{Production cost} = \frac{CQ}{T} = \frac{DCQ}{Q(1-x)} = \frac{DC}{(1-x)} \tag{3.253}$$

The disposal cost per time unit for the scrapped items is determined by:

$$\text{Disposal cost} = \frac{C_S x Q}{T} = \frac{DC_S x Q}{Q(1-x)} = \frac{DC_S x}{(1-x)} \tag{3.254}$$

The setup cost per time unit is obtained with:

$$\text{Setup cost} = \frac{K}{T} = \frac{DK}{Q(1-x)} = \frac{DK}{(1-x)}\left(\frac{1}{Q}\right) \quad (3.255)$$

Based on Fig. 3.19, the holding cost per unit of finished product is given by:

$$\text{Holding cost for finished product} = \frac{h}{2T}[I \times (t_2 + t_3)] \quad (3.256)$$

Substituting I, t_2, t_3, and T from Eqs. (3.245), (3.246), (3.248), and (3.251), respectively, thus (see Appendix C of for detail calculations),

$$\text{Holding cost for finished product} = \frac{h}{2}$$
$$\times \left[((1-x)P - D)\frac{Q}{P} + \frac{P}{((1-x)P - D)}\left(\frac{B^2}{Q}\right)\right]$$
$$- hB$$
$$(3.257)$$

Based on Fig. 3.19, the backorder cost per unit of finished product is computed by:

$$\text{Backordering cost} = \frac{C_b}{2T}[B \times (t_1 + t_4)] \quad (3.258)$$

Substituting t_1 and t_4 from Eqs. (3.244) and (3.249) respectively, hence,

$$\text{Backordering cost} = \frac{C_b}{2T}\left[B \times \left(\frac{B}{(1-x)P - D} + \frac{B}{D}\right)\right] \quad (3.259)$$

From Eq. (3.251),

$$\text{Backordering cost} = \frac{C_b D}{2(1-x)Q}\left[B \times \left(\frac{B}{(1-x)P - D} + \frac{B}{D}\right)\right]$$
$$= \frac{C_b P}{2((1-x)P - D)}\left(\frac{B^2}{D}\right) \quad (3.260)$$

The total ordering cost per for time unit is determined by:

$$\text{Ordering cost} = \sum_{j=1}^{n}\frac{O_j}{T_j} = \sum_{j=1}^{n}\frac{DO_j}{(1-x)}\left(\frac{1}{M_j Q}\right) \quad (3.261)$$

The total purchasing cost per time unit is calculated by:

3.3 Fully Backordered

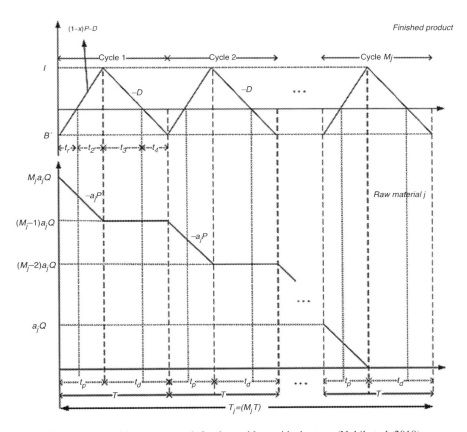

Fig. 3.19 The on-hand inventory graph for the problem with shortage (Nobil et al. 2018)

$$\text{Raw material purchasing cost} = \sum_{j=1}^{n} \frac{M_j C_j^{\text{Raw}} \alpha_j Q}{T_j} = \sum_{j=1}^{n} \frac{C_j^{\text{Raw}} \alpha_j D}{(1-x)} \quad (3.262)$$

From Fig. 3.19, the area of the raw material j in this figure is equal to:

$$\left\{ \frac{\alpha_j Q t_P}{2} + \frac{3\alpha_j Q t_P}{2} + \frac{5\alpha_j Q t_P}{2} + \cdots + \frac{(2M_j - 3)\alpha_j Q t_P}{2} + \frac{(2M_j - 1)\alpha_j Q t_P}{2} \right\}$$
$$+ \left\{ \alpha_j Q t_d + 2\alpha_j Q t_d + 3\alpha_j Q t_d + \cdots + (M_j - 2)\alpha_j Q t_d + (M_j - 2)\alpha_j Q t_d \right\}$$
$$(3.263)$$

Because the uptime period and downtime period, denoted by t_P and t_d, are same as the model without shortage, in other words, $t_P = \frac{Q}{P}$ and $t_d = \left(\frac{(1-x)P-D}{DP} \right) Q$. Therefore, the area of the raw material j in Fig. 3.19 is obtained as follows:

Area of raw material $j = \left[\dfrac{\alpha_j}{2P} + \dfrac{\alpha_j}{2}\left(\dfrac{(1-x)P - D}{DP}\right)Q^2 M_j^2\right] - \dfrac{\alpha_j}{2}$

$$\times \left(\dfrac{(1-x)P - D}{DP}\right)Q^2 M_j \qquad (3.264)$$

Thus, the total holding cost per time unit is computed as follows:

Holding cost of raw materials $= \displaystyle\sum_{j=1}^{n}\left\{\dfrac{Dh_j}{(1-x)}\left[\dfrac{\alpha_j}{2P} + \dfrac{\alpha_j}{2}\left(\dfrac{(1-x)P - D}{DP}\right)QM_j\right]\right.$

$$\left.-\dfrac{Dh_j \alpha_j}{2(1-x)}\left(\dfrac{(1-x)P - D}{DP}\right)Q\right\} \qquad (3.265)$$

Now, based on Eqs. (3.253)–(3.255), (3.257), (3.260)–(3.262), and (3.265), the total cost for the IPPI model with shortage is given by:

$$TC = \sum_{j=1}^{n}\left\{\dfrac{D\left(C_j^{\text{Raw}}\alpha_j + C_S x + C\right)}{(1-x)} + \dfrac{DK}{(1-x)}\left(\dfrac{1}{Q}\right) + \dfrac{DO_j}{(1-x)}\left(\dfrac{1}{M_j Q}\right)\right.$$

$$+ \left[\dfrac{Dh_j}{(1-x)}\left[\dfrac{\alpha_j}{2P} + \dfrac{\alpha_j}{2}\left(\dfrac{(1-x)P - D}{DP}\right)\right]\right]QM_j$$

$$\left. + \left[\dfrac{h}{2}\left(1 - x - \dfrac{D}{P}\right) - \dfrac{Dh_j \alpha_j}{2(1-x)}\left(\dfrac{(1-x)P - D}{DP}\right)\right]Q\right\} - hB + \dfrac{(h + C_b)P}{2((1-x)P - D)}\left(\dfrac{B^2}{Q}\right)\right\} \qquad (3.266)$$

Thus, the optimization problem is stated as follows (Nobil et al. 2018):

$$TC = \sum_{j=1}^{n}\left[\Delta_1^j + \Delta_2\left(\dfrac{1}{Q}\right) + \Delta_3^j\left(\dfrac{1}{M_j Q}\right) + \Delta_4^j(M_j Q) + \Delta_5^j Q - hB + \Delta_6\left(\dfrac{B^2}{Q}\right)\right]$$

s.t. $QK > 0$

$M_j \geq 1$ & integer; $j = 1, 2, \ldots, n$

$$(3.267)$$

3.3 Fully Backordered   217

$$\Delta_1^j = \frac{D\left(C_j^{\text{Raw}}\alpha_j + C_S x + C\right)}{(1-x)} > 0; j = 1, 2, \ldots, n$$

$$\Delta_2 = \frac{DK}{(1-x)} > 0$$

$$\Delta_3^j = \frac{DO_j}{(1-x)} > 0; j = 1, 2, \ldots, n \quad (3.268)$$

$$\Delta_4^j = \frac{Dh_j}{(1-x)}\left[\frac{\alpha_j}{2P} + \frac{\alpha_j}{2}\left(\frac{(1-x)P-D}{DP}\right)\right] > 0; j = 1, 2, \ldots, n$$

$$\Delta_5^j = \frac{h}{2}\left(1-x-\frac{D}{P}\right) - \frac{Dh_j}{2(1-x)}\left(\frac{(1-x)P-D}{DP}\right); j = 1, 2, \ldots, n$$

$$\Delta_6 = \frac{(h+C_b)P}{2((1-x)P-D)} > 0$$

They have used metaheuristic algorithms to solve their models, so for more detailed information, readers can refer to Nobil et al. (2018).

3.3.1.5 Service Level Constraint

Chiu (2006) studied the effect of service level constraint on EPQ model with random defective rate. In the realistic inventory control and management, due to certain internal orders of parts/materials and other operating considerations, the planned backlogging is the strategy to effectively minimize overall inventory costs. While allowing backlogging, abusive shortage in an inventory model, however, may cause an unacceptable service level and turn into possible loss of future sales (because of the loss of customer goodwill). Therefore, the maximal allowable shortage level per cycle is always set as an operating constraint of the business in order to achieve minimal service level while deriving the optimal lot size decision (Chiu 2006).

Some notations which are specifically used to model this problem are shown in Table 3.14.

The EPQ model assumes that the production rate P must always be greater than or equal to the demand rate D. The production rate of perfect-quality items must always be greater than or equal to the sum of the demand rate and the production rate of defective items ($P - d - D \geq 0$ or $1 - x - \frac{D}{P} \geq 0$). Figure 3.20 depicts the on-hand inventory level and allowable backorder level for the EPQ model with backlogging permitted. For the following derivation, they employ the solution procedures used by Hayek and Salameh (2001). From Fig. 3.20, one can obtain the cycle length T, production uptime t_1, the maximum level of on-hand inventory H_1, production downtime t_2, shortage permitted time t_3, and t_4 as follows (Chiu 2006):

$$T = \sum_{i=1}^{4} t_i = \frac{Q_b(1-x)}{D}, \qquad (3.269)$$

$$t_1 = \frac{H_1}{P-d-D}, \qquad (3.270)$$

$$H_1 = (P-d-D)\frac{Q_b}{P} - B, \qquad (3.271)$$

$$t_2 = \frac{H_1}{D}, \qquad (3.272)$$

$$t_3 = \frac{B}{D}, \qquad (3.273)$$

$$t_4 = \frac{B}{P-d-D}, \qquad (3.274)$$

$$(t_1 + t_4) = \frac{Q}{P}. \qquad (3.275)$$

The scrap items built up randomly during production uptime ($t_1 + t_4$) are (Chiu 2006):

$$d \cdot (t_1 + t_4) = x \cdot Q. \qquad (3.276)$$

The inventory cost per cycle is (Chiu 2006):

$$\begin{aligned}\text{TC}(Q,B) = {}& CQ + C_S x Q + K + h\left[\frac{H_1}{2}(t_1+t_2)\right] \\ & + C_b\left[\frac{B}{2}(t_3+t_4)\right] + h\left[\frac{d(t_1+t_4)}{2}(t_1+t_4)\right]. \end{aligned} \qquad (3.277)$$

Since scrap items are produced randomly during a regular production run, the cycle length T is a variable (see Fig. 3.20). One may employ the renewal reward theorem (Zipkin 2000) to cope with the variable cycle length. By substituting variables from Eqs. (3.269) to (3.276) in Eq. (3.277), the expected cost $E[\text{TCU}(Q,B)] = \frac{E[\text{TC}(Q_b, B)]}{E(T)}$ can be obtained as follows (Chiu and Chiu 2003):

$$\begin{aligned} E[\text{TCU}(Q,B)] = {}& D\left[\frac{C}{1-E(x)} + C_S\frac{E(x)}{1-E(x)}\right] + \frac{KD}{Q}\frac{1}{1-E(x)} \\ & + \frac{h}{2}\left[\left(1-\frac{D}{P}\right)Q_b - 2B\right]\frac{1}{1-E(x)} + \frac{(C_b+h)}{1-E(x)} E\left(\frac{1-x}{1-x-\frac{D}{P}}\right)\frac{B^2}{2Q} \\ & + h\left[B - \left(1-\frac{D}{P}\right)Q\right]\frac{E(x)}{1-E(x)} + \frac{hQ}{2}\frac{E(x^2)}{1-E(x)} \end{aligned} \qquad (3.278)$$

3.3 Fully Backordered

Table 3.14 Notation of given problem (Chiu 2006)

t_1	Production uptime (time)
t_2	Production downtime (time)
t_3	Time shortage permitted (time)
t_4	Time needed to satisfy all the backorders by the next production (time)

Fig. 3.20 On-hand inventory of the EPQ model with random defective rate and backlogging permitted (Chiu 2006)

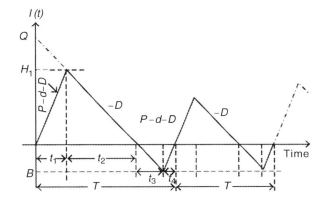

For the proof of convexity of $E[\mathrm{TCU}(Q, B)]$, one can utilize the Hessian matrix equation (Rardin and Rardin 1998):

$$[Q \ \ B] \begin{pmatrix} \dfrac{\partial^2 E[\mathrm{TCU}(Q,B)]}{\partial Q^2} & \dfrac{\partial^2 E[\mathrm{TCU}(Q,B)]}{\partial Q \partial B} \\ \dfrac{\partial^2 E[\mathrm{TCU}(Q,B)]}{\partial Q \partial B} & \dfrac{\partial^2 E[\mathrm{TCU}(Q,B)]}{\partial B^2} \end{pmatrix} \begin{bmatrix} Q \\ B \end{bmatrix}$$

$$= \frac{2KD}{Q} \frac{1}{1 - E(x)} > 0. \tag{3.279}$$

Equation (3.279) is positive, because all parameters are positives. Hence, the expected inventory cost function $E[\mathrm{TCU}(Q, B)]$ is a strictly convex function for all Q and B different from zero (Chiu 2006).

Hence, it follows that for the optimal production lot size Q and the maximal level of backorder B, one can differentiate $E[\mathrm{TCU}(Q, B)]$ with respect to Q and with respect to B and solve the linear system of Eq. (3.280) by letting these partial derivatives equal to zero (Chiu 2006):

$$\frac{\partial E[\text{TCU}(Q,B)]}{\partial Q} = -\frac{KD}{Q^2}\frac{1}{1-E(x)} + \frac{h}{2}\left(1-\frac{D}{P}\right)\frac{1}{1-E(x)}$$
$$-\frac{B^2}{2Q^2}\frac{(C_b+h)}{1-E(x)}E\left(\frac{1-x}{1-x-\frac{\lambda}{P}}\right) - h\left(1-\frac{D}{P}\right)\frac{E(x)}{1-E(x)} + \frac{h}{2}\frac{E(x^2)}{1-E(x)} = 0$$

(3.280)

$$\frac{\partial E[(\text{TCU}(Q,B))]}{\partial B} = -h\frac{1}{1-E(x)} + \frac{B}{Q}\frac{(C_b+h)}{1-E(x)}E\left(\frac{1-x}{1-x-\frac{D}{P}}\right)$$
$$+ h\frac{E(x)}{1-E(x)}$$
$$= 0$$

(3.281)

Hence, one derives the optimal production policy, Q^*_b and B^*, as shown below (Chiu 2006):

$$Q^* = \sqrt{\frac{2KD}{h\left(1-\frac{D}{P}\right) - \frac{h^2}{C_b+h}\frac{\{1-E(x)\}^2}{E\left(\frac{1-x}{1-x-\frac{D}{P}}\right)} - 2h\left(1-\frac{D}{P}\right)E(x) + hE(x^2)}}$$

(3.282)

$$B^* = \frac{h}{(C_b+h)}\frac{1-E(x)}{E\left(\frac{1-x}{1-x-\frac{D}{P}}\right)}Q^*$$

(3.283)

Now for the EPQ model with random defective rate when shortage is not permitted, the cycle length $T = t_1 + t_2$ (see Fig. 3.20). The expected annual cost $E[\text{TCU}(Q)] = \frac{E(\text{TC}(Q))}{E(T)}$ can be obtained as follows (Chiu et al. 2003):

$$E[\text{TCU}(Q)] = D\left[\frac{C}{1-E(x)} + C_S\frac{E(x)}{1-E(x)}\right] + \frac{KD}{Q}\frac{1}{1-E(x)} + \frac{hQ}{2}\left(1-\frac{D}{P}\right)\frac{1}{1-E(x)}$$
$$- hQ\left(1-\frac{D}{P}\right)\frac{E(x)}{1-E(x)} + \frac{hQ}{2}\frac{E(x^2)}{1-E(x)}$$

(3.284)

Differentiating $E[\text{TCU}(Q)]$ with respect to Q twice, we find that $E[\text{TCU}(Q)]$ is convex, and by minimizing the expected annual cost $E[\text{TCU}(Q)]$, one can derive the optimal production quantity Q^* as shown in Eq. (3.285) (Chiu 2006):

$$Q^* = \sqrt{\frac{2KD}{h\left(1-\frac{D}{P}\right) - 2h\left(1-\frac{D}{P}\right)E(x) + hE(x^2)}}.$$

(3.285)

Now effects of backlogging and service level constraint on the EPQ model will be considered.

3.3 Fully Backordered

The expected annual cost per when backlogging is not permitted is always greater than or equal to that of the EPQ model with allowed backlogging. That is, $E(\text{TCU}(Q)) \geq E(\text{TCU}(Q, B))$, for any given $Q = Q_{\text{backlogging}}$ (Chiu 2006), because:

$$E(\text{TCU}(Q)) - E(\text{TCU}(Q, B)) = \frac{hB}{1 - E(x)} - \frac{(C_b + h)}{1 - E(x)} E\left(\frac{1 - x}{1 - x - \frac{D}{P}}\right)$$
$$\times \frac{B^2}{2Q} - h \frac{E(x)}{1 - E(x)} B \qquad (3.286)$$

Substituting B, one has

$$E(\text{TCU}(Q)) - E(\text{TCU}(Q, B)) = \frac{h^2}{2(C_b + h)} \frac{1 - E(x)}{E\left(\frac{1-x}{1-x-\frac{D}{P}}\right)} Q \geq 0 \qquad (3.287)$$

Since parameters h and C_b are nonnegative numbers, the random defective rate x and, $1 - x - \frac{D}{P} \geq 0$ and the production lot size $Q \geq 0$, hence Eq. (3.287) ≥ 0.

So it is better (in terms of total inventory costs) to permit shortage and have them backordered for the EPQ model with random defective rate. While allowing backlogging, abusive shortage in an inventory model, however, may cause an unacceptable service level and turn into possible loss of future sales. Hence, the maximal allowable shortage level per cycle is always set as an operating constraint for the business in order to attain the minimal service level. Suppose that set α to be the maximum proportion of shortage permitted per cycle (i.e., the service level $= (1 - \alpha)\%$), then (Chiu 2006):

$$\alpha = \frac{t_3 + t_4}{T}, \qquad (3.288)$$

$$\frac{\alpha}{1 - \alpha} = \frac{t_3 + t_4}{t_1 + t_2}. \qquad (3.289)$$

Substituting t_1, t_2, t_3, and t_4, one obtains (Chiu 2006):

$$\frac{\alpha}{1 - \alpha} = \frac{B}{\left(1 - x - \frac{D}{P}\right)Q - B}, \qquad (3.290)$$

Substituting B, one has the following (Chiu 2006):

$$\frac{1}{\alpha} = \frac{1}{1 - E(x)} \frac{(C_b + h)}{h} E\left(\frac{1 - x}{1 - x - \frac{D}{P}}\right)\left(1 - x - \frac{D}{P}\right), \qquad (3.291)$$

$$C_{\rm b} = h \cdot \left\{ \frac{1}{\alpha} \left[\frac{1 - E(x)}{(1 - x - D/P)} \cdot E\left(\frac{1 - x}{1 - x - D/P} \right)^{-1} \right] - 1 \right\}. \qquad (3.292)$$

Assume that (Chiu 2006):

$$f(\alpha, x) = h \cdot \left\{ \frac{1}{\alpha} \left[\frac{1 - E(x)}{(1 - x - \frac{D}{P})} \cdot E\left(\frac{1 - x}{1 - x - \frac{D}{P}} \right)^{-1} \right] - 1 \right\}. \qquad (3.293)$$

Equation (3.293) represents the relationship between the imputed backorder cost f (α, x) and the maximum proportion of shortage permitted time α. In other words, when the service level $(1 - \alpha)\%$ of the EPQ model is set, the corresponding imputed backorder cost $f(\alpha, x)$ can be obtained. Hence, one can utilize this information to determine whether or not the service level is achievable. For the computation of $E\left(\frac{1-x}{1-x-\frac{D}{P}} \right)$, one can refer to Chiu (2006).

Let C_{bi} be the tangible backorder cost per item. If $C_{bi} > f(\alpha, x)$, then the service level $(1 - \alpha)\%$ is achievable. Otherwise, the tangible backorder cost should be increased to $f(\alpha, x)$, and then use it to derive the new optimal operating policy (in terms of Q^* and B^*), so that the overall inventory costs can be minimized and the service level constraint will be attained. Letting C_{bi} be the adjustable intangible backorder cost (per item per unit time), then b_i should satisfy the following condition in order to attain the $f(\alpha, x)$ service level (Chiu 2006):

$$b_i \geq [f(\alpha, x) - C_{bi}]. \qquad (3.294)$$

Therefore, by using $C_b = f(\alpha, x)$, one can derive the new optimal production lot size Q^* and the optimal backorder level B^* that minimizes the expected annual inventory costs as well as achieves the minimal service level $(1 - \alpha)\%$ (Chiu 2006).

Example 3.8 Chiu (2006) considered a company which produces a product for several regional clients. He assumed $D = 4000$ units per year, $P = 10,000$ units per year, $C_s = \$0.3$ per scrap item, $K = \$450$ per setup, $C = \$2$ per item, $x \sim U[0, 0.1]$, $C_{bt} = \$0.2$ per item backordered per unit time, $\alpha = 0.3$, and $h = \$0.6$ per item per year.

First let $C_b = C_{bt}$. From Eqs. (3.278), (3.282), and (3.283), one obtains the overall costs $E[TCU(Q_b^*, B^*)] = \$9087$, the optimal production quantity $Q^* = 6284$, and the optimal backorder level $B^* = 2589$.

For EPQ model with backlogging not allowed, from Eqs. (3.286) to (3.287), the total cost $E[TCU(Q^*)] = \$9625$ and the optimal production quantity $Q^* = 3323$ are obtained. One notices that the EPQ model with backlogging permitted has a lower overall cost than that of the EPQ model with no shortage allowed.

Another research to investigate the effects of service level constraint on order and shortage quantities is performed by Shyu et al. (2009). Readers for more detailed information can refer to this work.

3.4 Partial Backordered

3.4.1 Continuous Delivery

In this section, the EPQ models with scrap and continuous delivery policy with partial backordered shortage are presented.

3.4.1.1 Multi-product Single-Machine System

Taleizadeh et al. (2010) developed a multi-product single-machine production system under economic production quantity (EPQ) model in which the existence of only one machine causes a limited production capacity for the common cycle length of all products, the production defective rates are random variables, shortages are allowed and take a combination of backorder and lost sale, and there is a service rate constraint for the company.

Imperfect production processes, due to process deterioration or some other factors, may randomly generate X percent of defective items at a rate d. The inspection cost per item is involved when all items are screened. All defective items are assumed to be scrapped; i.e., no rework is allowed. The annual constant production rate (P) is much larger than the annual constant demand rate (D) as the basic assumption of the finite production model. In other words, the expected production rate of the scrapped items θ can be expressed as $d = PE[X]$. Also, Taleizadeh et al. (2010) assumed that there is a real constant production capacity limitation on a single machine on which all products are produced and that the setup cost is nonzero.

Some notations which are specifically used to model this problem are shown in Table 3.15. Index $j = 1, 2, \ldots, n$ refers to the number of products.

The production rate P_j is always assumed to be greater than or equal to the demand rate D_j. Furthermore, the production rate of the perfect-quality items is assumed to be greater than or equal to the sum of the demand rate and the production rate of defective items. In other words, $P_j - D_j - d_j \geq 0$ or $1 - E(x) - \frac{D_j}{P_j} \geq 0$..

Figure 3.21 depicts the on-hand inventory level and allowable backorder level of the EPQ model with permitted backlogging. To model the problem, a part of modeling procedure used in Hayek and Salameh (2001) is applied. Since all products are manufactured on a single machine with a limited capacity, the cycle length for all of them is equal ($T_1 = T_2 = \ldots = T_n = T$). Then, based on Fig. 3.21, for $j = 1, 2, \ldots, n$, one obtains (see Appendix A of Taleizadeh et al. 2010):

$$T = \sum_{i=1}^{4} t^i_j = \frac{Q^B_j\left(1 - E(x_j)\right) + (1 - \xi_j)B_j}{D_j} \qquad (3.295)$$

$$t_j^1 = \frac{I_j^1}{P_j - D_j - d_j} \tag{3.296}$$

$$I_j^1 = (P_j - D_j - d_j)\frac{Q_j}{P_j} - \xi_j B_j \tag{3.297}$$

$$t_j^2 = \frac{I_j^1}{D_j} \tag{3.298}$$

$$t_j^3 = \frac{\xi_j B_j}{\xi_j D_j} = \frac{B_j}{D_j} \tag{3.299}$$

$$t_j^4 = \frac{\xi_j B_j}{P_j - D_j - d_j} \tag{3.300}$$

$$t_j^1 + t_j^4 = \frac{Q_j}{P_j} \tag{3.301}$$

The objective function of the model is the summation of the expected annual production, holding, shortage, disposal, and setup costs as:

$$\begin{aligned}Z =\ & \text{Production cost} + \text{Holding cost} + \text{Backordering cost} \\ & + \text{Disposal cost} + \text{Setup cost}\end{aligned} \tag{3.302}$$

In the following subsections, different parts of the objective function are described.

The production cost per unit and the production quantity per period of the jth product are C_j and Q_j, respectively. Hence, the production cost of the jth product per period is $C_j \cdot Q_j$. While the total annual production cost of the jth product in a disjoint production policy (each product is ordered separately) is $NC_j \cdot Q_j$, this cost for the joint policy (all products have a unique ordering cycle) is $\frac{C_j Q_j}{T}$. Furthermore, since the shortages are in combinations of backorders and lost sales, based on Eq. (3.307), one obtains:

$$Q_j = \frac{T \times D_j - (1 - \xi_j)B_j}{E(1 - x_j)} = \frac{T \times D_j - (1 - \xi_j)B_j}{1 - E(x_j)} \tag{3.303}$$

Hence, the expected annual production cost will be:

$$\sum_{j=1}^{n} \frac{C_j \left[\frac{T \times D_j - (1-\xi_j)B_j}{E(1-x_j)}\right]}{T} = \sum_{j=1}^{n} \frac{C_j D_j}{1 - E(x_j)} - \sum_{j=1}^{n} \left[\frac{C_j(1-\xi_j)}{1 - E(x_j)}\right]\frac{B_j}{T} \tag{3.304}$$

3.4 Partial Backordered

Table 3.15 Notation of given problem (Taleizadeh et al. 2010)

$f_{x_j}(x_j)$	The probability density function of x_j
ξ_j	The fraction of jth product shortage that is backordered
α	The safety factor of total allowable shortages
I_j^1	The maximum units of on-hand inventory level, when the regular production process stops, (unit)
N	The number of cycles per year
t_j^1	The production uptime of the jth product, (time)
t_j^2	The production downtime of the jth product, (time)
t_j^3	The permitted shortage time of the jth product, (time)
t_j^4	The time needed to satisfy all backorders in the next production of the jth product, (time)
Z	The annual expected total costs, ($/year)

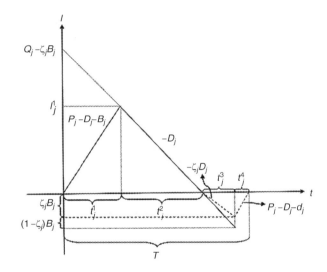

Fig. 3.21 A production–inventory cycle (Taleizadeh et al. 2010)

The holding cost per unit of the jth product per unit time for both the healthy and the scrapped items is h_j. According to Fig. 3.21, the total holding costs of healthy items per cycle and per year are shown in Eqs. (3.305) and (3.306), respectively:

$$\sum_{j=1}^{n} h_j \left[\frac{I_j^1}{2} \left(t_j^1 + t_j^2 \right) \right] \tag{3.305}$$

$$N \sum_{j=1}^{n} h_j \left[\frac{I_j^1}{2} \left(t_j^1 + t_j^2 \right) \right] \tag{3.306}$$

However, Eq. (3.306) for the joint production policy in which $N = \frac{1}{T}$ becomes:

$$\frac{1}{T}\sum_{j=1}^{n}h_j\left[\frac{I_j^1}{2}\left(t_j^1+t_j^2\right)\right] \qquad (3.307)$$

Finally, the expected total annual holding cost of healthy items is (see Appendix B of Taleizadeh et al. 2010):

$$\sum_{j=1}^{n}h_j(P_j-d_j)\left[\frac{(P_j-D_j-d_j)D_j}{2(P_j)^2(1-E(x_j))^2}T-\frac{(P_j-D_j-d_j)(1-\xi_j)+\xi_j P_j(1-E(x_j))}{(P_j)^2(1-E(x_j))^2}B_j\right.$$
$$\left.+\frac{(P_j-D_j-d_j)^2(1-\xi_j)^2+2\xi_j(1-\xi_j)P_j(1-E(x_j))(P_j-D_j-d_j)+(\xi_j P_j)^2(1-E(x_j))^2}{2D_j(P_j)^2(1-E(x_j))^2(P_j-D_j-d_j)}\frac{(B_j)^2}{T}\right] \qquad (3.308)$$

Since the scrap items of each product are assumed to be held until the end of its production time, based on Fig. 3.21, the total holding costs of the scrapped items per cycle and per year are shown in Eqs. (3.309) and (3.310), respectively:

$$\sum_{j=1}^{n}h_j\left[\frac{d_j(t_j^1+t_j^4)}{2}\left(t_j^1+t_j^4\right)\right] \qquad (3.309)$$

$$N\sum_{j=1}^{n}h_j\left[\frac{d_j(t_j^1+t_j^4)}{2}\left(t_j^1+t_j^4\right)\right] \qquad (3.310)$$

Again, for the joint production policy, Eq. (3.310) becomes:

$$\frac{1}{T}\sum_{j=1}^{n}h_j\left[\frac{d_j(t_j^1+t_j^4)}{2}\left(t_j^1+t_j^4\right)\right]=\frac{1}{T}\sum_{j=1}^{n}h_j\left[\frac{d_j}{2}\left(\frac{Q_j}{P_j}\right)^2\right] \qquad (3.311)$$

Hence, the expected total annual holding cost of scrapped items according to Eq. (3.303) is:

$$\sum_{j=1}^{n}h_j d_j\left[\frac{(D_j)^2 T-2D_j(1-\xi_j)B_j}{2(P_j)^2(1-E(x_j))^2}+\frac{(1-\xi_j)^2}{2(P_j)^2(1-E(x_j))^2}\left(\frac{B_j^2}{T}\right)\right] \qquad (3.312)$$

Finally, the expected total annual holding cost of healthy and scrapped items is:

3.4 Partial Backordered

$$\sum_{j=1}^{n} h_j(P_j - d_j) \left[\frac{(P_j - D_j - d_j)D_j}{2(P_j)^2(1-E(x_j))^2} T - \frac{(P_j - D_j - d_j)(1-\xi_j) + \xi_j P_j(1-E(x_j))}{(P_j)^2(1-E(x_j))^2} B_j \right.$$
$$\left. + \frac{(P_j - D_j - d_j)^2(1-\xi_j)^2 + 2\xi_j(1-\xi_j)P_j(1-E(x_j))(P_j - D_j - d_j) + (\xi_j P_j)^2(1-E(x_j))^2}{2D_j(P_j)^2(1-E(x_j))^2(P_j - D_j - d_j)} \frac{(B_j)^2}{T} \right]$$
$$+ \sum_{j=1}^{n} h_j d_j \left[\frac{(D_j)^2 T - 2D_j(1-\xi_j)B_j}{2(P_j)^2(1-E(x_j))^2} + \frac{(1-\xi_j)^2}{2(P_j)^2(1-E(x_j))^2} \left(\frac{B_j^2}{T} \right) \right] \qquad (3.313)$$

Based on Fig. 3.21, the backordered and lost sale costs per cycle are shown in Eqs. (3.314) and (3.315), respectively:

$$\sum_{j=1}^{n} C_{bj} \xi_j \left[\frac{B_j}{2} \left(t_j^3 + t_j^4 \right) \right] \qquad (3.314)$$

$$\sum_{j=1}^{n} \widehat{\pi}_j (1 - \xi_j) B_j \qquad (3.315)$$

These costs for a year become:

$$N \sum_{j=1}^{n} C_{bj} \xi_j \left[\frac{B_j}{2} \left(t_j^3 + t_j^4 \right) \right] \qquad (3.316)$$

$$N \sum_{j=1}^{n} \widehat{\pi}_j (1 - \xi_j) B_j \qquad (3.317)$$

Because of the joint production policy, Eqs. (3.316) and (3.317) will change to Eqs. (3.318) and (3.319), respectively:

$$\frac{1}{T} \sum_{j=1}^{n} C_{bj} \xi_j \left[\frac{B_j}{2} \left(t_j^3 + t_j^4 \right) \right] \qquad (3.318)$$

$$\frac{1}{T} \sum_{j=1}^{n} \widehat{\pi}_j (1 - \xi_j) B_j \qquad (3.319)$$

Finally, the expected annual backordered and lost sale costs are (see Appendix C of Taleizadeh et al. 2010):

$$\frac{1}{T} \sum_{j=1}^{n} C_{bj} \xi_j \left[\frac{(P_j - (1-\xi_j)D_j - d_j)(B_j)^2}{2D_j(P_j - D_j - d_j)} \right] \qquad (3.320)$$

$$\frac{1}{T}\sum_{j=1}^{n}\widehat{\pi}_j(1-\xi_j)B_j \qquad (3.321)$$

Since the quantity of scrapped items is $E(x_j)Q_j$, the expected total disposal cost per cycle is $\sum_{j=1}^{n}Cs_jE(x_j)Q_j$. This quantity per year becomes:

$$N\sum_{j=1}^{n}Cs_jE(x_j)Q_j \qquad (3.322)$$

$$\frac{1}{T}\sum_{j=1}^{n}Cs_jE(x_j)Q_j \qquad (3.323)$$

Since $Q_j = \frac{T \times D_j - (1-\xi_j)B_j}{1-E(x_j)}$ the annual expected total scrapped item cost is:

$$\begin{aligned}\frac{1}{T}\sum_{j=1}^{n}Cs_jE(x_j)&\frac{T \times D_j - (1-\xi_j)B_j}{1-E(x_j)}\\&=\sum_{j=1}^{n}\frac{Cs_jE(x_j)D_j}{1-E(x_j)} - \sum_{j=1}^{n}\frac{Cs_jE(x_j)(1-\xi_j)}{1-E(x_j)}\frac{B_j}{T}\end{aligned} \qquad (3.324)$$

The cost of a setup is K which occurs N times per year. So, the annual setup cost will be:

$$NK = \frac{K}{T} \qquad (3.325)$$

As a result, the objective function of the model becomes (Taleizadeh et al. 2010):

3.4 Partial Backordered

$$\text{Min} Z = \sum_{j=1}^{n} \frac{C_j D_j}{1-E(x_j)} - \sum_{j=1}^{n} \left[\frac{C_j(1-\xi_j)}{1-E(x_j)}\right] \frac{B_j}{T}$$

$$+ \sum_{j=1}^{n} h_j(P_j - d_j) \left[\frac{(P_j - D_j - d_j)D_j}{2(P_j)^2(1-E(x_j))^2} T - \frac{(P_j - D_j - d_j)(1-\xi_j) + \xi_j P_j(1-E(x_j))}{(P_j)^2(1-E(x_j))^2} B_j \right.$$

$$+ \left. \frac{(P_j - D_j - d_j)^2(1-\xi_j)^2 + 2\xi_j(1-\xi_j)P_j(1-E(x_j))(P_j - D_j - d_j) + (\xi_j P_j)^2(1-E(x_j))^2 (B_j)^2}{2D_j(P_j)^2(1-E(x_j))^2(P_j - D_j - d_j)} \frac{}{T}\right]$$

$$+ \sum_{j=1}^{n} h_j d_j \left[\frac{(D_j)^2 T - 2D_j(1-\xi_j)B_j}{2(P_j)^2(1-E(x_j))^2} + \frac{(1-\xi_j)^2}{2(P_j)^2(1-E(x_j))^2}\left(\frac{B_j^2}{T}\right)\right]$$

$$+ \frac{1}{T}\sum_{j=1}^{n} C_{bj}\xi_j \left[\frac{(P_j - (1-\xi_j)D_j - d_j)(B_j)^2}{2D_j(P_j - D_j - d_j)}\right] + \frac{1}{T}\sum_{j=1}^{n} \hat{\pi}_j(1-\xi_j)B_j$$

$$+ \sum_{j=1}^{n} \frac{Cs_j E(x_j) D_j}{1-E(x_j)} - \sum_{j=1}^{n} \frac{Cs_j E(x_j)(1-\xi_j)}{1-E(x_j)} \frac{B_j}{T} + \frac{K}{T}$$

$$= \sum_{j=1}^{n} \left[\frac{h_j d_j(1-\xi_j)^2}{2(P_j)^2(1-E(x_j))^2} + \frac{C_{bj}\xi_j(P_j - (1-\xi_j)D_j - d_j)}{2D_j(P_j - D_j - d_j)}\right.$$

$$+ \left. \frac{(P_j - D_j - d_j)^2(1-\xi_j)^2 + 2\xi_j(1-\xi_j)P_j(1-E(x_j))(P_j - D_j - d_j) + (\xi_j P_j)^2(1-E(x_j))^2 (B_j)^2}{2D_j(P_j)^2(1-E(x_j))^2(P_j - D_j - d_j)} \frac{}{T}\right]$$

$$- \sum_{j=1}^{n} \left[\frac{h_j d_j D_j(1-\xi_j)}{2(P_j)^2(1-E(x_j))^2} + \frac{h_j(P_j - d_j)\left[(P_j - D_j - d_j)(1-\xi_j) + \xi_j P_j(1-E(x_j))\right]}{2(P_j)^2(1-E(x_j))^2}\right] B_j$$

$$- \sum_{j=1}^{n} \left(\frac{C_j + \hat{\pi}_j + Cs_j E(x_j)(1-\xi_j)}{1-E(x_j)} \frac{B_j}{T} + \sum_{j=1}^{n}\left(\frac{h_j(P_j - d_j)(P_j - D_j - d_j)D_j}{2(P_j)^2(1-E(x_j))^2} + \frac{h_j d_j(D_j)^2}{2(P_j)^2(1-E(x_j))^2}\right)T\right)$$

$$+ \sum_{j=1}^{n} \left[\frac{(C_j + Cs_j E(x_j))D_j}{1-E(x_j)}\right] + \frac{K}{T} \quad (3.326)$$

To make sure that all of the n products will be produced by a single machine, a capacity limitation should be considered as explained in the next subsection.

The maximum capacity of the single machine and the minimum service rate are the two constraints of the model that are described in the two following subsections. Since $t_j^1 + t_j^4$ and ts_j are the production time and setup time of the jth product, respectively, the summation of the total production and setup time (for all products) will be $\sum_{j=1}^{n}\left(t_j^1 + t_j^4\right) + \sum_{j=1}^{n} \text{ts}_j$ in which it should be smaller or equal to the period length (T). So the capacity constraint of the model is (Taleizadeh et al. 2010):

$$\sum_{j=1}^{n}\left(t_j^1 + t_j^4\right) + \sum_{j=1}^{n} \text{ts}_j \leq T \quad (3.327)$$

Then, based on the derivation in Appendix D of Taleizadeh et al. (2010), one obtains:

$$\frac{\sum_{j=1}^{n} \text{ts}_j - \sum_{j=1}^{n} \frac{(1-\xi_j)B_j}{P_j(1-E(x_j))}}{1 - \sum_{j=1}^{n} \frac{D_j}{P_j(1-E(x_j))}} \leq T \qquad (3.328)$$

Since the shortage quantity of the jth product per period is B_j, the annual demand of the jth product is D_j, the number of periods in each year is N, and the safety factor of allowable shortage is α, the service rate constraint becomes:

$$\sum_{j=1}^{n} \frac{N \times B_j}{D_j} \leq T \qquad (3.329)$$

According to Appendix D of Taleizadeh et al. (2010), the service rate constraint is:

$$T^{\text{SL}} = \frac{\sum_{j=1}^{n} \frac{C_j^2}{2C_j^1 D_j}}{\alpha - \sum_{j=1}^{n} \frac{C_j^3}{2C_j^1 D_j}} \leq T \qquad (3.330)$$

Based on the objective function in Eq. (3.326) and the constraints in Eqs. (3.328) and (3.330), the final model is:

3.4 Partial Backordered

$$\text{Min } Z = \sum_{j=1}^{n} \left[\frac{\hat{\pi}_j(1-\xi_j)}{2D_j} + \frac{h_j d_j(1-\xi_j)^2}{2(P_j)^2(1-E(x_j))^2} + \frac{C_{bj}\xi_j(P_j-(1-\xi_j)D_j-d_j)}{2D_j(P_j-D_j-d_j)} \right]$$

$$\text{Min } Z = \sum_{j=1}^{n} \left[\frac{\hat{\pi}_j(1-\xi_j)}{2D_j} + \frac{h_j d_j(1-\xi_j)^2}{2(P_j)^2(1-E(x_j))^2} + \frac{C_{bj}\xi_j(P_j-(1-\xi_j)D_j-d_j)}{2D_j(P_j-D_j-d_j)} \right.$$

$$+ \left. \frac{(P_j-D_j-d_j)^2(1-\xi_j)^2 + 2\xi_j(1-\xi_j)P_j(1-E(x_j))(P_j-D_j-d_j) + (\xi_j P_j)^2(1-E(x_j))^2}{2D_j(P_j)^2(1-E(x_j))^2(P_j-D_j-d_j)} \right] \frac{(B_j)^2}{T}$$

$$- \sum_{j=1}^{n} \left[\frac{h_j d_j D_j(1-\xi_j)}{2(P_j)^2(1-E(x_j))^2} + \frac{h_j(P_j-d_j)[(P_j-D_j-d_j)(1-\xi_j) + \xi_j P_j(1-E(x_j))]}{2(P_j)^2(1-E(x_j))^2} \right] B_j$$

$$- \sum_{j=1}^{n} \frac{(C_j + Cs_j E(x_j))(1-\xi_j) B_j}{1-E(x_j)} \frac{1}{T}$$

$$+ \sum_{j=1}^{n} \left(\frac{h_j(P_j-d_j)(P_j-D_j-d_j)D_j}{2(P_j)^2(1-E(x_j))^2} + \frac{h_j d_j(D_j)^2}{2(P_j)^2(1-E(x_j))^2} \right) T + \sum_{j=1}^{n} \left[\frac{(C_j + Cs_j E(x_j))D_j}{1-E(x_j)} \right] + \frac{K}{T}$$

$$\text{s.t.: } \frac{\sum_{j=1}^{n} ts_j - \sum_{j=1}^{n} \frac{(1-\xi_j)B_j}{P_j(1-E(x_j))}}{1-\sum_{j=1}^{n} \frac{D_j}{P_j(1-E(x_j))}} \leq T$$

$$T^{SL} = \frac{\sum_{j=1}^{n} \frac{C_j^2}{2C_j^1 D_j}}{\alpha - \sum_{j=1}^{n} \frac{C_j^3}{2C_j^1 D_j}} \leq T$$

$$T, B_j \geq 0 \quad \forall j, j = 1, 2, \ldots, n.$$

(3.331)

Example 3.9 Consider a multi-product inventory control problem with five products in which their general and specific data are given and in Tables 3.16 and 3.18, respectively. Two numerical examples are given. In the first example, the probability distribution of x_j is uniform, and in the second example, the distribution for X_j is normal. The setup cost is $K = \$100,000$, and the safety factor of total allowable shortages is $\alpha = 0.35$.

Based on data of Table 3.17, the problem is solved using the proposed algorithm, and the optimal results are given in Tables 3.18 and 3.19 for the uniform and normal distributions, respectively.

Table 3.16 General data (Taleizadeh et al. 2010)

Product	D_j	P_j	ts_j	ζ_j	$\wedge \pi_j$	C_j	h_j	C_{bj}	Cs_j
1	800	10,000	0.01	0.75	1000	500	15	350	80
2	900	11,000	0.015	0.80	900	400	12	300	70
3	1000	12,000	0.02	0.85	800	300	9	250	60
4	1100	13,000	0.025	0.90	700	200	6	200	50
5	1200	14,000	0.03	0.95	600	100	3	150	40

Table 3.17 Specific data (Taleizadeh et al. 2010)

Product	$x_j \sim U[a_j, b_j]$				$x_j \sim N[\mu_j, \sigma_j^2]$		
	a_j	b_j	$E[x_j]$	d_j	$\mu_j = E[x_j]$	σ_j^2	d_j
1	0	0.1	0.05	500	0.25	0.01	2500
2	0	0.15	0.075	825	0.28	0.02	3080
3	0	0.2	0.1	1200	0.33	0.03	3960
4	0	0.25	0.125	1625	0.38	0.04	4940
5	0	0.3	0.15	2100	0.42	0.05	5880

Table 3.18 The optimal results of Example 1 (uniform distribution) (Taleizadeh et al. 2010)

Product	Uniform						
	T_{Min}	T^{SL}	T	T^*	B_j	Q_j^B	Z
1					268.5	2531.2	
2					254.6	2951.2	
3	0.1578	3.0897	1.9841	3.0897	223.6	3395.8	1,625,500
4					172.6	3864.5	
5					98.9	4356.2	

Table 3.19 The optimal results of Example 2 (normal distribution) (Taleizadeh et al. 2010)

Product	Uniform						
	T_{Min}	T^{SL}	T	T^*	B_j	Q_j^B	Z
1					349.3	4280.6	
2					334.7	5059.8	
3	0.2044	4.1222	1.6771	4.1222	302.5	6084.9	2,246,700
4					239.2	7275.1	
5					137.2	8517	

3.5 Conclusion

This report provided a comprehensive review of all EPQ models which considered a scrapped item. A brief introduction and literature review of mentioned problem is provided. For each problem the framework and mathematical model are presented separately, and then a numerical example of each study is reported. In total, ten studies were reviewed and analyzed. The difference between reviewed works was the stochastic parameters which were considered in those problems. Finally, these studies can be extended by incorporating other features in future studies.

References

Akbarzadeh, M., Esmaeili, M., & Taleizadeh, A. A. (2015). EPQ model with scrap and backordering under vendor managed inventory policy. *Journal of Industrial and Systems Engineering, 8*(1), 85–102.

Akbarzadeh, M., Taleizadeh, A. A., & Esmaeili, M. (2016). Developing economic production quantity model with scrap, rework and backordering under vendor managed inventory policy. *International Journal of Advanced Logistics, 5*(3–4), 125–140.

Chiu, S. W. (2008). Production lot size problem with failure in repair and backlogging derived without derivatives. *European Journal of Operational Research, 188*(2), 610–615.

Chiu, S. W., Wang, S. L., & Chiu, Y. S. P. (2007). Determining the optimal run time for EPQ model with scrap, rework, and stochastic breakdowns. *European Journal of Operational Research, 180*(2), 664–676.

Chiu, S. W., Lin, H. D., Wu, M. F., & Yang, J. C. (2011b). Determining replenishment lot size and shipment policy for an extended EPQ model with delivery and quality assurance issues. *Scientia Iranica, 18*(6), 1537–1544.

Chiu, S. W., Chiu, Y. P., & Wu, B. P. (2003). An economic production quantity model with the steady production rate of scrap items. *The Journal of Chaoyang University of Technology, 8*(1), 225–235.

Chiu, Y. P., & Chiu, S. W. (2003). A finite production model with random defective rate and shortages allowed and backordered. *Journal of Information & Optimization Sciences, 24*(3), 553–567.

Chiu, Y.-S.P., Wang, S.-S., Ting, C.-K., Chuang, H.-J., Lien, Y.-L. (2008). Optimal run time for EMQ model with backordering, failure-in- rework and breakdown happening in stock-piling time. *WSEAS Transactions on Information Science and Applications, 5*(4), 475–486.

Chiu, Y. S. P., Liu, S. C., Chiu, C. L., & Chang, H. H. (2011a). Mathematical modeling for determining the replenishment policy for EMQ model with rework and multiple shipments. *Mathematical and Computer Modelling, 54*(9–10), 2165–2174.

Chiu, Y. S. P., Chen, K. K., & Ting, C. K. (2012). Replenishment run time problem with machine breakdown and failure in rework. *Expert Systems with Applications, 39*(1), 1291–1297.

Chiu, Y. S. P. (2006). The effect of service level constraint on EPQ model with random defective rate. *Mathematical Problems in Engineering*, 98502. https://doi.org/10.1155/MPE/2006/98502.

Chiu, Y. S. P., Chiu, S. W., Li, C. Y., & Ting, C. K. (2009). Incorporating multi-delivery policy and quality assurance into economic production lot size problem. *Journal of Scientific and Industrial Research, 68*(6), 505–512.

Chiu, S., Cheng, C. B., Wu, M. F., & Yang, J. C. (2010). An algebraic approach for determining the optimal lot size for EPQ model with rework process. *Mathematical and Computational Applications, 15*(3), 364–370.

Chiu, Y. S. P., Lin, H. D., Hwang, M. H., & Pan, N. (2011). Computational optimization of manufacturing batch size and shipment for an integrated EPQ model with scrap. *American Journal of Computational Mathematics, 1*(3), 202.

Chung, K. J. (1997). Bounds for production lot sizing with machine breakdowns. *Computers & Industrial Engineering, 32*(1), 139–144.

Grubbström, R. W., & Erdem, A. (1999). The EOQ with backlogging derived without derivatives. *International Journal of Production Economics, 59*(1–3), 529–530.

Ghorpade, S. R., & Limaye, B. V. (2010). *A course in multivariable calculus and analysis.* New York: Springer Science & Business Media.

Hayek, P. A., & Salameh, M. K. (2001). Production lot sizing with the reworking of imperfect quality items produced. *Production Planning & Control, 12*(6), 584–590.

Hillier, F. S., & Lieberman, G. J. (1995). *Introduction to operations research* (pp. 424–469). New York: McGraw Hill.

Hsu, L. F., & Hsu, J. T. (2016). Economic production quantity (EPQ) models under an imperfect production process with shortages backordered. *International Journal of Systems Science, 47*(4), 852–867.

Huang, Y.-F. (2006). Algebraic improvement on effects of random defective rate and imperfect rework process on Economic Production Quantity model. *Journal of Applied Sciences, 6*(5), 1082–1084.

Nobil, A. H., Sedigh, A. H. A., & Cárdenas-Barrón, L. E. (2016). A multi-machine multi-product EPQ problem for an imperfect manufacturing system considering utilization and allocation decisions. *Expert Systems with Applications, 56*, 310–319.

Nahmias, S., & Cheng, Y. (2005). *Production and operations analysis* (Vol. 6). New York: McGraw-Hill.

Nobil, A. H., Cárdenas-Barrón, L. E., & Nobil, E. (2018). Optimal and simple algorithms to solve integrated procurement-production-inventory problem without/with shortage. *RAIRO-Operations Research, 52*(3), 755–778.

Rosenblatt, M. J., & Lee, H. L. (1986). Economic production cycles with imperfect production processes. *IIE Transactions, 18*(1), 48–55.

Rardin, R. L., & Rardin, R. L. (1998). *Optimization in operations research* (Vol. 166). Upper Saddle River, NJ: Prentice Hall.

Sarkar, B., Cárdenas-Barrón, L. E., Sarkar, M., & Singgih, M. L. (2014). An economic production quantity model with random defective rate, rework process and backorders for a single stage production system. *Journal of Manufacturing Systems, 33*(3), 423–435.

Shyu, M. L., Hsu, K. H., Tu, Y. C., & Huang, Y. F. (2009). The EPQ model with random defective rate under service constraint without calculus. *Journal of Information and Optimization Sciences, 30*(2), 245–251.

Silver, E. A., Pyke, D. F., & Peterson, R. (1998). *Inventory management and production planning and scheduling* (Vol. 3, p. 30). New York: Wiley.

Shyu, M. L., Hsu, K. H., Tu, Y. C., & Huang, Y. F. (2014). The EPQ model with random defective rate under service constraint without calculus. *Journal of Information and Optimization Sciences, 30*, 245–251.

Taleizadeh, A. A., Niaki, S. T. A., & Najafi, A. A. (2010). Multiproduct single-machine production system with stochastic scrapped production rate, partial backordering and service level constraint. *Journal of Computational and Applied Mathematics, 233*(8), 1834–1849.

Taleizadeh, A. A., Samimi, H., Sarkar, B., & Mohammadi, B. (2017). Stochastic machine breakdown and discrete delivery in an imperfect inventory-production system. *Journal of Industrial & Management Optimization, 13*(3), 1511–1535.

Zipkin, P. H. (2000). *Foundations of inventory management* (1st ed.). Boston, MA: McGraw-Hill/Irwin.

Chapter 4
Rework

4.1 Introduction

Scrap and rework costs are a manufacturing reality impacting organizations across all industries and product lines. Scrap and rework costs are caused by many things—when the wrong parts are ordered, when engineering changes are not effectively communicated, or when designs are not properly executed on the manufacturing line. No matter why scrap and rework occurs, its impact on an organization is always the same—wasted time and money. And while no one, especially an operations manager, wants to admit it, these expenses add up quickly and negatively impact the bottom line.

Although it is near impossible to eliminate scrap and rework completely, you can reduce the amount of scrap and rework in your organization by optimizing the way you document product data, review manufacturing processes, and communicate manufacturing and engineering changes throughout your supply chain. If priority is given to evaluating and improving your manufacturing processes, it becomes much easier to reduce the amount of scrap and rework in your organization.

When imperfect-quality items are produced in the finite production model, as described in real-life situations, one cannot depend on the classical EPQ model to determine the optimal replenishment policy. The effect of defective items on the finite production model must be studied in order to minimize overall inventory costs. This chapter considers the EPQ model with the rework process of imperfect-quality items and the assumption that not all of the defective are repairable; a portion θ of them are scrap and will not be reworked. Mathematical modeling and analysis is employed in this chapter, and the disposal cost for each scrap item and the repairing and holding costs for each reworked items are included in the cost analysis. The renewal reward theorem is utilized to deal with the variable cycle length, and the optimal lot size that minimizes the overall costs for the imperfect-quality EPQ model

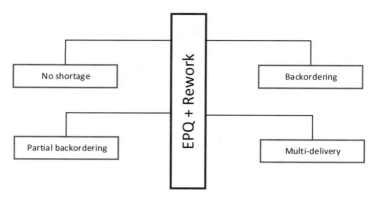

Fig. 4.1 Categories of EPQ model considering rework

is derived where backorders are permitted. A numerical example has been used to illustrate the proposed methodology. For future research, one interesting and realistic consideration will be that when the rework process itself is imperfect.

This chapter studied an imperfect production system with a set of new working assumptions such as process compressibility, reworking, and inspection simultaneously. Also regular production process and rework process can be carried out with different process rates in their corresponding upper and lower bounds by paying their related costs and different stochastic percent of defectives. Next an integrated model is presented which simultaneously determines production lot size, backlog, rate of regular production, and rate of rework with the objective of minimizing the total costs. While the model is nonlinear and could not be easily solved within a closed-form solution, a simple algorithm is developed to obtain optimal solution. The mentioned problem based on the several features is organized. So four categorizes, no shortage, partial backordering, full backordering, and multi-delivery, are provided which is shown in Fig. 4.1. All categories are investigated in the next sections.

The common notations of EPQ models are shown in Table 4.1. To integrate the model of this chapter, these notations are considered for all models.

4.2 Literature Review

In today's manufacturing environment, most firms are confronted with fierce competition both domestic and offshore, in terms of quality, on-time delivery, and price. Customers demand quality and expect delivery on time in full (DOTIF) and usually negotiate for yearly price decreases. Consequently, the primary goal of firms today is to reduce costs while improving overall quality. In fact, reducing the inventory level is the most effective way of controlling product costs, quality, and delivery time.

4.2 Literature Review

Table 4.1 Notations

i	Index of product, $i = 1, 2, \ldots$
P, P_1 or P_{1i}, P_i	Production rate of product or product i (units per unit time)
P_2 or P_{2i}	Rework rate of non-conforming item or item i (units per unit time)
Q or Q_i	Production quantity of product or product i (unit)
D or D_i	Demand rate of product or product i (units per unit time)
R_s	Rate of screening (unit per year)
C	Manufacturing or production cost per unit (\$/unit)
C_J	Reject cost per unit (\$/unit)
C_R	Rework cost per unit (\$/unit)
C_d	Disposal cost per unit (\$/unit)
C_q	Quality improvement cost per unit (\$/unit)
C_T	Delivery cost per unit (\$/unit)
C_I	Inspection or screening cost per unit (\$/unit)
d	Defective rate (units per unit time)
θ	Portion of the imperfect-quality items cannot be reworked and are scrapped (%)
I, I_1, H, H_1	Inventory level (unit)
K	Fixed cost (\$/setup)
K_S	Fixed delivery cost per shipment (\$/shipment)
x	Proportion of defective (%)
B	Size number of backordered (unit)
C_b	Cost of backordered per unit per time (\$/unit/unit time)
$\widehat{C}_b$	Cost of backordered per unit (\$/unit)
$\widehat{\pi}$	Cost of lost sale per unit (\$/unit)
β	Proportion of backordering (%)
$1 - \beta$	Proportion of lost sale (%)
ts	Setup time of machine to produce product (time)
SL	Safety factor of total allowable shortages (%)
F	Proportion of time during which inventory level is positive (%)
T	Period of time (time)
Q^T or Q_i^T	Number of end product or product i to be transported in each shipment (unit)
s	Unit selling price (\$/unit)
v	Unit discounted selling price of defective items (\$/unit)
N	Number of cycle
h	Holding cost per unit per time for healthy item (\$/unit/unit time)
h_1	Holding cost for each imperfect-quality items being reworked or not reworked per unit time (\$/item/unit time)
$E[\cdot]$	Expected value operator

The classical EPQ model has been used for a long time and is widely accepted and implemented. Regardless of its simplicity, the EOQ and EPQ model is still applied industrywide today (Osteryoung 1986; Zipkin 2000). However, finding an economic order quantity has been based on some unrealistic assumptions. One of unrealistic assumption in EPQ model is that all produced items are healthy. The classical EPQ model shows that the optimal lot size will generate minimum manufacturing cost, thus producing minimum total setup cost and inventory cost. However, this is only true if all manufactured products are of perfect quality. In reality this is not the case; therefore, it is necessary to look at and allow cost for carrying imperfect-quality items, because this cost can influence the decision for selecting the economic lot size. A number of works have been published to address this unrealistic assumption. A brief discussion of these works is given below.

Porteus (1986) was one of a group of people who formulated the relationship between process quality improvement and setup cost reduction. In the model, he found that there is a significant relationship between them. Lowering the setup cost alone makes the lot size smaller and the defective items fewer and produces lower annual cost. He illustrated that the annual cost can be further reduced when a joint investment in both process quality improvement and setup reduction is optimally made. From his model, it is assumed that once the process is out of control, it continues to produce defective items until the next setup is adjusted. However, this particular assumption is not realistic in the case of dynamic process control and when the product type is expensive.

Many researchers have extended Porteus' studies; for example, Chand (1989) validates Porteus's model by including the learning effects on setup frequency and process quality. Tapiero et al. (1987) have presented a theoretical framework to examine the trade-offs between pricing, reliability, design, and quality control issues in manufacturing operations. Cheng (1989) has proposed an EPQ model with a flexible and imperfect process. A geometric programming (GP) approach has been developed for solving this model. The investment costs of this model tend toward infinity when the setup cost is close to zero. Cheng (1991) proposed an EOQ model with demand-dependent unit cost and imperfect production processes. He formulated the optimization problem as a GP, and it is solved to obtain a closed-form optimal solution.

Salameh and Jaber (2000) hypothesized a production–inventory situation where items are not of perfect quality. The imperfect-quality items could be used in another production–inventory situation. Their work also considered that the imperfect items can be sold as a single batch at a lower price by the end of 100% inspection. It shows that the economic lot size quantity tends to increase as the average percentage of imperfect-quality items increases. However, it does not include the impact of the reject and the rework on their model and ignore the factor of when to sell. Furthermore, their work only considered the EOQ model. Goyal and Cárdenas-Barrón (2001) presented a simple approach for determining the economic production quantity for an item with imperfect quality. It is suggested that this simple approach is comparable to the optimal method of Salameh and Jaber. Some related research can be found in Chan et al. (2003), Cárdenas-Barrón et al. (2012, 2015), Taleizadeh and Noori-Daryan (2016), Taleizadeh and Heydaryan (2017), Taleizadeh et al. (2019, 2020), Shafiee-Gol et al. (2016), Moshtagh and Taleizadeh (2017),

Keshavarz et al. (2019), Rosenblatt & Lee (1986), Aggarwal & Aneja (2016), Glock & Jaber (2013), Tersine & Tersine (1988), Krishnamoorthi & Panayappan (2012), Taft (1918), Taleizadeh et al. (2015) and Alizadeh-Basbam and Taleizadeh (2020).

4.3 No Shortage

4.3.1 Imperfect Item Sales

Chan et al. (2003) developed an imperfect inventory system that a process produces a single product in a batch size of Q. In addition, storage and withdrawals are uniform and continuous. The demand rate for the product is deterministic and constant over a planning horizon of 1 year. The production process produces this item with finite production rate P units per year. The manufacturing cost of each unit is $C\$$, and the inventory holding cost per unit per year and the setup cost per batch are donated by h and K, respectively. In this problem, it is assumed that each lot produced contains p_1 percent imperfect-quality items, with a known probability distribution function $f(p_1)$. Items of imperfect quality detected by the inspection process are sold at a lower price. It is assumed that the inspection cost for each unit is a fixed constant and the detection of defectives is achieved by nondestructive and error-free testing. The lot also contains a percentage of defectives, p_2, with a known probability distribution function $f(p_2)$. It is assumed that these defective items can be reworked instantaneously at a cost and kept in stock. After the rework process, these items are assumed to be of good quality. Each lot reworked also contains a percentage of defectives, p_3, with a known probability distribution function $f(p_3)$. These units are rejected with an associated cost. Three different approaches are developed to comply with three different situations/cases as below.

Some new notations which are specially used for this problem presented in Table 4.2.

Case I Imperfect-quality items are sold at a discounted price when identified and are not counted into the inventory. The production rate of good items per year (including the items that can be reworked) is represented as $P_p = P[1 - p_1 - p_3]$. The amount of good-quality items added to stock, as the result of a single production run, is $(P_p - D)(Q/P) = Q[1 - p_1 - p_3 - (D/P)]$ (Chan et al. 2003) (see Fig. 4.2).

Case II After the inspection process, the imperfect items are kept in stock and sold at the end of the production period within each cycle as a single batch at a reduced price per unit. As a result, the production rate during the production period will be expressed as $P_g = P[1 - p_3]$ (Chan et al. 2003) (see Fig. 4.3).

Case III These imperfect-quality items are kept in stock and sold at the end of the cycle (just before next production run). The production rate is the same as Case II. The main difference between Case II and Case III is the time factor for the sale of the imperfect-quality items (Chan et al. 2003) (see Fig. 4.4).

Table 4.2 New notations of given problem

P_p	Production rate of good items per year (including the items that can be reworked) is represented as $P_p = P[1 - p_1 - p_3]$ (units per unit time)
p_1	Percentage of imperfect-quality items, with a known probability distribution function $f(p_1)$ (units per unit time)
p_2	Percentage of rework items, with a known probability distribution function $f(p_2)$ (units per unit time)
p_3	Percentage of reject items from rework process, with a known probability distribution function $f(p_3)$ (units per unit time)
P_g	Production rate of good items and imperfect-quality items per unit time (include the rework items) (units per unit time)
Q^*	Lot size in number of units per lot (optimal value for the new model) (unit)
Q^{*I}	Lot size in number of units per lot for Case I (unit)
Q^{*II}	Lot size in number of units per lot for Case II (unit)
Q^{*III}	Lot size in number of units per lot for Case III (unit)
Q'	Optimum lot size in number of units per lot for the classical EPQ model (unit)

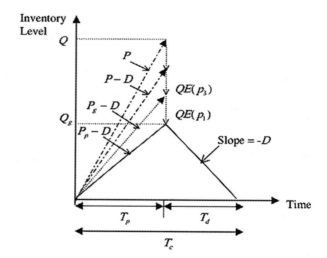

Fig. 4.2 The behavior of the inventory level per cycle for Case I (Chan et al. 2003)

Define $E(P_p) = [1 - E(P_1) - E(P_3)]$ as expected proportion of good items used. The amount of good items available for use in the lot per each single production run is $Q[E(P_p) - D/P]$. To avoid shortage, it is assumed that:

$$E(P_p) \geq D/P \qquad (4.1)$$

Consider a firm having an expected total profit (ETP) as follows:

$$\text{ETP}(Q) = \text{ETR}(Q) - \text{ETC}(Q) \qquad (4.2)$$

4.3 No Shortage

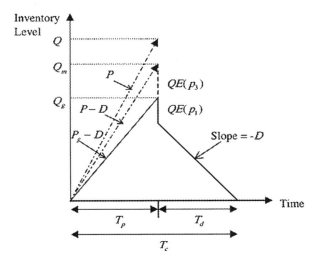

Fig. 4.3 The behavior of the inventory level per cycle for case II (Chan et al. 2003)

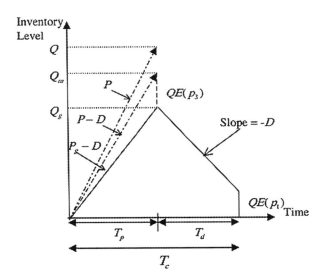

Fig. 4.4 The behavior of the inventory level per cycle for Case III (Chan et al. 2003)

where ETR(Q) and ETC(Q) are the expected total revenue and expected total cost per year. The magnitude of these is obtained as follows:

ETR(Q) = (Good quality items) × full price per item + (imperfect quality items)

× lower price per item

$$= Ds + \frac{D \cdot E(p_1) \cdot v}{E(P_p)} = Ds + Dv\xi$$

where:

$$\xi = E(p_1)/E(P_p) \qquad (4.3)$$

ETC(Q) is the sum of setup cost, rejection cost, rework cost, inspection cost, manufacturing cost, and inventory holding cost, in which only the inventory holding cost varies on the different cases. The inventory holding cost is obtained as the average inventory times holding cost per item per year, giving for Case I, Case II, and Case III, respectively. For the purposes of brevity, the derivation of the inventory holding cost is omitted. The detail derivation is, however, available from the authors:

$$\frac{hQ[E(P_p) - D/P]}{2}, \quad \frac{hQ}{2} \times \left\{ E(P_p) - \frac{D}{P}[1 - \xi] \right\}, \quad hQ\left\{ E(P_p) + 2E(p_1) - \frac{D}{P}[1 + \xi] \right\},$$

The total number of units produced per year is $D/E(P_p)$; the number of setups is the number of units produced per year divided by the lot size, Q. Therefore, the annual setup cost is given as $DK/[E(P_p)Q]$. The manufacturing cost and the inspection cost per unit is constant, and then the annual manufacturing cost and the inspection cost can be written as $CD/E(P_p)$ and $C_1D/E(P_p)$, respectively. In addition, an average of $[D/E(P_p)]E(P_3)$ units are rejected per year with an associated cost; C_J and $[D/E(P_p)]E(P)$ units are reworked per year with a cost C_R; the annual rejection cost and rework cost can be represented as $DC_JE(p_3)/E(P_p)$ and $DC_RE(p_2)/E(P_p)$, respectively. The values of $E(P_1)$, $E(P_2)$, and $E(P_3)$ follow from the assumed distribution of the quality characteristic being measured. By substituting all the cost components into Eq. (4.2), the ETP(Q) for the three different approaches can be expressed as:

Case I Imperfect items sold as identified which are not kept in stock (see Fig. 4.2) (Chan et al. 2003):

$$\text{ETP}(Q)^{\text{I}} = \left[\underbrace{Ds + Dv\xi}_{\text{Revenue}} - \left\{ \underbrace{\frac{hQ[E(P_p) - D/P]}{2}}_{\text{Holding Cost}} + \underbrace{\left[\frac{K}{Q} + C_JE(p_3) + C_RE(p_2) + C_1 + C \right]\left[\frac{D}{E(P_p)} \right]}_{\text{Setup, Rejection, Rework, Inspection and Production Costs}} \right\} \right]$$

$$(4.4)$$

4.3 No Shortage

Case II Imperfect items sold at the end of production period (see Fig. 4.3) (Chan et al. 2003):

$$\text{ETP}(Q)^{\text{II}} = \overbrace{[Ds + Dv\beta]}^{\text{Revenue}}$$
$$-\left\{\underbrace{\frac{hQ[E(P_{\text{p}}) - (D/P)(1-\beta)]}{2}}_{\text{Holding Cost}} + \underbrace{\left[\frac{K}{Q} + C_{\text{J}}E(p_3) + C_{\text{R}}E(p_2) + C_{\text{I}} + C\right]\left[\frac{D}{E(P_{\text{p}})}\right]}_{\text{Setup, Rejection, Rework, Inspection and Production Costs}}\right\} \quad (4.5)$$

Case III Imperfect items sold just before the next batch (see Fig. 4.4) (Chan et al. 2003):

$$\text{ETP}(Q)^{\text{III}} = $$
$$\overbrace{[Ds + Dv\beta]}^{\text{Revenue}} - \left\{\underbrace{\frac{hQ[E(P_{\text{p}}) + 2E(p_1) - (D/P)(1+\beta)]}{2}}_{\text{Holding Cost}} + \underbrace{\left[\frac{K}{Q} + C_{\text{J}}E(p_3) + C_{\text{R}}E(p_2) + C_{\text{I}} + C\right]\left[\frac{D}{E(P_{\text{p}})}\right]}_{\text{Setup, Rejection, Rework, Inspection and Production Costs}}\right\} \quad (4.6)$$

Differentiating and equating $d[\text{ETP}(Q)]/dQ = 0$, the optimal Q^* (lot size) which generates minimum expected total cost can be given by:

$$Q^{*\text{I}} = \sqrt{\frac{2DK}{h[E(P_{\text{p}}) - (D/P)][E(P_{\text{p}})]}} \quad (4.7)$$

$$Q^{*\text{II}} = \sqrt{\frac{2DK}{h\left[E(P_{\text{p}}) - \frac{D}{P}[1-\xi]\right][E(P_{\text{p}})]}} \quad (4.8)$$

$$Q^{*\text{III}} = \sqrt{\frac{2DK}{h\left[E(P_{\text{p}}) + 2E(p_1) - \frac{D}{P}[1+\xi]\right][E(P_{\text{p}})]}} \quad (4.9)$$

The second derivates of Eqs. (4.4)–(4.6) give $\text{ETP}''(Q)^{\text{I}} = \text{ETP}''(Q)^{\text{II}} = \text{ETP}''(Q)^{\text{III}} = -2DK/\{[1 - E(p_1) - E(p_3)]Q^3\} \leq 0$ and are negative for all values of positive Q, which implies that there exist unique values of $Q^{*\text{I}}$, $Q^{*\text{II}}$, and $Q^{*\text{III}}$ that maximize Eqs. (4.4)–(4.6), respectively. Note that when $P_1 = P_1 = P_1 = 0$ (i.e., all products produced are of perfect quality),

Table 4.3 Optimal values (Chan et al. 2003)

Case	Order size	Average inventory cost	Setup cost	ETP(Q)
I	1850	$1696.08	$1694.68	$777,166.60
II	1801	$1742.13	$1740.79	$777,074.45
II	1607	$1951.36	$1950.94	$776,655.06

Eqs. (4.7)–(4.9) reduce to the classical EPQ model, $Q' = (2DC_d/h(1 - D/P))^{1/2}$. If the time for producing a batch of units is zero, or approaches zero, the arrival rate is infinite. When this occurs, the batch model reverts to the classical EOQ model, $Q' = (2DC_d/h)^{1/2}$.

Example 4.1 Chan et al. (2003) developed a numerical illustration which is provided to illustrate the usefulness of the models developed in previous section. Suppose that an electronic company producing high-voltage transformers for Model: XTA with a production capacity of 960 units per 8-h shift. Items in range of 93–107 V are used directly by the company for Model: XTA. Items outside this range and in the range of 90–110 V may be used for various activities or sold to another producer. If the quality characteristic exceeds the upper limit of 110 V, the transforms are rejected; if it is lower than the lower limit of 90 V, the manufacturer can adjust the voltage in the plant by changing a resistor at a cost and kept in stock instantaneously. Thus, the parameters needed for analyzing for this situation are given as:

$C_J = \$15$ per unit, $C_R = \$8$ per unit, $D = 60,000$ units per year, $h = \$3$ per unit per year
$C_I = \$0.5$ per unit, $C = \$15$ per unit, $K = \$45$ per lot, $s = \$30$ per unit, $v = \$12$ per unit

Since the plant works one shift per day, 5 days a week, and 50 weeks a year, the annual production capacity is $P = 960(5)(50) = 240,000$ units/year. Then, the expected value of $E(p_1)$, $E(p_2)$, and $E(p_3)$ can be obtained from the normal distribution, $x \sim N(\mu = 100, \sigma = 5)$, as 0.1160, 0.0228, and 0.0228, respectively. Substituting these values into Eqs. (4.7)–(4.9), the optimal values of Q^{*I}, Q^{*II}, and Q^{*III} can be obtained. Table 4.3 illustrates the optimal lot sizes, average inventory cost, setup cost, and the expected total profit for the three different cases (Chan et al. 2003).

4.3.2 Rework Policy

Chiu and Chiu (2003) studied the effect of the reworking of repairable defective items on EPQ model. They assumed that x percent of defective items were generated randomly by an imperfect process, at a production rate d, and not all of the defective items produced are repairable. There is a portion $\theta\%$ of the imperfect-quality items cannot be repaired, are scrap items. Furthermore, the proposed EPQ model does not

4.3 No Shortage

allow backorders when excessive demand occurred, the optimal production quantity derived, must be able to satisfy demand at all times. The production rate P is constant and is much larger than the demand rate D. The production rate d of the defective items could be expressed as the production rate times the defective rate: $d = P \cdot x$. All repairable defective items are reworked at a steady rate P_1, and the following notations are used in our analysis: The production rate of perfect-quality items must always be greater than or equal to the sum of the demand rate and the production rate of defective items (Chiu and Chiu 2003).

The production rate of perfect-quality items must always be greater than or equal to the sum of the demand rate and the production rate of defective items, so $P - d - D \geq 0$ must be satisfied:

$$0 \leq x \leq \left(1 - \frac{D}{P}\right) \tag{4.10}$$

For the following derivation, they employed the solution procedures used by Hayek and Salameh (2001), referring to Fig. 4.5:

$$T = \frac{Q(1 - \theta \cdot x)}{D} \tag{4.11}$$

where $0 \leq \theta \leq 1$ and $\theta \cdot x \cdot Q$ are scrap items randomly produced by the regular production process. Hence, the cycle length T is a variable, not a constant:

$$T = t_1 + t_2 + t_3 \tag{4.12}$$

The production uptime t_1 is:

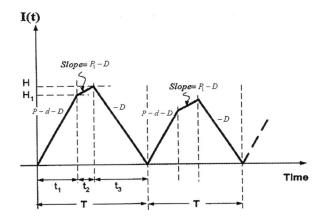

Fig. 4.5 On-hand inventory of perfect-quality items (Chiu and Chiu 2003)

Fig. 4.6 On-hand inventory of defective items (including scrap items) (Chiu and Chiu 2003)

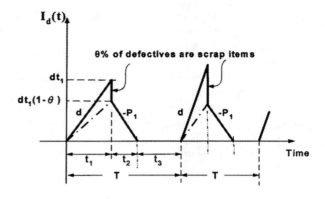

$$t_1 = \frac{Q}{P} \tag{4.13}$$

and

$$H_1 = (P - d - D)t_1 = Q\left(1 - x - \frac{D}{P}\right) \tag{4.14}$$

The total defective items produced during the regular production uptime t_1, as illustrated in Fig. 4.6, are:

$$d \cdot t_1 = x \cdot Q \tag{4.15}$$

The repairable portion $(1 - \theta)\%$ of imperfect-quality items are reworked immediately when the regular production ends. The time t_2 needed for the rework is computed in (4.16), and the maximum level of on-hand inventory when rework process finished is calculated in (4.17):

$$t_2 = \frac{x \cdot Q(1 - \theta)}{P_1} = \frac{d \cdot Q(1 - \theta)}{P_1 P} \tag{4.16}$$

$$H = H_1 + (P_1 - D)t_2 = Q\left(1 - \frac{D}{P} - \frac{Dd(1 - \theta)}{P_1 P} - \frac{d\theta}{P}\right) \tag{4.17}$$

The production downtime t_3 is:

$$t_3 = \frac{H}{D} = Q\left(\frac{1}{D} - \frac{P_1 + d(1 - \theta)}{P_1 P} - \frac{d\theta}{DP}\right) \tag{4.18}$$

Solving the inventory cost per cycle, TC(Q) is (Chiu and Chiu 2003):

4.3 No Shortage

$$\text{TC}(Q) = \overbrace{K}^{\text{Setup Cost}} + \overbrace{CQ}^{\text{Production Cost}} + \overbrace{C_R xQ(1-\theta)}^{\text{Rework Cost}} + \overbrace{C_d(\theta xQ)}^{\text{Disposal Cost}}$$

$$+ \overbrace{h\left[\frac{H_1 + dt_1}{2}(t_1) + \frac{(H_1 + H)}{2}(t_2) + \frac{H}{2}(t_3)\right]}^{\text{Holding Cost}} + \underbrace{h_1 \frac{P_1 t_2}{2}(t_2)}_{\text{Holding Cost of Reworked Item}}$$

(4.19)

In this model, and as in real-life situation, the percentage of defective items is considered to be a random variable with a known probability density function. Thus, to take the randomness of imperfect production quality into account, one can utilize the expected values of x in the inventory cost analysis. Since $\theta \cdot x \cdot Q$ are scraps produced randomly during a regular production run, it follows that the cycle length is a variable. They employed the renewal reward theorem approach to cope with the variable cycle length, that is, to compute the $E[T]$ first. Then in the expected annual inventory cost $E[\text{TCU}(Q)] = E[\text{TC}(Q)]/E[T]$, from Eqs. (4.11) through (4.19), one obtains that (Chiu and Chiu 2003):

$$E[\text{TCU}(Q)] = D\left[C\frac{1}{1-\theta E[x]} + C_R(1-\theta)\frac{E[x]}{1-\theta E[x]} + C_d\theta\frac{E[x]}{1-\theta E[x]}\right]$$
$$+ \frac{KD}{Q}\frac{1}{1-\theta E[x]} + \frac{hQ}{2}\left(1 - \frac{D}{P}\right)\frac{1}{1-\theta E[x]} + \frac{DQ(1-\theta)^2}{2P_1}$$
$$\times (h_1 - h)\frac{E[x^2]}{1-\theta \cdot E[x]} - hQ\theta\left(1 - \frac{D}{P}\right)\frac{E[x]}{1-\theta E[x]} + \frac{hQ\theta^2}{2}\frac{E[x^2]}{1-\theta E[x]}$$

(4.20)

Suppose that:

$$E_0 = \frac{1}{1-\theta E[x]}; \quad E_1 = \frac{E[x]}{1-\theta E[x]}; \quad E_2 = \frac{E[x]^2}{1-\theta E[x]},$$

Then, Eq. (4.20) becomes (Chiu and Chiu 2003):

$$E[\text{TCU}(Q)] = D[CE_0 + C_R(1-\theta)E_1 + C_d\theta E_1] + \frac{KD}{Q}E_0 + \frac{hQ}{2}\left(1 - \frac{D}{P}\right)E_0$$
$$+ \frac{DQ(1-\theta)^2}{2P_1}(h_1 - h)E_2 - hQ\theta\left(1 - \frac{D}{P}\right)E_1 + \frac{hQ\theta^2}{2}E_2$$

(4.21)

Differentiating $E[\text{TCU}(Q)]$ with respect to Q, the first and the second derivatives of $E[\text{TCU}(Q)]$ are shown in Eqs. (4.22) and (4.23):

$$\frac{dE[\text{TCU}(Q)]}{dQ} = \frac{-KD}{Q^2}E_0 + \frac{h}{2}\left(1 - \frac{D}{P}\right)E_0 + \frac{D(1-\theta)^2}{2P_1}(h_1 - h)E_2 - h$$
$$\cdot \theta\left(1 - \frac{D}{P}\right)E_1 + \frac{h \cdot \theta^2}{2}E_2 \qquad (4.22)$$

$$\frac{d^2 E[\text{TCU}(Q)]}{dQ^2} = \frac{2KD}{Q^3}E_0 \qquad (4.23)$$

From Eq. (4.23), since E_0, K, D, and Q are all positive numbers, the second derivative of $E[\text{TCU}(Q)]$ with respect to Q is greater than zero; the expected total inventory cost function $E[\text{TCU}(Q)]$ is a convex, for all Q is different from zero. The optimal production quantity Q^* can be obtained by setting the first derivative of $E[\text{TCU}(Q)]$ equal to zero; refer to Eq. (4.24) (Chiu and Chiu 2003):

$$Q^* = \sqrt{\frac{2KD}{\left(h\left(1 - \frac{D}{P}\right) + \frac{D(1-\theta)^2}{P_1}(h_1 - h)E[x^2] - 2h\theta\left(1 - \frac{D}{P}\right)E[x] + h\theta^2 E[x^2]\right)}} \qquad (4.24)$$

Supposing that all of the defective items are not repairable, meaning that they are all scrap items ($\theta = 1$), from Eq. (4.24), one obtains:

$$Q^* = \sqrt{\frac{2KD}{\left(h\left(1 - \frac{D}{P}\right) - 2h\left(1 - \frac{D}{P}\right)E[x] + hE[x^2]\right)}} \qquad (4.25)$$

Further, if the process produces all perfect-quality items, i.e., $x = 0$, it follows that Eq. (4.24) will give the same result as that of the classical EPQ model (Nahmias and Cheng 2005; Silver et al. 1998; Tersine 1994) as below:

$$Q^* = \sqrt{\frac{2KD}{h\left(1 - \frac{D}{P}\right)}} \qquad (4.26)$$

Example 4.2 Chiu and Chiu (2003) presented an example for a local company which manufactures a product for several regional industrial clients. It has experienced a relatively flat demand of 4000 units per year. This item is produced at a rate of 10,000 units per year. The accounting department has estimated that it costs $450 to initiate a production run, each unit costs the company $2 to manufacture, and the cost of holding is $0.6 per item per year. The defective items are reworked at a rate of 600 units per year, each repairable defective item costs the company $0.5 to rework, and there is a disposal cost of $0.3 for each scrap item inspected and identified prior to starting the reworking process. In addition, there is a holding cost of $0.8 per year,

4.3 No Shortage

per unit of the items being reworked. Supposing that the percentage of defective items produced is uniformly distributed over the interval [0, 0.2], and not all of the imperfect-quality items are repairable, there is a θ % of them that are scrap items. Chiu and Chiu (2003) used the following parameters as $P = 10,000$ units per year, $D = 4000$ units per year, $P_1 = 600$ units per year, $x =$ Uniform [0. 0.2], $\theta = 0.3$ (scrap rate out of the imperfect-quality items), $K = \$450$ for each production run, $C = \$2$ per item, $C_R = \$0.5$ repaired cost for each item reworked, $C_d = \$0.3$ disposal cost for each item reworked, $h = \$0.6$ per item per unit time, and $h_1 = \$0.8$ per item reworked per unit (Chiu and Chiu 2003).

The optimal production lot size can be computed from Eq. (4.24). For example, when $\theta = 0.1$, the value of $Q^* = 3162$ units, and if $\theta = 0.3$, $Q^* = 3200$ units. One notices that as x increases, the value of Q^* decreases. For different θ values, if θ increases, then Q^* increases. The optimal total inventory costs can also be obtained from Eq. (4.21). For example, when $\theta = 0.1$, the value of $E[TCU(Q^*)] = \$9281$, and if $\theta = 0.3$, then $E[TCU(Q^*)] = \$9354$. One notices that as x increases, $E[TCU(Q^*)]$ increases. For different θ values, if θ increases, then $E[TCU(Q^*)]$ increases.

4.3.3 Imperfect Rework

Chiu et al. (2004) examined the EPQ model with the random defective rate and an imperfect rework process. Consider that a practical production process generates randomly x percent of imperfect-quality items at a production rate P. The basic assumption of the finite production model with imperfect-quality items produced is that the production rate P must always be greater than or equal to the sum of the demand rate D and the production rate of defective items d. In order to model the problem, the following specific notations are used for this problem as presented in Table 4.4.

Hence, the following condition must hold (Chiu et al. 2004):

$$P - d - D \geq 0$$

$$0 \leq x \leq \left(1 - \frac{D}{P}\right) \tag{4.27}$$

For the following derivation, the solution procedures are those used by Hayek and Salameh (2001). The production cycle length (T) is the summation of the production uptime (t_1), the reworking time (t_2), and the production downtime (t_3), referring to Fig. 4.7 (Chiu et al. 2004):

$$T = \sum_{i=1}^{3} t_i \tag{4.28}$$

Table 4.4 New notations of given problem (Chiu et al. 2004)

θ_1	The proportion of reworked items that fail (become scraps), θ_1 is assumed to be a random variable with known probability density function
d_1	The production rate of scrap items (during the rework process), in units per unit time

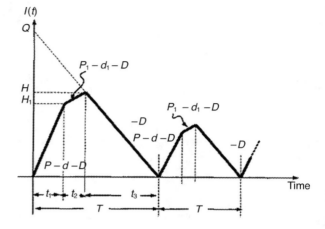

Fig. 4.7 On-hand inventory of perfect-quality item (Chiu et al. 2004)

The production uptime t_1 needed to accumulate H_1 units of perfect-quality items is (Chiu et al. 2004):

$$t_1 = \frac{Q}{P} = \frac{H_1}{P-d-D} \qquad (4.29)$$

and

$$H_1 = (P-d-D) \cdot \frac{Q}{P} \qquad (4.30)$$

The total defective items produced during the regular production uptime t_1, as illustrated in Fig. 4.8, are (Chiu et al. 2004):

$$d \cdot t_1 = x \cdot Q \qquad (4.31)$$

The repairable portion $(1 - \theta)$ of defective items is reworked right after the regular production process ends. The time t_2 needed for the reworking is computed in Eq. (4.32). The maximum level of on-hand inventory H is obtained in Eq. (4.33) (Chiu et al. 2004):

4.3 No Shortage

Fig. 4.8 On-hand inventory of defective items (Chiu et al. 2004)

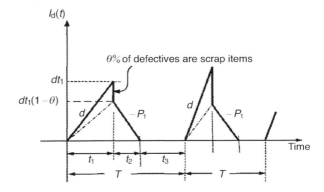

$$t_2 = \frac{xQ(1-\theta)}{P_1} = \frac{dQ(1-\theta)}{P_1 P} \quad (4.32)$$

$$H = H_1 + (P_1 - d_1 - D)t_2$$
$$= Q\left(1 - \frac{d\theta}{P} - \frac{D(P_1 + d)}{P_1 P} - \frac{d_1 d}{P_1 P} + \frac{d_1 d\theta}{P_1 P} + \frac{Dd\theta}{P_1 P}\right) \quad (4.33)$$

Referring to Fig. 4.9, since the rework process itself is assumed to be imperfect either, a random portion θ_1 of the reworked items becomes scrap items. The production rate d_1 in producing scrap items during the rework process can be written as:

$$d_1 = P_1 \cdot \theta_1, \quad \text{where } 0 \leq \theta_1 < 1 \quad (4.34)$$

The total scrap items produced when the rework process ends can be computed in Eq. (4.35):

$$d_1 \cdot t_2 = \theta_1[x(1-\theta)Q] \quad (4.35)$$

The production downtime "t_3" is:

$$t_3 = \frac{H}{D} = Q\left(\frac{H}{D} - \frac{d\theta}{DP} - \frac{(P_1 + d)}{P_1 P} - \frac{d_1 d}{DP_1 P} + \frac{d_1 d\theta}{DP_1 P} + \frac{d\theta}{P_1 P}\right) \quad (4.36)$$

The cycle length T can be obtained (Chiu et al. 2004):

$$T = \frac{Q[1 - \theta x - (1-\theta)x\theta_1]}{D} = \frac{Q\{1 - x[\theta + (1-\theta)\theta_1]\}}{D} \quad (4.37)$$

where $0 < \theta < 1$ and $\theta \cdot x \cdot Q$ are scrap items randomly produced during the regular production process and $[(1-\theta) \cdot x \cdot \theta_1] \cdot Q$ are scrap items produced during the rework

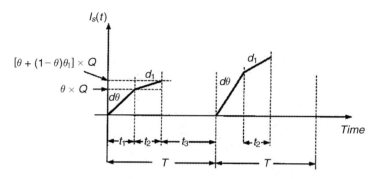

Fig. 4.9 On-hand inventory level of scrap items (Chiu et al. 2004)

process. Hence, the cycle length T is not a constant. Letting 'φ' denotes the overall scrap rate, then $\varphi = [\theta + (1-\theta) \cdot \theta_1]$. Then, the cycle length can be rewritten as:

$$T = \sum_{i=1}^{3} t_i = \frac{Q[1 - \varphi \cdot x]}{D} \tag{4.38}$$

Solving the total inventory cost per cycle, TC(Q) is (Chiu et al. 2004):

$$\text{TC} = \overbrace{K}^{\text{Setup Cost}} + \overbrace{CQ}^{\text{Production Cost}} + \overbrace{C_R[x(1-\theta)Q]}^{\text{Rework Cost}} + \overbrace{C_d(\varphi x Q)}^{\text{Disposal Cost}}$$
$$+ \underbrace{h\left[\frac{H_1 + d \cdot t_1}{2}(t_1) + \frac{H_1 + H}{2}(t_2) + \frac{H}{2}(t_3)\right]}_{\text{Holding Cost}} + \underbrace{h_1 \frac{P_1 t_2}{2}(t_2)}_{\text{Holding Cost of Reworked Items}}$$

(4.39)

In this model, as in real life, the proportion of imperfect-quality items is considered to be a random variable. Thus, to take the randomness of imperfect production quality into account, one can utilize the expected values of x, θ, and θ_1 in the inventory cost analysis. Let $E[x]$, θ, φ, and θ_1 represent the expected values of x, θ, φ, and θ_1, respectively. And to cope with the variable cycle length, the renewal reward theorem approach is employed to compute $E[T]$. Hence, it follows that in the expected total inventory cost $E[\text{TCU}(Q)] = E[\text{TC}(Q)]/E[T]$, from Eqs. (4.29) through (4.39), one obtains (Chiu et al. 2004):

4.3 No Shortage

$$E[TCU(Q)] = D\left[C\frac{1}{1-\varphi E[x]} + C_R(1-\theta)\frac{E[x]}{1-\varphi E[x]} + C_d\varphi\frac{E[x]}{1-\varphi E[x]}\right]$$
$$+ \frac{KD}{Q}\frac{1}{1-\varphi E[x]} + \frac{hQ}{1}\left(1-\frac{D}{P}\right)\frac{1}{1-\varphi \cdot E[x]}$$
$$+ \frac{DQ(1-\theta)^2}{2P_1}[h_1 - h(1-\theta_1)]\frac{E[x^2]}{1-\varphi E[x]} \quad (4.40)$$
$$- hQ\varphi\left(1-\frac{D}{P}\right)\frac{E[x]}{1-\varphi E[x]} + \frac{hQ\varphi^2}{2}\frac{E[x^2]}{1-\varphi E[x]}$$

where:

$$\varphi = [\theta + (1-\theta)\theta_1]$$

For simplicity assume;

$$E_0 = \frac{1}{1-\varphi E[x]}; \quad E_1 = \frac{E[x]}{1-\varphi E[x]}; \quad E_2 = \frac{E[x^2]}{1-\varphi E[x]}$$

Then, Eq. (4.40), the expected inventory cost per unit time becomes:

$$E[TCU(Q)] = D[CE_0 + C_R(1-\theta)E_1 + C_d\varphi E_1] + \frac{KD}{Q}E_0 + \frac{hQ}{1}\left(1-\frac{D}{P}\right)E_0$$
$$+ \frac{DQ(1-\theta)^2}{2P_1}[h_1 - h(1-\theta_1)]E_2 - hQ\varphi\left(1-\frac{D}{P}\right)E_1 + \frac{hQ\varphi^2}{2}E_2$$
$$(4.41)$$

The optimal production lot size can be obtained by minimizing the cost function $E[TCU(Q)]$. Differentiating $E[TCU(Q)]$ with respect to Q, the first and the second derivatives of $E[TCU(Q)]$ are shown in Eqs. (4.42) and (4.43):

$$\frac{dE[TCU(Q)]}{dQ} = \frac{-KD}{Q^2}E_0 + \frac{h}{2}\left(1-\frac{D}{P}\right)E_0 + \frac{D(1-\theta)^2}{2P_1}[h_1 - h(1-\theta_1)]E_2$$
$$- h\varphi\left(1-\frac{D}{P}\right)E_1 + \frac{h\varphi^2}{2}E_2 \quad (4.42)$$

$$\frac{d^2 E[TCU(Q)]}{dQ^2} = \frac{2KD}{Q^3}E_0 \quad (4.43)$$

Equation (4.43) is positive, because E_0, K, D, and Q are all positive. The second derivative of $E[TCU(Q)]$ with respect to Q is greater than zero; hence, the expected inventory cost function $E[TCU(Q)]$ is a convex function for all Q different from zero (Chiu et al. 2004). The optimal production quantity can be obtained by setting the first derivative of $E[TCU(Q)]$ equal to zero (referring to Eq. (4.42)), from: $\frac{dE[TCU(Q)]}{dQ} = 0$

$$Q^* = \sqrt{\frac{2KD}{h\left(1 - \frac{D}{P}\right) + \frac{D(1-\theta)^2}{P_1}[h_1 - h(1 - \theta_1)]E[x^2] - 2h\varphi\left(1 - \frac{D}{P}\right)E[x] + h\varphi^2 E[x^2]}} \quad (4.44)$$

Example 4.3 Chiu et al. (2004) proposed an example for a regional firm which produces a product for several industrial clients. This item has experienced a relatively flat demand of 4000 units per year, and it is produced at a rate of 10,000 units per year. The accounting department has estimated that it costs the company $450 to initiate a production run, and each unit costs $2 to manufacture, in which inspection cost per item is included. The cost of holding is $0.6 per item per year, and the disposal cost is $0.3 for each scrap item. The defective rate x is uniformly distributed over the interval [0, 0.2]. The rate of rework is 600 units per year. Each defective item costs the company $0.5 to repair plus an additional holding cost of $0.8 per item reworked per year. The scrap rates, 0 and 9, are both assumed to follow the uniform distribution over the range [0, 0.1]. Summary of parameters used is as follows: $P = 10,000$ units per year, $D = 4000$ units per year, $P_1 = 600$ units per year, $x \sim$ Uniform [0, 0.2], $\theta \sim$ Uniform [0, 0.1], $\theta_1 \sim$ Uniform [0, 0.1], $K = \$450$ for each production run, $C = \$2$ per item, $C_R = \$0.5$ repaired cost for each item reworked, $C_d = \$0.3$ disposal cost for each item reworked, $h = \$0.6$ per item per unit time, and $h_1 = \$0.8$ per item reworked per unit.

Hence, it follows that the overall scrap rate, $\varphi = [\theta + (1 - \theta)\theta_1] = 0.0975$, and the optimal production lot size can be obtained from Eq. (4.44). For example, if $\varphi = 0.1$, then the value of optimal lot size $Q^* = 3113$. One notices that as x increases, the value of Q^* decreases, and for different φ values, as φ increases, the value of Q^* increases. The optimal inventory costs can be calculated from Eq. (4.40). For example, if $\varphi = 0.1$, then the value of $E[TCU(Q^*)] = \$9453$. One notices that as x increases, the value of $E[TCU(Q^*)]$ increases, and for different φ values, as φ increases, the optimal cost function $E[TCU(Q^*)]$ increases (Chiu et al. 2004).

4.3.4 Quality Screening

Moussawi-Haidar et al. (2016) studied a production quantity model, in which production occurs at a rate P and demand occurs at rate D units per unit time, $P > D$. The inventory builds up at the rate $P - D$. During production, a random proportion P of defective items is produced, with a known probability density function $f(P)$. Demand during production is met from non-defective items only, which requires the units demanded to be screened before they are sold to customers. During this process, if an item is found to be defective, it is replaced with a

4.3 No Shortage

non-defective item. The number of defective items accumulated when production stops, and before screening is conducted, is equal to the total number of items screened from the total demand. As soon as production stops, screening the remaining units of the produced lot is conducted at the rate x per unit per unit time, where $x > D$. The screening cost during production is higher than that after production, i.e., $C_{I1} > C_{I2}$.

In order to model the problem, the following specific notations are used for this problem as presented in Table 4.5.

They analyzed two models that differently address the defective items identified during production and screening. The first model assumes that defective items are sold at a discount at the end of the production cycle. The second model assumes that defective items are reworked at a constant rate.

4.3.4.1 Salvaging of Defective Items

The defective items accumulated at the end of the screening period are sold at a discounted price v. The following notation will be used to develop the mathematical model (Moussawi-Haidar et al. 2016). According to Fig. 4.10, demand during production is met using good items only. Therefore, in $[0, t_1]$, a number of units are screened before they are sold to customers. To be able to satisfy demand from good items only, more than the demand is screened. The total number of units screened can be computed as follows:

$$\left[D + Dp + Dp^2 + \cdots \right] t_1 = \frac{D}{1-p} t_1 \quad (4.45)$$

At the end of t_1, the number of defective items identified is the total number of units screened during the interval $[0, t_1]$, as given in (4.45), less the demand during this period. This is written as:

$$\left[\frac{D}{1-p} - D \right] t_1 = \frac{pD}{1-p} \frac{y}{P} \quad (4.46)$$

The on-hand inventory not screened at the end of t_1 is equal to the maximum inventory level, $y(1 - D/P)$, less the number of defective items identified at the end of t_1, as given in (4.46). This is depicted in Fig. 4.10:

Table 4.5 New notations of given problem (Moussawi-Haidar et al. 2016)

C_{I1}	Screening cost per item during production ($/item)
C_{I2}	Screening cost per item after production stops ($/item)
y	Total number of items produced during a production cycle (item)

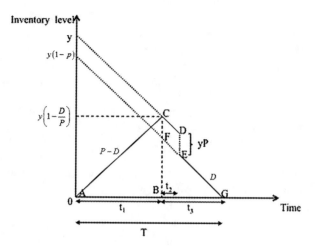

Fig. 4.10 The behavior of the inventory level over a production cycle (Moussawi-Haidar et al. 2016)

$$y\left(1 - \frac{D}{P}\right) - \frac{pD}{1-p}\frac{y}{P} \qquad (4.47)$$

At t_1, the on-hand inventory not screened in $[0, t_1]$ is screened at the rate x. It can be easily checked that the total number of defective items in a cycle, yp, is the summation of the defective items found during the interval $[0, t_1]$, $pP/(1-p)\cdot y/P$, and those found during the screening period t: $p\left[y\left(1 - \frac{D}{P}\right) - \frac{pDy}{(1-p)P}\right]$.

Two conditions are required. First, to avoid shortages during production, the number of good items produced should meet demand during production, i.e., $N(y, p) \geq Dt_1$, which implies the following condition on p:

$$p \leq 1 - D/P \qquad (4.48)$$

The on-hand inventory not screened at the end of production is expressed in (4.47) and requires t_2 units of time to be screened at rate x per unit per unit time. Thus, t_2 can be written as:

$$t_2 = \frac{y(1 - D/P) - (pD/(1-p))(y/P)}{x} \qquad (4.49)$$

They let t_3 be the time from when production stops until the end of the cycle, i.e., $t_3 = T - t_1$. Then t_3 can be written as:

$$t_3 = \frac{y(1 - D/P) - yp}{D} \qquad (4.50)$$

4.3 No Shortage

The second condition is a limit on the screening time t_2. Naturally, since t_2 should be less than t_3, after some term arrangement, the following condition on the screening rate, R_s, is derived:

$$R_s > \frac{D(1 - D/P) - pD^2/(1-p)}{1 - D/P - p} \tag{4.51}$$

Let TR(y) be the total revenue per cycle. TR(y) is the summation of the selling price of good-quality items and the discounted selling price of defective items. Thus, it is written as:

$$\text{TR}(y) = sy(1-p) + vyp \tag{4.52}$$

Also, let TC(y) be the total cost per cycle. TC(y) is the summation of the production setup cost, unit production cost, screening cost during and after production, and inventory holding cost. The number of units screened at the end of production, i.e., at time t_1, is equal to the inventory level at t_1 less the total number of defectives identified during production and given in (4.46). To compute the holding cost expression, they refer to Fig. 4.10, in which the average inventory is the summation of the three areas, ABC, CDEF, and BGF. From Fig. 4.10, the cycle time T can be found as $T = y(1 - P)/D$. Computing the areas of the three triangles, the total cost per cycle, TC(y), can be written as follows (Moussawi-Haidar et al. 2016):

$$\text{TC}(y) = \underbrace{K}_{\text{Setup Cost}} + \underbrace{Cy}_{\text{Production Cost}} + \underbrace{C_{11} \frac{D}{(1-p)} \frac{y}{P}}_{\text{Screening cost during production}} + \underbrace{C_{12} y \left[(1 - D/P) - \frac{pD}{P(1-p)} \right]}_{\text{Screening cost after production}}$$

$$+ \underbrace{h \left[\frac{y^2(1 - D/P - p)^2}{2D} + \frac{y^2(1 - D/P)}{2P} + \frac{y^2 p \left(1 - D/P - \frac{pD}{P(1-p)} \right)}{R_s} \right]}_{\text{Holding Cost}}$$

$$\tag{4.53}$$

The total profit per cycle is the total revenue less the total cost and is given as:

$$\text{TP}(y) = sy(1-p) + vyp - \left[K + Cy + C_{11} \frac{D}{(1-p)} \frac{y}{P} + C_{12} y \left[(1 - D/P) - \frac{pD}{P(1-p)} \right] \right.$$

$$\left. + h \left[\frac{y^2(1 - D/P - p)^2}{2D} + \frac{y^2(1 - D/P)}{2P} + \frac{y^2 p \left(1 - D/P - \frac{pD}{P(1-p)} \right)}{R_s} \right] \right]$$

$$\tag{4.54}$$

Taking the expected value of the total profit per cycle ETP(y) with respect to p, they got the following expression for the expected total profit per cycle ETP(y):

$$\text{ETP}(y) = sy(1 - E(p)) + vyE(p) - \left[K + Cy + C_{11}DE\left(\frac{1}{1-p}\right)\frac{y}{P} + C_{12}y\left[(1 - D/P) - \frac{D}{P}E\left(\frac{p}{1-p}\right)\right]\right]$$

$$+ h\left[\frac{y^2 E\left[(1 - D/P - p)^2\right]}{2D} + \frac{y^2(1 - D/P)}{2P} + \frac{y^2 E(p)\left(1 - D/P - \frac{D}{P}E\left(\frac{p}{1-p}\right)\right)}{R_s}\right]$$

(4.55)

Using the renewal reward theorem (see Ross 1996, Theorem 4.6.1), they found the expected profit per unit time as follows:

$$\text{ETPU}(y) = \frac{\text{ETP}(y)}{E(T)},$$

where the expected duration of the production cycle is $E(T) = [y(1 - E(p))]/D$. This gives the following expression for the expected profit per unit time, ETPU(y) (Moussawi-Haidar et al. 2016):

$$\text{ETPU}(y) = sD + \frac{vDE(p)}{1 - E(p)} - \frac{C_{11}D^2}{P(1 - E(p))}E\left(\frac{1}{1-p}\right) - \frac{C_{12}D}{1 - E(p)}\left(1 - D/P - \frac{D}{P}E\left(\frac{p}{1-p}\right)\right)$$

$$- \frac{CD}{1 - E(p)} - K\frac{D}{y(1 - E(p))} - h\frac{y}{1 - E(p)}\left[\frac{E\{(1 - D/P - p)^2\}}{2} + \frac{D(1 - D/P)}{2P} + \frac{DE(p)\left(1 - D/P - \frac{D}{P}E\left(\frac{p}{1-p}\right)\right)}{R_s}\right]$$

(4.56)

Setting the unit discounted price v to zero results in a model that allows scrapping the defective items, rather than selling them at discount. Thus, their model in this case reduces to one that allows scrapping defective items. All the mathematical expressions will remain unchanged.

The first-order conditions of ETPU(y) with respect to y give the optimal production lot size as (Moussawi-Haidar et al. 2016):

$$y^* = \sqrt{\frac{KD}{h\left[\frac{E\left[(1 - D/P - p)^2\right]}{2} + \frac{D(1 - D/P)}{2P} + \frac{DE(p)\left(1 - D/P - \frac{D}{P}E(p/(1-p))\right)}{R_s}\right]}}$$

(4.57)

4.3 No Shortage

4.3.4.2 Reworking of Defective Items

In this section Moussawi-Haidar et al. (2016) considered that all defective items will be reworked. Using previous models and also Figs. 4.11 and 4.12, the new total profit becomes:

$$\text{ETPU}(y) = \overbrace{sD}^{\text{Revenue}} - \overbrace{CD}^{\text{Production Cost}} - \overbrace{C_R pD}^{\text{Rework Cost}} - \overbrace{\frac{DK}{y}}^{\text{Setup Cost}} - \overbrace{C_{11} \frac{D^2}{P} E\left(\frac{1}{1-p}\right)}^{\text{Screening cost during production}}$$

$$- \overbrace{C_{12} DJ}^{\text{Screening cost after production}} - \overbrace{h_1 y \frac{Dd^2}{2P^2 P_1}}^{\text{Holding Cost of Reworked Items}}$$

$$- hy \underbrace{\left[\frac{D\widetilde{J}}{2D} + \frac{DJ}{x}\left(\widetilde{J} - \frac{DJ}{2x}\right) + \left(\widetilde{J} - \frac{DJ}{x}\right)\frac{dD}{PP_1} + \left(\widetilde{J} - \frac{DJ}{x} - \frac{Dd}{PP_1}\right)^2 / 2 + \frac{dD}{2P^2} + \frac{dJD}{Px} \right]}_{\text{Holding Cost}}$$

(4.58)

where $J = 1 - \frac{D}{P}\left[1 - E\left(\frac{p}{1-p}\right)\right]$ and $\widetilde{J} = 1 - \frac{D}{P} - \frac{d}{P} = 1 - \frac{D}{P} - E(p)$.

And after proving the concavity and setting the first derivative respect to y equal to zero and some simplification,

$$y^* = \sqrt{\frac{KD}{h\left[\frac{D\widetilde{J}}{2P} + \frac{DJ}{x}\left(\widetilde{J} - \frac{DJ}{2x}\right) + \left(\widetilde{J} - \frac{DJ}{x} + \frac{1}{2} + \frac{J}{P_1}\right)\frac{DE(p)}{P_1} + h_1 \frac{DE(p)^2}{2P_1} + \left(\widetilde{J} - \frac{DJ}{x} - \frac{DE(p)}{P_1}\right)^2 / 2\right]}}$$

(4.59)

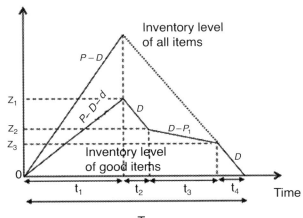

Fig. 4.11 The behavior of the inventory level over a production cycle when defective items are reworked (Moussawi-Haidar et al. 2016)

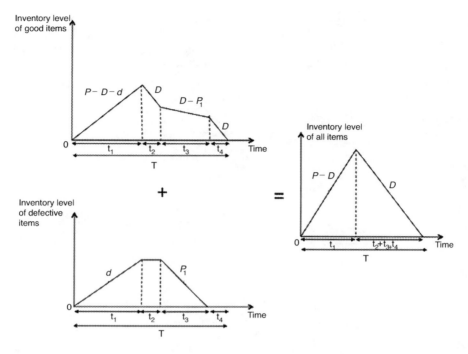

Fig. 4.12 The inventory of good, defective, and all items (Moussawi-Haidar et al. 2016)

Example 4.4 Moussawi-Haidar et al. (2016) analyzed how the optimal production quantity and optimal profit vary with the model parameters, for each of the models. Note that in each model, the conditions on the expected proportion of defective items and the screening rate should be held. So they developed numerical results similar to those in Hayek and Salameh (2001). This illustrates the application of their model and allows comparing their results with those of Hayek and Salameh (2001) for the model with rework. They set $D = 1200$ units/year, $P = 1600$ units/year, $P_1 = 100$ units/year, $C = \$104$, $s = \$200$/unit, $v = \$80$/unit, $R_s = 175{,}200$ units/year, $C_{I1} = \$0.5$/unit, $C_{I2} = \$0.6$/unit, $K = \$1500$, $h = \$20$/unit/year, $h_1 = \$22$/unit/year, $C_R = \$8$/unit, $p \sim U[0, 0.1]$ with probability distribution function:

$$f(p) = \begin{cases} 10 & 0 \le p \le 0.1 \\ 0 & \text{otherwise} \end{cases}$$

Then,

$$E(p) = 0.05$$
$$E\left(\frac{1}{1-p}\right) = 1.0536$$
$$E\left(\frac{p}{1-p}\right) = 0.0536$$

When the defective items are salvaged, and $p = 0$, the economic production quantity and related profits are, respectively, 848.52 units and $110,332. For more detailed information, readers can refer to Moussawi-Haidar et al. (2016).

4.4 Backordering

4.4.1 Simple Rework

Cárdenas-Barrón (2009) developed an EPQ inventory model in an imperfect production system environment with rework process and planned backorders. In addition to common assumptions of the EPQ model, he assumed that the proportion of defective products is known, the products are 100% screened, and the screening cost is not considered. All defective products are reworked and converted into good-quality products. Scrap is not generated at any cycle; backorders are allowed, and all backorders are satisfied. Production and reworking are done in the same manufacturing system at the same production rate. Two types of backorder costs are considered: linear backorder cost (backorder cost is applied to average backorders) and fixed backorder cost (backorder cost is applied to maximum backorder level allowed). The new notation which is used in this model is $\widehat{C}_b$ presenting the unit backorder cost independent of time.

Figure 4.13 shows the inventory behavior for the EPQ inventory model with rework at the same cycle and planned backorders. Using Fig. 4.13, the maximum inventory I_{max} can be calculated as the sum of $I_1 + I_2$ (Cárdenas-Barrón 2009).

According to triangle (146), it can be concluded that (Cárdenas-Barrón 2009):

$$\tan\theta_1 = P(1-x) - D = \frac{I_1 + B}{T_1 + T_2}$$

Also, it is obvious that $T_1 + T_2 = TP$ is the production time of manufacturing Q units; therefore $T_1 + T_2$ is equal to Q/P. Then according to Fig. 4.13,

$$\frac{Q}{P}[P(1-x) - D] = I_1 + B$$
$$I_1 = Q[(1-x) - D/P] - B$$

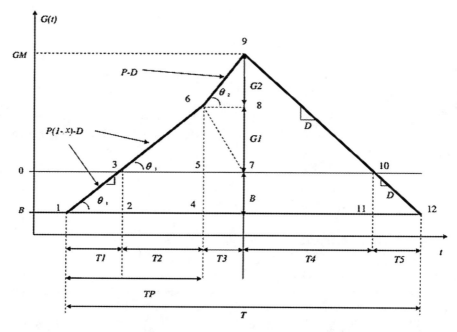

Fig. 4.13 Inventory behavior for the EPQ inventory model with rework (Cárdenas-Barrón 2009)

$$P - D = \frac{I_2}{T_3}$$

where T_3 is the production time of manufacturing the defective products (xQ); therefore T_3 is equal to xQ/P. Then,

$$\frac{xQ}{P}(P - D) = I_2$$

$$I_2 = xQ(1 - D/P)$$

Adding I_1 and I_2, the maximum inventory I_{max} is obtained (Cárdenas-Barrón 2009):

$$I_{max} = I_1 + I_2 = Q[(1 - x) - D/P] - B + xQ(1 - D/P)$$

$$I_{max} = Q[1 - (1 + x)(D/P)] - B$$

According to triangle (356), one can obtain T_2 as:

4.4 Backordering

$$T_2 = \frac{Q[(1-x) - D/P] - B}{[P(1-x) - D]}$$

Cárdenas-Barrón (2009) presented that:

$$\text{Area of triangle } (356) = \frac{T_2 I_1}{2} = \frac{\{Q[(1-x) - D/P] - B\}^2}{2[P(1-x) - D]}$$

$$\text{Area of triangle } (567) = \frac{T_3 I_1}{2} = \frac{xQ\{Q[(1-x) - D/P] - B\}}{2P}$$

$$\text{Area of triangle } (679) = \frac{T_3 I_{\max}}{2} = \frac{xQ\{Q[1 - (1+x)(D/P)] - B\}}{2P}$$

According to triangle (7910), T_4 can be obtained as:

$$T_4 = \frac{Q[1 - (1+x)(D/P)] - B}{D}$$

where T_4 is the time needed for maximum consumption at hand inventory level $I_{\max}$; then,

$$\text{Area of triangle } (7910) = \frac{T_4 I_{\max}}{2} = \frac{\{Q[1 - (1+x)(D/P)] - B\}^2}{2D}$$

Then, the inventory average which can be computed summing the area of triangles (356), (567), (679), and (7910) divided by T gives:

$$I = \frac{\frac{\{Q[(1-x) - D/P] - B\}^2}{2[P(1-x) - D]} + \frac{\{Q[1 - (1+x)(D/P)] - B\}^2}{2D} + \frac{xQ\{Q[1 - \frac{x}{2} - (1 + \frac{x}{2})(D/P)] - B\}}{P}}{T}$$

After some simplifications,

$$I = \frac{(Q^2 + B^2)(1-x) + Q^2(1 + x + x^2)\left(\frac{D^2}{P^2}\right) + Q^2(x^3 - 2)\left(\frac{D}{P}\right) + 2BQ\left(\frac{D}{P} + (x-1)\right)}{2Q[(1-x) - D/P]}$$

In order to simplify the mathematical expression, Cárdenas-Barrón (2009) defined $Z = 1 - x$, $E = 1 - x - D/P$, $I = (1 + x + x^2)\left(\frac{D^2}{P^2}\right)$, $O = (x^3 - 2)(D/P)$, $U = D/P + x - 1$.

Then,

$$I = \frac{Q^2(Z+I+O) + B^2Z - 2BQE}{2QE}$$

$$I = \frac{Q}{2}\left(\frac{Z+I+O}{E}\right) + \frac{B^2Z}{2QE} - B$$

After some algebra,

$$I = \frac{Q}{2}\left[1 - (1+x+x^2)(D/P)\right] + \frac{B^2(1-x)}{2Q(1-x-D/P)} - B$$

assume:

$$L = 1 - (1+x+x^2)(D/P)$$

Finally, the inventory average is given by

$$I = \frac{Q}{2}L + \frac{B^2Z}{2QE} - B$$

And easily T_1 and T_5 can be expressed by:

$$T_1 = \frac{B}{[P(1-x) - D]}$$

$$T_5 = \frac{B}{D}$$

So according to Fig. 4.12 and using T_1 and T_5, the average backorders J can be computed as below:

$$J = \frac{\frac{B^2}{2[P(1-x)-D]} + \frac{B^2}{2D}}{T}$$

$$J = \frac{B^2(1-x)}{2Q[(1-x) - D/P]} \quad (4.60)$$

$$J = \frac{B^2Z}{2QE} \quad (4.61)$$

Cárdenas-Barrón (2009) assumed that all defective products have the same manufacturing cost when they are reworked. Consequently, the total cost function (TC) is given as:

4.4 Backordering

$$\text{TC} = \overbrace{\frac{KD}{Q}}^{\text{Setup Cost}} + \overbrace{h\bar{I}}^{\text{Holding cost}} + \overbrace{\frac{\widehat{C_b}BD}{Q} + \widehat{C_b}J}^{\text{Shortage Cost}} + \overbrace{CD(1+x)}^{\text{Production Cost}} \quad (4.62)$$

Substituting $I = \frac{Q}{2}L + \frac{B^2 Z}{2QE} - B$ and $J = \frac{B^2 Z}{2QE}$ into the second and fourth terms of Eq. (4.62), respectively, Eq. (4.62) can be rewritten as:

$$\text{TC}(Q,B) = \frac{KD}{Q} + \frac{hQL}{2} + \frac{hB^2 Z}{2QE} - hB + \frac{\widehat{C_b}DB}{Q} + \frac{C_b B^2 Z}{2QE} + CD(1+x) \quad (4.63)$$

The problem is to find the lot size (Q) and the size of backorders (B) that minimize the total cost inventory system (4.63). Assuming Q and B are continuous, let us take the first partial derivatives with respect to Q and B of Eq. (4.63), with these derivatives expressed by Eqs. (4.64) and (4.65), respectively:

$$\frac{\partial \text{TC}(Q,B)}{\partial Q} = -\frac{KD}{Q^2} + \frac{hL}{2} - \frac{hB^2 Z}{2Q^2 E} - \frac{\widehat{C_b}BD}{Q^2} - \frac{C_b B^2 Z}{2Q^2 E} \quad (4.64)$$

$$\frac{\partial \text{TC}(Q,B)}{\partial B} = \frac{hBZ}{QE} - h + \frac{\widehat{C_b}D}{Q} + \frac{C_b BZ}{QE} \quad (4.65)$$

In order to verify that Eq. (4.63) is convex in Q and B, they must show that the following two well-known conditions hold:

$$\frac{\partial^2 \text{TC}(Q,B)}{\partial Q^2} > 0, \quad \frac{\partial^2 \text{TC}(Q,B)}{\partial B^2} > 0 \quad (4.66)$$

$$\left(\frac{\partial^2 \text{TC}(Q,B)}{\partial Q^2}\right)\left(\frac{\partial^2 \text{TC}(Q,B)}{\partial B^2}\right) - \left(\frac{\partial^2 \text{TC}(Q,B)}{\partial Q \partial B}\right)^2 > 0 \quad (4.67)$$

First, they proved condition (4.66). By taking the second partial derivatives with respect to Q and B of Eq. (4.63), it can be easily shown that condition (4.66) is satisfied. The second partial derivatives are given by Eqs. (4.68) and (4.69). Both equations must be greater than zero. Since x is on interval (0, 1), thus K is greater than zero, and E also will be greater than zero if and only if x is less than $1 - D/P$. Notice that the former analysis is valid for any finite, nonzero value of Q and B. Therefore, both second derivatives (Eqs. 4.68 and 4.69) are greater than zero (Cárdenas-Barrón 2009):

$$\frac{\partial^2 \text{TC}(Q,B)}{\partial Q^2} = \frac{2KD}{Q^3} + \frac{hB^2 K}{Q^3 E} + \frac{2\widehat{C_b}BD}{Q^3} + \frac{C_b B^2 K}{Q^3 E} > 0 \quad (4.68)$$

$$\frac{\partial^2 TC(Q,B)}{\partial Q^2} = \frac{hK}{QE} + \frac{C_b K}{QE} > 0 \qquad (4.69)$$

After that, they proved condition (4.67). For the sake of brevity, they provided only the final results. Taking the partial derivatives with respect to Q and B of Eq. (4.66) yields (Cárdenas-Barrón 2009),

$$\frac{\partial^2 TC(Q,B)}{\partial Q \partial B} = -\frac{1}{Q^2}\left[\frac{BK}{E}(h+C_b) + \widehat{C_b}D\right] \qquad (4.70)$$

Then,

$$\left(\frac{\partial^2 TC(Q,B)}{\partial Q \partial B}\right)^2 = \left\{-\frac{1}{Q^2}\left[\frac{BK}{E}(h+C_b) + \widehat{C_b}D\right]\right\}^2 \qquad (4.71)$$

Developing the squared binomial, one obtains (Cárdenas-Barrón 2009):

$$\left(\frac{\partial^2 TC(Q,B)}{\partial Q \partial B}\right)^2 = \frac{B^2 K^2 (h+C_b)^2}{Q^4 E^2} + \frac{2\widehat{C_b}BDK(h+C_b)}{Q^4 E} + \frac{\left(\widehat{C_b}D\right)^2}{Q^4} \qquad (4.72)$$

On the other hand,

$$\left(\frac{\partial^2 TC(Q,B)}{\partial Q^2}\right)^2 \left(\frac{\partial^2 TC(Q,B)}{\partial B^2}\right)^2 = \frac{2KD(H+W)}{Q^4 E} + \frac{B^2 K^2 (h+C_b)^2}{Q^4 E^2}$$
$$+ \frac{2\widehat{C_b}BDK(h+C_b)}{Q^4 E} \qquad (4.73)$$

Substituting Eqs. (4.72) and (4.73) into (4.67), and after simplifying, one obtains (Cárdenas-Barrón 2009):

$$\left(\frac{\partial^2 TC(Q,B)}{\partial Q^2}\right)\left(\frac{\partial^2 TC(Q,B)}{\partial B^2}\right) - \left(\frac{\partial^2 TC(Q,B)}{\partial Q \partial B}\right)^2$$
$$= \frac{1}{Q^4 E}\left[2KDZ(h+C_b) - E\left(\widehat{C_b}D\right)^2\right] \qquad (4.74)$$

Equation (4.74) should be greater than zero if and only if the following condition is satisfied (Cárdenas-Barrón 2009):

4.4 Backordering

$$2KDZ(h + C_b) - E(\widehat{C}_bD)^2 > 0 \tag{4.75}$$

Finally, it can be concluded that total cost function (4.63) is convex in Q and B, if and only if condition (4.75) is satisfied. Then Cárdenas-Barrón (2009) using the first-order derivative derived the optimal values as below:

$$Q = \sqrt{\frac{2KDZ(h + C_b) - E(\widehat{C}_bD)^2}{h[Z(h + C_b)L - Eh]}} \tag{4.76}$$

$$B = \frac{E(hQ - \widehat{C}_bD)}{Z(h + C_b)} \tag{4.77}$$

Substituting Eqs. (4.76) and (4.77) into Eq. (4.63) and after some algebraic steps,

$$\text{TC}^* = \left\{\frac{1}{Z(h + C_b)}\right\}$$
$$\times \left\{\sqrt{\left[2KDZ(h + C_b) - E(\widehat{C}_bD)^2\right][Z(h + C_b)L - hE]h} + hE\widehat{C}_bD\right\}$$
$$+ CD(1 + x) \tag{4.78}$$

Equation (4.78) also can be written as (Cárdenas-Barrón 2009):

$$\text{TC}(Q) = \left\{\frac{h}{Z(h + C_b)}\right\}\left\{Q[Z(h + C_b)L - hE] + E\widehat{C}_bD\right\} + CD(1 + x) \tag{4.79}$$

Another alternative mathematical expression for the total cost was proposed by Goyal and Cárdenas-Barrón (2003), which is (Cárdenas-Barrón 2009):

$$\text{TC}(Q, B) = \frac{D}{Q}\left(2K + \widehat{C}_bB\right) + CD(1 + x) \tag{4.80}$$

It is important to mention that when x is equal to $1 - D/P$, then E is zero. Then,

$$Q = \sqrt{\frac{2KD}{hL}} \tag{4.81}$$

$$B = 0 \tag{4.82}$$

The reader can remember that L is equal to $1 - (1 + x + x^2)(D/P)$, and one may obtain a negative value under the radical in Eq. (4.81) when L is less than zero.

Therefore, $1 - (1 + x + x^2)(D/P)$ must be greater than zero to avoid a negative value for L. Thus, it is easy to see that the valid interval for R can be determined solving the following expression (Cárdenas-Barrón 2009):

$$1 - (1 + x + x^2)(D/P) > 0$$

$$x < \frac{-1 + \sqrt{1 - 4(1 - P/D)}}{2}$$

Then,

$$0 < x < \frac{-1 + \sqrt{1 - 4(1 - P/D)}}{2} \tag{4.83}$$

The reader can note that the right side of condition (4.83) in some cases may be greater than one, which has no practical meaning because the percentage of defective products x should be on interval $(0, 1)$. Therefore, when the right side of condition (4.83) is greater than one, then the valid interval for x is $(0, 1)$, otherwise given by condition (4.83) (Cárdenas-Barrón 2009).

Example 4.5 Cárdenas-Barrón (2009) presented an example with parameters values of $D = 300$ units per year, $P = 550$ units per year, $K = \$120$ per lot size, $h = \$0.5$ per unit per year, $\widehat{C}_b = \$1$ per unit short, $C_b = \$10$ per unit short per year, and $C = \$7$ per unit. The different values are considered for x, and results are presented in Table 4.6.

Example 4.6 The second example of Cárdenas-Barrón (2009) is presented using $D = 4800$ units per year, $P = 24{,}000$ units per year, $K = \$50$ per lot size, $h = \$50$ per unit per year, $\widehat{C}_b = \$0.1$ per unit shortage, $C_b = \$14.4$ per unit shortage per year, and $C = \$3$ per unit. The results are presented in Table 4.7.

4.4.2 Defective Product

Wee et al. (2013) revisited the work of Cárdenas-Barrón (2009) by changing the decision variables. Cárdenas-Barrón (2009) considered order quantity and backordering level as decision variables, while the optimal solution condition in the work of Wee et al. (2013) is analyzed using the production time and the time to eliminate backorders as decision variables instead of the classical decision variables of lot and backorder quantities. They developed two different models which are presented in this section, and assumptions are as same as those used in the work of Cárdenas-Barrón (2009).

4.4 Backordering

Table 4.6 Solution to Example 4.5 (Cárdenas-Barrón 2009)

x	Q	B	TC(Q, B)
0	87.97727	31.052	2546.884
0.01	88.52751	30.88151	2564.528
0.05	90.9152	30.13339	2634.411
0.1	94.21531	28.95957	2720.633
0.15	97.58277	27.34423	2806.496
0.2	100.4781	25.05102	2893.368
0.25	102.0588	21.83155	2983.122
0.3	101.2903	17.53177	3078.104
0.35	97.36266	12.24558	3180.858
0.4	90.24759	6.382393	3293.635
0.45	80.85083	0.515502	3417.967
1−300/550	79.92997	0	3429.874

4.4.2.1 Model I: EPQ Model for Production Time Greater Than or Equal to the Time to Eliminate Backorders

Figure 4.14 depicts the production–inventory model with rework at the same cycle. During interval t_1, there is no production, and maximal backordering level is achieved. Production and screening simultaneously start with the same rate. The total production size during interval $t_2 + t_3$ is $P(t_2 + t_3)$ units. One can see that the perfect items increase at a rate of $(1 - x)P - D$ during interval $t_2 + t_3$. To eliminate backorders and build a positive stock, $(1 - x)P - D > 0$ is assumed. Since the production process contains p rate of defective items, $P(t_2 + t_3)p$ units of defective items are reworked to perfect items during interval t_4. Wee et al. (2013) assumed the defective items are 100% reworked to perfect items and then the inventory increases at a rate of $P - D$ during interval t_4. Afterward, during interval t_5, the inventory is depleted without production.

Referring to Fig. 4.14, the backordering level B is as follows:

$$B = ((1 - x)P - D)t_2 \qquad (4.84)$$

During the interval t_3, the inventory level increases from zero to I_1 with a rate of $(1 - x)P - D$. So:

$$I_1 = ((1 - x)P - D)t_3 \qquad (4.85)$$

Since $B = kt_1$, then:

$$t_1 = \frac{((1 - x)P - D)t_2}{D} \qquad (4.86)$$

They know that the interval t_4 is assumed to rework the defective units generated during the total production time $(t_2 + t_3)$. Thus,

Table 4.7 Solution to Example 4.6 (Cárdenas-Barrón 2009)

x	Q	B	TC(Q, B)
0	87.97727	31.052	2546.884
0.01	1574.694	24.72761	15,283.11
0.05	1583.218	24.7332	15,855.13
0.1	1594.994	24.73314	16,569.7
0.15	1608.08	24.71774	17,283.76
0.2	1622.556	24.67667	17,997.29
0.25	1638.51	24.59631	18,710.28
0.3	1656.044	24.45839	19,422.72
0.35	1675.265	24.23812	20,134.6
0.4	1696.296	23.90122	20,845.89
0.45	1719.262	23.39939	21,556.59
0.5	1744.292	22.66302	22,266.68
0.55	1771.505	21.589	22,976.14
0.6	1800.982	20.01965	23,684.99
0.65	1832.718	17.70373	24,393.21
0.7	1866.505	14.22006	25,100.85
0.75	1901.682	8.813453	25,808

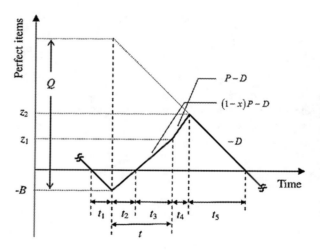

Fig. 4.14 Inventory of perfect-quality items for EPQ model (Wee et al. 2013)

$$t_4 = P(t_2 + t_3)$$

and

$$t_4 = x(t_2 + t_3) \tag{4.87}$$

From Fig. 4.12, it is shown that the maximum inventory level I_2 is:

4.4 Backordering

$$I_2 = I_1 + (P-D)t_4 = (P(1-x)-D)t_3 + (P-D)t_4$$
$$= (xt_2 + t_3)P - (x(t_2+t_3)+t_3)D \quad (4.88)$$

Since z_2 units is consumed at a rate of k during the interval t_5, one has:

$$t_5 = \frac{I_2}{D} = \frac{(xt_2+t_3)P - (x(t_2+t_3)+t_3)D}{D} \quad (4.89)$$

The total inventory cost per cycle, TC, is as below:

$$TC = \overbrace{CP(t_2+t_3)}^{\text{Production Cost}} + \overbrace{C_R P(t_2+t_3)x}^{\text{Rework Cost}} + \overbrace{K}^{\text{Setup Cost}} +$$
$$\underbrace{\left[C_b \frac{(t_1+t_2)B}{2} + \widehat{C}_b B\right]}_{\text{Shortage Cost}} + h \underbrace{\left[\frac{t_3 I_1}{2} + \frac{t_4(I_1+I_2)}{2} + \frac{t_5 I_2}{2}\right]}_{\text{Holding Cost}} \quad (4.90)$$

From Eqs. (4.86), (4.87), and (4.89), the total cycle time can be rewritten as

$$T = \sum_{i=1}^{5} t_i = \frac{(t_2+t_3)P}{D} \quad (4.91)$$

For ease of notation and analysis, t ($t = t_2 + t_3$) is set as decision variable. Moreover, the rework and production unit costs are the same ($C = C_R$). Then the total cost per unit time is expressed as (Wee et al. 2013):

$$TCU(t_2,t) = \frac{TC}{T} = C(1+x)D + \frac{KD}{tP} + \frac{C_b t_2^2 (1-x)}{2t}((1-x)P-D) + \frac{\widehat{C}_b t_2 D}{tP}((1-x)P-D)$$
$$+ \frac{h}{2t}((t-(1-x)t_2)^2 P + 2t \cdot t_2 D - (1+x+x^2)t^2 D - (1-x)t_2^2 D) \quad (4.92)$$

Obviously, the objective is to minimize TCU(t_2, t) subject to $t \geq t_2 \geq 0$. The production time t is greater than zero.

Lemma 4.1 For any given $t > 0$, the following cases that minimize TCU($t_2|t$) exist (Wee et al. 2013):

Case I. For a given $t \leq \frac{\widehat{C}_b D}{hP}$, $t_2^* = 0$;

Case II. If $hP - (C_b+h)(1-x)P \geq 0$, for a given $\frac{\widehat{C}_b D}{hP} < t \leq \frac{\widehat{C}_b D}{hP-(C_b+h)(1-x)P}$, $0 < t_2^* \leq t$;

If $hP - (C_b+h)(1-x)P < 0$, for a given $\frac{\widehat{C}_b D}{hP} < t$, $0 < t_2^* \leq t$;

Case III. If $hP - (C_b + h)(1-x)P > 0$, for a given $t > \frac{\widehat{C_b}D}{hP-(C_b+h)(1-x)P}$, $t_2^* > t$.

Proof See Appendix of Wee et al. (2013).

Lemma 4.1 shows that the optimal time to eliminate backorders, t_2^*, is dependent on the production time, t. The time $t_2^* = 0$ implies that shortage is not allowed. If the production time is predetermined, Lemma 4.1 shows the decision whether to schedule shortage period. If the production time is greater than $\frac{\widehat{C_b}D}{hP}$, the optimal policy is to allow a shortage period. Lemma 4.1 Case III shows if production time is greater than $\frac{\widehat{C_b}D}{hP-(C_b+h)(1-x)P}$, the optimal time to eliminate backorders is greater than the production time, which is an infeasible solution for Model I (Fig. 4.14). Taking the first derivatives of TCU(t_2, t) with respect to t_2 and t, one has (Wee et al. 2013):

$$\frac{\partial}{\partial t_2}\text{TCU}(t_2,t) = \frac{((1-x)P-D)\left(((C_b+h)(1-x)t_2 - ht)P + \widehat{C_b}D\right)}{Pt} \quad (4.93)$$

$$\frac{\partial}{\partial t}\text{TCU}(t_2,t) = \frac{\left\{\begin{array}{l}-2KD - 2\widehat{C_b}Dt_2((1-x)P-D) + t^2hP(P-D(1+x+x^2)) \\ -(C_b+h)(1-x)t_2^2P((1-x)P-D)\end{array}\right\}}{2Pt^2} \quad (4.94)$$

Let $\frac{\partial}{\partial t_2}\text{TCU}(t_2,t) = 0$ and $\frac{\partial}{\partial t}\text{TCU}(t_2,t) = 0$, so:

$$t_2(t) = \frac{hPt - \widehat{C_b}D}{(C_b+h)(1-x)P} \quad (4.95)$$

$$t(t_2) = \sqrt{\frac{2KD + t_2((1-x)P-D)\left((C_b+h)(1-x)Pt_2 + 2\widehat{C_b}D\right)}{hP(P-D(1+x+x^2))}} \quad (4.96)$$

Let $(\widehat{t_2}, \widehat{t})$ be the solution satisfying the first-order conditions for TCU(t_2, t). Substituting Eq. (4.95) into Eq. (4.96), they derived the production time and time $(\widehat{t})$ to eliminate backorders $(\widehat{t_2})$ as follows:

$$\widehat{t} = \frac{1}{P}\sqrt{\frac{D}{h}\frac{2K(C_b+h)(1-x)P - \widehat{C_b}^2D((1-x)P-D)}{hD+(1-x)C_bP-(1-x^3)(C_b+h)D}} \quad (4.97)$$

$$\widehat{t_2} = \frac{hP\widehat{t} - \widehat{C_b}D}{(C_b+h)(1-x)P} \quad (4.98)$$

From the denominator of Eq. (4.97),

4.4 Backordering

$$hD + (1-x)C_bP - (1-x^3)(C_b+h)D$$
$$> (C_b+h)D - (1-x^3)(C_b+h)D((1-x)P > D) = (C_b+h)Dx^3 \geq 0 \quad (4.99)$$

Thus, $\widehat{t}$ and $\widehat{t}_2$ in Eqs. (4.97) and (4.98) are real solutions. Based on the above analysis and results, Wee et al. (2013) proposed the following theorem:

Theorem 4.1

(i) If $2hKP - \widehat{C}_b^2 D(P - (1+x+x^2)D) > 0$, i.e., $hP\widehat{t} - \widehat{C}_b D > 0$, then $TCU(t_2, t)$ is convex in t_2 and t. If $\widehat{t} \geq \widehat{t}_2$ the optimal solution (t_2^*, t^*) of $TCU(t_2, t)$ is $(\widehat{t}_2, \widehat{t})$.

(ii) If $2hKP - \widehat{C}_b^2 D(P - (1+x+x^2)D) \leq 0$, i.e., $hP\widehat{t} - \widehat{C}_b D \leq 0$, then the optimal solution (t_2^*, t^*) of $TCU(t_2, t)$ is $\left(0, \sqrt{\frac{2DK}{hP(P-D(1+x+x^2))}}\right)$.

Proof See Appendix of Wee et al. (2013).

Substituting (t_2, t) by (t_2^*, t^*) into Eq. (4.92), the optimal total cost per unit time can be derived, i.e., $TCU(t_2^*, t^*)$. From t^* and t_2^*, the optimal batch size and optimal maximal backordering level are as follows (Wee et al. 2013):

$$Q^* = Pt^* \quad (4.100)$$

$$B^* = ((1-x)P - D)t_2^* \quad (4.101)$$

The constraint of $2hKP - \widehat{C}_b^2 D(P - (1+x+x^2)D) > 0$ in Theorem 4.1(i) is equivalent to Eq. (4.101) in Cárdenas-Barrón (2009). Replacing t^* and t_2^* by $\widehat{t}$ and $\widehat{t}_2$, respectively, it can be concluded that Q^* and B^* are the same as Eqs. (4.102) and (4.103) in Cárdenas-Barrón (2009). Similar to the result in Cárdenas-Barrón (2009), Theorem 4.1(ii) derives the optimal solution as in Jamal et al. (2004). Substituting t^* by $\sqrt{\frac{2DK}{hP(P-D(1+x+x^2))}}$ into Eq. (4.100), it is shown that the resulting Q^* is identical to Eq. (4.94) in Jamal et al. (2004).

4.4.2.2 Model II: EPQ Model for Production Time Less than the Time to Eliminate Backorders

In this subsection, Model II inventory model is developed. Figure 4.15. describes the inventory behavior for which the production run time is less than the time to eliminate backorders. The production starts at the beginning of interval s_2. Here, the reworking period is $s_3 + s_4$. They used s_2 and s_3 to obtain the following expressions for the objective function. From Fig. 4.15, it can be seen that (Wee et al. 2013):

$$B_1 = (P - D)s_3 \qquad (4.102)$$

The maximum backordering level (B_2) is:

$$B_2 = B_1 + ((1 - x)P - D)s_2 = (P - D)s_3 + ((1 - x)P - D)s_2 \qquad (4.103)$$

Since $B_2 = Ds_1$,

$$s_1 = \frac{(P - D)s_3 + ((1 - x)P - D)s_2}{D} \qquad (4.104)$$

The interval $s_3 + s_4$ is the time for reworking the defective units produced during the total production time (s_2). Thus, one has $P(s_3 + s_4) = Ps_2 x$ and:

$$s_4 = xs_2 - s_3 \qquad (4.105)$$

From Fig. 4.14, the maximum inventory level (I) is:

$$I = (P - D)s_4 = (P - D)(xs_2 - s_3) \qquad (4.106)$$

Since I units with a rate of k is consumed during interval s_5, one has:

$$s_5 = \frac{I}{D} = \frac{(P - D)(xs_2 - s_3)}{D} \qquad (4.107)$$

The total inventory cost per cycle, TC', is:

$$\mathrm{TC}' = \overbrace{CPs_2 + CPs_2 x}^{\text{Production Cost}} + \overbrace{K}^{\text{Setup Cost}} + \overbrace{h \frac{I(s_4 + s_5)}{2}}^{\text{Holding Cost}} + \\ \overbrace{\left[C_b \left(\frac{B_2 s_1}{2} + \frac{(B_1 + B_2)s_2}{2} + \frac{B_1 s_3}{2} \right) + \widehat{C}_b B_2 \right]}^{\text{Shortage Cost}} \qquad (4.108)$$

From Eqs. (4.104), (4.105), and (4.107), the total cycle time S is rewritten as:

$$S = \sum_{i=1}^{5} s_i = \frac{s_2 P}{D} \qquad (4.109)$$

For simplicity of notation, they defined s_2 and s ($s = s_2 + s_3$) as decision variables of the inventory model. Then the total cost per unit time is simplified as:

4.4 Backordering

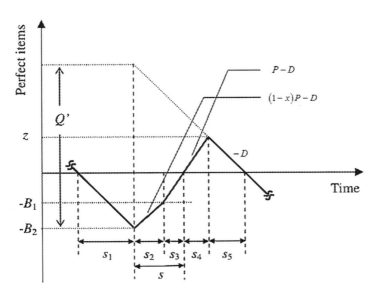

Fig. 4.15 Inventory of perfect-quality items for EPQ model II (Wee et al. 2013)

$$\text{TCU}'(s_2, s) = \frac{TC'}{S} = C(1+x)D + \frac{KD}{s_2 P} + \frac{h}{2s_2}(s - s_2 - s_2 x)^2 (P - D)$$
$$+ \frac{C_b}{2s_2}\Big[(s - s_2 x)^2 P - ((1-x)s_2^2 + 2(1-x)(s - s_2)s_2 \quad (4.110)$$
$$+ (s - s_2)^2)D\Big] + \frac{\widehat{C}_b D}{s_2 P}[(s - s_2 x)P - sD]$$

Evidently, the objective is to minimize $\text{TCU}'(s_2, s)$ subject to $s > s_2 > 0$. Taking the first derivatives of $\text{TCU}'(s_2, s)$ with respect to s and s_2, one has:

$$\frac{\partial}{\partial s}\text{TCU}(s_2, s) = \frac{(P - D)\Big[\widehat{C}_b D - P((h + (C_b + h)x)s_2 - (C_b + h)s)\Big]}{Ps_2} \quad (4.111)$$

$$\frac{\partial}{\partial s_2}\text{TUC}(s_2, s) = \frac{-2KD - 2\widehat{C}_b D(P - D)s - h\big(s^2 - (1+x)^2 s_2^2\big)P(P - D)}{2Ps_2^2}$$
$$+ \frac{C_b\big(xs_2^2(xP - D) - s^2(P - D)\big)}{2s_2^2}$$
$$(4.112)$$

Using $\frac{\partial}{\partial s}\text{TCU}'(s_2, s) = 0$ and $\frac{\partial}{\partial s_2}\text{TCU}'(s_2, s) = 0$, one obtains:

$$s(s_2) = \frac{(h + (C_b + h)x)Ps_2 - \widehat{C}_b D}{(C_b + h)P} \qquad (4.113)$$

$$s_2(t) = \sqrt{\frac{2KD + s(P-D)\left(s(C_b+h)P + 2\widehat{C}_b D\right)}{P\left(h(P-D)(1+x)^2 + C_b x(xP-D)\right)}} \qquad (4.114)$$

Let $(\widehat{s}_2, \widehat{s})$ be the solution satisfying the first-order conditions for TCU'(s_2, s). Substituting Eq. (4.113) into Eq. (4.114), it can be derived that (Wee et al. 2013):

$$\widehat{s}_2 = \frac{1}{P}\sqrt{\frac{D}{C_b} \frac{2K(C_b+h)P - \widehat{C}_b^2 D(P-D)}{h(P - (1 + (1-x)x)D) - C_b x(1-x)D}} \qquad (4.115)$$

$$\widehat{s} = \frac{(h + (C_b + h)x)P\widehat{s}_2 - \widehat{C}_b D}{(C_b + h)P} \qquad (4.116)$$

Using the previous analysis and results, they proposed the following theorem for Model II:

Theorem 4.2 If $h(P - (1 + (1-x)x)D) - C_b x(1-x)D > 0$ and

$$2KP(C_b+h)(hx - C_b(1-x))^2 > \widehat{C}_b^2(C_b+h)D\left(hx^2(P-D) + C_b(1-x)((1-x)P) - D\right)$$

i.e., $(hx - C_b(1-x))P\widehat{s}_2 - \widehat{C}_b D > 0$, then TCU'$(s_2, s)$ is convex in s_2 and s. The optimal solution (s_2, s) of TCU'(s_2^*, s^*) is $(\widehat{s}_2, \widehat{s})$ (Wee et al. 2013).

Proof See Appendix of Wee et al. (2013).

Substituting (s_2, s) by (s_2^*, s^*) into Eq. (4.110), the optimal total cost per unit time can be obtained. The optimal lot size and the optimal backordering level can be expressed as:

$$Q'^* = Ps_2^* \qquad (4.117)$$

$$B_2^* = ((1-x)P - D)s_2^* + (P-D)(s^* - s_2^*) \qquad (4.118)$$

Example 4.7 Wee et al. (2013) to verify their results used the parameter values of Cárdenas-Barrón (2009) as $D = 300$ unit per year, $P = 550$ units per year, $K = \$50$ per lot size, $h = \$50$ per unit per year, $\widehat{C}_b = \$1$ per unit shortage, $C_b = \$10$ per unit shortage per year, and $C = \$7$ per unit. Firstly, they applied Model I to find the optimal solutions. Since $2hKP - \widehat{C}_b^2 D(P - (1 + x + x^2)D) > 0$ for interval $p \in [0, 1 - \frac{300}{550}]$, they derived $(\widehat{t}_2, \widehat{t})$ from Eqs. (4.97) and (4.98). Table 4.8 illustrates the optimal policy of this example for Model I with different defective rates. When $p > 0.20$, t_3 and I_1 are negative values (infeasible optimal solutions for Model I).

4.4 Backordering

When $p > 0.20$, the optimal policy of the example is Lemma 4.1 Case III (Wee et al. 2013). To obtain feasible policies, Wee et al. (2013) applied Model II when Model I leads to infeasible policies. Table 4.9 illustrates the optimal policy for Model II with different defective rates. They had shown that when $p = 0.20$, Model II results in an infeasible optimal solution. Table 4.9 indicates that the optimal solutions are derived from Model II when $p > 0.20$.

4.4.3 Random Defective Rate: Same Production and Rework Rates

Sarkar et al. (2014) developed the work of Cárdenas-Barrón (2009) by considering that proportion of defective products in each cycle is a stochastic variable and follows a probability distribution (uniform, triangular, and beta). Other assumptions are as same as Cárdenas-Barrón's (2009) work. Basically, three different inventory models are developed for three different distribution density functions such as uniform, triangular, and beta.

Case A The proportion of defective products follows a uniform distribution.

In order to take the randomness of proportion of defective products into account, the expected value of R is used in the development and analysis of inventory model. The inventory behavior through time is represented in Fig. 4.16. According to Fig. 4.16, the maximum inventory I_{max} is simply computed as the sum of $I_1 + I_2$. From to triangle (146), it is easy to see that (Sarkar et al. 2014):

$$\tan \theta_1 = P(1 - E[x]) - D = \frac{I_1 + B}{T_1 + T_2} \quad (4.119)$$

In this case, it is assumed that the x follows a uniform distribution with range $[a, b]$, $0 < a < b < 1$. For a uniform distribution, it is well-known that the expected value for x is given as $E[x] = (a + b)/2$. Furthermore, the production time of producing Q units is $T_P = T_1 + T_2$. Therefore, $T_1 + T_2$ must be equal to Q/P. Substituting the expected value $E[x]$ and $T_1 + T_2$, one obtains:

$$P\left(1 - \frac{a+b}{2}\right) - D = \frac{I_1 + B}{T_1 + T_2} \quad (4.120)$$

or

$$I_1 = Q\left[\frac{2-a-b}{2} - \frac{D}{P}\right] - B = Q\left(1 - \frac{a+b}{2} - \frac{D}{P}\right) - B \quad (4.121)$$

According to triangle (689), it is easy to see that:

Table 4.8 The results of Example 1 for Model I with different p values (Wee et al. 2013)

p	0.2	0.25	0.30	0.35	0.40	0.45	1–300/550
TCU*	2893.37	2983.12	3078.1	3180.86	3293.64	3417.97	3429.87
Q^*	100.48	102.06	101.29	97.36	90.25	80.85	79.93
t_1^*	0.53	−0.96	−1.88	−2.07	−1.46	−0.15	0
t_2^*	9.66	10.64	11.93	13.42	14.95	16.39	16.51
B^*	25.05	21.83	17.53	12.25	6.38	0.52	0
t_1^*	0.0835	0.07277	0.05844	0.04082	0.02127	0.00172	0
t_2^*	0.17894	0.19406	0.20626	0.21297	0.21275	0.2062	0.205361
t_3^*	0.00375	−0.00850	−0.02209	−0.03594	−0.04866	−0.05920	−0.06003
t_4^*	0.03654	0.04639	0.05525	0.06196	0.06563	0.06615	0.06606
t_5^*	0.0322	0.03547	0.03978	0.04474	0.04983	0.05463	0.05505

4.4 Backordering

Table 4.9 The results of Example 1 for Model II with different p values (Wee et al. 2013)

p	0.2	0.25	0.30	0.35	0.40	0.45	1–300/550
TCU*	2893.26	2982.45	3072.88	3164.63	3257.76	3352.35	3361.02
Q^*	100.29	102.97	105.33	107.29	108.76	109.67	109.72
I^*	9.87	10.07	10.25	10.4	10.51	10.58	10.58
B_1^*	−0.75	1.63	4.11	6.67	9.26	11.85	12.08
B_2^*	24.78	22.69	20.39	17.88	15.19	12.35	12.08
s_1^*	0.08258	0.07563	0.06797	0.05962	0.05065	0.04116	0.04028
s_2^*	0.18234	0.18722	0.19152	0.19507	0.19774	0.19939	0.19949
s_3^*	−0.00301	0.00651	0.01644	0.02667	0.03705	0.0474	0.04834
s_4^*	0.03948	0.04029	0.04101	0.0416	0.04205	0.04232	0.04234
s_5^*	0.0329	0.03358	0.03418	0.03467	0.03504	0.03527	0.03528

$$\tan \theta_2 = P - D = \frac{I_2}{T_3} \tag{4.122}$$

$$T_3 = \frac{E[x]Q}{P} = \frac{Q}{P}\frac{(a+b)}{2} \tag{4.123}$$

where T_3 is the production time of producing the defective products. Therefore, T_3 must be equal to $E[x]Q/P$. Thus,

$$I_2 = T_3(P - D) = \frac{Q(a+b)(P-D)}{2P} = \frac{Q(a+b)(1-D/P)}{2} \tag{4.124}$$

Now, the maximum inventory $I_{\max}$ can be obtained as summing I_1 and I_2; hence (Sarkar et al. 2014),

$$\begin{aligned}I_{\max} = I_1 + I_2 &= Q\left(1 - \frac{a+b}{2} - \frac{D}{P}\right) - B + \frac{Q(a+b)(1-D/P)}{2} \\ &= Q\left[1 - \frac{D}{P}\left(1 + \frac{a+b}{2}\right)\right] - B\end{aligned} \tag{4.125}$$

From Fig. 4.16, T is the sum of T_1, T_2, T_3, T_4, and T_5. It is well-known that T is the time between runs. And T_2 and T_4 are:

$$T_2 = \frac{Q[(1-E[x]) - D/P] - B}{[P(1-E[x]) - D]} = \frac{Q\left[\left(1 - \frac{a+b}{2}\right) - D/P\right] - B}{\left[P\left(1 - \frac{a+b}{2}\right) - D\right]} \tag{4.126}$$

$$T_4 = \frac{Q\left[1 - (1+E[x])\frac{D}{P}\right] - B}{D} = \frac{Q\left[1 - \left(1 + \frac{a+b}{2}\right)\frac{D}{P}\right] - B}{D} \tag{4.127}$$

where T_2 is the time needed to build up I_1 units in inventory and T_4 is the time needed for consumption at hand maximum inventory level $I_{\max}$, then (Sarkar et al. 2014). As it was stated before, T_3 is equal to $E[x]Q/P = ((a+b)/2)Q/P$. In order to calculate the average inventory, the area of above horizontal line (time) should be calculated:

$$\frac{T_2 I_1}{2} = \frac{\left\{Q\left[\left(1 - \frac{a+b}{2}\right) - \frac{D}{P}\right] - B\right\}^2}{2\left[P\left(1 - \frac{a+b}{2}\right) - D\right]} \tag{4.128}$$

$$\frac{T_3 I_1}{2} = \frac{\left(\frac{a+b}{2}\right)Q\left\{Q\left[1 - \left(1+\frac{a+b}{2}\right)\frac{D}{P}\right] - B\right\}}{2P} \tag{4.129}$$

$$\frac{T_4 I_{\max}}{2} = \frac{\left\{Q\left[1 - \left(1+\frac{a+b}{2}\right)\frac{D}{P}\right] - B\right\}^2}{2D} \tag{4.130}$$

Finally, the cyclic inventory average I can be calculated as below (Sarkar et al. 2014):

4.4 Backordering

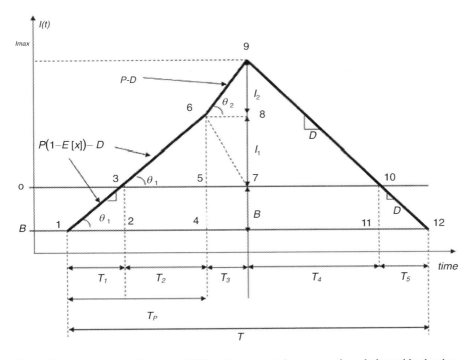

Fig. 4.16 Inventory behavior for the EPQ with rework at the same cycle and planned backorders (Sarkar et al. 2014)

$$\bar{I} = \frac{1}{T} \times \left[\frac{\{Q[(1-\frac{a+b}{2}) - \frac{D}{P}] - B\}^2}{2\left[P\left(1 - \frac{a+b}{2}\right) - D\right]} + \frac{\left(\frac{a+b}{2}\right)Q\left\{Q\left[1 - \frac{a+b}{4} - \left(1 + \frac{a+b}{4}\right)\frac{D}{P}\right] - B\right\}}{P} \right. $$
$$\left. + \frac{\{Q[1 - (1+\frac{a+b}{2})\frac{D}{P}] - B\}^2}{2D} \right] \quad (4.131)$$

After some simplifications,

$$\bar{I} = \frac{1}{2Q\left[\left(1-\frac{a+b}{2}\right)-\frac{D}{P}\right]}\left[(Q^2+B^2)\left(1-\frac{a+b}{2}\right)+\frac{Q^2D^2}{P^2}\left(1+\frac{a+b}{2}+\left(\frac{a+b}{2}\right)^2\right)\right.$$
$$\left. +\frac{Q^2D}{P}\left(\left(\frac{a+b}{2}\right)^3-2\right)+2BQ\left(\frac{D}{P}+\frac{a+b}{2}-1\right)\right] \quad (4.132)$$

In order to express the above mathematical equation in a more compact expression, the following symbols were defined:

$$\gamma = 1 - \frac{a+b}{2} \quad (4.133)$$

$$E = 1 - \frac{a+b}{2} - \frac{D}{P} \quad (4.134)$$

$$I = \left[1 + \frac{a+b}{2} + \left(\frac{a+b}{2}\right)^2\right]\left(\frac{D^2}{P^2}\right) \quad (4.135)$$

$$O = \left[\left(\frac{a+b}{2}\right)^3 - 2\right]\left(\frac{D}{P}\right) \quad (4.136)$$

$$U = \frac{D}{P} + \frac{a+b}{2} - 1 = -E \quad (4.137)$$

Then,

$$\bar{I} = \frac{1}{2QE}\left[Q^2(\gamma + I + O) + B^2\gamma - 2BQE\right] \quad (4.138)$$

$$\bar{I} = \frac{Q}{2}\left(\frac{\gamma + I + O}{E}\right) + \frac{B^2\gamma}{2QE} - B \quad (4.139)$$

With further rearrangement:

$$\bar{I} = \frac{Q}{2}\left[1 - \left(1 + \frac{a+b}{2} + \left(\frac{a+b}{2}\right)^2\right)\left(\frac{D}{P}\right)\right] + \frac{B^2\left(1-\frac{a+b}{2}\right)}{2Q\left(1-\frac{a+b}{2}-\frac{D}{P}\right)} - B \quad (4.140)$$

If L is defined as follows,

$$L = 1 - \left(1 + \frac{a+b}{2} + \left(\frac{a+b}{2}\right)^2\right)\left(\frac{D}{P}\right) \quad (4.141)$$

finally, the inventory average is given as follows:

4.4 Backordering

$$\bar{I} = \frac{Q}{2}L + \frac{B^2\gamma}{2QE} - B \tag{4.142}$$

With regard to inventory average of backorders B, it can be determined by the sum of the area of triangles: under time line and divided by T. T_1 and T_5 can be calculated as:

$$T_1 = \frac{B}{[P(1-E(x))-D]} = \frac{B}{\left[P\left(1-\frac{a+b}{2}\right)-D\right]} \tag{4.143}$$

$$T_5 = \frac{B}{D} \tag{4.144}$$

where T_1 is the time needed to satisfy the backorder level once production process is started again and T_5 is the time needed to build up the backorder level of B units. Thus, for average shortage (Sarkar et al. 2014):

$$\frac{T_1 B}{2} = \frac{B^2}{2\left[P\left(1-\frac{a+b}{2}\right)-D\right]} \tag{4.145}$$

$$\frac{T_5 B}{2} = \frac{B^2}{2D} \tag{4.146}$$

So cyclic backordering can be determined by dividing the average of shortage by T, as below:

$$\bar{B} = \frac{1}{T}\left[\frac{B^2}{2\left[P\left(1-\frac{a+b}{2}\right)-D\right]} + \frac{B^2}{2D}\right] \tag{4.147}$$

$$\bar{B} = \frac{B^2\left(1-\frac{a+b}{2}\right)}{2Q\left[\left(1-\frac{a+b}{2}\right)-\frac{D}{P}\right]} \tag{4.148}$$

$$\bar{B} = \frac{B^2\gamma}{2QE} \tag{4.149}$$

Therefore, the total cost of the system is (Sarkar et al. 2014):

$$\text{TC} = \overbrace{\frac{KD}{Q}}^{\text{Setup Cost}} + \overbrace{h\bar{G}}^{\text{Holding Cost}} + \overbrace{\frac{\bar{C}_b BD}{Q} + C_b \bar{B}}^{\text{Shortage Cost}} + \overbrace{CD(1+E[x])}^{\text{Production Cost}} \tag{4.150}$$

After, substituting the value of above expressions in one obtains (Sarkar et al. 2014):

$$\mathrm{TC}(Q,B) = \frac{KD}{Q} + \frac{hQL}{2} + \frac{hB^2\gamma}{2QE} - hB + \frac{\widehat{C}_b BD}{Q} + \frac{C_b B^2 \gamma}{2QE} + CD(2-\gamma) \quad (4.151)$$

The cost equation consists of two decision variables as Q and B. In order to derive the optimal values of both decision variables, the second-order Hessian matrix should be positive definite. So one should perform:

Necessary conditions

$$\frac{\partial \mathrm{TC}(Q,B)}{\partial Q} = 0 \quad \text{and} \quad \frac{\partial \mathrm{TC}(Q,B)}{\partial B} = 0 \quad (4.152)$$

Sufficient conditions

$$\frac{\partial^2 \mathrm{TC}(Q,B)}{\partial Q^2} > 0 \quad \text{and} \quad \frac{\partial^2 \mathrm{TC}(Q,B)}{\partial Q^2} \frac{\partial^2 \mathrm{TC}(Q,B)}{\partial B^2} - \left(\frac{\partial^2 \mathrm{TC}(Q,B)}{\partial Q \partial B}\right)^2$$

$$> 0 \quad (4.153)$$

For minimization of the cost equation, the condition is $((2KD(C_b + h)\gamma)/Q^4 E) - \left(\widehat{C}_b^2 D^2/Q^4\right) > 0$, i.e., if the expression $((2KD(C_b + h)\gamma)/Q^4 E) - \left(\widehat{C}_b^2 D^2/Q^4\right)$ is greater than 0, the sufficient condition of the optimality criteria is satisfied. Therefore, it can be concluded that the cost equation is convex when the expression $((2KD(C_b + h)\gamma)/Q^4 E) - \left(\widehat{C}_b^2 D^2/Q^4\right) > 0$. The optimal values are as follows:

$$\frac{\partial \mathrm{TC}}{\partial Q} = 0 \quad (4.154)$$

$$Q = \sqrt{\frac{2KD\gamma(C_b + h) - E\widehat{C}_b^2 D^2}{h[\gamma L(C_b + h) - Eh]}} \quad (4.155)$$

$$\frac{\partial \mathrm{TC}}{\partial B} = 0 \quad (4.156)$$

or

$$B = \frac{\left(hQ - \widehat{C}_b D\right)E}{(C_b + h)\gamma} \quad (4.157)$$

However, the solution (Q, B) does not necessarily exist although $\mathrm{TC}(Q, B)$ is convex as it was shown in Chung (2011). He proved that $\mathrm{TC}(Q, B)$ exists if and only

4.4 Backordering

if $2KDh \geq \widehat{C}_b^2 D^2 L$ and if $2KDh < \widehat{C}_b^2 D^2 L$, then $B^* = 0$ and $Q^* = \sqrt{2KD/hL}$. Substituting the above optimal values in the cost equation and after some simplification, the minimum cost is obtained as follows:

$$TC^* = \frac{\sqrt{\left(2KD\gamma(C_b\pi + h) - E\widehat{C}_b^2 D^2\right)(h[\gamma L(C_b + h) - Eh]) + DE\widehat{C}_b h}}{\gamma(C_b + h)}$$
$$+ CD(2 - \gamma) \tag{4.158}$$

Case B The proportion of defective products follows a triangular distribution.

In this case, it is assumed that s follows a triangular distribution with parameters $[a, b, c]$; $0 < a < b < c < 1$ where parameters a and c are the inferior and superior limits, respectively, and b is the mode of the triangular distribution. For a triangular distribution, it is well-known that the expected value for x is given as $E[x] = (a + b + c)/3$. For this case, from Fig. 4.16, the maximum inventory I_{max} is also computed as the sum of $I_1 + I_2$. So (Sarkar et al. 2014):

$$\tan\theta_1 = P(1 - E[x]) - D = \frac{I_1 + B}{T_1 + T_2} \tag{4.159}$$

Similar to previous case, $T_1 + T_2 = Q/P$. Substituting the expected value $E[x]$ and $T_1 + T_2$, one obtains:

$$P\left(1 - \frac{a+b+c}{3}\right) - D = \frac{I_1 + B}{T_P} \tag{4.160}$$

or

$$I_1 = Q\left[1 - \frac{a+b+c}{3} - \frac{D}{P}\right] - B \tag{4.161}$$

Also similar to previous case:

$$T_1 = \frac{B}{P(1 - E[x]) - D} \tag{4.162}$$

$$T_2 = \frac{Q[(1 - E[x]) - D/P] - B}{[P(1 - E[x]) - D]} = \frac{Q\left[\left(1 - \frac{a+b+c}{3}\right) - D/P\right] - B}{\left[P\left(1 - \frac{a+b+c}{3}\right) - D\right]} \tag{4.163}$$

$$T_3 = \frac{E[x]Q}{P} = \frac{Q(a+b+c)}{3P} \tag{4.164}$$

$$T_4 = \frac{Q[1-(1+E[x])D/P]-B}{D} = \frac{Q\left[1-\frac{D}{P}\left(1+\frac{a+b+c}{3}\right)\right]-B}{D} \quad (4.165)$$

$$T_5 = \frac{B}{D} \quad (4.166)$$

Since, T_3 is equal to $E[x]Q/P$. Hence,

$$I_2 = T_3(P-D) = Q\left(\frac{a+b+c}{3}\right)\left[1-\frac{D}{P}\right] \quad (4.167)$$

Therefore, the maximum inventory $I_{\max}$ can be found as:

$$I_{\max} = I_1 + I_2 = Q\left[1-\frac{D}{P}\left(1+\frac{a+b+c}{3}\right)\right] - B \quad (4.168)$$

In order to calculate the average inventory, Sarkar et al. (2014) presented the following equations:

$$\frac{1}{2}T_2 I_1 = \frac{\left[Q\left(1-\frac{a+b+c}{3}-\frac{D}{P}\right)-B\right]^2}{2\left[P\left(1-\frac{a+b+c}{3}\right)-D\right]} \quad (4.169)$$

$$\frac{I_1 T_3}{2} = \frac{Q\frac{a+b+c}{3}\left[Q\left(1-\frac{a+b+c}{3}-\frac{D}{P}\right)-B\right]}{2P} \quad (4.170)$$

$$\frac{I_{\max} T_3}{2} = \frac{Q\frac{a+b+c}{3}\left[Q\left[1-\frac{D}{P}\left(1+\left(\frac{a+b+c}{3}\right)\right)\right]-B\right]}{2P} \quad (4.171)$$

$$\frac{T_4 I_{\max}}{2} = \frac{\left\{Q\left[1-\frac{D}{P}\left(1+\frac{a+b+c}{3}\right)\right]-B\right\}^2}{2D} \quad (4.172)$$

Finally, the inventory average $\bar{I}_{\text{Tri}}$ can be calculated as (Sarkar et al. 2014):

$$\bar{I}_{\text{Tri}} = \frac{D}{Q}$$

$$\times \left[\begin{array}{l} \dfrac{\left[Q\left(1-\frac{a+b+c}{3}-\frac{D}{P}\right)-B\right]^2}{2\left[P\left(1-\frac{a+b+c}{3}\right)-D\right]} + \dfrac{Q\frac{a+b+c}{3}\left[Q\left(1-\frac{a+b+c}{3}-\frac{D}{P}\right)-B\right]}{2P} \\ + \dfrac{Q\frac{a+b+c}{3}\left[Q\left[1-\frac{D}{P}\left(1+\frac{a+b+c}{3}\right)\right]-B\right]}{2P} + \dfrac{\left[Q\left\{1-\frac{D}{P}\left(1+\frac{a+b+c}{3}\right)\right\}-B\right]^2}{2D} \end{array} \right]$$

$$(4.173)$$

4.4 Backordering

In order to express the above mathematical equation in a more compact expression, the following symbols were defined:

$$K_{Tri} = 1 - \frac{a+b+c}{3} \tag{4.174}$$

$$E_{Tri} = 1 - \frac{a+b+c}{3} - \frac{D}{P} \tag{4.175}$$

$$I_{Tri} = \left(1 + \frac{a+b+c}{3} + \left(\frac{a+b+c}{3}\right)^2\right)\left(\frac{D}{P}\right)^2 \tag{4.176}$$

$$O_{Tri} = \left(\left(\frac{a+b+c}{3}\right)^3 - 2\right)\left(\frac{D}{P}\right) \tag{4.177}$$

$$U_{Tri} = \frac{D}{P} + \frac{a+b+c}{3} - 1 = -E_{Tri} \tag{4.178}$$

Then,

$$\bar{I}_{Tri} = \frac{Q}{2}\left[\frac{K_{Tri} + I_{Tri} + O_{Tri}}{E_{Tri}}\right] + \frac{B^2 K_{Tri}}{2QE_{Tri}} - B \tag{4.179}$$

Assume L_{Tri} is defined as follows:

$$L_{Tri} = 1 - \left(1 + \frac{a+b+c}{3} + \left(\frac{a+b+c}{3}\right)^2\right)\left(\frac{D}{P}\right) \tag{4.180}$$

Also to calculate the average shortage using Fig. 4.16 (Sarkar et al. 2014),

$$\frac{T_1 B}{2} = \frac{B^2}{2\left[P\left(1 - \frac{a+b+c}{3}\right) - D\right]} \tag{4.181}$$

$$\frac{T_5 B}{2} = \frac{B^2}{2D} \tag{4.182}$$

Thus, the inventory average of backorders $\bar{B}$ divided by T results in cyclic average backordering as below:

$$\bar{B} = \frac{1}{T}\left[\frac{B^2}{2\left[P\left(1 - \frac{a+b+c}{3}\right) - D\right]} + \frac{B^2}{2D}\right] = \frac{B^2 K_{Tri}}{2QE_{Tri}} \tag{4.183}$$

Therefore, the total cost of the system is as follows (Sarkar et al. 2014):

$$\text{TC}(Q,B) = \overbrace{\frac{AD}{Q}}^{\text{Setup Cost}} + \overbrace{\frac{hQL_{\text{Tri}}}{2} + \frac{hB^2 K_{\text{Tri}}}{2QE_{\text{Tri}}} - hB}^{\text{Holding Cost}} + \overbrace{\frac{\widehat{C_b}BD}{Q} + \frac{C_b B^2 K_{\text{Tri}}}{2QE_{\text{Tri}}}}^{\text{Shortage Cost}} +$$

$$\overbrace{CD(2 - K_{\text{Tri}})}^{\text{Production Cost}} \qquad (4.184)$$

In order to derive the optimal values of decision variables, similar to previous case,

Necessary conditions

$$\frac{\partial \text{TC}(Q,B)}{\partial Q} = 0 \quad \text{and} \quad \frac{\partial \text{TC}(Q,B)}{\partial B} = 0 \qquad (4.185)$$

Sufficient conditions

$$\frac{\partial^2 \text{TC}(Q,B)}{\partial Q^2} > 0 \quad \text{and} \quad \frac{\partial^2 \text{TC}(Q,B)}{\partial Q^2} \frac{\partial^2 \text{TC}(Q,B)}{\partial B^2} - \left(\frac{\partial^2 \text{TC}(Q,B)}{\partial Q \partial B}\right)^2$$
$$> 0 \qquad (4.186)$$

For minimization of the cost function, the condition presented in Eq. 4.186, should be satisfied. Therefore, it can be concluded that the cost equation is convex when the expression $\left((2KD(C_b + h)K_{\text{Tri}})/Q^4 E_{\text{Tri}}\right) - \left(\widehat{C}_b^2 D^2/Q^4\right) > 0$. The optimal values are as follows (Sarkar et al. 2014):

$$Q = \sqrt{\frac{2KDK_{\text{Tri}}(C_b + h) - E_{\text{Tri}}\widehat{C}_b^2 D^2}{h[K_{\text{Tri}}L_{\text{Tri}}(C_b + h) - E_{\text{Tri}}h]}} \qquad (4.187)$$

$$B = \frac{\left(hQ - \widehat{C}_b D\right) E_{\text{Tri}}}{(C_b + h)K_{\text{Tri}}} \qquad (4.188)$$

It is to be noted that $\text{TC}(Q, B)$ exists if $2KDh \geq \widehat{C}_b^2 D^2 L_{\text{Tri}}$. On the other hand, if $2KDh < \widehat{C}_b^2 D^2 L_{\text{Tri}}$, then $B^* = 0$ and $Q^* = \sqrt{2KD/hL_{\text{Tri}}}$. Substituting the above optimal values in the cost equation and after some simplification, the minimum cost is obtained as follows (Sarkar et al. 2014):

4.4 Backordering

$$TC^* = \frac{\sqrt{\left(2KDK_{Tri}(C_b+h) - E_{Tri}\widehat{C}^2{}_b D^2\right)\left(h[K_{Tri}L_{Tri}(C_b+h) - E_{Tri}h]\right) + DE_{Tri}\widehat{C}_b h}}{K_{Tri}(C_b+h)}$$
$$+ CD(2 - K_{Tri})$$
(4.189)

Case C The proportion of defective products follows a beta distribution.

In this case, it is assumed that x follows a beta distribution with range $[\alpha, \beta]$; $0 < \alpha < \beta < 1$ where parameters α and β are the inferior and superior limits, respectively, of the beta distribution. For beta distribution, it is well-known that the expected value for x is given as $E[x] = \alpha/(\alpha + \beta)$.

Similar as previous cases, they followed the same procedure to obtain the decision variable Q and B. From Fig. 4.16, the maximum inventory I_{max} is calculated as the sum of $I_1 + I_2$:

$$P(1 - E[x]) - D = \frac{I_1 + B}{T_1 + T_2} \quad (4.190)$$

As the production time of producing Q units is $T_P = T_1 + T_2$. Hence, $T_1 + T_2 = Q/P$. Substituting the expected value $E[x]$ and $T_1 + T_2$, one obtains:

$$P\left(1 - \frac{\alpha}{\alpha + \beta}\right) - D = \frac{I_1 + B}{T_P} \quad (4.191)$$

or

$$I_1 = Q\left[1 - \frac{\alpha}{\alpha + \beta} - \frac{D}{P}\right] - B \quad (4.192)$$

From Fig. 4.16, one has:

$$T_1 = \frac{B}{P(1 - E[x]) - D} \quad (4.193)$$

$$T_2 = \frac{Q\left[1 - \frac{\alpha}{\alpha+\beta} - \frac{D}{P}\right] - B}{P\left(1 - \frac{\alpha}{\alpha+\beta}\right) - D} \quad (4.194)$$

$$T_3 = \frac{E[x]Q}{P} = \frac{Q\alpha}{P(\alpha + \beta)} \quad (4.195)$$

$$T_4 = \frac{Q\left[1 - \frac{D}{P}\left(1 + \frac{\alpha}{\alpha+\beta}\right) - B\right]}{D} \tag{4.196}$$

$$T_5 = \frac{B}{D} \tag{4.197}$$

Using the above equations,

$$I_2 = T_3(P - D) = Q\left(\frac{\alpha}{\alpha+\beta}\right)\left[1 - \frac{D}{P}\right] \tag{4.198}$$

$$I_{\max} = I_1 + I_2 + Q\left[1 - \frac{D}{P}\left(1 + \frac{\alpha}{\alpha+\beta}\right)\right] - B \tag{4.199}$$

In order to calculate the average inventory,

$$\frac{1}{2}T_2 I_1 = \frac{\left[Q\left(1 - \frac{\alpha}{\alpha+\beta} - \frac{D}{P}\right) - B\right]^2}{2\left[P\left(1 - \frac{\alpha}{\alpha+\beta}\right) - D\right]} \tag{4.200}$$

$$\frac{T_3 I_1}{2} = \frac{Q\frac{\alpha}{\alpha+\beta}\left[Q\left(1 - \frac{\alpha}{\alpha+\beta} - \frac{D}{P}\right) - B\right]}{2P} \tag{4.201}$$

$$\frac{I_{\max}T_3}{2} = \frac{Q\frac{\alpha}{\alpha+\beta}\left[Q\left(1 - \left(\frac{\alpha}{\alpha+\beta} + 1\right)\frac{D}{P}\right) - B\right]}{2P} \tag{4.202}$$

$$\frac{T_4 I_{\max}}{2} = \frac{\left[Q\left\{1 - \frac{D}{P}\left(1 + \frac{\alpha}{\alpha+\beta}\right)\right\} - B\right]^2}{2D} \tag{4.203}$$

Finally, as before, the inventory average I_{beta} can be calculated summing the area presented in Eqs. (4.200)–(4.203) and divided by T. Hence, one obtains I_{beta} as (Sarkar et al. 2014):

$$\bar{I}_{\text{beta}} = \frac{D}{Q}$$

$$\times \left[\frac{\left[Q\left(1 - \frac{\alpha}{\alpha+\beta} - \frac{D}{P}\right) - B\right]^2}{2\left[P\left(1 - \frac{\alpha}{\alpha+\beta}\right) - D\right]} + \frac{Q\frac{\alpha}{\alpha+\beta}\left[Q\left(1 - \frac{\alpha}{\alpha+\beta} - \frac{D}{P}\right) - B\right]}{2P} \right.$$
$$\left. + \frac{Q\frac{\alpha}{\alpha+\beta}\left[Q\left(1 - \frac{D}{P}\left(1 + \frac{\alpha}{\alpha+\beta}\right)\right) - B\right]}{2P} + \frac{\left[Q\left\{1 - \frac{D}{P}\left(1 + \frac{\alpha}{\alpha+\beta}\right)\right\} - B\right]^2}{2D}\right]$$

$$\tag{4.204}$$

4.4 Backordering

In order to express the above mathematical equation in a more compact expression, the following were defined symbols:

$$K_{beta} = 1 - \frac{\alpha}{\alpha + \beta} \tag{4.205}$$

$$E_{beta} = 1 - \frac{\alpha}{\alpha + \beta} - \frac{D}{P} \tag{4.206}$$

$$I_{beta} = \left(1 + \frac{\alpha}{\alpha + \beta} + \left(\frac{\alpha}{\alpha + \beta}\right)^2\right)\left(\frac{D}{P}\right)^2 \tag{4.207}$$

$$O_{beta} = \left(\left(\frac{\alpha}{\alpha + \beta}\right)^3 - 2\right)\left(\frac{D}{P}\right) \tag{4.208}$$

$$U_{beta} = \frac{D}{P} + \frac{\alpha}{\alpha + \beta} - 1 = -E_{beta} \tag{4.209}$$

Then,

$$\bar{I}_{beta} = \frac{Q}{2}\left[\frac{K_{beta} + I_{beta} + O_{beta}}{E_{beta}}\right] + \frac{B^2 K_{beta}}{2QE_{beta}} - B \tag{4.210}$$

Simplifying, one obtains:

$$\bar{I}_{beta} = \frac{Q}{2}\left[1 - \left(1 + \frac{\alpha}{\alpha + \beta} + \left(\frac{\alpha}{\alpha + \beta}\right)^2\right)\left(\frac{D}{P}\right)\right] + \frac{B^2\left(1 - \frac{\alpha}{\alpha+\beta}\right)}{2Q\left(1 - \frac{\alpha}{\alpha+\beta} - \frac{D}{P}\right)} - B \tag{4.211}$$

If L_{beta} is defined as follows,

$$L_{beta} = 1 - \left(1 + \frac{\alpha}{\alpha + \beta} + \left(\frac{\alpha}{\alpha + \beta}\right)^2\right)\left(\frac{D}{P}\right) \tag{4.212}$$

finally, the inventory average is given by:

$$\bar{I}_{beta} = \frac{Q}{2} L_{beta} + \frac{B^2 A_{beta}}{2QE_{beta}} - B \tag{4.213}$$

In order to calculate the average shortage, one has:

$$\frac{T_1 B}{2} = \frac{B^2}{2\left[P\left(1 - \frac{\alpha}{\alpha+\beta}\right) - D\right]} \tag{4.214}$$

$$\frac{T_5 B}{2} = \frac{B^2}{2D} \tag{4.215}$$

Thus, the inventory average of backorders $\overline{B}$ can be calculated adding the area presented in Eqs. (4.215) and (4.216), and divided by T, one obtains (Sarkar et al. 2014):

$$\overline{B} = \frac{1}{T}\left[\frac{B^2}{2\left[P\left(1 - \frac{\alpha}{\alpha+\beta}\right) - D\right]} + \frac{B^2}{2D}\right] = \frac{B^2 K_{beta}}{2QE_{beta}} \tag{4.216}$$

Therefore, the total cost of the system is (Sarkar et al. 2014):

$$\text{TC}(Q, B) = \overbrace{\frac{KD}{Q}}^{\text{Setup Cost}} + \overbrace{\frac{hQL_{beta}}{2} + \frac{hB^2 K_{beta}}{2QE_{beta}} - hB}^{\text{Holding Cost}} + \overbrace{\frac{\widehat{C}_b BD}{Q} + \frac{C_b B^2 K_{beta}}{2QE_{beta}}}^{\text{Shortage Cost}} +$$

$$\overbrace{CD(2 - K_{beta})}^{\text{Production Cost}} \tag{4.217}$$

In order to solve the problem, the necessary and sufficient conditions should be evaluated:

Necessary conditions

$$\frac{\partial \text{TC}(Q, B)}{\partial Q} = 0 \quad \text{and} \quad \frac{\partial \text{TC}(Q, B)}{\partial B} = 0 \tag{4.218}$$

Sufficient conditions

$$\frac{\partial^2 \text{TC}(Q, B)}{\partial Q^2} > 0 \quad \text{and} \quad \frac{\partial^2 \text{TC}(Q, B)}{\partial Q^2} \frac{\partial^2 \text{TC}(Q, B)}{\partial B^2} - \left(\frac{\partial^2 \text{TC}(Q, B)}{\partial Q \partial B}\right)^2$$

$$> 0 \tag{4.219}$$

For minimization of the cost equation, the condition is $((2KD(C_b + h)K_{beta})/Q^4 E_{beta}) - \left(\widehat{C}_b^2 D^2/Q^4\right) > 0$. That is, if this expression is greater than 0, the sufficient condition of the optimality criteria is satisfied. It can be concluded that the cost equation is convex when the expression is greater than 0. The optimal values are as follows (Sarkar et al. 2014):

4.4 Backordering

$$Q = \sqrt{\frac{2KDK_{\text{beta}}(C_b + h) - E_{\text{beta}}\widehat{C}_b^2 D^2}{h[K_{\text{beta}}L_{\text{beta}}(C_b + h) - E_{\text{beta}}h]}} \qquad (4.220)$$

$$B = \frac{\left(hQ - \widehat{C}_b D\right)E_{\text{beta}}}{(C_b + h)K_{\text{beta}}} \qquad (4.221)$$

Example 4.8 Sarkar et al. (2014) considered numerical experiments based on Cárdenas-Barrón's (2009) data as $D = 300$ units/year, $a = 0.03$, $b = 0.07$, $P = 550$ units/year, $C_b = \$10$/unit/year, $h = \$50$/unit/year, $\widehat{C}_b = \$1$/unit short, $K = \$50$/lot size, and $C = \$7$/unit. Then, the optimal solution is TC = $\$2634.41$/year, $Q = 90.92$ units, $B = 30.13$ units.

Example 4.9 The values of the following parameters are to be taken in appropriate units: $D = 300$ units/year, $a = 0.03$, $b = 0.04$, $c = 0.07$, $P = 550$ units/year, $C_b = \$10$/unit/year, $h = \$50$/unit/year, $C'_b = \$1$/unit short, $K = \$50$/lot size, and $C = \$7$/unit. Then, the optimal solution is TC = $\$2628.63$/year, $Q = 90.71$ units, $B = 30.21$ units (Sarkar et al. 2014).

Example 4.10 The values of the following parameters are to be taken in appropriate units: $D = 300$ units/year, $\alpha = 0.03$, $\beta = 0.07$, $P = 550$ units/year, $C_b = \$10$/unit/year, $h = \$50$/unit/year, $\widehat{C}_b = \$1$/unit short, $K = \$50$/lot size, and $C = \$7$/unit. Then, the optimal solution is TC = $\$3078.10$/year, $Q = 101.29$ units, $B = 17.53$ units.

Sarkar et al. (2014) compared numerical outcomes of the three models in Table 4.10.

4.4.4 Rework Process and Scraps

Sivashankari and Panayappan (2014) developed different modeling of an EPQ model with shortage, defective items, and reworking. Figure 4.17 shows the inventory behavior for the production inventory model during one cycle with planned backorders in which rework is not presented.

4.4.4.1 Without Rework

Times t_1 and t_4 are needed to build up B units of items, so:

$$t_1 = \frac{B}{P - D - d}; \quad t_4 = \frac{B}{D} \qquad (4.222)$$

Time t_2 is needed to build up Q_1 units of items; therefore:

Table 4.10 Optimum cost, order quantity and backorder for three distribution functions (Sarkar et al. 2014)

Example no.	Total cost (per year)	Order quantity (units)	Backorder quantity (units)
1	2634.41	90.92	30.13
2	2628.63	90.71	30.21
3	3078.1	101.29	17.53

$$t_2 = \frac{I_{\text{Max}}}{P - D - d} \qquad (4.223)$$

The production phase occurs during time:

$$t_1 + t_2 = \frac{Q}{P} \rightarrow t_2 = \frac{Q}{P} - \frac{B}{P - D - d} \qquad (4.224)$$

Time t_3 is needed to consume all units I_{Max} at demand rate D:

$$t_3 = \frac{I_{\text{Max}}}{D} \qquad (4.225)$$

but,

$$Q_1 = (P - D - d)t_2 - B = (P - D - d)\left(\frac{Q}{P}\right) - B \qquad (4.226)$$

The production cycle time T (from Eqs. 4.222–4.226) is given by:

$$T = \sum_{i=1}^{4} t_i = \frac{Q}{P}\left[1 + \frac{P - D - d}{D}\right] = \frac{Q}{D}(1 - x) \qquad (4.227)$$

Backordering occurs during time $t_1 + t_4$. The average shortage during $t_1 + t_4$ is:

$$\frac{1}{T}\left[\frac{1}{2}Bt_1 + \frac{1}{2}Bt_4\right] = \frac{B}{2T}\left[\frac{B}{D} + \frac{B}{P - D - d}\right] = \frac{PB^2}{2Q(P - D - d)} \qquad (4.228)$$

Positive inventory occurs during the time. Therefore, the average inventory during time is:

4.4 Backordering

$$\frac{1}{T}\left[\frac{1}{2}Qt_2 + \frac{1}{2}I_{\text{Max}}t_3\right] = \frac{I_{\text{Max}}}{2T}\left[\frac{I_{\text{Max}}}{D} + \frac{I_{\text{Max}}}{P-D-d}\right] = \frac{(P-d)I_{\text{Max}}^2}{2Q(P-D-d)}$$

$$= \frac{P}{2Q(P-D-d)}\left[\frac{Q(P-D-d)}{P} - B\right]^2 \quad (4.229)$$

$$= \frac{Q(P-D-d)}{2P} + \frac{PB^2}{2Q(P-D-d)} - B$$

The total cost function TC(Q, B) = setup cost + production cost + holding cost + shortage cost is (Sivashankari and Panayappan 2014):

$$\text{TC}(Q,B) = \overbrace{\frac{DK}{Q(1-x)}}^{\text{Setup Cost}} + \overbrace{\frac{CD}{1-x}}^{\text{Production Cost}} +$$

$$\overbrace{\frac{hQ(P-D-d)}{2P} + \frac{hPB^2}{2Q(P-D-d)} - hB}^{\text{Holding Cost}} + \overbrace{\frac{C_bPB^2}{2Q(P-D-d)}}^{\text{Shortage Cost}} \quad (4.230)$$

The necessary conditions for having a minimum are:

$$\frac{\partial \text{TC}(Q,B)}{\partial Q} = 0 \quad \text{and} \quad \frac{\partial \text{TC}(Q,B)}{\partial B} = 0 \quad (4.231)$$

Partially differentiating Eq. (4.230) with respect to Q gives:

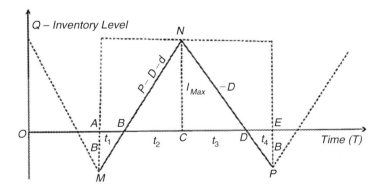

Fig. 4.17 Production inventory model in one cycle with defective item (Sivashankari and Panayappan 2014)

$$\frac{\partial TC}{\partial Q} = \frac{-DK}{Q^2(1-x)} + \frac{h(P-D-d)}{2P} - \frac{PB^2h}{2Q^2(P-D-d)}$$
$$- \frac{C_b PB^2}{2Q^2(P-D-d)}$$
$$= 0 \qquad (4.232)$$

So this yields to:

$$Q^2 = \frac{2PD(P-D-d)K + PB^2(h+C_b)(1-x)}{h(P-D-d)^2(1-x)}$$
$$= \frac{2PDK + PB^2}{h(P-D-d)^2(1-x)} + \frac{P^2(h+C_b)B^2}{h(P-D-d)^2} \qquad (4.233)$$

Partially differentiating Eq. (4.230) with respect to B gives:

$$\frac{\partial TC}{\partial B} = \frac{PBh}{Q(P-D-d)} - h + \frac{C_b PB}{Q(P-D-d)} = 0 \qquad (4.234)$$

Therefore,

$$B = \frac{hQ(P-D-d)}{P(h+C_b)} \qquad (4.235)$$

Substituting Eq. (4.235) into (4.233) and simplifying yields to:

$$Q = \sqrt{\frac{2PDK(h+C_S)}{hC_b(P-D-d)(1-x)}} \qquad (4.236)$$

Sivashankari and Panayappan (2014) proved that the derived optimal solutions are global one and its related Hessian matrix is positive definite.

Example 4.11 Sivashankari and Panayappan (2014) presented an example with parameter values $P = 5000$ units, $D = 4500$ units, $K = 100$, $h = 10$, $C = 100$, $C_b = h = 10, 100$, and $x = 0.01$. The optimal solutions are $Q^* = 1421.34$, $B^* = 63.96$, and total cost $= 455{,}185.06$.

4.4.4.2 Rework Case and Quality Improvement

Sivashankari and Panayappan (2014) and Krishnamoorthi and Panayappan (2012) developed case Sect. 4.4.4.1 under rework policy as presented in Fig. 4.18.

4.4 Backordering

H_1 represents the quantity of good items remaining after consumption at the end of time t_1:

$$t_1 = \frac{Q}{P} \tag{4.237}$$

$$I_1 = (P - D - \lambda)t_1 = (P - D - d)\left(\frac{Q}{P}\right) - B \tag{4.238}$$

Time t_1 needed to build up Q_1 units of item; therefore:

$$t_1 = \frac{I_1}{P - D - d} = \frac{(P - D - d)(Q/P) - B}{P - D - d} = \frac{Q}{P} - \frac{B}{P - D - d} \tag{4.239}$$

Time t_2 needed to rework the defective items:

$$t_2 = \frac{MS}{P} = \frac{OJ - JK}{P} = \frac{xQ - x\theta Q}{P} = \frac{xQ(1 - \theta)}{P} \tag{4.240}$$

H_2 represents the quantity of items that should remain after consumption:

$$\begin{aligned} I_2 &= I_1 + NS = I_1 + (P - D)t_2 = (P - D - d)(Q/P) - B + \frac{(P - D)xQ(1 - \theta)}{P} \\ &= (P - D)\left(\frac{Q}{P}\right) - Qx - B + \frac{(P - D)xQ(1 - \theta)}{P} \end{aligned} \tag{4.241}$$

Time t_3 needed to build up H_2 units of items; therefore:

$$t_3 = \frac{1}{D}\left[(P - D)\left(\frac{Q}{P}\right) - Qx - B + \frac{(P - D)xQ(1 - \theta)}{P}\right] \tag{4.242}$$

Shortages time:

$$t_4 = \frac{B}{D}$$

$$t_5 = \frac{B}{P - D - d} \tag{4.243}$$

So the period length is:

$$T = t_1 + t_2 + t_3 + t_4 + t_5 = \frac{Q}{D}\left[\frac{D}{P} + \frac{xD(1-\theta)}{P} + \frac{(P-D)}{P} - x + \frac{(P-D)x(1-\theta)}{P}\right]$$
$$= \frac{Q}{D}[1 - x - x(1-\theta)] = \frac{Q}{D}(1 - x\theta) \qquad (4.244)$$

The average inventory is calculated as:

$$\bar{I} = \frac{1}{T}\left[\frac{1}{2}I_1 t_1 + I_1 t_2 + \frac{1}{2}(I_2 - I_1)t_2 + \frac{1}{2}t_3 I_2\right] \qquad (4.245)$$

And after some simplifications,

$$\bar{I} = \frac{Q}{2P(1-x\theta)}\left[P(1-x\theta)^2 - D\left(1 + x - 2x\theta + x^2(1-\theta)^2\right)\right]$$
$$- \frac{DB}{2P(1-x\theta)}[1 + 2x(1-\theta)] \qquad (4.246)$$

The average shortage is as follows:

$$\bar{B} = \frac{1}{T}\left[\frac{1}{2}Bt_4 + \frac{1}{2}Bt_5\right] = \frac{B^2(P-D)}{2T(P-D-d)} = \frac{B^2 P(1-x)}{2Q(P-D-d)(1-x\theta)} \qquad (4.247)$$

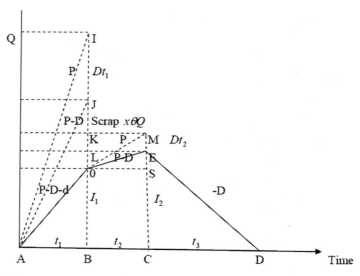

Fig. 4.18 On-hand inventory of EPQ model with the rework and shortages permitted (Krishnamoorthi and Panayappan 2012)

4.4 Backordering

The total cost of the system TC(Q, B) is the accumulation of the setup cost, production cost, holding cost, shortage cost, reworking cost, rejection cost, and quality cost for defective items (Krishnamoorthi and Panayappan 2012).

$$TC = \underbrace{\overbrace{\frac{DK}{Q(1-x\theta)}}^{\text{Setup Cost}} + \overbrace{\frac{CD}{1-x\theta}}^{\text{Production Cost}} + \overbrace{\frac{C_R Dx(1-\theta)}{1-x\theta}}^{\text{Rework Cost}} + \overbrace{\frac{C_J Dx\theta}{1-x\theta}}^{\text{Rejection Cost}} + \overbrace{\frac{C_q Dx}{1-x\theta}}^{\substack{\text{Quality}\\\text{Improvement}\\\text{Cost}}} + \overbrace{\frac{C_b PB^2(1-x)}{2Q(P-D-d)(1-x\theta)}}^{\text{Shortage Cost}} \\ + \underbrace{\frac{hQ}{2P(1-x\theta)}\left(P(1-x\theta)^2 - D\left(1+x-2x\theta+x^2(1-\theta)^2\right)\right) - \frac{hDB}{2P(1-x\theta)}(1+2x(1-\theta))}_{\text{Holding Cost}}}$$

(4.248)

Partially derivative TC(Q, B) with respect to Q and B (Krishnamoorthi and Panayappan 2012),

$$\frac{\partial TC}{\partial Q} = \left[\frac{-DK}{Q^2(1-x\theta)} + \frac{h\left(P(1-x\theta)^2 - D\left(1+x-2x\theta+x^2(1-\theta)^2\right)\right)}{2P(1-x\theta)} - \frac{C_b PB^2(1-x)}{2Q^2(P-D-d)(1-x\theta)}\right] = 0$$

(4.249)

Let:

$$E = P(1-x\theta)^2 - D\left(1+x-2x\theta+x^2(1-\theta)^2\right)$$ (4.250)

Then,

$$Q^2 = \frac{2PD(P-D-d)K + P^2 B^2 C_b(1-x)}{h(P-D-d)(E)}$$ (4.251)

And:

$$\frac{\partial TC}{\partial B} = \frac{-Dh(1-2x(1-\theta))}{2P(1-\theta x)} + \frac{2BPC_b(1-x)}{2Q(P-D-d)(1-\theta x)} = 0$$ (4.252)

So,

$$B = \frac{DQh(P-D-d)(1+2x-2x\theta)}{2P^2 C_b(1-x)}$$ (4.253)

Substitute the value of (4.253) in (4.251), after simplification:

$$Q^2\left[1 - \frac{D^2h(P-D-d)(1+2x-2x\theta)^2}{4P^2(1-x)C_b(E)}\right] = \frac{2PDK}{h(E)} \qquad (4.254)$$

Therefore, the optimum lot size is:

$$Q = \sqrt{\frac{8P^3D(1-x)C_bK}{4P^2(1-x)C_bh\left[P(1-x\theta)^2 - D\left(1+x-2x\theta+x^2(1-\theta)^2\right)\right] - D^2h^2(P-D-d)(1+2x-2x\theta)^2}} \qquad (4.255)$$

Example 4.12 Krishnamoorthi and Panayappan (2012) to support their proposed model presented an example with parameter values. Let $P = 5000$ units; $D = 4500$ units; $K = 100$; $h = 10$; $C_R = 5$; $C_J = 1$; $C_q = 5$; $C = 100$; $x = 0.01$ and $\theta = 0.1$; and $C_b = 10$ (Krishnamoorthi and Panayappan 2012). *The optimal solutions are $Q^* = 1120.76$; $B^* = 46.67$; and total cost $= 451{,}686.71$.*

4.4.5 Rework and Preventive Maintenance

Chen et al. (2010) considered a single-item production process. At the beginning of a production cycle, the production system is assumed to be in an in-control state in which the process only produces acceptable items. After a period of production time, process may shift to out-of-control state to produce some non-conforming items. The elapsed time for a process to shift is a random variable that follows a general distribution with increasing hazard rate. In practical production systems, some non-conforming items may be reworked. Hence a percentage of the non-conforming items are scrapped items; they are discarded before the rework process starts. Other non-conforming items are reworked, and the rework process starts immediately when the regular production ends. The state of a process is observed by inspection. The process is inspected at time $t_1, t_2, \ldots, t_m$, and PM is carried out right after each inspection. The production cycle ends either when the system is in the out-of-control state or the last inspection is completed. Then to renew a production cycle, extra works on the system are needed and should be ensured that the next cycle begins with in-control state. Therefore, a PM does not be performed in the last inspection when the system is in the out-of-control state (Chen et al. 2010). The behavior of inventory system is presented in Fig. 4.19. Also some new notations which are specifically used for the proposed model are presented in Table 4.11.

It should be noticed that unlike common imperfect inventory control systems, in this work order quantity and backordering level are not decision variables. In this model the optimal length of the inspection intervals, the optimal cost of PM, and the number of inspections, based on the integrated model, are decision variables.

4.4 Backordering

The expected regular production cycle time is given by (Chen et al. 2010):

$$E(T) = \sum_{i=1}^{m} h_i \cdot \prod_{j=1}^{i-1} (1 - p_j) \qquad (4.256)$$

When the general production cycle ends, the total expected number of non-conforming items is $E(N)$. And in realistic production systems, the imperfect-quality items may be reworked; hence the rework process could eliminate waste and affect the cost of manufacturing. Then, a percentage $(1 - x_1)$ of the non-conforming items are reworked and the number of non-conforming items which can be reworked is $(1 - x_1)E(N)$. The rework rate of non-conforming items in units per unit time is P_r. Then, the expected rework cycle time can be obtained as the following formula shows (Chen et al. 2010):

$$E(T_r) = (1 - x_1)E(N)/P_r \qquad (4.257)$$

The total expected number of non-conforming items per production cycle is given by:

$$E(N) = \sum_{j=1}^{m} p_j E(N_j) \cdot \prod_{i=1}^{j-1} (1 - p_i) \qquad (4.258)$$

where $E(N_j)$ is the expected number of non-conforming items produced due to out of control during the period (t_{j-1}, t_j) and is given by:

$$E(N_j) = \int_{a_{j-1}}^{b_j} \theta P(b_j - t) \frac{f(t)}{\overline{F}(a_{j-1})} dt =$$

$$= \frac{\theta P}{\overline{F}(a_{j-1})} \left\{ b_j [F(b_j) - F(a_{j-1})] - \int_{a_{j-1}}^{b_j} tf(t)dt \right\} \qquad (4.259)$$

Then from Fig. 4.19, the expected inventory cycle length is presented as follows (Chen et al. 2010):

$$E(\text{Cycle Length}) = \frac{1}{D}[P \cdot E(T)] \qquad (4.260)$$

The expected inventory holding cost is calculated by multiplying the inventory holding cost per unit of time by the expected inventory over the course of the inventory cycle (area under the curve in Fig. 4.19) from the time at which shortages begin to be filled ($t = B/(P - D)$) to the time when inventory is depleted ($t = E$

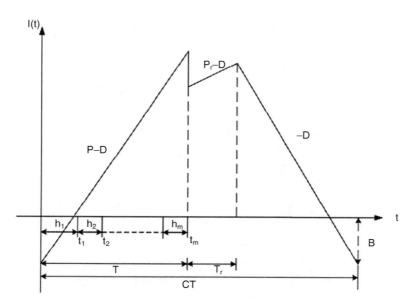

Fig. 4.19 Inventory cycle (Chen et al. 2010)

Table 4.11 New notations of given problem

P_r	Rework rate of non-conforming items in units per unit time (units per unit time)
T_r	The time of reworking non-conforming items for each cycle (time)
m	Number of inspections carried out during each production run
h_j	Length of the jth inspection interval
t_j	Time for the jth inspection $t_j = \sum_{j=1}^{m} h_j$
θ	Non-conforming rate when system is in the out-of-control state
x_1	The percentage of non-conforming items that are scrapped and will not be reworked, $0 \leq x_1 \leq 1$
x_2	The percentage of scrapped items produced during the rework process $0 \leq x_2 \leq 1$
η	Imperfect factor
γ_k	Imperfectness coefficient at the kth PM
b_j	Actual age of system instantly before the jth PM
a_j	Actual age of system instantly after the jth PM with $a_0 = 0$
C_{PM}^a	Cost of the actual PM activities ($/PM activity)
C_{PM}^{max}	Cost of the maximum PM level ($/PM activity)
$R(T)$	Restoration cost ($/restoration)
p_j	The conditional probability that the process shifts to the out-of-control state during the time interval (t_{j-1}, t_j)
N_j	Number of non-conforming items produced within (t_{j-1}, t_j)
$f(t)$	Probability density function of the time to shift
$F(t)$	Cumulative distribution of $f(t)$, $\overline{F}(t) = 1 - F(t)$, $r(t) = f(t)/\overline{F}(t)$

4.4 Backordering

$(CT) - B/D$). Hence, the expected inventory holding cost is determined by the following formula (Chen et al. 2010):

$$E(\text{Holding Cost}) = \left(\frac{h}{2}\right)$$
$$\times \left\{\frac{P}{D(P-D)}[(P-D)E(T)-B]^2 - \frac{1}{P_r}[(1-x_1)E(N)]^2\right\} \quad (4.261)$$

The integrated model they developed is also allowing shortages for the determination of EPQ. So from Fig. 4.19, the expected shortage cost is given by (Chen et al. 2010):

$$E(\text{Shortage Cost}) = \left(\frac{C_b}{2}\right) B^2 \frac{P}{(P-D)D} \quad (4.262)$$

When the regular production ends, a percentage x_1 of the non-conforming items are scrapped, and the others are reworked. Restated, the quantity that can be reworked is $(1 - x_1)E(N)$. And the repair cost for each non-conforming item reworked is C_R. Then, the expected reworking cost is given by (Chen et al. 2010):

$$E(\text{Rework Cost}) = C_R(1-x_1)E(N) \quad (4.263)$$

Also, when regular production ends, the number of non-conforming items that are scrapped is $x_1 E(N)$. The rework process is assumed to be imperfect; a percentage x_2 of the reworked items fail the restoring process and become scrap items. The quantity of the reworked items that become scrap is $(1 - x_1)x_2 E(N)$ when the rework process ends. The disposal cost for each scrapped item is C_d. Therefore, the expected disposal cost for scrap items is given by (Chen et al. 2010):

$$E(\text{Disposal Cost}) = C_d[x_1 + (1-x_1)x_2]E(N) \quad (4.264)$$

The system cannot be as good as new after implementing PM, but it will be younger, according to the level of PM activities. The reduction in the used age of the equipment is a function of the PM cost. Assume after each PM the system recovers. The parameter is a degradation factor which impacts the influence of PM activities on the used age of the process. Let γ_k be the imperfect coefficient at the kth PM, then (Chen et al. 2010):

$$\gamma_k = \eta^{k-1} \frac{C_{PM}^a}{C_{PM}^{max}} \quad (4.265)$$

Ben-Daya (1999) considered both linear and nonlinear relationships between the age reduction and PM cost. Here, they assumed the relationship is linear as follows (Chen et al. 2010):

$$a_k = (1 - \gamma_k)b_k \tag{4.266}$$

Note that the actual age of a production system at time t_j is given by:

$$\begin{aligned} b_1 &= h_1 \\ b_j &= a_{j-1} + h_j \quad \text{for } j = 2, 3, \ldots, m \end{aligned} \tag{4.267}$$

Usually, the PM activity is implemented after each inspection unless the system is ceased, and no PM is being implemented at the end of the production cycle. These give the expected maintenance cost E(PM) in the following lemma. The expected PM cost per regular production cycle is given by (Chen et al. 2010):

$$E(\text{PM Cost}) = C_{\text{PM}}^a \underbrace{\sum_{j=1}^{m-1} \prod_{i=1}^{j} (1 - p_i)}_{n_{\text{PM}}=\text{expected number of PM per production cycle}} \tag{4.268}$$

There might be errors in inspections, but for simplicity, such errors are not considered. The expected number of inspection in the production cycle is one more than the expected number of PM because there is one inspection at the end of the production cycle. This gives:

$$E(\text{Inspection Cost}) = (n_{\text{PM}} + 1)C_{\text{I}} = C_{\text{I}} \left[1 + \sum_{j=1}^{m-1} \prod_{i=1}^{j} (1 - p_i) \right] \tag{4.269}$$

The production system in out-of-control state must be terminated for maintenance. When this happens, the charge for restoration is needed. Therefore, the restoration cost (RC) should be included in the total cost. Assume that the out-of-control state occurs at the time t in the period $(a_j - 1, b_j)$ and the detection delay time is $b_j - t$. According to Ben-Daya (2002), when the restoration cost is assumed to change linearly with the detection delay, $\text{RC}_j = R(b_j - t) = r_0 + r_1(b_j - t)$ where r_0 and r_1 are constants. The restoration cost per production cycle is given by (Chen et al. 2010):

4.4 Backordering

$$E(\text{Restoration Cost}) = \sum_{j=1}^{m} p_j \prod_{i=1}^{j-1} (1 - p_i)$$
$$\times \left((r_0 + r_1 b_j) \frac{F(b_j) - F(a_{j-1})}{\overline{F}(a_{j-1})} - r_1 \int t \frac{f(t)}{\overline{F}(a_{j-1})} dt \right)$$
(4.270)

The expected total cost (ETC) per each cycle is summation of setup (setup cost is K), holding, shortage, preventive maintenance, restoration, disposal, inspection, and rework costs. Moreover, after the production cycle ends, the total amount of production minus the quantity which is disposal leaves quantity that can be sold. Therefore, the expected total revenue is:

$$\text{Expected Total Revenue} = S[E(Q) - (x_1 + (1 - x_1)x_2)E(N)] \quad (4.271)$$

where $E(Q)$ is the expected total production quantity and $E(Q) = P \cdot E(T)$.

Then, the expected total profit per unit time can be obtained obtain given by:

$$\text{ETP} = \text{Expected Total Profit}$$
$$= \frac{\text{Expected Total Revenue} - \text{Expected Total Cost}}{E(\text{Cycle Length})} \quad (4.272)$$

The next problem is optimizing the decision variables for the above integrated profit model. Methods of adjusting the number of inspections and optimizing the solution are also discussed. Using the above formulas, one determines simultaneously the optimal length of the inspection intervals, $h_1, h_2, \ldots, h_m$; the optimal cost of PM, C_{PM}^a; and the number of inspections, m, based on the integrated model.

As equipment breaks down, an inspection schedule should be arranged to find the fault quickly and then propose the correct measure. In practice, most of the mechanical malfunctions follow the non-Markov shock model with IFR function. As the production process continues, the optimal inspection interval is progressively reduced. Each inspection interval has the same cumulative hazard rate. Restated (Chen et al. 2010),

$$\int_{t_j}^{t_{j+1}} r(t)dt = \int_0^{t_1} r(t)dt \quad \text{for } j = 2, 3, \ldots, m \quad (4.273)$$

Since the hazard rate is reduced at the end of each inspection interval due to the PM activities, condition (4.301) becomes (Chen et al. 2010):

$$\int_{a_{j-1}}^{b_j} r(t)dt = \int_0^{h_1} r(t)dt \quad \text{for } j = 2, 3, \ldots, m \tag{4.274}$$

Usually researchers assume that the time of process staying in the in-control state is a random variable, following Weibull probability function. So its probability distribution function is given by (Chen et al. 2010):

$$f(t) = \lambda v t^{v-1} e^{-\lambda t^v} \quad t > 0, v \geq 1, \lambda > 0 \tag{4.275}$$

So hazard rate of the Weibull distribution is applied in Eq. (4.273) to obtain the length of the inspection intervals (Chen et al. 2010):

$$h_j = \left[(a_{j-1})^v + (h_1^v)\right]^{\frac{1}{v}} - a_{j-1} \quad \text{for } j = 2, 3, \ldots, m \tag{4.276}$$

This means, when h_1 is determined, so are other h_j's. Therefore, to maximum expected total profit, the value of decision variables $h_1, \ldots, _m$ and C_{PM}^a are needed to be determined. They implemented the stepwise partial particularization procedure to achieve the goal. Since the characteristics of the profit function, some modifications to the standard method have to be made to account for the inherent internality constraint on the number of inspections. The optimal value of $m \geq 2$ could be determined by choosing m that satisfies two inequalities (Chen et al. 2010):

$$\text{ETP}(m-1) \leq \text{ETP}(m) \quad \text{and} \quad \text{ETP}(m+1) \leq \text{ETP}(m) \tag{4.277}$$

Therefore, the optimal value of the number of inspections, m^*, and the length of the first inspection interval, h_1^*, can be obtained by the following procedure if the PM level is determined (Chen et al. 2010).

Step 1: Estimate m_0. The maximum number of inspection existed during each production run according to historical experience and the condition of production.
Step 2: First setup $m = 1$. One can search an optimal value h_1 and calculate the expected total cost $\text{ET}\pi_1$ under this condition.
Step 3: Repeat Step 2 for $m = 2, 3, \ldots, m_0$. One has the optimal value h_1 and the expected total profit form ETP_2 to ETP_{m_0} under different m, respectively.
Step 4: The optimal values m^* and h_1^* must meet the following condition:

$$\text{ETP}(h_1^*, m^*, C_{PM}^a) = \text{Max}\{\text{ETP}_j, j = 1, 2, \ldots, m_0\}$$

Example 4.13 Chen et al. (2010) presented several examples to illustrate the important aspects of the integrated profit model. The time that the process remains in the in-control state is assumed to follow a Weibull distribution with scale and shape parameters $\lambda = 5$ and $v = 2.5$. The following parameters are fixed $D = 500$,

4.4 Backordering

$P = 1000$, $P_r = 700$, $K = \$150$, $h = \$0.5$, $C_b = \$8$, $C_1 = \$10$, $C_d = \$20$, $C_{PM}^{Max} = \$30$, $C_R = \$5$, $s = \$10$, $\eta = 0.99$, $r_0 = 50$, $r_1 = 0.5$, and $x_2 = 0.1$.

The results show clearly that the expected total profit increases when the actual implemented PM level increases. The optimum PM level when PM $C_{PM}^a = \$30$ is obtained when max $C_{PM}^{Max} = \$30$, leading to a total profit of \$4728.41 much more than without PM (\$4672.17). Using comprehensive numerical analysis, Chen et al. (2010) found the optimal number of inspections, the optimal length of first inspection interval, the EPQ and the expected total profit per unit time under different values of x_1 ($x_1 = 0.1$, 0.5 and 1.0), B ($B = 0$, 50 and 100) and (0.2 and 0.4) as presented in Table 4.12.

4.4.6 Random Defective Rate: Different Production and Rework Rates

Chiu (2003) studies the effect of the reworking of defective items on the finite production model. He assumed production process may generate randomly x percent of defective items at a production rate d. The inspection cost per item is involved when all items are screened. Not all of the defective items produced are reworked. A portion θ of the imperfect-quality items are scrap and must be discarded before the rework process starts. The production rate P is a constant and is much larger than the demand rate D. When regular production ends, the reworking of defective items starts immediately at a constant rate P_1. The production rate d of the imperfect-quality items can be expressed as the product of the production rate times the percentage of defective items produced.

The production rate P must always be greater than or equal to the sum of the demand rate D and the defective rate d. Therefore, the following condition must hold:

$$P - D - d \geq 0; \quad 0 \leq x \leq 1 - \frac{D}{P} \qquad (4.278)$$

For the following derivation, the solution procedures are those used by Hayek and Salameh (2001). Referring to Fig. 4.20,

$$T = \frac{Q(1 - \theta x)}{D} \qquad (4.279)$$

where $0 \leq \theta \leq 1$ and $\theta x Q$ are scrap items randomly produced by the regular production process. Hence, the cycle length T is a variable, not a constant:

$$T = \sum_{i=1}^{5} t_i, \quad t_1 + t_5 = \frac{Q}{P} \tag{4.280}$$

Therefore, the production uptime t_1 is:

$$t_1 = \frac{I_1}{P - D - d} \tag{4.281}$$

$$I_1 = (P - D - d)\frac{Q}{P} - B \tag{4.282}$$

The time t_2 needed to rework $(1 - \theta)$ imperfect-quality items is computed in Eq. (4.283), and the maximum level of on-hand inventory when the rework process finished is calculated in Eq. (4.284):

$$t_2 = \frac{xQ(1 - \theta)}{P_1} = \frac{dQ(1 - \theta)}{P_1 P} \tag{4.283}$$

$$I = I_1 + (P_1 - D)t_2 = Q\left(1 - \frac{D}{P} - \frac{d\theta}{P} - \frac{Dd(1 - \theta)}{P_1 P}\right) - B \tag{4.284}$$

Using Fig. 4.20,

$$t_3 = \frac{I}{D} \tag{4.285}$$

$$t_4 = \frac{B}{D} \tag{4.286}$$

$$t_5 = \frac{B}{P - D - d} \tag{4.287}$$

The defective items produced during the regular production uptime t_1, as depicted in Fig. 4.21, are (Chiu 2003):

$$d(t_1 + t_5) = xQ \tag{4.288}$$

The reworking of $(1 - \theta)$ imperfect-quality items starts immediately, when the regular production time t_1 ends (Chiu 2003). Solving the inventory cost per cycle, TC(Q, B) is (Chiu 2003):

$$\begin{aligned}\text{TC}(Q,B) = &\overbrace{CQ}^{\text{Production Cost}} + \overbrace{C_R xQ(1-\theta)}^{\text{Rework Cost}} + \overbrace{C_d xQ\theta}^{\text{Disposal Cost}} + K + \overbrace{h_1 \frac{P_1 t_2}{2}(t_2)}^{\text{Holding Cost of Reowrked Item}}\\ &+ \underbrace{h\left[\frac{I_1}{2}(t_1) + \frac{I_1 + I}{2}(t_2) + \frac{I}{2}(t_3) + \frac{d(t_1 + t_5)}{2}(t_1 + t_5)\right]}_{\text{Holding Cost}} + \underbrace{C_b \frac{B}{2}(t_4 + t_5)}_{\text{Shortage Cost}}\end{aligned}$$

$$\tag{4.289}$$

4.4 Backordering

Table 4.12 Optimal values under various conditions (Chen et al. 2010)

		$B=0$				$B=50$				$B=100$			
		m^*	h_1^*	Q^*	ETP	m^*	h_1^*	Q^*	ETP	m^*	h_1^*	Q^*	ETP
$\theta=0.2$	$d_1=0.1$	3	0.29	702	4741.53	4	0.29	845	4737.7	6	0.29	1051	4694.67
	$d_1=0.5$	3	0.26	668	4732.98	4	0.26	815	4729.45	6	0.26	1047	4686.05
	$d_1=1$	3	0.24	640	4725.49	4	0.24	788	4722.37	6	0.24	1033	4678.79
$\theta=0.4$	$d_1=0.1$	3	0.26	664	4731.89	4	0.26	810	4728.41	6	0.25	1045	4684.98
	$d_1=0.5$	3	0.24	625	4721.17	4	0.23	773	4718.31	6	0.24	1022	4674.68
	$d_1=1$	4	0.21	729	4713.48	4	0.21	743	4709.62	7	0.22	1098	4666.39

Fig. 4.20 On-hand inventory of non-defective items (Chiu 2003)

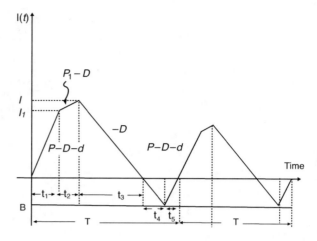

Using the renewal reward theorem in dealing with the variable cycle length, that is to compute $E[T]$ first. Then the expected annual cost $E[TCU(Q, B)] = E[TC(Q, B)]/E[T]$, so (Chiu 2003):

$$E[TCU(Q,B)] = D\left[C\frac{1}{1-\theta E[x]} + C_R(1-\theta)\frac{E[x]}{1-\theta E[x]} + C_d\theta\frac{E[x]}{1-\theta E[x]}\right]$$
$$+ \frac{KD}{Q}\frac{1}{1-\theta E[x]} + \frac{h}{2}\left[\left(1-\frac{D}{P}\right)Q - 2B\right]\frac{1}{1-\theta E[x]}$$
$$+ \frac{DQ(1-\theta)^2}{2P_1}(h_1-h)\frac{E[x^2]}{1-\theta E[x]}$$
$$+ \frac{B^2}{2Q}(C_b+h)E\left(\frac{1-x}{1-x-D/P}\right)\frac{1}{1-\theta E[x]}$$
$$+ h\theta\left[B-\left(1-\frac{D}{P}\right)Q\right]\frac{E[x]}{1-\theta E[x]} + \frac{hQ\theta^2}{2}\frac{E[x^2]}{1-\theta E[x]}$$
(4.290)

Let:

$$E_0 = \frac{1}{1-\theta E[x]}; \quad E_1 = \frac{E[x]}{1-\theta E[x]}; \quad E_2 = \frac{E[x^2]}{1-\theta E[x]}; \quad \text{and} \quad E_3$$
$$= E\left(\frac{1-x}{1-x-D/P}\right)\frac{1}{1-\theta E[x]} \tag{4.291}$$

Then Eq. (4.290), the expected annual cost, becomes (Chiu 2003):

4.4 Backordering

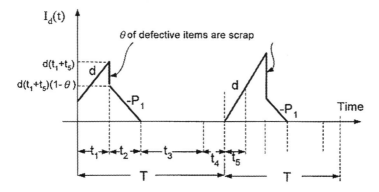

Fig. 4.21 On-hand inventory of defective items (Chiu 2003)

$$E[\text{TCU}(Q,B)] = D\left[CE_0 + C_R(1-\theta)E_1 + C_d\theta\frac{E[x]}{1-\theta E[x]}E_1\right] + \frac{KD}{Q}E_0 + \frac{h}{2}\left[\left(1-\frac{D}{P}\right)Q - 2B\right]E_0$$
$$+ h\theta\left[B - \left(1-\frac{D}{P}\right)Q\right]E_1 + \frac{DQ(1-\theta)^2}{2P_1}(h_1-h)E_2 + \frac{B^2}{2Q}(C_b+h)E_3 + \frac{hQ\theta^2}{2}E_2$$
(4.292)

For the proof of convexity of $E[\text{TCU}(Q, B)]$, one can utilize the Hessian matrix equation and obtain the following (Chiu 2003):

$$[Q\ B]\begin{bmatrix} \frac{\partial^2 E[\text{TCU}(Q,B)]}{\partial Q^2} & \frac{\partial^2 E[\text{TCU}(Q,B)]}{\partial Q \partial B} \\ \frac{\partial^2 E[\text{TCU}(Q,B)]}{\partial B \partial Q} & \frac{\partial^2 E[\text{TCU}(Q,B)]}{\partial B^2} \end{bmatrix}\begin{bmatrix} Q \\ B \end{bmatrix} = \frac{2KD}{Q}E_0 > 0 \quad (4.293)$$

Since the Hessian matrix is positive definite, the first derivative respect to decision variables yields to optimal values:

$$\frac{\partial E[\text{TCU}(Q,B)]}{\partial Q} = \frac{-KD}{Q^2}E_0 + \frac{h}{2}\left(1-\frac{D}{P}\right) + \frac{D(1-\theta)^2}{2P_1}(h_1-h)E_2$$
$$- \frac{B^2}{2Q^2}(C_b+h)E_3 - h\theta\left(1-\frac{D}{P}\right)E_1 + \frac{h\theta^2}{2}E_2 \quad (4.294)$$

$$\frac{\partial E[\text{TCU}(Q,B)]}{\partial B} = -hE_0 + \frac{B}{Q}(C_b+h)E_3 + h\theta E_1 \quad (4.295)$$

Setting partial derivatives presented in Eqs. (4.294) and (4.295) equal to zero yields to:

$$Q^* = \sqrt{\frac{2KD}{h(1-D/P) + \left[\left(D(1-\theta)^2\right)/p_1\right](h_1-h)E[x^2] - \left(h^2\{1-\theta E[x]\}^2\right)/\{(C_b+h)E[(1-x)/(1-x-D/P)]\} - 2h\theta(1-D/P)E[x] + h\theta^2 + E[x^2]}} \quad (4.296)$$

$$B^* = \left(\frac{h}{C_b+h}\right)\left(\frac{1}{E_3}\right)Q^* = \left(\frac{h}{C_b+h}\right)\frac{1-\theta E[x]}{E[(1-x)/(1-x-D/P)]}Q^* \quad (4.297)$$

Example 4.14 Chiu (2003) presented an example in which a company produces a product for several industrial clients under the following parameters: $P = 10{,}000$ units per year, $D = 4000$ units per year, $P_1 = 600$ units per year, $x = U[0, 0.2]$, $\theta = 0.1$, $K = \$450$ for each production run, $C = \$2$ per item (inspection cost per item is included), $C_R = \$0.5$ per item, $C_d = \$0.3$ for each scrap item, $h = \$0.6$ per item per unit time, $h_1 = \$0.8$ per item reworked per unit time, and $C_b = \$0.2$ per item backordered per unit time.

The optimal production lot size can be obtained from Eq. (4.296) as $Q^* = 5929$ units. The optimal backorder quantity can be computed from Eq. (4.297) as optimal $B^* = 2558$ units.

4.4.7 Imperfect Rework Process

Chiu (2007) derived the optimal replenishment policy for imperfect-quality economic manufacturing quantity (EMQ) model with rework and backlogging. He used a random defective rate, and all items produced are inspected, and the defective items are classified as scrap and repairable. A rework process is involved in each production run when regular manufacturing process ends, and a rate of failure in repair is also assumed. This assumption is the main difference between this work and previous ones. Some new notations which are used in this model are presented in Table 4.13.

According to previous models and referring to Fig. 4.22,

$$T = \sum_{i=1}^{5} t_i \quad (4.298)$$

The production uptime t_1 needed to accumulate I_1 units of perfect-quality items is (Chiu 2007):

$$t_1 = \frac{I_1}{P - D - d} \quad (4.299)$$

4.4 Backordering

$$I_1 = (P - D - d)\frac{Q}{P} - B \qquad (4.300)$$

The basic assumption of the imperfect-quality EMQ model is that the production rate P of perfect-quality items must always be greater than or equal to the sum of the demand rate D and the production rate of defective items d. Hence, the following condition must hold:

$$0 \leq x \leq 1 - \frac{D}{P} \qquad (4.301)$$

The time t_2 needed to rework of the repairable defective items is computed as

$$t_2 = \frac{xQ(1-\theta)}{P_1} \qquad (4.302)$$

Since the rework process is assumed to be imperfect, the production rate of the scrap items, d_1, can be written as:

$$d_1 = P_1 \cdot \theta_1 \qquad (4.303)$$

The maximum level of on-hand inventory, when the rework process ends, is (Chiu 2007):

$$I = I_1 + (P_1 - d_1 - D)t_2$$
$$= Q\left(1 - \frac{d\theta}{P} - \frac{D(P_1 + d)}{P_1 P} - \frac{d_1 d}{P_1 P} + \frac{d_1 d\theta}{P_1 P} + \frac{D d\theta}{P_1 P}\right) - B \qquad (4.304)$$

Similar to previous model,

$$t_3 = \frac{I}{D} \qquad (4.305)$$

$$t_4 = \frac{B}{D} \qquad (4.306)$$

$$t_5 = \frac{B}{P - D - d} \qquad (4.307)$$

The defective items produced during the regular production uptime $t_1 + t_5$, as illustrated in Fig. 4.23, are:

$$d(t_1 + t_5) = xQ \qquad (4.308)$$

Among the defective items, a random portion y of the imperfect-quality items is scrap; the reworking of $(1 - \theta)$ of defective items starts immediately, when the

Table 4.13 Notations

d_1	Defective production rate during rework process (units per unit time)
θ_1	Portion of the defective produced during rework process (%)

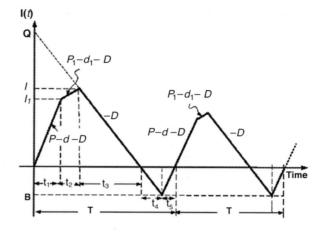

Fig. 4.22 On-hand inventory of perfect-quality items (Chiu 2007)

regular production time t_1 ends. Since the rework process is assumed to be imperfect either, a random portion y_1 of the reworked items fail the repairing and become scrap (refer to Fig. 4.24); they are calculated as follows:

$$d_1 t_2 = \theta_1 [x(1-\theta)Q] \tag{4.309}$$

Hence, the cycle length T becomes (Chiu 2007):

$$T = \frac{Q[1 - \theta \cdot x - (1-\theta)x\theta_1]}{D} \tag{4.310}$$

where $0 \leq \theta \leq 1$ and $[\theta x Q]$ are scrap items randomly produced during the regular production process and $[(1-\theta)x\theta_1]Q$ are scrap items randomly generated during the rework process. Hence, the cycle length T is not a constant. If φ is used to denote the total scrap rate, then $\varphi = [\theta + \theta_1(1-\theta)]$; one notes that the mean of random variable φ will follow the standard normal distribution (based on the Central Limit theorem). Therefore, Eq. (4.309) can be rewritten as:

$$T = \frac{Q[1 - x\varphi]}{D} \tag{4.311}$$

The total inventory cost per cycle, TC, is:

4.4 Backordering

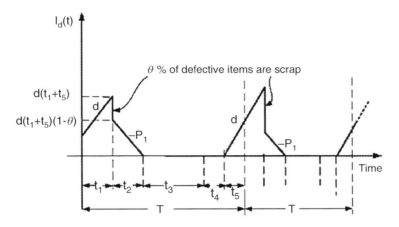

Fig. 4.23 On-hand inventory of defective items (Chiu 2007)

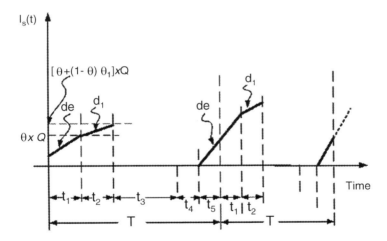

Fig. 4.24 On-hand inventory of scrap items (Chiu 2007)

$$TC(Q,B) = \overbrace{K}^{Setup\,Cost} + \overbrace{CQ}^{Production\,Cost} + \overbrace{C_R[x\cdot(1-\theta)\cdot Q]}^{Rework\,Cost} + \overbrace{C_d(\varphi\cdot x\cdot Q)}^{Disposal\,Cost} + \overbrace{h_1\frac{P_1\cdot t_2}{2}(t_2)}^{Holding\,Cost\,of\,Reworked\,Items}$$
$$+ \underbrace{h\left[\frac{I_1}{2}(t_1) + \frac{I_1+I}{2}(t_2) + \frac{I}{2}(t_3) + \frac{d(t_1+t_5)}{2}(t_1+t_5)\right]}_{Holding\,Cost} + \underbrace{C_b\frac{B}{2}(t_4+t_5)}_{Backordering\,Cost}$$

(4.312)

Chiu (2007) assumed that the proportion of defective items x and the ratio of scrap items θ and θ_1 are random variables with known probability density functions.

Thus, to take the randomness of imperfect production quality into account, one can utilize the expected values of x, θ, and θ_1 in the inventory cost analysis. Let $E[x]$, θ, and θ_1 represent the expected values of x, θ, and θ_1, respectively. Since the production cycle length is not a constant, one may employ the renewal theorem approach to cope with the variable cycle length, that is, to compute the $E[T]$ first. Then the expected annual inventory cost function $E[TCU(Q, B)]/E[T]$ becomes (Chiu 2007):

$$E[TCU(Q,B)] = D\left[C\frac{1}{1-\varphi E[x]} + C_R(1-\theta)\frac{E[x]}{1-\varphi E[x]} + C_d\varphi\frac{E[x]}{1-\varphi E[x]}\right] + \frac{KD}{Q}\frac{1}{1-\varphi E[x]}$$
$$+ \frac{h}{2}\left[Q\left(1-\frac{D}{P}\right) - 2B\right]\frac{1}{1-\varphi E[x]} + \frac{DQ(1-\theta)^2}{2P_1}[h_1 - h(1-\theta_1)]\frac{E[x^2]}{1-\varphi E[x]}$$
$$+ \frac{B^2}{2Q}(C_b + h)E\left(\frac{1-x}{1-x-\frac{D}{P}}\right)\frac{1}{1-\varphi E[x]} + h\varphi\left[B - \left(1-\frac{D}{P}\right)Q\right]\frac{E[x]}{1-\varphi E[x]}$$
$$+ \frac{hQ\varphi^2}{2}\frac{E[x^2]}{1-\varphi E[x]}$$

(4.313)

where $\varphi = [\theta + (1-\theta)\theta_1]$, and if

$$E_0 = \frac{1}{1-\varphi \cdot E[x]}; \quad E_1 = \frac{E[x]}{1-\varphi \cdot E[x]}; \quad E_2 = \frac{E[x^2]}{1-\varphi \cdot E[x]} \quad \text{and} \quad E_3$$
$$= E\left(\frac{1-x}{1-x-D/P}\right)\frac{1}{1-\varphi E[x]}$$

Then Eq. (4.340), the expected annual cost, becomes (Chiu 2007):

$$E[TCU(Q,B)] = D[C \cdot E_0 + C_R \cdot (1-\theta) \cdot E_1 + C_d \cdot \varphi \cdot E_1] + \frac{KD}{Q}E_0$$
$$+ \frac{h \cdot Q}{1}\left(1 - \frac{D}{P}\right) \cdot E_0 + \frac{DQ(1-\theta)^2}{2P_1}[h_1 - h(1-\theta_1)] \cdot E_2$$
$$+ \frac{B^2}{2Q}(C_b + h)E_3 + h \cdot \varphi \cdot \left[B - \left(1-\frac{D}{P}\right)Q\right] \cdot E_1 + \frac{h \cdot Q \cdot \varphi^2}{2} \cdot E_2$$

(4.314)

Similar to previous model, to derive the optimal values, using Hessian matrix,

4.4 Backordering

$$[Q \ B] \begin{bmatrix} \dfrac{\partial^2 E[\text{TCU}(Q,B)]}{\partial Q^2} & \dfrac{\partial^2 E[\text{TCU}(Q,B)]}{\partial Q \partial B} \\ \dfrac{\partial^2 E[\text{TCU}(Q,B)]}{\partial B \partial Q} & \dfrac{\partial^2 E[\text{TCU}(Q,B)]}{\partial B^2} \end{bmatrix} \begin{bmatrix} Q \\ B \end{bmatrix} = \dfrac{2KD}{Q} E_0 > 0 \quad (4.315)$$

So $E[\text{TCU}(Q, B)]$ is strictly convex. Therefore,

$$\frac{\partial E[\text{TCU}(Q,B)]}{\partial Q} = \frac{-KD}{Q^2} E_0 + \frac{h}{2}\left(1 - \frac{D}{P}\right) + \frac{D(1-\varphi)^2}{2P_1}(h_1 - h)E_2 \\ - \frac{B^2}{2Q^2}(C_b + h)E_3 - h\varphi\left(1 - \frac{D}{P}\right)E_1 + \frac{h\varphi^2}{2} E_2 \quad (4.316)$$

$$\frac{\partial E[\text{TCU}(Q,B)]}{\partial B} = -hE_0 + \frac{B}{Q}(C_b + h)E_3 + h\varphi E_1 \quad (4.317)$$

Setting partial derivatives presented in Eqs. (4.316) and (4.317) equal to zero yields to:

$$Q^* = \sqrt{\frac{2KD}{h(1 - D/P) + \left[\left(D(1-\varphi)^2\right)/p_1\right](h_1 - h)E[x^2] - \left(h^2\{1 - \varphi E[x]\}^2\right)/\{(C_b + h)E[(1-x)/(1-x-D/P)]\} - 2h\varphi(1 - D/P)E[x] + h\varphi^2 + E[x^2]}} \quad (4.318)$$

$$B^* = \left(\frac{h}{C_b + h}\right)\left(\frac{1}{E_3}\right) Q^* = \left(\frac{h}{C_b + h}\right) \frac{1 - \varphi E[x]}{E[(1-x)/(1-x-D/P)]} Q^* \quad (4.319)$$

Example 4.15 Chiu (2007) supposed a supplier produces a product with the parameters values of $K = \$450$ for each production run, $C = \$2$ per item (inspection cost per item is included), $h = \$0.6$ per item per unit time, $C_R = \$0.5$ repaired cost for each item reworked, $h_1 = \$0.8$ per item reworked per unit time, $C_d = \$0.3$ for each scrap item, $C_b = \$0.2$ per item per unit time, $D = 4000$ units per year, $P = 10,000$ units per year, $P_1 = 600$ units per year, $x =$ the proportion of imperfect-quality items produced is uniformly distributed over the interval [0, 0.2], θ is uniformly distributed over the range [0, 0.1]; its expected value 0.05, and θ_1 is uniformly distributed within [0, 0.1] where its expected value is 0.05.

Hence, from the definition of φ, the overall scrap $\varphi = [\theta + \theta_1(1 - \theta)] = 0.0975$ and the optimal replenishment policy can be calculated from Eqs. (4.318) to (4.319). Then the value of optimal lot size $Q^* = 5305$ and the optimal allowable backorder level is $B^* = 2175$.

4.5 Partial Backordering

4.5.1 Immediate Rework

In this section, an EMQ model with production capacity limitation, imperfect production processes, immediate rework, and partial backordered quantities in a multi-product single-machine manufacturing system is developed. The defective items of n different types of products are generated at a rate x_i; $i = 1, 2, \ldots, n$ per cycle. So the good item quantities are $(1-x_i)P_i$. The production and demand rates of the ith product per cycle are P_i and D_i, respectively. In this production system, each cycle consists of three parts: production uptime, rework time, and production downtime. Since all of the products are manufactured on a single machine with a limited capacity, a unique cycle length for all items is considered, that is, $T_1 = T_2 = \cdots = T_n = T$. They assumed that the total scrapped items are reworkable and no imperfect items are produced at the end of the rework process. Also, the producer has to use the same resource for production and rework processes simultaneously. Because a single machine has a limited joint production system capacity, shortage is allowed with a certain fraction of it to be backordered. In this work, they extended Jamal et al. (2004) to consider a more realistic inventory control problem in which multi-product single-machine strategy is used to produce several items under immediate rework, partial backordering, and capacity constraints (Taleizadeh and Wee 2015).

Figure 4.25 shows the inventory control problem under study. First, a single-product problem which consists of ith product is first developed. The fundamental assumption of an economic manufacturing model with rework process is:

$$(1 - x_i)P_i - D_i \geq 0 \tag{4.320}$$

Figure 4.25 shows that T_i^1 and T_i^5 are the production uptimes for non-defective and defective items, respectively. T_i^2 is the reworking time and T_i^3 and T_i^4 are the production downtimes, respectively. Finally, the cycle length is:

$$T = \sum_{j=1}^{5} T_i^j \tag{4.321}$$

In this model, a part of the modeling procedure is adopted from Jamal et al. (2004). As noted before, since all products are manufactured on a single machine with a limited capacity, the cycle length for all products is equal ($T_1 = T_2 = \cdots = T_n = T$), based in Fig. 4.25. One has:

4.5 Partial Backordering

$$T_i^1 = \frac{Q_i}{P_i} - \frac{\beta_i B_i}{(1-x_i)P_i - D_i} \tag{4.322}$$

$$T_i^2 = \frac{Q_i}{P_i} x_i \tag{4.323}$$

$$T_i^3 = \frac{\left(1 - \frac{D_i}{P_i} - x_i \frac{D_i}{P_i}\right)Q_i}{D_i} - \frac{\beta_i B_i}{D_i} \tag{4.324}$$

$$T_i^4 = \frac{B_i}{D_i} \tag{4.325}$$

$$T_i^5 = \frac{\beta_i B_i}{(1-x_i)P_i - D_i} \tag{4.326}$$

It is evident from Fig. 4.25 that:

$$I_i = ((1-x_i)P_i - D_i)\frac{Q_i}{P_i} - \beta_i B_i \tag{4.327}$$

$$I_i' = I_i + x_i(P_i - D_i)\frac{Q_i}{P_i} \tag{4.328}$$

Hence, using the equation the cycle length for a single product problem is:

$$T = \sum_{j=1}^{5} T_i^j = \frac{Q_i + (1-\beta_i)B_i}{D_i} \tag{4.329}$$

And the order quantity for the ith product is:

$$Q_i = D_i T - (1-\beta_i)B_i \tag{4.330}$$

The elements of the cost function are the setup cost, the holding cost, the processing cost, the rework cost, and the shortage cost which are expressed as:

$$TC = C_A + C_P + C_{Re} + C_H + C_B + C_L \tag{4.331}$$

$$C_A = \frac{\sum_{i=1}^{n} K_i}{T} \tag{4.332}$$

The production cost per unit is c_i, and the production quantity of ith product per period is Q_i. So, the production cost of ith product per period is $C_i Q_i$. Hence, the annual production cost for ith product is $NC_i Q_i$, and the following cost is the joint policy cost:

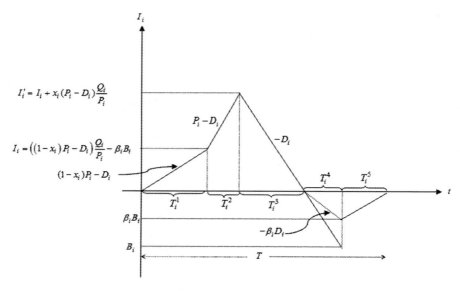

Fig. 4.25 On-hand inventory of perfect-quality items (Taleizadeh and Wee 2015)

$$C_P = \frac{\sum_{i=1}^{n} C_i Q}{T} \qquad (4.333)$$

The rework cost per unit of ith product is C_{R_i}, and the quantity of ith product that needs to be reworked per period is $x_i Q_i$. So, the rework cost of ith product per period is $C_{R_i} x_i Q_i$. Hence, the rework cost for ith product per year is $N C_{R_i} x_i Q_i$, and the annual rework cost for the joint policy is:

$$C_{Re} = \frac{\sum_{i=1}^{n} C_{R_i} x_i Q_i}{T} \qquad (4.334)$$

From Fig. 4.25, Eq. (4.335) shows the inventory holding cost of the system for an independent and joint production policy, respectively:

$$\begin{aligned} C_H &= N \sum h_i \left[\frac{I_i}{2} (t_i^1) + \frac{I_i + I_i'}{2} (t_i^2) + \frac{I_i'}{2} (t_i^3) \right] \\ &= \frac{1}{T} \sum h_i \left[\frac{I_i}{2} (t_i^1) + \frac{I_i + I_i'}{2} (t_i^2) + \frac{I_i'}{2} (t_i^3) \right] \end{aligned} \qquad (4.335)$$

Also, from Fig. 4.25, Eqs. (4.335) and (4.336) show the annual backordered and the lost sale costs in the joint policy production, respectively:

4.5 Partial Backordering

$$C_B = \frac{\sum_{i=1}^{n} C_{b_i}\beta_i\left(t_i^4 + t_i^5\right)B_i}{2T} \qquad (4.336)$$

$$C_L = \frac{\sum_{i=1}^{n} \widehat{\pi}_i(1-\beta_i)B_i}{2T} \qquad (4.337)$$

where $C_{b_i}\beta_i B_i$ and $\widehat{\pi}_i(1-\beta_i)B_i$ are the backordered and the lost sale cost of ith product per period, respectively. Consequently, one has:

$$\begin{aligned} TC &= C_A + C_P + C_R + C_H + C_B + C_L \\ &= \frac{\sum_{i=1}^{n} K_i}{T} + \frac{\sum_{i=1}^{n} C_i Q_i}{T} + \frac{\sum_{i=1}^{n} C_{R_i} x_i Q_i}{T} + N \sum h_i \left[\frac{I_i}{2}(t_i^1) + \frac{I_i + I_i'}{2}(t_i^2) + \frac{I_i'}{2}(t_i^3)\right] \\ &\quad + \frac{\sum_{i=1}^{n} C_{b_i}\beta_i\left(t_i^4 + t_i^5\right)B_i}{2T} + \frac{\sum_{i=1}^{n} \widehat{\pi}_i(1-\beta_i)B_i}{2T} \end{aligned}$$
$$(4.338)$$

In the joint production systems with reworks, the total production, rework, and setup times should be smaller than the cycle length. In our problem, $\sum_{i=1}^{n}\left(t_i^1 + t_i^2 + t_i^5\right) + \sum_i ts_i$ must be smaller or equal to $T(T_1 = T_2 = \cdots = T_n = T)$. Hence, the capacity constraint is:

$$\sum_{i=1}^{n}\left(t_i^1 + t_i^2 + t_i^5\right) + \sum_i ts_i \leq T$$

From Eqs. (4.350), (4.351), and (4.354), the capacity constraint model becomes:

$$\sum_i \frac{(1+x_i)}{P_i}(D_i T - (1-\beta_i)B_i) + \sum_i ts_i \leq T \qquad (4.339)$$

The final model of the joint production system is:

$$\text{Min}: \ TC(T,B_i) = \frac{\alpha_1}{T} + \alpha_2 T - \sum_{i=1}^{n} \alpha_{3i} B_i - \sum_{i=1}^{n} \alpha_{4i}\frac{B_i}{T} + \sum_{i=1}^{n} \alpha_{5i}\frac{B_i^2}{T} + \sum_{i=1}^{n}(C_i + C_{R_i} x_i)D_i$$

$$\text{s.t.}: \ T \geq \frac{\sum_{i=1}^{n} ts_i - \sum_{i=1}^{n}\frac{(1+x_i)}{P_i}(1-\beta_i)B_i}{\left(1 - \sum_{i=1}^{n}\frac{(1+x_i)D_i}{P_i}\right)} = T_{\text{Min}}^{\text{Production}}$$

$$T, B_i \ \forall i; \ i = 1, 2, \ldots, n$$

$$(4.340)$$

where:

$$\alpha_1 = \sum_{i=1}^{n} K_i > 0 \qquad (4.341a)$$

$$\alpha_2 = \sum_{i=1}^{n} h_i \left(\begin{array}{c} \dfrac{(D_i)^2((1-x_i)P_i - D_i) + 4x_i(D_i)^2((1-x_i)P_i - D_i) + 2x_i^2(D_i)^2(P_i - D_i)}{2(P_i)^2} \\ + (((1-x_i)P_i - D_i) + x_i(P_i - D_i))\left(D_i - \dfrac{1}{P_i} - \dfrac{x_i}{P_i}\right)\dfrac{(D_i)^2}{P_i} \end{array} \right)$$
$$> 0$$
$$(4.341b)$$

$$\alpha_{3i} = h_i \left(\begin{array}{c} \dfrac{((1-x_i)P_i - D_i)D_i(1-\beta_i) + 2\beta_i P_i D_i}{2(P_i)^2} \\ + \dfrac{(1-\beta_i)D_i(4x_i((1-x_i)P_i - D_i) + x_i^2(P_i - D_i)) + x_i P_i D_i \beta_i + x_i^2 D_i(1-\beta_i)(P_i - D_i)}{(P_i)^2} \\ + \left(\dfrac{((1-x_i)P_i - D_i) + x_i(P_i - D_i)}{P_i D_i} + \left(D_i - \dfrac{1}{P_i} - \dfrac{\alpha_i}{P_i}\right)\right)\beta_i D_i \end{array} \right) > 0$$
$$(4.341c)$$

$$\alpha_{4i} = \left(h_i(((1-x_i)P_i - D_i) + x_i(P_i - D_i))\left(D_i - \dfrac{1}{P_i} - \dfrac{x_i}{P_i}\right)\dfrac{(D_i(1-\beta_i))}{P_i} - \dfrac{(\widehat{\pi}_i - C_i - C_{R_i}x_i)(1-\beta_i)}{2} \right) > 0$$
$$(4.341d)$$

$$\alpha_{5i} = \left(\begin{array}{c} \dfrac{(1+x_i)h_i\beta_i(1-\beta_i)}{P_i} + \dfrac{h_i\beta_i^2}{2(1-x_i)P_i - 2D_i} + \dfrac{h_i((1-x_i)P_i - D_i)(1-\beta_i)^2}{2(P_i)^2} \\ + \dfrac{h_i(1-\beta_i)^2(2x_i((1-x_i)P_i - D_i) + x_i^2(P_i - D_i))}{(P_i)^2} + \dfrac{C_{b_i}\beta_i((1-x_i)P_i - (1-\beta_i)D_i)}{2((1-x_i)P_i - D_i)D_i} \\ + \dfrac{\beta_i(1-\beta_i)h_i(((1-x_i)P_i - D_i) + \beta_i(P_i - D_i) + P_i - (1+x_i)D_i)}{P_i D_i} \\ + \dfrac{h_i\beta_i^2}{D_i} - h_i(((1-x_i)P_i - D_i) + x_i(P_i - D_i))\left(D_i - \dfrac{1+x_i}{P_i}\right)\dfrac{(1-\beta_i)^2}{P_i} \end{array} \right) > 0$$
$$(4.341e)$$

Since the Hessian matrix of objective function is positive for all nonzero B_i and T, $TC(T, B_i)$ is convex:

$$[T, B_1, B_2, \ldots, B_n] \times \mathbf{H} \times \begin{bmatrix} T \\ B_1 \\ B_2 \\ \vdots \\ B_n \end{bmatrix} = \dfrac{2\alpha_1 + \sum_{i=1}^{n} \alpha_{4i} B_i}{T} > 0 \qquad (4.342)$$

4.5 Partial Backordering

To derive the optimal values of the decision variables, take the partial differentiations of TC(T, B_i) with respect to T and B_i (for details, see Appendix 2 of Taleizadeh and Wee (2015)):

$$\frac{\partial \text{TC}(T, B_i)}{\partial T} = \frac{-\sum_{i=1}^{n} \alpha_{5i} B_i^2 - \alpha_1 + \sum_{i=1}^{n} \alpha_{4i} B_i}{T^2} + \alpha_2 \rightarrow T$$

$$= \sqrt{\frac{\alpha_1 - \sum_{i=1}^{n}\left(\frac{\alpha_{4i}^2}{4\alpha_{5i}}\right)}{\left(\alpha_2 - \sum_{i=1}^{n}\left(\frac{\alpha_{3i}^2}{4\alpha_{5i}}\right)\right)}} \quad (4.343)$$

$$\frac{\partial \text{TC}(T, B_i)}{\partial B_i} = \frac{2\alpha_{5i} B_i - \alpha_{4i}}{T} - \alpha_{3i} \rightarrow B_i = \frac{\alpha_{3i} T + \alpha_{4i}}{2\alpha_{5i}} \quad (4.344)$$

To ensure feasibility, both $\alpha_1 - \sum_{i=1}^{n}\left(\frac{\alpha_{4i}^2}{4\alpha_{5i}}\right)$ and $\alpha_2 - \sum_{i=1}^{n}\left(\frac{\alpha_{3i}^2}{4\alpha_{5i}}\right)$ should simultaneously be positive or negative. In order to solve the above problem, Taleizadeh and Wee (2015) introduced the following solution procedures:

Step 1: Check for feasibility:
If $(1 - x_i)P_i - D_i \geq 0$, $\sum_{i=1}^{n}(1 + x_i)\frac{D_i}{P_i} < 1$ and both $\alpha_1 - \sum_{i=1}^{n}\left(\frac{\alpha_{4i}^2}{4\alpha_{5i}}\right)$ and $\alpha_2 - \sum_{i=1}^{n}\left(\frac{\alpha_{3i}^2}{4\alpha_{5i}}\right)$ be either positive or negative simultaneously, go to *Step* 2.

Step 2: Find a solution:
Using Eqs. (4.343) and (4.344), calculate T and B_i.

Step 3: Check the constraints.

Step 4: Derive the optimal solution:
Based on the derived value of T^*, then B_i^* can be derived from Eq. (4.344). For $Q_i^* = D_i T^* - (1 - \beta_i) B_i^*$, the optimal values of the order quantity can be obtained. Calculate the objective function using Eq. (4.339), and then go to *Step* 5.

Step 5: Terminate the procedure.

Taleizadeh and Wee (2015) considered a production system with production capacity limitation, imperfect production processes, immediate rework, and partial backordered quantity. The defective items of n different types of products are generated at a rate x_i; $i = 1, 2, \ldots, n$ per cycle. The production and demand rates of the ith item per cycle are P_i and D_i, respectively. So, the perfect item quantities are $(1-x_i)P_i$. They assumed that the total scrapped items are reworkable and no imperfect items occur at the end of the rework process. Also, the producer has to use the same resource for production and rework processes simultaneously. Shortage is allowed with certain fraction of it to be backordered because the single machine has limited joint production system capacity. Two multi-product problems with immediate rework and capacity constraint with partial backordering are considered for 15 products.

Example 4.16 The general and the specific data of mentioned examples are given in Table 4.14. The best results using the proposed methodology are shown in Table 4.15. Since $T = 2.5619$ is greater than its lower bound $T_{\text{Min}}^{\text{Production}} = 0.1159$, so $T^* = T = 2.5619$.

Example 4.17 The general and the specific data of mentioned examples are given in Table 4.16. The best results using the proposed methodology are shown in Table 4.17. Since $T = 2.5619$ is smaller than $T_{\text{Min}}^{\text{Production}} = 2.6247$, so $T^* = T_{\text{Min}}^{\text{Production}}$.

4.5.2 Repair Failure

Material is considered as one of the most important resources in any production system, and management of inventory is playing an important role in increasing the profitability of an organization. In the last decades, there have been tremendous efforts by industries to reduce the cost of inventory. The primary concern on inventory management is to reduce the costs of setup and holding. Inventory management has direct relationship with maintaining market share since customers may switch to different vendors due to the shortage. When goods are produced internally, the economic production quantity (EPQ) model is employed to determine the optimal production lot size. The traditional EPQ model assumption does not consider defective items. Due to imperfect quality of raw materials and/or production facilities, rework and repair of the defective items are considered in this study. This study is significant because a number of production units such as printed circuit board assembly in the PCBA manufacturing, metal components, and plastic injection molding have rework items (Taleizadeh et al. 2010).

As studied previous models presented in this chapter, the imperfect-quality EPQ model considers a manufacturing process with a constant production rate P and demand rate D, where $P > D$. This process randomly generates x percent of defective items at a rate d. Taleizadeh et al. (2010) assumed that all items produced are screened and the inspection cost per item is included in the unit production cost C. All defective items produced are reworked at a rate of P^1 at the end of each production cycle.

Also some new notations which are specifically used for the proposed model are presented in Table 4.18.

They assume an imperfect rework process where a random portion θ of the items is scrapped. Let d be the production rate of the defective items during the regular manufacturing process (it can be expressed as the product of production rate P and the defective percentage x) where $d = Px$. Let d_1 be the production rate of scrapped items during the rework which could be expressed as the product of the reworking rate and the percentage of scrapped items produced during the rework process with $d^1 = P^1\theta$. A real constant production capacity limitation on a single machine in which all products are produced and the setup cost is considered to be nonzero are

4.5 Partial Backordering

assumed. Since all products are manufactured on a single machine with a limited capacity, the cycle length for each is equal, i.e., $T_1 = \cdots = T_n = T$.

They first presented the problem statement for a single product case and then they changed it to a multi-product case. The basic assumption of EPQ model with imperfect-quality items produced is that P_i must always be greater than or equal to the sum of demand rate D_i and the production rate of defective items d_i. Therefore, one has (Taleizadeh et al. 2010):

$$P_i - D_i - d_i \geq 0, \quad 0 \leq x_i \leq 1 - \frac{D_i}{P_i} \tag{4.345}$$

The production cycle length (see Fig. 4.26) is the summation of the production uptime, the reworking time, the production downtime, and the shortage permitted time:

$$T = \sum_{j=1}^{5} t_i^j \tag{4.346}$$

The modeling procedure is adopted from Hayek and Salameh (2001). Since all products are manufactured on a single machine with a limited capacity, the cycle length of each product will be equal. One has (Taleizadeh et al. 2010):

$$t_i^1 = \frac{I_i}{P_i - D_i - d_i} \tag{4.347}$$

$$t_i^5 = \frac{\beta_i B_i}{P_i - D_i - d_i} \tag{4.348}$$

$$t_i^2 = \frac{x_i Q_i}{P_1} \tag{4.349}$$

$$t_i^3 = \frac{I_i^{\text{Max}}}{D_i} \tag{4.350}$$

$$t_i^4 = \frac{B_i}{D_i} \tag{4.351}$$

$$I_i = (P_i - D_i - d_i)\frac{Q_i}{P_i} - \beta_i B_i \tag{4.352}$$

$$I_i^{\text{Max}} = I_i + (P_{1i} - d_{1i} - D_i)t_i^2 = Q_i\left(1 - \frac{D_i}{P_{1i}P} - \frac{d_{1i}d_i}{P_{1i}P} - \frac{D_i d_i}{P_{1i}P_i}\right) - \beta_i B_i \tag{4.353}$$

Note that $t_i^1 - t_i^5$ are the production uptimes, t_i^2 is the reworking time, and t_i^3 and t_i^4 are the production downtimes. Also t_i^4 is the permitted shortage time, and t_i^5 is the time needed to satisfy all the backorders for the next production. During the rework

Table 4.14 General data for Example 4.16 (Taleizadeh and Wee 2015)

Product	D_i	P_i	ts_i	K_i	C_{R_i}	C_i	h_i	C_{b_i}	$\hat{\pi}_i$	x_i	β_i
1	150	5000	0.0025	500	15	34	2	5	1	0.05	0.5
2	200	5500	0.003	600	14	32	4	7	3	0.1	0.5
3	250	6000	0.0035	700	13	30	6	9	5	0.15	0.5
4	300	6500	0.004	800	12	28	8	11	7	0.2	0.5
5	350	7000	0.0045	900	11	26	10	13	9	0.25	0.5
6	400	7500	0.0025	1000	10	24	12	15	11	0.05	0.6
7	450	8000	0.003	1100	9	22	14	17	13	0.1	0.6
8	500	8500	0.0035	1200	8	20	16	19	15	0.15	0.6
9	550	9000	0.004	1300	7	18	18	21	17	0.2	0.6
10	600	9500	0.0045	1400	6	16	20	23	19	0.25	0.6
11	650	10,000	0.0025	1500	5	14	22	25	21	0.05	0.7
12	700	10,500	0.003	1600	4	12	24	27	23	0.1	0.7
13	750	11,000	0.0035	1700	3	10	26	29	25	0.15	0.7
14	800	11,500	0.004	1800	2	8	28	31	27	0.2	0.7
15	850	12,000	0.0045	1900	1	6	30	33	29	0.25	0.7

Table 4.15 The best results for Example 4.16 (Taleizadeh and Wee 2015)

Product	$T_{\text{Min}}^{\text{Production}}$	T	T^*	Q_i^*	B_i^*	TC^*
1	0.1159	2.5619	2.5619	380.1	8.33	932,400
2				505.6	13.59	
3				630.7	19.63	
4				755.4	26.3	
5				879.9	33.5	
6				1006.40	45.89	
7				1130.90	54.78	
8				1255.30	64.02	
9				1379.60	73.58	
10				1503.80	83.42	
11				1631.20	113.48	
12				1755.50	125.98	
13				1879.80	138.75	
14				2004.20	151.77	
15				2128.10	165.03	

process, the production rate of scrap items can be written as in Eq. (4.354) and calculated as in Eq. (4.355):

$$d_{1i} = P_{1i}\theta_i \tag{4.354}$$

$$d_{1i}t_i^2 = x_i\theta_i Q_i \tag{4.355}$$

Hence, it follows that the cycle length in single product state is:

4.5 Partial Backordering

Table 4.16 General data for Example 1 (Taleizadeh and Wee 2015)

Product	D_i	P_i	ts_i	K_i	C_{R_i}	C_i	h_i	C_{b_i}	$\hat{\pi}_i$	x_i	β_i
1	200	5000	0.0005	500	15	34	2	5	1	0.05	0.5
2	250	5500	0.001	600	14	32	4	7	3	0.1	0.5
3	300	6000	0.0015	700	13	30	6	9	5	0.15	0.5
4	350	6500	0.002	800	12	28	8	11	7	0.2	0.5
5	400	7000	0.0025	900	11	26	10	13	9	0.25	0.5
6	450	7500	0.0005	1000	10	24	12	15	11	0.05	0.6
7	500	8000	0.001	1100	9	22	14	17	13	0.1	0.6
8	550	8500	0.0015	1200	8	20	16	19	15	0.15	0.6
9	600	9000	0.002	1300	7	18	18	21	17	0.2	0.6
10	650	9500	0.0025	1400	6	16	20	23	19	0.25	0.6
11	700	10,000	0.0005	1500	5	14	22	25	21	0.05	0.7
12	750	10,500	0.001	1600	4	12	24	27	23	0.1	0.7
13	800	11,000	0.0015	1700	3	10	26	29	25	0.15	0.7
14	850	11,500	0.002	1800	2	8	28	31	27	0.2	0.7
15	900	12,000	0.0025	1900	1	6	30	33	29	0.25	0.7

Table 4.17 The best results for Example 2 (Taleizadeh and Wee 2015)

Product	$T_{\text{Min}}^{\text{Production}}$	T	T^*	Q_i^*	B_i^*	TC^*
1	2.6247	2.5619	2.6247	389.5	8.48	981,350
2				518	13.83	
3				646.2	19.98	
4				774	26.77	
5				901.6	34.1	
6				1031.20	46.79	
7				1158.80	55.86	
8				1286.20	65.28	
9				1413.60	75.03	
10				1540.80	85.07	
11				1671.30	115.88	
12				1798.70	128.65	
13				1926.00	141.69	
14				2053.30	154.99	
15				2180.50	168.55	

$$T = \frac{Q_i[1 - \theta_i \cdot x_i]}{D_i} \tag{4.356}$$

$$Q_i = \frac{D_i T - (1 - \beta_i) B_i}{1 - \theta_i x_i} \tag{4.357}$$

Table 4.18 New notations of given problem

P_{1i}	Rework rate of non-conforming items in units per unit time ith item (units per unit time)
d_{1i}	The production rate of scrapped items during the rework process of ith item (units per unit time)
h_{1_i}	Unit holding cost for each scrap of ith item (units per unit time)

Since both the random defective rate and the scrap rate are in [0, 1] and [0, $Q_i\theta_i x_i$] are the scrap items randomly produced during the imperfect rework process, it follows that the production cycle length T is not a constant. Solving the total inventory cost per year TC(Q, B) yields:

$$TC(Q,B) = \underbrace{NC_iQ_i}_{\text{Production Cost}} + \underbrace{NC_{R_i}x_iQ_i}_{\text{Rework Cost}} + \underbrace{NC_{d_i}x_iQ_i\theta_i}_{\text{Disposal Cost}} + \overbrace{Nh_i\left[\frac{I_i}{2}(t_i^1) + \frac{I_i + I_i^{Max}}{2}(t_i^2) + \frac{I_i^{Max}}{2}(t_i^3) + \frac{d_i(t_i^1 + t_i^5)}{2}(t_i^1 + t_i^5)\right]}^{\text{Holding Cost of Perfect Quality Items}}$$

$$+ \underbrace{Nh_{1_i}\left[\frac{P_{1i}t_i^2}{2}(t_i^2)\right]}_{\text{Holding Cost of Imperfect Quality Items}} + \underbrace{NC_{b_i}\frac{\beta_iB_i}{2}(t_i^4 + t_i^5)}_{\text{Back Ordered Cost}} + \underbrace{N\hat{\pi}_i\frac{(1-\beta_i)B_i}{2}(t_i^4)}_{\text{Lost Sale Cost}} + \underbrace{NK_i}_{\text{Setup Cost}}$$

(4.358)

The maximum capacity of the single machine and the service rate are two constraints of the model that are described in the following subsections. Since $t_i^1 + t_i^2 + t_i^5$ and S_i are the production uptimes, the rework time and the setup time of the ith product, respectively, the summation of the total production uptimes, rework and setup time (for all products) is $\sum_{i=1}^{n}(t_i^1 + t_i^2 + t_i^5) + \sum_{i=1}^{n} ts_i$ which is smaller or equal to the period length (T). Therefore, one has:

$$\sum_{i=1}^{n}(t_i^1 + t_i^2 + t_i^5) + \sum_{i}^{n} ts_i \leq T \quad (4.359)$$

$$\sum_{i}^{n}\frac{D_i(P_{1i} + d_i)}{P_iP_{1i}(1 - \theta_ix_i)} + \sum_{i}^{n} ts_i \leq T$$

$$T \geq \frac{\sum_{i}^{n} ts_i}{1 - \sum_{i}^{n}\frac{D(P_{1i}+d_i)}{P_iP_{1i}(1-\theta_ix_i)}} = T_{\text{Min}} \quad (4.360)$$

Since the shortage quantity of the ith product per period is B_j, the annual demand of the jth product is D_j, the number of periods in each year is N, and the safety factor of allowable shortage is SL. Therefore, the service level constraint is as follows:

$$\sum_{i}^{n} \frac{NB_i}{D_i} \leq \text{SL}$$

4.5 Partial Backordering

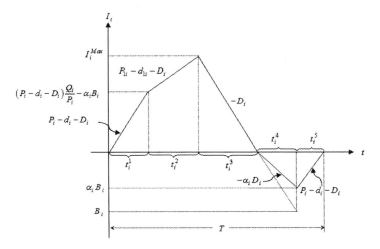

Fig. 4.26 EPQ inventory system (Taleizadeh et al. 2010)

$$T \geq \frac{\sum_{j}^{n}\frac{C_j^2}{2C_j^1 D_j}}{\mathrm{SL} - \sum_{j}^{n}\frac{C_j^3}{2C_j^1 D_j}} = T_{\mathrm{SL}} \quad (4.361)$$

Finally, the final model is:

$$\text{Min}: \ \mathrm{TC}(T,B_i) = \sum_{i=1}^{n} C_i^1 \frac{(B_i)^2}{T} - \sum_{i=1}^{n} C_i^2 \frac{B_i}{T} - \sum_{i=1}^{n} C_i^3 B_i + \sum_{i=1}^{n} C_i^4 T + \sum_{i=1}^{n} C_i^5 + \sum_{i=1}^{n} \frac{K_i}{T}$$

$$\text{s.t.}: \ T \geq \frac{\sum_{i=1}^{n} t s_i}{\left[1 - \sum_{i=1}^{n} \frac{D(P_{1i}+d_i)}{P_i P_{1i}(1-\theta_i x_i)}\right]}$$

$$T \geq \frac{\sum_{j=1}^{n}\frac{C_j^2}{2C_j^1 D_j}}{\left(\mathrm{SL} - \sum_{j=1}^{n}\frac{C_j^3}{2C_j^1 D_j}\right)} = T_{\mathrm{SL}}$$

$$T, B_i \geq 0 \ \ \forall i, i = 1,2,\ldots,n$$

(4.362)

In which:

$$C_i^1 = h_i \left(\frac{J_i(1-\beta_i)^2}{(1-\theta_i x_i)^2} + \frac{(1-\beta_i)R_i}{(1-\theta_i x_i)} + \frac{\beta_i^2(P_i - d_i)}{2(P_i - d_i - D_i)} \right)$$

$$+ h_{1i} \left(\frac{1}{2P_i^1} \left(\frac{d_i(1-\beta_i)}{(1-\theta_i x_i)P_i} \right)^2 \right) + C_{bi} \left(\frac{\beta_i}{2} \left(\frac{P_i - d_i - (1-\beta_i)D_i}{D_i(P_i - d_i - D_i)} \right) \right)$$

$$+ \frac{(1-\beta_i)\hat{\pi}_i}{2} \left(\frac{P_i - d_i - (1-\beta_i)D_i}{D_i(P_i - d_i - D_i)} \right) \tag{4.363a}$$

$$C_i^2 = \left(\frac{C_i(1-\beta_i)}{(1-\theta_i x_i)} \right) + \left(\frac{C_{R_i} x_i (1-\beta_i)}{(1-\theta_i x_i)} \right) + \left(\frac{C_{d_i} x_i \theta_i (1-\beta_i)}{(1-\theta_i x_i)} \right) > 0 \tag{4.363b}$$

$$C_i^3 = h_i \left(\frac{2J_i D_i (1-\beta_i)}{(1-\theta_i x_i)} + \frac{D_i R_i}{(1-\theta_i x_i)} \right)$$

$$+ h_{1i} \left(\frac{1}{2P_{1i}} \left(\frac{d_i}{P_i} \right)^2 \left(\frac{2D_i(1-\beta_i)}{(1-\theta_i x_i)} \right) \right) \tag{4.363c}$$

$$C_i^4 = h_i \frac{J_i D_i^2}{(1-\theta_i x_i)^2} + h_{1i} \left(\frac{1}{2P_{1i}} \left(\frac{d_i}{P_i} \right)^2 \left(\frac{D_i}{(1-\theta_i x_i)} \right)^2 \right) > 0 \tag{4.363d}$$

$$C_i^5 = \frac{C_i D_i}{(1-\theta_i x_i)} + \frac{C_{R_i} x_i D_i}{(1-\theta_i x_i)} + \frac{C_{d_i} x_i \theta_i D_i}{(1-\theta_i x_i)} > 0 \tag{4.363e}$$

$$\mu_i = 1 - \frac{D_i}{P_i} - \frac{d_i(P_{1i} - d_{1i} - D_i)}{2P_{1i}P_i}, \quad U_i = 1 - \frac{D_i}{P_i} - \frac{d_i(d_{1i} + D_i)}{P_{1i}P_i}, \quad J_i$$

$$= \frac{(P_i - D_i)}{2P_i^2} + \frac{\mu_i d_i}{P_{1i}P_i} + \frac{U_i}{2D_i}, \quad R_i = \frac{\beta_i d_i}{P_{1i}P_i} + \frac{\beta_i U_i}{D_i} + \frac{\beta_i}{P_i} \tag{4.363f}$$

Since the Hessian matrix of objective function is positive for all nonzero B_i and T, $X \times \mathbf{H} \times X^T = \left(2\sum_{i=1}^n K_i + \sum_{i=1}^n C_i^2 B_i \right)/T \geq 0$, $TC(T, B_i)$ is convex. To derive the optimal values of the decision variables, take the partial differentiations of $TC(T, B_i)$ with respect to T and B_i (for details, see Appendix 2 of Taleizadeh and Wee (2015)):

$$\frac{\partial TC(T, B_i)}{\partial T} = 0 \rightarrow T = \sqrt{\frac{\sum_{i=1}^n K_i - \sum_{i=1}^n \left[\frac{(C_i^2)^2}{4C_i^1} \right]}{\sum_{i=1}^n C_i^4 - \sum_{i=1}^n \left[\frac{(C_i^3)^2}{4C_i^1} \right]}} \tag{4.364}$$

$$\frac{\partial TC(T, B_i)}{\partial B_i} = 0 \rightarrow B_i^* = \frac{C_i^3 T^* + C_i^2}{2C_i^1} \tag{4.365}$$

Then,

4.6 Multi-delivery

$$Q_i = \frac{D_i T - (1-\beta_i)B_i}{(1-\theta_i x_i)}, \quad 0 \le \theta_i \le 1 \tag{4.366}$$

To ensure production of all products by a machine and satisfy service level constraint of each, the following steps to derive T^*, B_i^*, and Q_i^* must be performed (Taleizadeh et al. 2010):

Step 1.
Check for feasibility. If $\sum_{i=1}^{n} \frac{D_i(P_i^1 + d_i)}{P_i P_i^1 (1-\theta_i x_i)} < 1$ and $\sum_{i=1}^{n} \frac{C_i^3}{2C_i^1 D_i} < SL$, go to step 2; else the problem is infeasible.

Step 2.
Calculate T using Eq. (4.364). If $T \ge 0$, go the Step 3; else the problem is infeasible.

Step 3.
Calculate T_{SL} using Eq. (4.358).

Step 4.
Calculate T_{Min} using Eq. (4.360).

Step 5.
If $T \ge$ Max $\{T_{Min}, T_{SL}\}$, then $T^* = T$; else $T = $ Max $\{T_{Min}, T_{SL}\}$.

Step 6.
Calculate $B_i^*, \forall i = 1, 2, \ldots, n$ using Eq. (4.365).

Step 7.
Calculate $Q_i^*, \forall i = 1, 2, \ldots, n$ using Eq. (4.366).

Step 8.
Terminate procedure.

Examples 4.18 and 4.19 Taleizadeh et al. (2010) considered five-product inventory control problem where the general and the specific data are given in Tables 4.19, 4.20, and 4.21. They considered two numerical examples with uniform and normal distributions for x_i and θ_i. Tables 4.22 and 4.23 show the best results for the two numerical examples. The safety level, *SL*, is 30%

4.6 Multi-delivery

4.6.1 Multi-delivery Policy and Quality Assurance

Chiu et al. (2009) developed a multi-delivery policy and quality assurance into an imperfect economic production quantity (EPQ) model with scrap and rework. A portion of non-conforming items produced is considered to be scrap, while the other is assumed to be repairable and is reworked in each cycle when regular production ends. Finished items can only be delivered to customers if whole lot is quality assured after rework. Fixed quantity multiple installments of finished batch are delivered by request to customers at a fixed interval of time. Expected integrated cost function per unit time is derived.

This paper incorporates a multi-delivery policy and quality assurance into an imperfect EPQ model with scrap and rework. Consider that during regular production time and x portion of defective items is produced randomly, at a production rate d. Among defective items, a portion is assumed to be scrap and the rest can be reworked and repaired at a rate P_1, in each cycle after a production run (see Fig. 4.27). Similar to previous case, $(P - d - D) > 0$ or $(1 - x - D/P) > 0$, where $d = Px$ (Chiu et al. 2009).

Also some new notations which are specifically used for this proposed model are presented in Table 4.24.

It is assumed that finished items can only be delivered to customers if whole lot is quality assured at the end of rework. Fixed quantity of n installments of finished batch is delivered by request to customers, at a fixed interval of time during production downtime t_3 (Fig. 4.27):

$$T = t_1 + t_2 + t_3 \quad (4.367)$$

$$t_1 = \frac{Q}{P} = \frac{H_1}{P - d} \quad (4.368)$$

$$t_2 = \frac{xQ(1 - \theta)}{P_1} \quad (4.369)$$

$$t_3 = nt_n = T - (t_1 + t_2) = Q\left(\frac{(1 - \theta x)}{D} - \frac{1}{P} - \frac{x(1 - \theta)}{P_1}\right) \quad (4.370)$$

$$H_1 = (P - d)t_1 = (P - d)\frac{Q}{P} = (1 - x)Q \quad (4.371)$$

$$H = H_1 + P_1 t_2 = (1 - \theta x)Q \quad (4.372)$$

where T, cycle length; H, maximum level of on-hand inventory when regular production process ends; H_1, maximum level of on-hand inventory when rework process finishes; Q, production lot size; t_1, production uptime for proposed EPQ model; t_2, time for reworking of defective items; t_3, time for delivering all quality assured finished products; t_n, fixed interval of time between each installment of finished products delivered during t_3; $I(t)$, on-hand inventory of perfect-quality items at time t; and $I_d(t)$, on-hand inventory of defective items at time t (Chiu et al. 2009). On-hand inventory of defective items during production uptime t_1 and reworking time t (Fig. 4.28) shows that maximum level of on-hand defective items is $d \cdot t_1$ and (Chiu et al. 2009):

$$d \cdot t_1 = P \cdot x \cdot t_1 = xQ \quad (4.373)$$

4.6 Multi-delivery

Table 4.19 General data for Examples 4.18 and 4.19 (Taleizadeh et al. 2010)

P	D_i	P_i	P_{1i}	ts_i	K_i	β_i	C_i	h_i	h_{1_i}	C_{d_i}	C_{b_i}	$\hat{\pi}_i$	C_{R_i}
1	600	4000	2000	0.003	500	0.3	15	5	2	3	10	12	1
2	700	4500	2500	0.004	450	0.4	12	4	2	3	8	10	2
3	800	5000	3000	0.005	400	0.5	10	3	2	3	6	8	3
4	900	5500	3500	0.006	350	0.6	8	2	2	3	4	6	4
5	1000	6000	4000	0.007	300	0.7	6	1	2	3	2	4	5

Table 4.20 Specific data for Example 4.18 (Taleizadeh et al. 2010)

	$X_i \sim U[a_i, b_i]$				$\theta_i \sim U[a_i, b_i]$			
Product	a_i	b_i	$E[X_i]$	$d_i = P_i E[X_i]$	a_i	b_i	$E[\theta_i]$	$d_{1i} = P_{1i} E[\theta_i]$
1	0	0.05	0.025	100	0	0.15	0.075	150
2	0	0.1	0.05	225	0	0.2	0.1	250
3	0	0.15	0.075	375	0	0.25	0.125	375
4	0	0.2	0.1	550	0	0.3	0.15	525
5	0	0.25	0.125	750	0	0.35	0.175	700

Table 4.21 Specific data for Example 4.19 (Taleizadeh et al. 2010)

	$X_i \sim U[a_i, b_i]$				$\theta_i \sim U[a_i, b_i]$			
Product	a_i	b_i	$E[X_i]$	$d_i = P_i E[X_i]$	a_i	b_i	$E[\theta_i]$	$d_{1i} = P_{1i} E[\theta_i]$
1	0.01	0.01	40	0.1	0.01	200	1	0.01
2	0.02	0.02	90	0.15	0.02	375	2	0.02
3	0.03	0.03	150	0.2	0.03	600	3	0.03
4	0.04	0.04	220	0.25	0.04	875	4	0.04
5	0.05	0.05	300	0.3	0.05	1200	5	0.05

A θ portion among non-conforming items is assumed to be scrap (Eq. 4.374). Other reparable portion $(1 - \theta)$ is reworked right after production uptime t_1 ends (Chiu et al. 2009):

$$\theta d \cdot t_1 = \theta \cdot P \cdot x \cdot t_1 = \theta x Q \qquad (4.374)$$

Total costs per cycle TC(Q) consist of setup cost, variable production cost, variable rework cost, disposal cost, fixed and variable delivery cost, holding cost during t_1 and t_2, variable holding cost for items reworked, and holding cost for finished goods during delivery time t_3 where n fixed-quantity installments of finished batch are delivered by request to customers at a fixed interval of time. Cost for each delivery is (Chiu et al. 2009):

$$K_S + C_T \left(\frac{H}{n}\right) \qquad (4.375)$$

Total delivery costs for n shipments in a cycle are:

Table 4.22 The best results for Example 4.18 (Taleizadeh et al. 2010)

Product	Uniform T_{Min}	T	T_{SL}	T^*	B_i	Q_i	Z
1	0.2617	0.5748	0.2574	0.5748	25.14	327.92	46,998
2					13.68	396.15	
3					9.02	459.66	
4					6.8	522.46	
5					6.52	585.68	

Table 4.23 The best results for Example 4.19 (Taleizadeh et al. 2010)

Product	Uniform T_{Min}	T	T_{SL}	T^*	B_i	Q_i	Z
1	0.1576	0.5778	0.2585	0.5778	25.2	329.34	46,017
2					13.74	397.37	
3					9.08	460.43	
4					6.86	522.46	
5					6.59	584.54	

$$n\left[K_S + C_T\left(\frac{H}{n}\right)\right] = nK_S + C_T H = nK_S + C_T Q(1 - \theta x) \tag{4.376}$$

Therefore, TC(Q) (Appendix A of Chiu et al. (2009)) is:

$$TC(Q) = \underbrace{K}_{\text{Setup Cost}} + \underbrace{CQ}_{\text{Production Cost}} + \underbrace{C_R[x(1-\theta)Q]}_{\text{Rework Cost}} + \underbrace{C_d[x\theta Q]}_{\text{Disposal Cost}} + \underbrace{C_T[Q(1-\theta x)]}_{\text{Delivery Cost}} + \underbrace{nK_S}_{\text{Fixed Delivery Cost}}$$
$$+ \underbrace{h_1 \frac{P_1 t_2}{2}(t_2)}_{\text{Holding Cost of Reworked Items}} + \underbrace{h\left[\frac{H_1 + dt_1}{2}(t_1) + \frac{H_1 + H}{2}(t_2)\right] + h\left(\frac{n-1}{2n}\right)Ht_3}_{\text{Holding Cost}}$$

(4.377)

Since x of defective items is assumed to be a random variable with a known probability density function, for randomness of defective rate, one can use values of x in inventory cost analysis. Substituting all related parameters from Eqs. (4.367) to (4.377) in TC(Q), one obtains expected production–inventory–delivery cost per unit time, $E[\text{TCU}(Q)]$ (Appendix B of Chiu et al. (2009)) as:

4.6 Multi-delivery

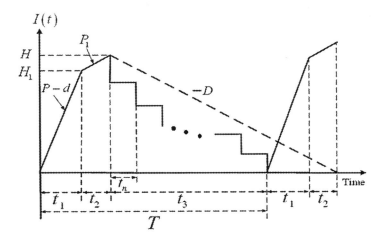

Fig. 4.27 On-hand inventory of perfect-quality items in EPQ model with multi-delivery policy, scrap, and rework (Chiu et al. 2009)

Table 4.24 New notations of given problem

P_1	Rework rate of non-conforming items in units per unit time (units per unit time)
t_1	Production uptime for product
t_2	The rework time for non-conforming product
n	Number of fixed quantity installments of the finished batch to be delivered to customers in each cycle, it is assumed to be a constant for all products
t_n	A fixed interval of time between each installment of finished products delivered.

$$E[\text{TCU}(Q)] = \frac{E[\text{TC}(Q)]}{E[T]} = \frac{CD}{1 - \theta E[x]} + \frac{(K + nK_S)}{Q(1 - \theta E[x])} + \frac{C_R E[x](1 - \theta)D}{(1 - \theta E[x])}$$
$$+ \frac{C_d E[x]\theta D}{(1 - \theta E[x])} + C_T D + \frac{hQD}{2P(1 - \theta E[x])} + \frac{h_1 (E[x])^2 QD(1 - \theta)^2}{2P_1 (1 - \theta E[x])}$$
$$+ \frac{hQD}{2P_1(1 - \theta E[x])}\left[\left(2E[x] - (E[x])^2 - \theta(E[x])^2\right)(1 - \theta)\right]$$
$$+ \left(\frac{n-1}{n}\right)\left[\frac{hQ(1 - \theta E[x])}{2} - \frac{hQD}{2P} - \frac{hQE[x](1 - \theta)D}{2P_1}\right]$$
(4.378)

Optimal production lot size can be obtained by minimizing expected cost function $E[\text{TCU}(Q)]$. Differentiating $E[\text{TCU}(Q)]$ with respect to the Q gives first and second derivative as (Chiu et al. 2009):

Fig. 4.28 On-hand inventory of defective items in EPQ model with multi-delivery policy, scrap and rework (Chiu et al. 2009)

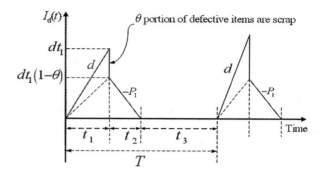

$$\frac{dE[\text{TCU}(Q)]}{dQ} = \frac{KD}{Q^2(1-\theta E[x])} + \frac{nK_S D}{Q^2(1-\theta E[x])} + \frac{hD}{2P(1-\theta E[x])}$$
$$+ \frac{hD}{2P_1(1-\theta E[x])}\left[\left(2E[x] - (E[x])^2 - \theta(E[x])^2\right)(1-\theta)\right]$$
$$+ \left(\frac{n-1}{n}\right)\left[\frac{h(1-\theta E[x])}{2} - \frac{hD}{2P} - \frac{hE[x](1-\theta)D}{2P_1}\right] + \frac{h_1(E[x])^2 D(1-\theta)^2}{2P_1(1-\theta E[x])}$$
(4.379)

$$\frac{d^2 E[\text{TCU}(Q)]}{dQ^2} = \frac{2(K + nK_S)D}{Q^3(1-\theta E[x])} \quad (4.380)$$

Equation (4.380) is positive because K, n, K_S, D, Q, and $(1 - \theta E[x])$ are all positive. Second derivative of $E[\text{TCU}(Q)]$ with respect to Q (Eq. 4.380) is greater than zero, and hence $E[\text{TCU}(Q)]$ is a convex function for all Q different from zero. Optimal production lot size Q^* can be obtained by setting first derivative (Eq. 4.379) of $E[\text{TCU}(Q)]$ equal to zero (Chiu et al. 2009):

$$Q^* = \sqrt{\frac{2(K+nK_S)D}{\left\{\begin{array}{l}\frac{hD}{P} + \frac{hD}{P_1}\left[2E[x] - (E[x])^2 - \theta(E[x])^2\right](1-\theta) \\ + \left(\frac{n-1}{n}\right)\left[h(1-\theta E[x])^2 - h\left(\frac{D}{P} + \frac{E[x](1-\theta)D}{P_1}\right)(1-\theta E[x])\right] + \frac{h_1(E[x])^2 D(1-\theta)^2}{P_1}\end{array}\right\}}}$$
(4.381)

Example 4.20 Chiu et al. (2009) presented an example in which it is assumed that a product can be manufactured at a rate of 60,000 units per year and this item has experienced a flat demand rate of 3400 units per year. During production uptime, random defective rate is assumed to be uniformly distributed over the interval [0, 0.3]. Among defective items, a portion $\theta = 0.1$ is considered to be scrap, and other portion can be reworked and repaired, at a rate $P_1 = 2100$ units per year. Additional parameters considered by this example are given as follows: $C_R = \$60$ per item reworked; $C_d = \$20$ per scrap item; $C = \$100$ per item; $K = \$20,000$ per production run; $h = \$20$ per item per year; $h_1 = \$40$ per item reworked per unit time (year); $n = 4$ installments of finished batch are delivered per cycle; $K_S = \$4350$ per shipment, a fixed cost; and $C_T = \$0.1$ per item delivered.

Optimal batch size $Q^* = 3495$ can be obtained from Eq. (4.381) and long-run average production–inventory delivery costs per year $E[TCU(Q^*)] = \$448,390$ from Eq. (4.378) (Chiu et al. 2009).

4.6.2 Multi-delivery and Partial Rework

Chiu et al. (2012) studied an extended EPQ model which incorporated quality assurance issue and a multi-delivery policy into the classic EPQ model. The quality assurance issue is in regard to the production system which has an x portion of random defective items produced at a production rate d, and among defective items, a θ portion is assumed to be scrap, and the other $(1 - \theta)$ portion can be reworked and repaired at a rate P_1, within the same cycle when regular production ends, while the multi-delivery policy is with regard to fixed quantity of n installments of the finished batch which are delivered to customer at a fixed interval of time during the production downtime (i.e., when the whole lot is quality assured at the end of rework) (Chiu et al. 2012). Figure 4.29 depicts the on-hand inventory of perfect-quality items of the proposed model. Figure 4.30 illustrates the expected reduction in inventory holding costs (in yellow/shade areas) of the proposed model (in blue). Based on the description of the proposed model and Fig. 4.29, the following expressions can be derived accordingly (Chiu et al. 2012):

$$t_1 = \frac{Q}{P} \tag{4.382}$$

$$t_2 = \frac{xQ(1-\theta)}{P_1} \tag{4.383}$$

$$t_3 = nt_n = T - (t_1 + t_2) \tag{4.384}$$

$$T = t_1 + t_2 + t_3 = \frac{Q}{D}(1 - \theta x) \tag{4.385}$$

$$t = \frac{D(t_1 + t_2)}{P - D} \tag{4.386}$$

$$H = (P - D)t = D(t_1 + t_2) \tag{4.387}$$

$$H_1 = Q(1 - x) - D(t_1 + t_2) \tag{4.388}$$

$$H_2 = H_1 + P_1 t_2 \tag{4.389}$$

The on-hand inventory of defective items during production uptime t_1 and reworking time t_2 is illustrated in Fig. 4.31. It is noted that the maximum level of

Fig. 4.29 On-hand inventory of perfect-quality items in EPQ model with $(n + 1)$ delivery policy and partial rework (Chiu et al. 2012)

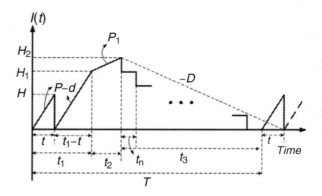

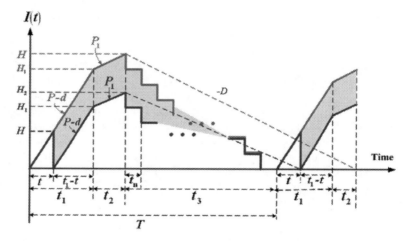

Fig. 4.30 Expected reduction in inventory holding costs (in yellow) of the proposed model in comparison with model of Chiu et al. (2012)

defective items is dt_1. A θ portion among non-conforming items is assumed to be scrap items as shown in Eq. (4.390). Other repairable portion $(1 - \theta)$ is reworked right after the production uptime t_1 ends (Chiu et al. 2012):

$$\theta D t_1 = \theta P x t_1 = \theta x Q \qquad (4.390)$$

Total production–inventory–delivery costs per cycle $TC(Q)$ consist of setup cost, variable production cost, variable rework cost, disposal cost, fixed and variable delivery cost, holding cost for perfect-quality items during production uptime t_1 and reworking time t_2, holding cost for defective items during uptime t_1, variable holding cost for items reworked during t_2, and holding cost for finished goods during

4.6 Multi-delivery

Fig. 4.31 On-hand inventory of non-conforming items in EPQ model with $(n + 1)$ delivery policy and partial rework (Chiu et al. 2012)

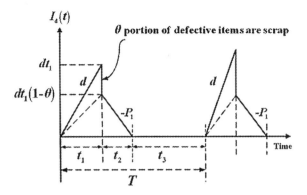

the delivery time t_3 where n fixed-quantity installments of the finished batch are delivered to customers at a fixed interval of time (for computation of the last term refer to Appendix of Chiu et al. (2009):

$$TC(Q,B) = \overbrace{K}^{Setup\,Cost} + \overbrace{CQ}^{Cost} + \overbrace{C_R[x(1-\theta)Q]}^{Rework\,Cost} + \overbrace{C_d(\theta x Q)}^{Disposal\,Cost} + \overbrace{(n+1)K_S}^{Cost} + \overbrace{C_T[x(1-\theta)Q]}^{Delivery\,Cost}$$

$$\underbrace{+h\left[\frac{H}{2}(t) + \frac{H_1+H_2}{2}(t_2) + \frac{H_1}{2}(t_1-t) + \frac{D(t_1)}{2}(t_1)\right] + h\frac{n-1}{2n}(H.t_3)}_{Holding\,Cost}$$

$$\underbrace{+h_1\frac{Dt_1(1-\theta)}{2}(t_2)}_{Holding\,Cost\,of\,Reworked\,Items}$$

(4.391)

Because x proportion of defective items is assumed to be a random variable with a known probability density function, one could use the expected values of x in the related cost analysis:

$$E[\text{TCU}(T)] = E[\text{TC}(T)]/E[T] \qquad (4.392)$$

Similar to previous case, Chiu et al. (2012) showed that the cost function is convex, so the root of first derivative of total cost with respect to Q yields to optimal values of production quantity as presented in Eq. (4.432):

$$Q^* = \sqrt{\dfrac{2[(n+1)K_S+K]D}{h\left\{\begin{array}{l}\dfrac{D}{P}\left[\dfrac{2D}{P^2}E\left(\dfrac{1}{1-x}\right)-\dfrac{1}{P}+\dfrac{(1-\theta)}{P_1}\right]\left[\dfrac{4D}{P}E\left(\dfrac{x}{1-x}\right)+\dfrac{2D(1-\theta)}{P_1}E\left(\dfrac{x^2}{1-x}\right)-2E(x)\right]\\ -\dfrac{DE(x)^2(1-\theta)^2}{P_1}\left[1+\dfrac{D}{P_1}\right]+[1-\theta E(x)]^2-\dfrac{D[1-2\theta E[x]]}{P}\\ -\dfrac{1}{n}[1-\theta E(x)]^2-\dfrac{D}{P}\left[2-\dfrac{D}{P}\right]+\dfrac{DE(x)(1-\theta)}{P_1}\left[-2[1-\theta E(x)]+\dfrac{2D}{P}+\dfrac{DE(x)(1-\theta)}{P_1}\right]\end{array}\right\}+\dfrac{h_1 DE(x)^2(1-\theta)^2}{P_1}}}$$

(4.393)

Example 4.21 Chiu et al. (2012) used again the data of their previous work (Chiu et al. 2009) here to demonstrate the aforementioned results derived by presented alternative approach. Using Eq. (4.393) the optimal production quantity is $Q^* = 4219$ and $E[TCU(Q^*)] = \$435{,}712$. It is noted that the overall reduction in production–inventory–delivery costs amounts to \$12,678, or 13.25% of total other related costs.

4.6.3 Multi-delivery Single Machine

Chiu et al. (2015) developed a multi-item EPQ model with scrap, rework, and multiple deliveries. Consider that L products are made in turn on a single machine with the purpose of maximizing the machine utilization. All items made are screened, and inspection cost for each item is included in the unit production cost C_i. During production process for each product i (where $i = 1, 2, \ldots, L$), an x_i portion of non-conforming items is produced randomly at a rate d_i. Among these non-conforming items, a θ_i portion is considered to be scrap items, and the other portion can be reworked and repaired at a rate of P_{2i} right after the end of regular production process in each cycle with an additional cost C_{Ri}. Under the normal operation, the constant production rate P_{1i} for product i must satisfy $(P_{1i} - d_i - D_i) > 0$, where D_i is the demand rate for product i per year and d_i can be expressed as $d_i = x_i P_{1i}$. Unlike classic EPQ model which assumes a continuous issuing policy for meeting product demands, this study adopts a multi-delivery policy. It is assumed that finished goods for each product i can only be delivered to customers if whole production lot is quality assured in the end of rework process for each product i. Fixed quantity n installments of the finished batch are delivered at a fixed interval of time during delivery time t_{3i} (refer to Fig. 4.32).

Also some new notations which are specifically used for this proposed model are presented in Table 4.25.

One can obtain the following formulas directly from Figs. 4.32 and 4.33 (Chiu et al. 2015):

4.6 Multi-delivery

$$t_{1i} = \frac{Q_i}{P_{1i}} = \frac{H_{1i}}{P_{1i} - d_i} \tag{4.394}$$

$$t_{2i} = \frac{x_i Q_i (1 - \theta_i)}{P_{2i}} \tag{4.395}$$

$$t_{3i} = n t_{ni} = T - (t_{1i} + t_{2i}) \tag{4.396}$$

$$T = t_{1i} + t_{2i} + t_{3i} = \frac{Q_i(1 - \theta_i x_i)}{D_i} \tag{4.397}$$

$$H_{1i} = (P_{1i} - d_i) t_{ii} \tag{4.398}$$

$$H_{2i} = H_{1i} + P_{2i} t_{2i} \tag{4.399}$$

$$d_i t_{1i} = x_i Q_i \tag{4.400}$$

Total delivery cost for product i (n shipments) in a cycle is:

$$n K_i^S + C_i^T Q_i (1 - \theta_i x_i) \tag{4.401}$$

Holding costs for finished products during the t_3, where n fixed-quantity installments of the finished batch are delivered to customers at a fixed interval of time, is:

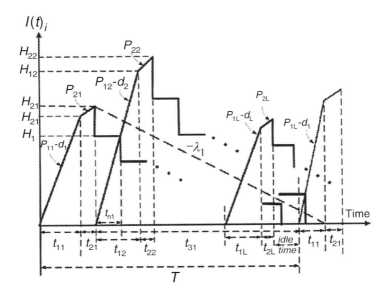

Fig. 4.32 On-hand perfect-quality inventory for product i in the proposed multi-item EPQ model under a common cycle policy (Chiu et al. 2015)

Table 4.25 New notations of given problem

P_{1i}	The production rate of ith item(units per unit time)
P_{2i}	Rework rate of non-conforming items in units per unit time ith item (units per unit time)
K_i^S	Fixed delivery cost per shipment for product I ($/shipment)
C_i^T	Unit shipping cost for product I ($/units)
t_{1i}	Production uptime for product i in the proposed EPQ model (time)
t_{2i}	The rework time for product i in the proposed EPQ model (time)
n	Number of fixed quantity installments of the finished batch to be delivered to customers in each cycle, it is assumed to be a constant for all products
t_{ni}	A fixed interval of time between each installment of finished products delivered during t_{2i}, for product i (time)
h_{1i}	Unit holding cost for each reworked item ($/units per unit time)

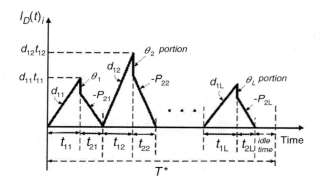

Fig. 4.33 On-hand inventory of defective items for product i in the proposed multi-item EPQ model under a common cycle policy (Chiu et al. 2015)

$$h_i\left(\frac{n-1}{2n}\right)H_{2i}t_{3i} \qquad (4.402)$$

Total production–inventory–delivery costs per cycle $TC(Q_i)$ for L products consist of the variable production cost, setup cost, rework cost, fixed and variable delivery cost, holding cost during production uptime t_{1i} and rework time t_{2i}, and holding cost for finished goods kept during the delivery time t_3. Therefore, total TC (Q_i) for L products are (Chiu et al. 2015):

4.6 Multi-delivery

$$\sum_{i}^{L} \mathrm{TC}(Q_i) = \sum_{i=1}^{L} \left\{ \begin{array}{c} \overbrace{C_i Q_i}^{\text{Production Cost}} + \overbrace{K_i}^{\text{Setup Cost}} + \overbrace{C_{R_i}[x_i(1-\theta_i)Q_i]}^{\text{Rework Cost}} + \overbrace{C_{d_i}[x_i\theta_i Q_i]}^{\text{Disposal Cost}} \\ \times \\ + \underbrace{nK_i^S}_{\text{Fixed Delivery Cost}} + \underbrace{C_i^T[Q_i(1-\theta_i x_i)]}_{\text{Shipment Cost}} + \overbrace{h_{1_i}\left[\frac{P_{2i}t_{2i}}{2}(t_{2i})\right]}^{\text{Holding Cost of Reworked Items}} \\ + \underbrace{h_i\left[\frac{H_{1i}+D_i t_{1i}}{2}(t_{1i}) + \frac{H_{1i}+H_{2i}}{2}(t_{2i}) + \frac{n-1}{2n}(I_{2i}t_{3i})\right]}_{\text{Holding Cost}} \end{array} \right\}$$

(4.403)

To take the randomness of defective rate x into account, by applying the expected values of x in the cost analysis and substituting all variables, the following expected $E[\mathrm{TCU}(Q)]$ can be obtained:

$$E[\mathrm{TCU}(Q)] = E\left[\sum_{i=1}^{L} \mathrm{TC}(Q_i)\right] \frac{1}{E[T]}$$

$$= \sum_{i=1}^{L} \left\{ \begin{array}{l} D_i\left[C_i \dfrac{1}{1-\theta_i \cdot E[x_i]} + C_{R_i}(1-\theta_i) \cdot \dfrac{E[x_i]}{1-\theta_i \cdot E[x_i]} + C_{d_i}\theta_i \dfrac{E[x_i]}{1-\theta_i \cdot E[x_i]}\right] + C_i^T D_i + \dfrac{K_i D_i}{Q_i} \dfrac{1}{1-\theta_i \cdot E[x_i]} \\ + \dfrac{nK_i^S D_i}{Q_i[1-\theta_i E(x_i)]} + \dfrac{h_{1_i}}{2}\left[\dfrac{Q_i D_i(1-\theta_i)^2}{P_i^2}\right] \dfrac{E(x_i)^2}{1-\theta_i \cdot E[x_i]} + E[x_i]\left(\dfrac{1-\theta_i}{P_{2i}} + \dfrac{1-\theta_i}{nP_{2i}}\right) + \dfrac{h_i Q_i D_i}{2} \dfrac{1}{1-\theta_i E(x_i)} \\ \times \left[\dfrac{1}{D_i} - \dfrac{1}{nD_i} + \dfrac{1}{nP_{1i}} + \left(1-\dfrac{1}{n}\right)E[x_i]\left(\dfrac{\theta_i}{P_{1i}} - \dfrac{2\theta_i}{D_i}\right) + E[x_i]^2\left(\dfrac{\theta_i^2}{D_i}\right) - E[x_i]^2\left(\dfrac{1-\theta_i}{P_{2i}} + \dfrac{\theta_i(1-\theta_i)}{nP_{2i}}\right)\right] \end{array} \right\}$$

(4.404)

where $E[T] = Q_i[1 - \theta_i E(x_i)]/D_i$. Replacing Q_i with T yields to:

$$E[\mathrm{TCU}(T)] = \sum_{i=1}^{L} \left\{ \begin{array}{l} \dfrac{C_i D_i}{1-\theta_i \cdot E[x_i]} + \dfrac{C_{R_i}(1-\theta_i)E[x_i]}{1-\theta_i \cdot E[x_i]} + \dfrac{C_{d_i}\theta_i C_i D_i}{1-\theta_i \cdot E[x_i]} + C_i^T D_i \\ + \dfrac{K_i}{T} + \dfrac{nK_i^S}{T} + \dfrac{h_{1_i} T D_i^2(1-\theta_i)^2}{2[1-\theta_i E(x_i)]^2}\left(\dfrac{E(x_i)^2}{P_{2i}}\right) \\ \times \\ + \dfrac{h_i T D_i^2}{2}\left[\dfrac{1}{D_i} - \dfrac{1}{D_i n} + \dfrac{1}{P_{1i}n[1-\theta_i E[x_i]]} + \dfrac{\theta_i E[x_i]}{P_{1i}n[1-\theta_i E[x_i]]^2}\right] \\ + \dfrac{h_i T D_i^2}{2}\left[\dfrac{(1-\theta_i)E[x_i][1-E[x_i]]}{P_{2i}n[1-\theta_i E[x_i]]^2} + \dfrac{(1-\theta_i)E[x_i]}{P_{2i}n[1-\theta_i E[x_i]]^2}\right] \end{array} \right\}$$

(4.405)

Let E_i^0, E_i^1 denote the following:

$$E_i^0 = \frac{1}{1 - \theta_i \cdot E[x_i]}$$

$$E_i^1 = \frac{E[x_i]}{1 - \theta_i \cdot E[x_i]} \quad (4.406)$$

Equation (4.405) becomes (Chiu et al. 2015):

$$E[\text{TCU}(T)] = \sum_{i=1}^{L} \left\{ \begin{array}{l} C_i D_i E_i^0 + C_{R_i}(1-\theta_i)E_i^1 + C_{d_i}\theta_i E_i^1 + C_i^T D_i + \dfrac{K_i D_i}{Q_i} \cdot \dfrac{1}{1-\theta_i \cdot E[x_i]} \\[6pt] \dfrac{K_i}{T} + \dfrac{nK_i^S}{T} + \dfrac{h_{1_i}TD_i^2(1-\theta_i)^2 \left(E_i^1\right)^2}{2P_{2i}} \\[6pt] + \dfrac{h_i TD_i^2}{2}\left[\dfrac{1}{D_i} - \dfrac{1}{D_i n} + \dfrac{E_i^0}{P_{1i}n} + \dfrac{\theta_i E_i^0 E_i^1}{P_{1i}n}\right] \\[6pt] + \dfrac{h_i TD_i^2}{2}\left[\dfrac{(1-\theta_i)E_i^1}{P_{2i}n} + \dfrac{(1-\theta_i)[1-E[x_i]]E_i^0 E_i^1}{P_{2i}n}\right] \end{array} \right\}$$

$$(4.407)$$

Chiu et al. (2015) developed an algebraic derivation using the optimal cycle length derived as below:

$$T^* = \sqrt{\dfrac{2\sum_{i=1}^{L}\left(K_i + nK_i^S\right)}{\sum_{i=1}^{L}\left(h_i D_i^2\left[\dfrac{1}{D_i} - \dfrac{1}{D_i n} + \dfrac{E_i^0}{P_{1i}n} + \dfrac{\theta_i E_i^0 E_i^1}{P_{1i}n} + \dfrac{(1-\theta_i)E_i^1}{P_{2i}n} + \dfrac{(1-\theta_i)[1-E[x_i]]E_i^0 E_i^1}{P_{2i}n}\right] + \dfrac{h_{1_i}TD_i^2(1-\theta_i)^2\left(E_i^1\right)^2}{2P_{2i}}\right)}}$$

$$(4.408)$$

Example 4.22 Chiu et al. (2015) considered a production schedule is to produce five products in turn on a single machine using a common production cycle policy. Production rate P_{1i} for each product is 58,000, 59,000, 60,000, 61,000, and 62,000, respectively, and annual demands D_i for five different products are 3000, 3200, 3400, 3600, and 3800, respectively. Random defective rates x_i during production uptime for each product follow the uniform distribution over the intervals of [0, 0.05], [0, 0.10], [0, 0.15], [0, 0.20], and [0, 0.25], respectively. Among the defective items, d_i portion is scrap items where D_i for five different products are 0, 0.025, 0.050, 0.075, and 0.100, respectively, and additional disposal costs are $20, $25, $30, $35, and $40 per scrapped item. The other portion of non-conforming items is assumed to be repairable at the reworking rates P_{2i} of 1800, 2000, 2200, 2400, and

4.6 Multi-delivery

2600, respectively, with additional reworking costs of $50, $55, $60, $65, and $70 per reworked item. Other parameters used include:

C_i = Unit manufacturing costs are $80, $90, $100, $110, and $120, respectively.
h_i = Unit holding costs are $10, $15, $20, $25, and $30, respectively.
K_i = Production setup costs are $3800, $3900, $4000, $4100, and $4200, respectively.
h_{1_i} = Unit holding costs per reworked are $30, $35, $40, $45, and $50, respectively.
K_i^S = The fixed delivery costs per shipment are $1800, $1900, $2000, $2100, and $2200.
C_i^T = Unit transportation costs are $0.1, $0.2, $0.3, $0.4, and $0.5, respectively.
n = Number of shipments per cycle, in this study it is assumed to be a constant 4.

The optimal common production cycle time $T^* = 0.6066$ (years) can be computed by Eq. (4.408), and applying Eq. (4.407), one obtains the expected production–inventory–delivery costs per unit time for L products, $E[TCU(T^* = 0.6066)] = \$2,015,921$.

4.6.4 Multi-product Two Machines

Chiu et al. (2018) developed a multi-product two-machine imperfect inventory system with discrete delivery. Assumed L diverse products (where $i = 1, 2, \ldots, L$) sharing a mutual part are to be produced using a two-machine fabrication scheme. Machine one (i.e., the stage 1) solely produces the common parts for all end products at a rate of $P_{1,0}$ (see Fig. 4.34). Then, machine two (i.e., the stage 2) fabricates L diverse products at annual rate of $P_{1,i}$, using a common cycle length strategy (see Fig. 4.35).

Also some new notations which are specifically used for this proposed model are presented in Table 4.26.

The objectives of the production–distribution plan are to meet annual demand rates D_i, shorten fabrication cycle length, and minimize overall relevant costs. The main purpose of this model is to determine the optimal values of number of shipments transported to sales offices per cycle and period length.

Under quality screening, random defective rate xi is observed in both production processes (where $i = 0, 1, 2, \ldots, L$; and $i = 0$ stands for its status of stage 1 when all *common parts* were produced by machine one). Defective items are produced at a rate of $d_{1,i}$. It is assumed that all defective items can be repaired by a follow-up rework process, at a rate of $P_{2,i}$, right after the end of regular production processes (see Figs. 4.34, 4.35, and 4.36). To disallow shortages, this study assumes $(P_{1,i} - d_{1,i} - D_i) > 0$.

The proposed two-machine multi-product fabrication scheme with postponement aims at releasing the production workload of common parts from machine two. Therefore, the proposed scheme should have a more efficient result on fabricating customized end products in the second stage. The proposed solution process starts with determining the optimal common production cycle time for machine two and then applying the obtained cycle length to machine one for production of all common parts in advance (see both Figs. 4.34 and 4.35). The following prerequisite condition must satisfy to ensure that machine two has *sufficient capacity* to fabricate and rework all L products under a common cycle length discipline:

$$\sum_{i=1}^{L}(t_{1,i}+t_{2,i}) < T \quad \text{or} \quad \sum_{i=1}^{L} Q_i \left[\frac{1}{P_{1,i}} + \frac{E[x_i]}{P_{2,i}}\right] < T \quad (4.409)$$

$$\sum_{i=1}^{L} D_i \left[\frac{1}{P_{1,i}} + \frac{E[x_i]}{P_{2,i}}\right] < 1 \quad (4.410)$$

In stage 2, the proposed fabrication must meet demand rate D_i of L diverse product i (where $i = 1, 2, \ldots, L$); the basic formulas displayed in Appendix B of Chiu et al. (2018) can be observed directly from Figs. 4.35, 4.36, 4.37 and 4.38. $TC_2(T, n)$ consists of fabrication variable and setup costs, reworking costs, the fixed and variable transportation costs, and total inventory holding costs for perfect and imperfect items in $t_{1,i}$ and $t_{2,i}$, for reworked items in $t_{2,i}$, for stocks stored at customers' side, and for safety stocks in the production cycle. So, $TC_2(T, n)$ is as follows (Chiu et al. 2018):

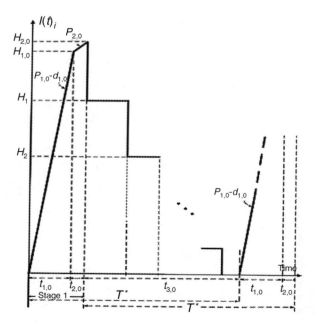

Fig. 4.34 On-hand inventory level of perfect-quality L customized end products in stage 2 of the proposed two-machine two-stage fabrication scheme (Chiu et al. 2018)

4.6 Multi-delivery

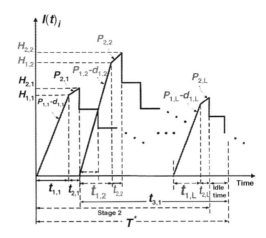

Fig. 4.35 On-hand inventory level of perfect-quality L customized end products in stage 2 of the proposed two-machine two-stage fabrication scheme (Chiu et al. 2018)

Table 4.26 New notations of given problem

i	Index of product
$P_{1,0}$	Annual rate of machine one (solely produces the common parts for all end products)
$P_{1,i}$	Annual rate of machine two (i.e., the stage 2) fabricates L diverse products
$P_{2,i}$	Rework rate of non-conforming items in units per unit time (units per unit time)
K_i^S	Fixed delivery cost per shipment ($/shipment)
C_i^T	Unit shipping cost ($/unit)
$h1,i$	Unit holding cost ($/units per unit time)
$h_{R,i}$	Holding cost per reworked item ($/units per unit time)
$h_{3,i}$	Unit stock holding cost at customer's side ($/units per unit time)
$h_{S,i}$	Safety stock's unit holding cost ($/units per unit time)
Q_i^T	Number of end product i to be transported in each shipment (unit)
t_{1i}	Fabrication time
t_{2i}	The rework time
t_{3i}	Delivery time
n	Number of delivery per cycle
t_{ni}	A fixed interval of time between each installment of finished products delivered
I_i	Leftover items of product i per shipment at the end tni

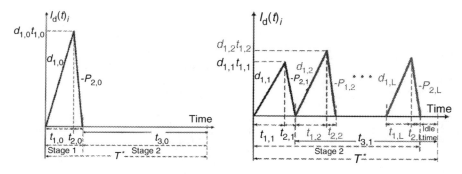

Fig. 4.36 On-hand inventory levels of defective *common intermediate parts* in stage 1 (left) and defective customized end products in stage 2 (right) (Chiu et al. 2018)

$$TC_2(T,n) = \sum_{i=1}^{L} \left\{ \begin{array}{c} \overbrace{C_i Q_i}^{\text{Production Cost}} + \overbrace{K_i}^{\text{Setup Cost}} + \overbrace{C_{R,i}[x_i Q_i]}^{\text{Rework Cost}} + \overbrace{nK_i^S}^{\text{Fixed Delivery Cost}} + \overbrace{C_i^T[Q_i]}^{\text{Delivery Cost}} + h_{R,i} \overbrace{\left[\frac{P_{2,i} t_{2,i}}{2}(t_{2,i})\right]}^{\text{Holding Cost of Reworked Items}} \\ + h_{1,i} \underbrace{\left[\frac{I_{1,i} + D_i t_{1,i}}{2}(t_{1,i}) + \frac{I_{1,i} + I_{2,i}}{2}(t_{1,2,i}) + \frac{n-1}{2n}(I_{2,i} t_{3,i}) + \frac{Q_i}{2} t_{1,i}\right]}_{\text{Holding Cost}} \\ + h_{1,i} \underbrace{\left[\frac{n(Q_i^T - I_i) t_{n,i}}{2} + \frac{n(n+1)}{2} I_i t_{n,i}^n + \frac{nI_i(t_{1,i} + t_{2,i})}{2}\right]}_{\text{Holding Cost at Customer Side}} + \underbrace{h_{S,i} x_i T Q_i}_{\text{Safety Stock Holding Cost}} \end{array} \right\}$$

(4.411)

Substitute Eqs. (B.1) to (B.12) from Appendix B of Chiu et al. (2018) in Eq. (4.411) and take into account randomness of x_i by using the expected values of x, and with further derivation, $E[TCU_2(T, n)]$ can be obtained as follows:

$$E[TCU_2(T,n)] = E[TC_2(T,n)]/E[T]$$

$$= \sum_{i=1}^{L} \left[\begin{array}{c} (C_i + C_i^T) D_i + \dfrac{K_i + nK_i^S}{T} + C_{R,i} D_i E[x_i] + \dfrac{h_{1,i} TD_i^2}{2}\left(\gamma_{1,i} - \dfrac{\gamma_{2,i}}{n}\right) \\ + \dfrac{h_{R,i} TD_i^2 (E[x_i])^2}{2P_{2,i}} + \dfrac{h_{3,i} TD_i^2}{2}\left(\dfrac{1}{\gamma_{1,i}} - \dfrac{E[x_i]}{\gamma_{2,i}} - \dfrac{\gamma_{1,i}}{n}\right) + h_{S,i} E[x_i] TD_i \end{array} \right]$$

(4.412)

$$\gamma_i^1 = \left[\frac{1}{D_i} - \frac{1}{P_{1,i}} - \frac{E[x_i]}{P_{2,i}}\right]$$

Fig. 4.37 On-hand inventory level of common parts waiting to be fabricated into customized end products in stage 2 of the proposed study (Chiu et al. 2018)

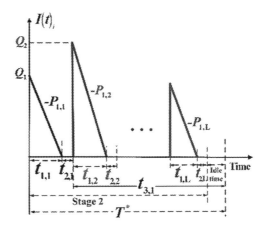

$$\gamma_i^2 = \left[\frac{1}{D_i} - \frac{E[x_i]^2}{P_{2,i}} + \frac{1}{P_{1,i}} + \frac{E[x_i]}{P_{2,i}}\right] \quad (4.413)$$

In stage 1, machine one has to make enough *common parts* in advance for the fabrication of L diverse end products. Hence, machine one must start producing common parts $(t_{1,0} + t_{2,0})$ ahead of time (see Fig. 4.35). The basic formulas displayed in Appendix C of Chiu et al. (2018) can also be observed directly from Figs. 4.35, 4.37, and 4.38. Similarly, machine one must have sufficient capacity to produce and rework all common intermediate parts. That is, the following prerequisite condition must satisfy (Chiu et al. 2018):

$$(t_{1,0} + t_{2,0}) < T \quad \text{or} \quad \left[\left(\frac{Q_0}{P_{1,0}} + \frac{E[x_0]}{P_{2,0}}\right)\right] < T \quad (4.414)$$

$$\left[\frac{D_0}{P_{1,0}} + \frac{E[x_0]D_0}{P_{2,0}}\right] < 1 \quad (4.415)$$

Total relevant fabrication costs per cycle for stage 1, $TC_1(T, n)$ consists variable production cost, setup cost, reworking cost, and total inventory holding costs for perfect and imperfect items in $t_{1,0}$ and $t_{2,0}$, for reworked items in $t_{2,0}$, and for the safety stocks. So, $TC_1(T, n)$ is (Chiu et al. 2018):

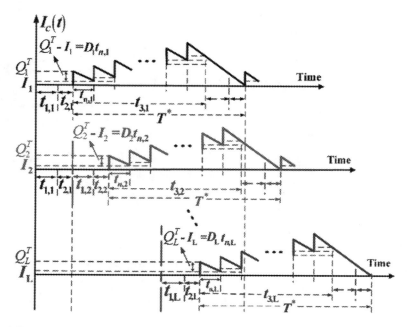

Fig. 4.38 On-hand inventory level of customized end products at the customer's side during the production cycle (Chiu et al. 2018)

$$\text{TC}_1(T,n) = \sum_{i=1}^{L} \left\{ \begin{array}{l} C_0 Q_0 + K_0 + C_{R_0}[x_0 Q_0] + h_{R,0}\left[\dfrac{D_{1,0} t_{1,0}}{2}(t_{2,0})\right] \\ + h_{1,0}\left[\dfrac{I_{1,0} + D_0 t_{1,0}}{2}(t_{1,0}) + \dfrac{I_{1,0} + I_{2,0}}{2}(t_{2,0}) + \sum_i I_i(t_{1,i} + t_{2,i})\right] + h_{S,0} x_0 T Q_0 \end{array} \right\} \tag{4.416}$$

Substitute previous in Eq. (4.416) and take into account the random defective rate x_0 by using the expected values of x_0, and with further derivation, $E[\text{TCU}_1(T,n)]$ can be derived as follows (Chiu et al. 2018):

$$E[\text{TCU}_1(T,n)] = E[\text{TC}_1(T,n)]/E[T] \tag{4.417}$$

Therefore, total relevant cost per unit time for the proposed study, $E[\text{TCU}(T,n)]$, is (Chiu et al. 2018):

4.6 Multi-delivery

$$E[\text{TCU}(T,n)] = E[\text{TCU}_1(T,n)] + E[\text{TCU}_2(T,n)] \tag{4.418}$$

To determine the optimal production–distribution policy, one must first prove the convexity of $E[\text{TCU}2(T, n)]$. Hessian matrix equations are employed to show if the following condition holds (see Appendix D of Chiu et al. 2018 for details). Since $[T,n] H \begin{bmatrix} T \\ n \end{bmatrix} = \sum_{i=1}^{L} \frac{2K_i}{T} \geq 0$, setting the partial derivative equal to zero gives:

$$T^* = \sqrt{\frac{\sum_{i=1}^{L} (K_i + n K_i^S)}{\sum_{i=1}^{L} \left\{ \frac{h D_i^2}{2} \left\{ \gamma_i^1 - \frac{\gamma_i^1}{n} \right\} + \frac{h_{R,i} D_i^2 (E[x_i])^2}{2 P_{2,i}^2} + \frac{h_{3,i} D_i^2}{2} \left[\frac{1}{P_{1,i}} + \frac{E[x_i]}{P_{2,i}} + \frac{\gamma_i^1}{n} \right] + h_{S,i} D_i E[x_i] \right\}}}$$

(4.419)

$$n^* = \sqrt{\frac{\left(\sum_{i=1}^{L} K_i\right) \sum_{i=1}^{L} \left[\frac{D_i^2}{2} (h_{3,i} - h_{1,i}) \gamma_i^1\right]}{\left(\sum_{i=1}^{L} K_i^S\right) \left\{\sum_{i=1}^{L} \frac{h_{1,i} D_i^2}{2} \gamma_i^2 + \frac{h_{R,i} D_i^2 (E[x_i])^2}{2 P_{2,i}^2} + \frac{h_{3,i} D_i^2}{2} \left[\frac{1}{P_{1,i}} + \frac{E[x_i]}{P_{2,i}} + \frac{\gamma_i^1}{n}\right] + h_{S,i} D_i E[x_i] \right\}}}$$

(4.420)

Example 4.23 Chiu et al. (2018) presented an example using the following values for parameters in stage 1: $K_0 = \$8500$, $C_0 = \$40$, $C_{R,0} = \$25$, $h_{1,0} = h_{S,0} = \$5$, $h_{R,0} = \$15$, and x_0 uniformly over the range [0, 0.04]. Consequently, the following values for parameters in stage 2, $C_i = \$80, \$70, \$60, \$50$, and $\$40$; $C_{R,i} = \$45, \$40, \$35, \$30$, and $\$25$; x_i over the ranges [0,0.21], [0, 0.16], [0, 0.11], [0, 0.06], and [0, 0.01]; and $K_i = \$10{,}500, \$10{,}000, \$9500, \$9000$, and $\$8500$, respectively, are considered. $P_{1,i} = 128{,}276, 124{,}068, 120{,}000, 116{,}066$, and $112{,}258$ (which also are based on the similar $1/\alpha$ relationship between $P_{1,i}$ and $P_{1,0}$; i.e., $P_{1,i} = 1/(1/P_{1,i} - 1/P_{1,0})$) and $P_{2,i} = 102{,}621, 99{,}254, 96{,}000, 92{,}852$, and $89{,}806$ (similarly they are calculated by $P_{2,i} = 1/(1/P_{2,i} - 1/P_{2,0})$), respectively. Also, $K_i^S = \$2200, \$2100, \$2000, \$1900$, and $\$1800$; $h_{1,i} = \$30, \$25, \$20, \$15$, and $\$10$; $h_{R,i} = \$50, \$45, \$40, \$35$, and $\$30$; $C_i^T = \$0.5, \$0.4, \$0.3, \$0.2$, and $\$0.1$; $h_{3,i} = \$90, \$85, \$80, \$75$, and $\$70$; and $h_{S,i} = \$30, \$25, \$20, \$15$, and $\$10$, respectively. Finally the optimal fabrication-distribution decisions $n^* = 3$, $T^* = 0.4453$, and $E[\text{TCU}(T^*, n^*)] = \$2{,}145{,}825$ can be obtained.

4.6.5 Shipment Decisions for a Multi-product

Chiu et al. (2016) developed an imperfect inventory system to simultaneously determine the production and shipment decisions for a multi-item vendor–buyer integrated inventory system with a rework process. Fabricating multi-products on a single machine with the aim of maximizing machine utilization is an operating goal of most manufacturing firms. In the proposed multi-product intra-supply chain system, the production rate is P_{1i} per year and the annual demand rate is D_i, where $i = 1, 2, \ldots, L$. All products made are checked for their quality, and the unit screening cost is included in the unit production cost C_i. It is also assumed that the production process can randomly produce x_i portion of non-conforming items at a rate d_i, where d_i can be expressed as $d_i = x_i P_{1i}$, and $(P_{1i} - d_i - D_i) > 0$ must be satisfied in order to sustain regular operations (i.e., avoid the occurrence of shortage). All defective items produced are reworked and fully repaired at the rate of P_{2i} at the end of each production cycle, with additional rework cost C_{Ri} per item. After the rework process, the entire quality assured lot of each product i is transported to sales offices/customers under a multi-delivery policy, in which n fixed quantity installments of the lot are shipped at fixed intervals of time in t_{3i} (Chiu et al. 2016). The schematic process of above description is presented in Figs. 4.39 and 4.40.

Also some new notations which are specifically used for this proposed model are presented in Table 4.26 in Sect. 4.6.4.

The on-hand inventory of product i stored at the sales offices/customers' side is illustrated in Fig. 4.41. Accordingly, the sales offices' holding costs along with delivery cost for all L products are included in the proposed cost analysis (Chiu et al. 2016).

Using Figs. 4.39 and 4.41, the following formulas can be obtained (Chiu et al. 2016):

$$H_{1i} = (P_{1i} - d_i)t_{1i} \quad (4.421)$$

$$H_{2i} = H_{1i} + P_{2i}t_{2i} \quad (4.422)$$

$$t_{1i} = \frac{Q_i}{P_{1i}} \quad (4.423)$$

$$t_{2i} = \frac{x_i Q_i}{P_{2i}} \quad (4.424)$$

$$t_{3i} = nt_{ni} = T - (t_{1i} + t_{2i}) \quad (4.425)$$

$$T = t_{1i} + t_{2i} + t_{3i} \quad (4.426)$$

$$d_i t_{1i} = x_i Q_i \quad (4.427)$$

Total delivery cost of n shipments of product i at t_{3i} is (Chiu et al. 2016):

$$nK_i^S + C_i^T Q_i \quad (4.428)$$

From Fig. 4.28, the holding cost of the finished items of product i at t_3 is (Chiu et al. 2016):

4.6 Multi-delivery

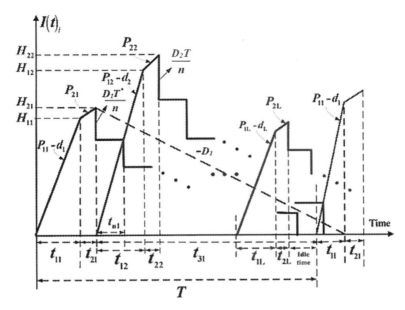

Fig. 4.39 On-hand inventory level of perfect-quality product i at time t in the proposed system (Chiu et al. 2016)

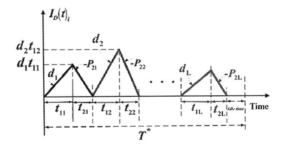

Fig. 4.40 On-hand inventory level of defective product i at time t in the proposed system (Chiu et al. 2016)

$$h_i\left(\frac{n-1}{2n}\right)H_{2i}t_{3i} \qquad (4.429)$$

According to the proposed multi-delivery policy, when n fixed quantity (i.e., D) installments of finished lot of product i are transported to sales offices at a fixed time interval t_{ni}, the following formulas are obtained (Chiu et al. 2016):

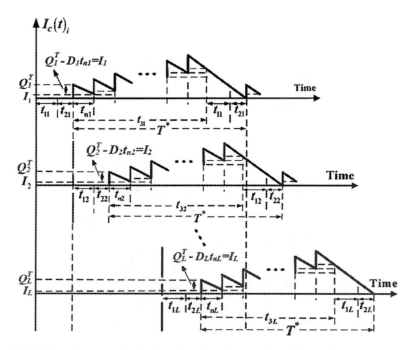

Fig. 4.41 On-hand inventory level of product i stored at the sales offices at time t in the proposed system (Chiu et al. 2016)

$$t_{ni} = \frac{t_{3i}}{n} \quad (4.430)$$

$$Q_i^T = \frac{H_{2i}}{n} \quad (4.431)$$

$$I_i = Q_i^T - D_i t_{ni} \quad (4.432)$$

The sales offices' stock holding cost of product i is (Chiu et al. 2016):

$$h_{2i}\left[n\frac{Q_i^T - I_i}{2}t_{ni} + \frac{nI_i}{2}(t_{1i} + t_{2i}) + \frac{n(n+1)}{2}I_i t_{ni}\right] \quad (4.433)$$

Therefore, TC(Q_i, n) for $i = 1, 2, \ldots, L$, which comprises the variable fabrication cost; setup cost; variable reworking cost; production units' inventory holding cost during the periods t_{1i}, t_{2i}, and t_{3i} (including holding cost of non-conforming items in

4.6 Multi-delivery

t_{1i}); inventory holding cost of reworked items in t_{2i}; fixed and variable transportation costs; and the stock holding cost from the sales offices/customers, is:

$$\sum_{i=1}^{L} \text{TC}(Q_i, n) = \sum_{i=1}^{L} \left\{ \begin{array}{l} \overbrace{C_i Q_i}^{\text{Production Cost}} + \overbrace{K_i}^{\text{Setup Cost}} + \overbrace{C_{R_i}[x_i Q_i]}^{\text{Rework Cost}} + \overbrace{nK_i^S}^{\text{Fixed Delivery Cost}} + \overbrace{C_i^T[Q_i]}^{\text{Shipment Cost}} + \overbrace{h_{1i}\left[\frac{d_{1i}t_{1i}}{2}(t_{2i})\right]}^{\text{Holding Cost of Reworked Items}} \\ + h_i \underbrace{\left[\frac{H_{1i} + d_i t_{1i}}{2}(t_{1i}) + \frac{H_{1i} + H_{2i}}{2}(t_{2i}) + \frac{n-1}{2n}(H_{2i}t_{3i})\right]}_{\text{Holding Cost}} \\ + h_{2i} \underbrace{\left[\frac{n(Q_i^T - I_i)t_{ni}}{2} + \frac{n(n+1)}{2}I_i t_{ni} + \frac{nI_i(t_{1i} + t_{2i})}{2}\right]}_{\text{Sales Offices' Stock Holding}} \end{array} \right\}$$

(4.434)

Substituting relevant parameters from Eq. (4.433) in Eq. (4.434), using the expected values of x to take randomness of defective rate into account, and applying the renewal reward theorem, $E[\text{TCU}(Q_i, n)]$ is obtained as follows (Chiu et al. 2016):

$$E[\text{TCU}(Q_i, n)] = \sum_{i=1}^{L} \left\{ \begin{array}{l} C_i D_i + \frac{K_i D_i}{Q_i} + C_{R_i}[E(x_i) D_i] + nK_i^S \frac{D_i}{Q_i} + C_i^T[D_i] + h_{1i}\left[\frac{Q_i D_i}{2P_{2i}}E(x_i)^2\right] \\ + \frac{h_i Q_i D_i}{2}\left[\frac{-1}{D_i n} + \frac{1}{P_{1i} n} + \frac{E(x_i)}{nP_{2i}} + \frac{1}{D_i} + \frac{E(x_i)}{P_{2i}} + \frac{E(x_i)^2}{P_{2i}}\right] \\ + \frac{h_{2i} Q_i D_i}{2}\left[\left[\frac{1}{D_i n} - \frac{1}{P_{1i} n} - \frac{E(x_i)}{nP_{2i}} + \frac{1}{P_{1i}} + \frac{E(x_i)}{P_{2i}}\right]\right] \end{array} \right\}$$

(4.435)

Since $Q_i = TD_i$,

$$E[\text{TCU}(T, n)] = \sum_{i=1}^{L}$$

$$\times \left\{ \begin{array}{l} C_i D_i + \frac{K_i}{T} + C_{R_i}[E(x_i) D_i] + \frac{nK_i^S}{T} + C_i^T[D_i] + h_{1i}\left[\frac{TD_i^2}{2P_{2i}}(E(x_i))^2\right] \\ + \frac{h_i TD_i^2}{2}\left[\frac{-1}{D_i n} + \frac{1}{P_{1i} n} + \frac{E(x_i)}{nP_{2i}} + \frac{1}{D_i} + \frac{E(x_i)}{P_{2i}} + \frac{(E(x_i))^2}{P_{2i}}\right] \\ + \frac{h_{2i} TD_i^2}{2}\left[\left[\frac{1}{D_i n} - \frac{1}{P_{1i} n} - \frac{E(x_i)}{nP_{2i}} + \frac{1}{P_{1i}} + \frac{E(x_i)}{P_{2i}}\right]\right] \end{array} \right\}$$

(4.436)

Chiu et al. (2016) used Hessian matrix to prove the convexity of the total cost function. Since $[T, n]H\begin{bmatrix}T\\n\end{bmatrix} = \sum_{i=1}^{L} \frac{2K_i}{T} \geq 0$ therefore, $E[\text{TCU}(T, n)]$ is strictly convex for all T and n not equal to zero, and $E[TCU(T, n)]$ has a minimum value. Then they set the first derivatives of $E[TCU(T, n)]$ with respect to T and with respect to n equal to zeros and solved the linear system and derived:

$$T^* = \sqrt{\frac{2\sum_i (K_i + nK_i^S)}{\sum_i \left\{ h_i D_i^2 \left\{ \frac{1}{D_i} + \frac{E[x_i]}{P_{2i}} - \frac{E[x_i]^2}{P_{2i}} \right\} + h_2 D_i^2 \left[\frac{1}{P_{1i}} + \frac{E[x_i]}{P_{2i}} \right] + \frac{h_{1i} D_i^2 E[x_i]^2}{P_{2i}} + \frac{D_i^2}{n} \left[\frac{1}{D_i} - \frac{1}{P_{1i}} - \frac{E[x_i]}{P_{2i}} \right] (h_{2i} - h_i) \right\}}} \tag{4.437}$$

$$n^* = \sqrt{\frac{2\sum_i (K_i) \cdot \sum_{i=1}^{L} D_i^2 \left[\frac{1}{D_i} - \frac{1}{P_{1i}} - \frac{E[x_i]}{P_{2i}} \right] (h_{2i} - h_i)}{\left(\sum_{i=1}^{L} K_i^S \right) \cdot \sum_{i=1}^{L} \left\{ h_i D_i^2 \left\{ \frac{1}{D_i} + \frac{E[x_i]}{P_{2i}} - \frac{E[x_i]^2}{P_{2i}} \right\} + h_2 D_i^2 \left[\frac{1}{P_{1i}} + \frac{E[x_i]}{P_{2i}} \right] + \frac{h_{1i} D_i^2 E[x_i]^2}{P_{2i}} \right\}}} \tag{4.438}$$

Example 4.24 Chiu et al. (2016) considered a five-product example being manufactured in sequence on a machine under the common cycle time policy in a multi-product inventory system with a rework process. Their annual production rates P_{1i} are 58,000, 59,000, 60,000, 61,000, and 62,000, respectively, and their annual demand rates D_i are 3000, 3200, 3400, 3600, and 3800, respectively. For each product, the production units have experienced the random non-conforming rates that follow the uniform distribution over intervals of [0, 0.05], [0, 0.10], [0, 0.15], [0, 0.20], and [0, 0.25], respectively. All non-conforming products are assumed to be repairable and are reworked at the end of the regular production, at annual rates P_{2i} of 46,400, 47,200, 48,000, 48,800, and 49,600, respectively. Additional costs for rework are $50, $55, $60, $65, and $70 per non-conforming product, respectively. Other values of system variables used in this example are listed below:

K_i = $17,000, $17,500, $18,000, $18,500, and $19,000.
C_i = $80, $90, $100, $110, and $120.
h_i = $10, $15, $20, $25, and $30.
h_{1i} = $30, $35, $40, $45, and $50.
K_i^S = $1800, $1900, $2000, $2100, and $2200.
h_{2i} = $70, $75, $80, $85, and $90.
C_i^T = $0.1, $0.2, $0.3, $0.4, and $0.5.

First, in order to determine the number of deliveries, using Eq. (4.437), $n^* = 4.4278$. Practically, n^* should be an integer number only, and to find the integer value of n^*, one can plug $n^+ = 5$ and $n^- = 4$ in Eq. (4.436) and obtain ($T = 0.6666$, $n^+ = 5$) and ($T = 0.6193$, $n^- = 4$). Then, using Eq. (4.435) with these two different values to obtain $E[TCU(0.6666, 5)] = \$2{,}229{,}865$ and $E[TCU(0.6193, 4)] = \$2{,}229{,}658$, respectively. By choosing a policy with minimum cost, the optimal production–shipment policy for the proposed system is determined as $n^* = 4$, $T^* = 0.6193$, and $E[TCU(T^*, n^*)] = \$2{,}229{,}658$.

4.6 Multi-delivery

4.6.6 Pricing with Rework and Multiple Shipments

Consider a situation in which a manufacturing system produces perfect and defective items. The perfect items are ready to cover the customer's demand. On the other hand, the defective items can be reworked after finishing the regular production process. At the end of the rework process, the manufacturer will deliver n equal size shipments to the customers during a specific time such that time between two consecutive deliveries during production downtime is equal (see Fig. 4.42). In addition to the aforementioned, it is important to point out that the manufacturing system randomly produces an x portion of defective items with a production rate d. Consequently, the production rate of defective items d can be expressed as $d = Px$. All items manufactured are screened, and inspection cost per item is included in the unit manufacturing cost. Furthermore, it is assumed that all defective items are reworkable at the end of the regular production. The reworkable items are recovered at a rate of P_1 in each cycle. Nonetheless, during the rework process, a θ_1 portion of reworked items fails and becomes scrap. If d_1 represents the production rate of scrap items during the rework process, then d_1 is calculated as $d_1 = p_1\theta_1$. The behavior of inventory level through time of proposed manufacturing problem is shown in Fig. 4.41. In the next section, the mathematical inventory model of manufacturing problem is presented (Taleizadeh et al. 2016).

This problem reexamines the research work of Chiu et al. (2014). According to Fig. 4.42, the following equations can be derived:

$$H_1 = (P - d)t_1 = (1 - x)Q \quad (4.439)$$

$$H = H_1 + (P_1 - d_1)t_2 = (1 - \theta_1 x)Q \quad (4.440)$$

$$t_1 = \frac{Q}{P} \quad (4.441)$$

$$t_2 = \frac{xQ}{P_1} \quad (4.442)$$

$$t_3 = T - t_1 - t_2 = Q\left(\frac{(1 - \theta_1 x)}{D} - \frac{1}{P} - \frac{x}{P_1}\right) \quad (4.443)$$

$$dt_1 = Pxt_1 = xQ \quad (4.444)$$

$$\theta_1 dt_1 = \theta_1 Pxt_1 = \theta_1 xQ \quad (4.445)$$

Thus, the profit function for each cycle production is given by (Taleizadeh et al. 2016):

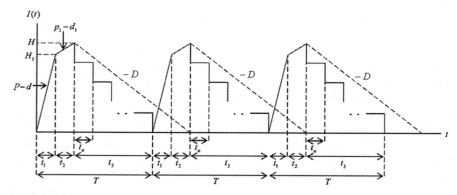

Fig. 4.42 Inventory level of perfect-quality items in the EPQ model with a multi-delivery policy and rework (Taleizadeh et al. 2016)

$$CP(s,Q) = \overbrace{sQ}^{Sale} - \begin{bmatrix} \overbrace{CQ}^{Production\,Cost} + \overbrace{K}^{Setup\,Cost} + \overbrace{C_R[xQ]}^{Rework\,Cost} + \overbrace{nK_S}^{Fixed\,Delivery\,Cost} + \overbrace{C_d[Qx\theta_1]}^{Disposal\,Cost} + \overbrace{C_T Q(1-x\theta_1)}^{Delivery\,Cost} \\ + \underbrace{h_1\left[\frac{P_1 t_2}{2}(t_2)\right]}_{Holding\,Cost\,of\,Reworked\,Items} + \underbrace{h\left[\frac{H_1+Dt_1}{2}(t_1) + \frac{H_1+H}{2}(t_2) + h\frac{n-1}{2n}(Ht_3)\right]}_{Holding\,Cost} \end{bmatrix}$$

(4.446)

Interested readers may see in Chiu et al. (2014) the details of the derivation of the second term in the above profit function. Dividing by T, the annual profit function is obtained as follows:

$$AP(s,Q) = \frac{CP(s,Q)}{T}$$
$$= \frac{sQ}{T} - \frac{1}{T} \times \begin{bmatrix} CQ + K + C_R[xQ] + C_d[Qx\theta_1] + nK_S + C_T Q(1-x\theta_1) \\ + h_1\left[\frac{P_1 t_2}{2}(t_2)\right] + h\left[\frac{H_1+Dt_1}{2}(t_1) + \frac{H_1+H}{2}(t_2)\right] + h\frac{n-1}{2n}(Ht_3) \end{bmatrix}$$

(4.447)

To incorporate the price in the above inventory model, the demand is considered as a linear function of price, and it is expressed as $D = a - bs$. Substituting D with $a - bs$ and $E(T)$ with $Q(1 - \theta_1 E(X))$, then the average annual profit becomes:

4.6 Multi-delivery

$$E[AP(s,Q)] = \frac{E[CP(s,Q)]}{E[T]}$$

$$= \frac{s(a-bs)}{1-\theta_1 E(x)} - \begin{bmatrix} \frac{(a-bs)}{1-\theta_1 E(x)}\left(C + \frac{K+nK_S}{Q} + C_R E[x] + C_d\theta_1 E[x] + h_1\left[\frac{E[x]^2 Q}{2P_1}\right]\right) \\ + \frac{hQ(a-bs)}{2P(1-\theta_1 E(x))} + \frac{hQ(a-bs)}{2P_1(1-\theta_1 E(x))}\left(2E(x) - E(x)^2 - \theta_1 E(x)^2\right) \\ + hQ\frac{n-1}{2n}\left(\frac{1-\theta_1 E(x)}{2} - \frac{a-bs}{2P} - \frac{E(x)(a-bs)}{2P_1}\right) + C_T Q(a-bs) \end{bmatrix}$$

(4.448)

From calculus, it is well-known that $E[AP(S,Q)]$ is concave if and only if the following equations are satisfied:

$$\frac{\partial^2 E[AP(s,Q)]}{\partial Q^2} = \frac{-2(K+nK_S)(a-bs)}{Q^3(1-\theta_1 E(X))} < 0 \quad (4.449)$$

$$\frac{\partial^2 E[AP(s,Q)]}{\partial s^2} = \frac{-2b}{1-\theta_1 E(X)} < 0 \quad (4.450)$$

$$[Q \quad s]\begin{bmatrix} \frac{\partial^2 E[AP(s,Q)]}{\partial Q^2} & \frac{\partial^2 E[AP(s,Q)]}{\partial Q \partial s} \\ \frac{\partial^2 E[AP(s,Q)]}{\partial Q \partial s} & \frac{\partial^2 E[AP(s,Q)]}{\partial s^2} \end{bmatrix}\begin{bmatrix} Q \\ s \end{bmatrix} < 0 \quad (4.451)$$

Equation (4.11) can be changed to Eq. (4.12) as below:

$$\left[Q^2\frac{\partial^2 E[AP(s,Q)]}{\partial Q^2} + 2Qs\frac{\partial^2 E[AP(s,Q)]}{\partial Q \partial s} + s^2\frac{\partial^2 E[AP(s,Q)]}{\partial s^2}\right] < 0 \quad (4.452)$$

The elements of Eq. (4.451) are derived as below:

$$\frac{\partial E(AP(s,Q))}{\partial Q} = (a-bs)\begin{pmatrix} \frac{K+nK_S}{Q^2(1-\theta_1 E(X))} - \frac{h}{2p_1(1-\theta_1 E(X))}\left(2E(X) - [E(x)]^2 - \theta_1[E(x)]^2\right) \\ -\frac{h_1[E(x)]^2}{2P_1(1-\theta_1 E(X))} + \left(\frac{n-1}{n}\right)\left(\frac{h(1-\theta_1 E(X))}{(a-bs)2} - \frac{h}{2P} - \frac{hE(X)}{2P_1}\right) \\ -\frac{h}{2P(1-\theta_1 E(X))} \end{pmatrix}$$

(4.453)

$$\frac{\partial E(\mathrm{AP}(s,Q))}{\partial s} = \frac{a-2bs}{1-\theta_1 E(X)}$$

$$+ b \begin{bmatrix} \left(C + \frac{(K+nK_S)}{Q} + C_R E(X) + C_d E(X)\theta_1 \right) \frac{1}{1-\theta_1 E(X)} \\ + \frac{hQ}{2P(1-\theta_1 E(X))} + \frac{hQ\left(2E(X)-[E(x)]^2 - \theta_1[E(x)]^2\right)}{2P_1(1-\theta_1 E(X))} \\ + \left(\frac{n-1}{n}\right)\left(\frac{-hQ}{2P} - \frac{hQE(X)}{2P_1}\right) + \frac{h_1[E(x)]^2 Q}{2P_1(1-\theta_1 E(X))} + C_T \end{bmatrix}$$

(4.454)

Thus,

$$\frac{\partial^2 E[\mathrm{AP}(s,Q)]}{\partial Q^2} = \frac{-2(K+nK_S)(a-bs)}{Q^3(1-\theta_1 E(X))} < 0 \qquad (4.455)$$

$$\frac{\partial^2 E[\mathrm{AP}(s,Q)]}{\partial s^2} = \frac{-2b}{1-\theta_1 E(X)} < 0 \qquad (4.456)$$

$$\frac{\partial^2 E[\mathrm{AP}(s,Q)]}{\partial Q \partial s} = \begin{pmatrix} \frac{-(K+nK_S)}{Q^2} + \frac{h}{2P} + \frac{h\left(2E(X)-[E(x)]^2-\theta_1[E(x)]^2\right)}{2P_1} + \\ \left(\frac{n-1}{n}\right)\left(\frac{-h}{2P} - \frac{hE(X)}{2P_1}\right)(1-\theta_1 E(X)) + \frac{h_1[E(x)]^2}{2P_1} \end{pmatrix}$$
$$\times \left(\frac{b}{(1-\theta_1 E(X))}\right)$$

(4.457)

If $\frac{\partial^2 E[\mathrm{AP}(s,Q)]}{\partial Q \partial s} < 0$, then $E[\mathrm{AP}(s,Q)]$ is strictly concave, because both $\frac{\partial^2 E[\mathrm{AP}(s,Q)]}{\partial s^2}$ and $\frac{\partial^2 E[\mathrm{AP}(s,Q)]}{\partial Q^2}$ are negative too. Moreover $\frac{\partial^2 E[\mathrm{AP}(s,Q)]}{\partial Q \partial s}$ is negative only if:

$$Q < Q_{\mathrm{Upper}}$$

$$= \sqrt{\frac{2(K+nK_S)}{\frac{h}{P} + \frac{h\left(2E(X)-[E(x)]^2-\theta_1[E(x)]^2\right)}{P_1} - \left(\frac{n-1}{n}\right)\left(\frac{h}{P}+\frac{hE(X)}{P_1}\right)(1-\theta_1 E(X))}}$$
$$+ \frac{h_1[E(x)]^2}{P_1}$$

(4.458)

4.6 Multi-delivery

Now, setting the first derivative of the objective function respect to Q, shown in Eq. (4.459), equal to zero gives:

$$Q^* = \sqrt{\dfrac{2(K+nK_S)}{\dfrac{h}{P}+\dfrac{h}{P_1}\left(2E(X)-[E(x)]^2-\theta_1[E(x)]^2\right)+\left(\dfrac{n-1}{n}\right)\left[\dfrac{h(1-\theta_1 E(X))^2}{(a-bs)}-h\left(\dfrac{1}{P}+\dfrac{E(X)}{P_1}\right)\times(1-\theta_1 E(X))\right]+\dfrac{h_1[E(x)]^2}{P_1}}}$$

(4.459)

It is worth remarking that the lot size Q^* is strictly less than Q_{Upper}. Therefore, the objective function is always concave. Then the optimal value of s is calculated as follows:

$$s^* = \dfrac{a}{b} - \dfrac{h\left(\dfrac{n-1}{n}\right)(1-\theta_1 E(X))^2 Q^{*2}}{2b(K+nK_S)-bQ^{*2}\left[\dfrac{h}{P}+\dfrac{h_1[E(x)]^2}{P_1}+\dfrac{h\left[2E(X)-[E(x)]^2-\theta_1[E(x)]^2\right]}{P_1}-h\left(\dfrac{n-1}{n}\right)(1-\theta_1 E(X))\left(\dfrac{1}{P}+\dfrac{E(X)}{P_1}\right)\right]}$$

(4.460)

4.6.6.1 Solution Procedure

Since Q is a function of s and s is a function of Q, then the following heuristic algorithm is proposed to obtain the solutions to the lot size and selling price.

Step 1: Set selling price equal to zero, and determine the upper level for the replenishment lot size Q_{Upper} using Eq. (4.458).
Step 2: Put $Q = 1$; calculate s and the long-run average profit with Eqs. (4.460) and (4.448), respectively.
Step 3: Put $Q = Q + 1$, calculate s and the long-run average profit with Eqs. (4.481) and (4.448), respectively.
Step 4: If $Q < Q_{\text{Upper}}$, then go to *Step 3*; otherwise go to *Step 5*.
Step 5: Select the solution with the maximum value of long-run average profit, and report the lot size and the selling price (Q and s).

Example 4.25 In this section, a numerical example is presented by Taleizadeh et al. (2016) to illustrate the inventory model and its solution through the proposed heuristic algorithm. Suppose that an item can be produced with a rate of 10,000 units per year with price sensitive demand function given by $D = 7000 - 250s$.

The random defective rate x is assumed to follow a uniform distribution between [0, 0.2] during the production uptime. In the next stage, all defective items are reworked at a rate of $p_1 = 1200$, but 10% of reworkable items is scrap; i.e., $\theta_1 = 0.1$. Other parameters are $K = \$15,000$ per production run; $K_S = \$3900$ per shipment, a fixed cost; $h_1 = \$35$ per unit per unit time; $h = \$20$ per item per unit time; $C = \$10$ per item; $C_R = \$5$ per item; $C_d = \$2$ per item; $C_T = \$0.5$ per item; and $n = 4$ shipments per cycle.

To solve this problem, it is required to apply the proposed heuristic algorithm that was developed in the previous section. *Step* 1, the upper level for the replenish lot size is $Q_{upper} = 2441$. *Step* 2, *Step* 3, and *Step* 4 for $Q = 1$ until $Q = 2441$, the selling price s and the long-run average profit are determined. *Step* 5, it is obtained that the maximum value of total profit is AT $P(s, Q) = 24,440$ such that the replenishment lot size is $Q^* = 2202$ and the selling price is $s^* = 14.812223$.

References

Aggarwal, K. K., & Aneja, S. (2016). An EOQ model with inspection error, rework and sales return. *International Journal of Advanced Operations Management, 8*(3), 185–199.

Alizadeh-Basbam, N., & Taleizadeh, A. A. (2020). A hybrid circular economy—Game theoretical approach in a dual-channel green supply chain considering sale's effort, delivery time, and hybrid re-manufacturing. *Journal of Cleaner Production, 250*, 119521.

Ben-Daya, M. (1999). Integrated production maintenance and quality model for imperfect processes. *IIE Transactions, 31*(6), 491–501.

Ben-Daya, M. (2002). The economic production lot-sizing problem with imperfect production processes and imperfect maintenance. *International Journal of Production Economics, 76*(3), 257–264.

Cárdenas-Barrón, L. E. (2009). Economic production quantity with rework process at a single-stage manufacturing system with planned backorders. *Computers & Industrial Engineering, 57*(3), 1105–1113.

Cárdenas-Barrón, L. E., Taleizadeh, A. A., & Treviño-Garza, G. (2012). An improved solution to replenishment lot size problem with discontinuous issuing policy and rework, and the multi-delivery policy into economic production lot size problem with partial rework. *Expert Systems with Applications, 39*(18), 13540–13546.

Cárdenas-Barrón, L. E., Treviño Garza, G., Taleizadeh, A. A., & Vasant, P. (2015). Determining replenishment lot size and shipment policy for an EPQ inventory model with delivery and rework. *Mathematical Problems in Engineering, 2015*, 595498, 8 p. https://doi.org/10.1155/2015/595498.

Chan, W. M., Ibrahim, R. N., & Lochert, P. B. (2003). A new EPQ model: integrating lower pricing, rework and reject situations. *Production Planning & Control, 14*(7), 588–595.

Chand, S. (1989). Lot sizes and setup frequency with learning in setups and process quality. *European Journal of Operational Research, 42*(2), 190–202.

Chen, J. M., Lin, Y. H., & Chen, Y. C. (2010). Economic optimisation for an imperfect production system with rework and scrap rate. *International Journal of Industrial and Systems Engineering, 6*(1), 92–109.

Cheng, T. C. E. (1989). An economic production quantity model with flexibility and reliability considerations. *European Journal of Operational Research, 39*(2), 174–179.

Cheng, T. C. E. (1991). An economic order quantity model with demand-dependent unit production cost and imperfect production processes. *IIE Transactions, 23*(1), 23–28.

References

Chiu, S. W., Gong, D. C., & Wee, H. M. (2004). Effects of random defective rate and imperfect rework process on economic production quantity model. *Japan Journal of Industrial and Applied Mathematics, 21*(3), 375.

Chiu, S. W., Chiu, Y. S. P., & Yang, J. C. (2012). Combining an alternative multi-delivery policy into economic production lot size problem with partial rework. *Expert Systems with Applications, 39*(3), 2578–2583.

Chiu, S. P., & Chiu, Y. P. (2003). An economic production quantity model with the rework process of repairable defective items. *Journal of Information and Optimization Sciences, 24*(3), 569–582.

Chiu, S. W. (2007). Optimal replenishment policy for imperfect quality EMQ model with rework and backlogging. *Applied Stochastic Models in Business and Industry, 23*(2), 165–178.

Chiu, S. W., Tseng, C. T., Wu, M. F., & Sung, P. C. (2014). Multi-item EPQ model with scrap, rework and multi-delivery using common cycle policy. *Journal of Applied Research and Technology, 12*(3), 615–622.

Chiu, Y. P. (2003). Determining the optimal lot size for the finite production model with random defective rate, the rework process, and backlogging. *Engineering Optimization, 35*(4), 427–437.

Chiu, Y. P., Chiang, K. W., Chiu, S. W., & Song, M. S. (2016). Simultaneous determination of production and shipment decisions for a multi-product inventory system with a rework process. *Advances in Production Engineering & Management, 11*(2), 141–151.

Chiu, Y. S., Wu, M. F., Chiu, S. W., & Chang, H. H. (2015). A simplified approach to the multi-item economic production quantity model with scrap, rework, and multi-delivery. *Journal of Applied Research and Technology, 13*(4), 472–476.

Chiu, Y.-S. P., Chiu, S. W., Li, C.-Y., & Ting, C.-K. (2009). Incorporating multi-delivery policy and quality assurance into economic production lot size problem. *Journal of Scientific and Industrial Research, 68*(6), 505–512.

Chiu, Y., Lin, H., Wu, M., & Chiu, S. (2018). Alternative fabrication scheme to study effects of rework of nonconforming products and delayed differentiation on a multiproduct supply-chain system. *International Journal of Industrial Engineering Computations, 9*(2), 235–248.

Chung, K. J. (2011). The economic production quantity with rework process in supply chain management. *Computers & Mathematics with Applications, 62*(6), 2547–2550.

Glock, C. H., & Jaber, M. Y. (2013). An economic production quantity (EPQ) model for a customer-dominated supply chain with defective items, reworking and scrap. *International Journal of Services and Operations Management, 14*(2), 236–251.

Goyal, S. K., & Cárdenas-Barrón, L. E. (2003). A simpler expression for the total average cost of inventory items with shortage permitted. *Industrial Engineering Journal, 32*(8), 24–27.

Goyal, S. K., & Cárdenas-Barrón, L. E. (2001). Note on: "An optimal batch size for a production system operating under a just-in-time delivery system". *International Journal of Production Economics, 72*(1), 99.

Hayek, P. A., & Salameh, M. K. (2001). Production lot sizing with the reworking of imperfect quality items produced. *Production Planning & Control, 12*(6), 584–590.

Jamal, A. M. M., Sarker, B. R., & Mondal, S. (2004). Optimal manufacturing batch size with rework process at a single-stage production system. *Computers & Industrial Engineering, 47*(1), 77–89.

Keshavarz, R., Makui, A., Tavakkoli-Moghaddam, R., & Taleizadeh, A. A. (2019). EPQ models with a production rate proportional to the power demand rate, rework process and scrapped items. *International Journal of Industrial Engineering: Theory, Applications and Practice, 26*(2), 173–198.

Krishnamoorthi, C., & Panayappan, S. (2012). An EPQ model for an imperfect production system with rework and shortages. *International Journal of Operational Research, 17*(1), 104–124.

Krishnamoorthi, C., & Panayappan, S. (2012). An EPQ model with imperfect production systems with rework of regular production and sales return. *American Journal of Operations Research, 2*(02), 225.

Moshtagh, S., & Taleizadeh, A. A. (2017). Stochastic integrated manufacturing and remanufacturing model with shortage, rework and quality based return rate in a closed loop supply chain. *Journal of Cleaner Production, 141*, 1548–1573.

Moussawi-Haidar, L., Salameh, M., & Nasr, W. (2016). Production lot sizing with quality screening and rework. *Applied Mathematical Modelling, 40*(4), 3242–3256.

Nahmias, S., & Cheng, Y. (2005). *Production and operations analysis* (Vol. 6). New York: McGraw-Hill.

Osteryoung, J. S. (1986). Use of the EOQ model for inventory analysis. *Production and Inventory Management, 27*, 39–45.

Porteus, E. L. (1986). Optimal lot sizing, process quality improvement and setup cost reduction. *Operations Research, 34*(1), 137–144.

Rosenblatt, M. J., & Lee, H. L. (1986). Economic production cycles with imperfect production processes. *IIE Transactions, 18*(1), 48–55.

Ross, S. M. (1996). *Stochastic processes* (2nd ed.). New York: Wiley.

Salameh, M. K., & Jaber, M. Y. (2000). Economic production quantity model for items with imperfect quality. *International Journal of Production Economics, 64*(1), 59–64.

Sarkar, B., Cárdenas-Barrón, L. E., Sarkar, M., & Singgih, M. L. (2014). An economic production quantity model with random defective rate, rework process and backorders for a single stage production system. *Journal of Manufacturing Systems, 33*(3), 423–435.

Sivashankari, C. K., & Panayappan, S. (2014). Production inventory model with reworking of imperfect production, scrap and shortages. *International Journal of Management Science and Engineering Management, 9*(1), 9–20.

Silver, E. A., Pyke, D. F., & Peterson, R. (1998). *Inventory management and production planning and scheduling* (Vol. 3, p. 30). New York: Wiley.

Shafiee-Gol, S., Nasiri, M. M., & Taleizadeh, A. A. (2016). Pricing and production decisions in multi-product single machine manufacturing system with discrete delivery and rework. *Opsearch, 53*, 873–888.

Taft, E. W. (1918). The most economical production lot. *Iron Age, 101*(18), 1410–1412.

Taleizadeh, A. A., & Heydaryan, H. (2017). Pricing, refund, and coordination optimization in a two stages supply chain of green and non-green products under hybrid production mode. *Journal of Remanufacturing, 7*, 49–76.

Taleizadeh, A. A., & Noori-Daryan, M. (2016). Pricing, replenishments and production policies in a supply chain of pharmacological product with rework process: A game theoretic approach. *Operational Research, An International Journal, 16*, 89–115.

Taleizadeh, A. A., Kalantari, S. S., & Cárdenas-Barrón, L. E. (2016). Pricing and lot sizing for an EPQ inventory model with rework and multiple shipments. *TOP, 24*(1), 143–155.

Taleizadeh, A. A., Karimi, M., & Torabi, S. A. (2020). A possibilistic closed-loop supply chain: Pricing, advertising and remanufacturing optimization. *Neural Computing and Applications, 32*, 1195–1215.

Taleizadeh, A. A., Wee, H. M., & Sadjadi, S. J. (2010). Multi-product production quantity model with repair failure and partial backordering. *Computers & Industrial Engineering, 59*(1), 45–54.

Taleizadeh, A. A., Alizadeh-Baban, N., & Niaki, S. T. A. (2019). A closed-loop supply chain considering carbon reduction, quality effort, and return policy under two remanufacturing scenarios. *Journal of Cleaner Production, 232*, 1230–1250.

Taleizadeh, A. A., & Wee, H. M. (2015). Manufacturing system with immediate rework and partial backordering. *International Journal of Advanced Operations Management, 7*(1), 41–62.

Taleizadeh, A. A., Kalantari, S. S., & Cárdenas-Barrón, L. E. (2015). Determining optimal price, replenishment lot size and number of shipments for an EPQ model with rework and multiple shipments. *Journal of Industrial & Management Optimization, 11*(4), 1059–1071.

Tapiero, C. S., Ritchken, P. H., & Reisman, A. (1987). Reliability, pricing and quality control. *European Journal of Operational Research, 31*(1), 37–45.

Tersine, R. J., & Tersine, M. G. (1988). *Instructor's manual to principles of inventory and materials management*. New York: North-Holland.

Tersine, R. J. (1994). *Principles of inventory and materials management* (4th ed.). Prentice Hall.

Wee, H. M., Wang, W. T., & Cárdenas-Barrón, L. E. (2013). An alternative analysis and solution procedure for the EPQ model with rework process at a single-stage manufacturing system with planned backorders. *Computers & Industrial Engineering, 64*(2), 748–755.

Zipkin, P. H. (2000). *Foundations of inventory management* (1st ed.). Boston, MA: McGraw-Hill/Irwin.

Chapter 5
Multi-product Single Machine

5.1 Introduction

The primary operation strategies and goals of most manufacturing firms are to seek a high satisfaction to customer's demands and to become a low-cost producer. To achieve these goals, the company must be able to effectively utilize resources and minimize costs.

The economic production quantity (EPQ) is a commonly used production model that has been studied extensively in the past few decades. One of the considered constraints in the EPQ inventory models is producing all items by a single machine. Since all of the products are manufactured on a single machine with a limited capacity, a unique cycle length for all items is considered. It is assumed there is a real constant production capacity limitation on the single machine on which all products are produced. If the rework is placed, both the production and rework processes are accomplished using the same resource, maybe the same cost and at the same or different speed. The first economic production quantity inventory model for a single-product single-stage manufacturing system was proposed by Taft (1918). Perhaps Eilon (1985) and Rogers (1958) were the first researchers who studied the multi-product single manufacturing system. Eilon (1985) proposed a multi-product lot-sizing problem classification for a system producing several items in a multi-product single-machine manufacturing system.

In this chapter, multi-product single-machine EPQ problems are presented in details. The presented models are classified into three categories in terms of the inclusion or noninclusion of the shortage and its types. The first category includes several models in which shortage is not allowed. In the second category of models, shortage is allowed and is back-ordered. Finally, the multi-product single-machine EPQ models with partial backordering are presented. The classification is shown in Fig. 5.1.

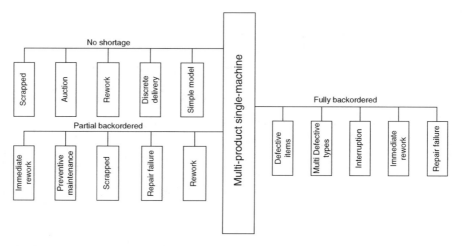

Fig. 5.1 Categories of multi-product single-machine EPQ

To provide a comprehensive introduction about the multi-product single-machine EPQ inventory management research status, in this chapter, the recent studies in relevant fields are reviewed. The literature review framework in this chapter provides a clear overview of the multi-product single-machine study field, which can be used as a starting point for further study. This survey presents a framework to classify different types of multi-product single-machine in terms of economic production quantity problem and reviews the literature based on the framework.

The common notations of multi-product single-machine EPQ problem are shown in Table 5.1. To integrate the models, these notations are used for all models. The main decision variable of this field is Q and T, but in some studies, other decision variables are considered too.

5.2 Literature Review

In this section, the background research from three perspectives are studied: multi-product single-machine EPQ models without shortage, models with shortage backordering, and models with partial backordering. Goyal (1984) developed a procedure for determining production quantities for two products manufactured on a single machine. This search procedure permits unequal batch quantities for the more frequently manufactured product. Wee et al. (2011) presented a multi-product single-machine EPQ model with multiple discrete deliveries and capacity and space constraints to determine the optimal period length, the optimal number of shipments, and the optimal order quantities. In order to solve the problem, they applied the

5.2 Literature Review

Table 5.1 Notations

n	Number of products (i is index of products $i = 1, 2, \ldots, n$)
Q_i	Production lot size of ith product for each cycle (unit)
T	Cycle length for all products (time)
B_i	The allowable backorder level of the ith product (unit)
D_i	Demand rate of ith product for each cycle (unit/year)
P_i or P_{1i}	Production rate of ith product for each cycle (unit/year)
P_{2i}	Rate of rework of all the defective items for ith product (unit/year)
d_i or d_{1i}	Production rate of defectives for ith product ($d_{1i} = xP_i$) (unit/year)
d_{2i}	Production rate of scrapped items during the reworking time for product i (unit/year)
K_i	Setup cost for each production run of ith product (\$/setup)
K_{S_i}	Fixed delivery cost for each shipment (\$/shipment)
ts_i	Setup time of machine to produce the ith product (time)
N	Number of cycles per year
I_i	Maximum level of on-hand inventory of ith product when regular production process stops (item)
H_i	Maximum level of on-hand inventory of ith product in units, when the reworking ends (item)
C_i	Production cost of ith product (\$/unit)
h_i	Holding cost of ith product (\$/unit/unit of time)
h_{1i}	Holding cost of each reworked item of ith product (\$/unit/unit of time)
α_i or x_i	Proportion of produced defective items of the ith item (constant or may be random variable) (%)
SL_i	Service factor for total allowable shortage of the ith product (%)
C_{Ri}	Rework cost of the ith product per unit (\$/unit)
θ_i	The scrapped quantity of the ith product (unit)
g	Cost of lost goodwill caused by unsatisfied demands (\$/unit)
C_{b_i}	The backordering cost per unit of the ith product per unit time (\$/unit/unit of time)
$\widehat{C}_{bi}$	The backordering cost per unit of the ith product (\$/unit)
$\widehat{\pi}_i$	The lost sale cost per unit of ith product per unit time (\$/unit)
s_i	The selling price of the ith product (\$/unit)
t_i^p	Production time in each cycle of ith product (time)
t_i^d	Downtime in each cycle of ith product (time)
t_i^r	Reworking time for ith product (time)
t_i^s	The permitted shortage time of the ith product (time)
t_i^b	The time needed to satisfy all backorders in the next production of the ith product (time)
C_{S_i}	The disposal cost per scrapped item of the ith product (\$/unit)
C_{T_i}	The transportation cost per item of the ith product (\$/unit)
CS	The annual expected total scrapped item cost (\$/year)
CP	Total production cost of all items (\$/year)
CH	Total holding cost of all items (\$/year)
CH_1	Total holding cost of all defective items (\$/year)
CA	Total setup cost of all items (\$/year)
CR	Total rework cost of all items (\$/year)

(continued)

Table 5.1 (continued)

CD	Total disposal cost of all items ($/year)
CC	Total warehouse construction cost of all items ($/year)
CB	Total backordering cost of all items ($/year)
W	Total available budget ($)
TC or ATC	Total cost ($/year)

extended cutting plane method, the particle swarm optimization, and harmony search algorithms. Taleizadeh et al. (2012) developed an EPQ model with production capacity limitation and breakdown with immediate rework for a single-stage production system with one machine. The aim of the problem is to determine both the optimal cycle length and the optimal production quantity for each product to minimize the expected total cost (holding, production, setup, rework costs). Nobil and Taleizadeh (2016) came up with the exact algorithm based on differentiation to solve a multi-product single-machine EPQ model for a defective production system. The faulty produced products are reworked or are put on auction as they are. Shafiee-Gol et al. (2016) presented pricing and production decisions in multi-product single-machine manufacturing system considering discrete delivery and rework.

Taleizadeh et al. (2010c) presented an economic production quantity model in which the production defective rate follows either a uniform or a normal probability distribution function. Backordering shortages are allowed, and the existence of only one machine causes a limited production capacity for the common cycle length of all products. The aim of their problem is to determine the optimal production quantity of each product such that the expected total cost is minimized. Two joint production systems in a form of multi-product single machine with and without rework are studied where shortage is allowed and back-ordered. For each system, the optimal cycle length and the backordered and production quantities of each product are determined such that the cost function is minimized (Taleizadeh et al. 2011). Taleizadeh et al. (2013b) developed a multi-product single-machine EPQ model with production capacity limitation and random defective production rate and failure during repair by considering shortage backordering. The objective was to determine the optimal period lengths, backordered quantities, and order quantities. Pasandideh et al. (2015) developed an economic production quantity inventory model for a multi-product single-machine lot-sizing problem with non-conforming items including scrap and rework, where reworks are classified into several groups based on failure severity. In this inventory model, shortage is allowed and is backordered. In their model, the available total budget is scarce and that there is a lower bound on the service level of each product. An EPQ inventory model with interruption in process, scrap and rework is analyzed by Taleizadeh et al. (2014). The shortages are permitted and fully back-ordered. In this EPQ inventory model, the decision variables are cycle length and backordered quantities of each product, and the main objective is to minimize the expected total cost.

5.3 No Shortage

A multi-product single-machine EPQ model with limited production capacity, random defective production rate, and repair failure is developed by Taleizadeh et al. (2010b). The aim of this problem is to minimize the expected total annual cost by optimizing the period length, the backordered quantities, and the rework items. When service level constraints and production capacity are considered, they have proved the proposed mathematical model to be convex. Taleizadeh et al. (2010a) extended a multi-product single-machine EPQ model with the production capacity limitation and random defective production rate. The shortage was assumed to occur in combination of backordered and lost sale, and there was a limitation on the service level. The aim of their problem is to determine the optimal solution of the period length, the shortage, and the order quantities such that the expected total cost, including holding, shortage, production, setup, and scrapped item costs, is minimized. Taleizadeh et al. (2013a) developed a multi-product single-machine EPQ model with partial backordering, rework, budget, and service level constraints. Their objective was to minimize the joint total cost of the system subject to service level and budget constraints. A multi-product single-machine economic production quantity model with preventive maintenance, scrap and rework is studied by Taleizadeh (2018). He assumed that preventive maintenance can occur when the inventory level is positive or negative. Indeed, two different scenarios are modeled, and according to the obtained results, a new one is selected. This scenario shows that the best time for the preventive maintenance is when all backordered demand is satisfied but inventory level is reached to zero. Taleizadeh and Wee (2015) developed a multi-product single-machine manufacturing system with rework, production capacity constraint, and partial backordering. The objective function of the proposed mathematical model is proved to be convex, and the minimum total cost is derived from the optimal period length, order, and backordered quantities. Taleizadeh et al. (2017) developed trade credit policy in multi-product single-machine model with repair failure and partial backordering. As a summary, features of reviewed studies are given in Table 5.2.

5.3 No Shortage

In this subsection, five different problems are presented in which shortage is not considered. The model development is investigated, and solution procedures to solve the optimization problem are presented. Also, numerical examples have been presented to illustrate the implementation of proposed method.

5.3.1 Simple Model

Goyal (1984) developed an EPQ mode for determining production quantities for two products manufactured on a single machine. The search procedure permits unequal

Table 5.2 Features of reviewed studies

Reference	Multi-product single machine	Shortage: No shortage	Shortage: Backordering	Shortage: Partial backordering	Defective	Scrapped	Rework	Defective after rework	Breakdown	Preventive maintenance	Constraint: Service level	Constraint: Capacity	Constraint: Budget	Constraint: Space	Decision variables: Number of shipment	Decision variables: Period length	Decision variables: Production quantity	Decision variables: Shortage quantity	Rework rate	Selling price	Number of shipment
Goyal (1984)	✓	✓															✓				
Wee et al. (2011)	✓				✓		✓					✓		✓	✓		✓	✓			✓
Taleizadeh et al. (2012)	✓	✓			✓		✓		✓			✓				✓	✓				
Shafiee-Gol et al. (2016)	✓	✓			✓	✓	✓					✓					✓			✓	
Nobil and Taleizadeh (2016)	✓	✓			✓											✓	✓		✓		
Pasandideh et al. (2015)	✓		✓			✓	✓				✓					✓	✓	✓			
Taleizadeh et al. (2011)	✓		✓			✓	✓				✓	✓				✓	✓	✓			
Taleizadeh et al. (2013b)	✓		✓				✓	✓				✓				✓	✓	✓			
Taleizadeh et al. (2014)	✓		✓				✓						✓			✓	✓	✓			
Taleizadeh et al. (2010c)	✓			✓						✓	✓	✓				✓	✓	✓			
Taleizadeh et al. (2013a)	✓			✓	✓		✓	✓			✓	✓				✓	✓	✓			
Taleizadeh et al. (2010b)	✓				✓			✓			✓	✓				✓	✓	✓			
Taleizadeh et al. (2010a)	✓			✓			✓					✓				✓	✓	✓			
Taleizadeh (2018)	✓					✓	✓					✓				✓	✓	✓			
Taleizadeh and Wee (2015)	✓					✓	✓					✓				✓	✓	✓			

5.3 No Shortage

batch quantities for the more frequently manufactured product. The total variable cost per unit of time for the ith product $V_i(N)$ is the sum of the cost of manufacturing setups and the stock holding cost. Hence, $V_i(N)$ is given by (Goyal 1984):

$$V_i(N) = NK_i + \frac{D_i h_i (1 - D_i/P_i)}{2N} \tag{5.1}$$

If it is possible to manufacture the ith product whenever producer intends to manufacture it, then the economic frequency of manufacturing setups is given by:

$$w = \sqrt{\frac{\frac{D_i h_i (1 - D_i/P_i)}{2N}}{K_i}} \tag{5.2}$$

If the second product is manufactured N times per unit of time, then the repetitive manufacturing cycle time is given by $1/N$. On the other hand, if the first product is manufactured NM times per unit of time, then the time interval between successive setups of the first product is given by $1/NM$. At time zero, the first setup for manufacturing the first product is undertaken. Once the production of the first product is over, the setup for manufacturing the second product is undertaken. The second setup for manufacturing the first product is undertaken at time $1/NM$, so the total time required to produce one batch of products 1 and 2 is given by:

$$ts_1 + \frac{D_1}{NMP_1} + ts_2 + \frac{D_2}{NP_2} \tag{5.3}$$

Hence, in order to ensure feasibility of the production schedule, the following condition should be met:

$$M \leq \frac{1 - \frac{D_1}{P_1}}{N\left(ts_1 + ts_2 + \frac{D_2}{NP_2}\right)} \tag{5.4}$$

If the setup times for both the products are assumed negligible, then the above expression reduces to:

$$M \leq x \tag{5.5}$$

where:

$$x = \frac{P_2(1 - D_1/P_1)}{D_2} \tag{5.6}$$

It can easily be proved that if for the first product the condition given by (5.5) is not satisfied for any given value of M, then the stock holding cost for the first product can be minimized by undertaking the first setup at the beginning of the manufacturing cycle and the second setup after an interval of $1/Nx$ time units. Hence, for the first product, one setup covers a time span of $1/\{N\text{Min}(M, x)\}$, and the remaining $(M - 1)$ setups, undertaken at equal time intervals, cover a time span of $/\{N\text{Min}(M, x) - 1\}/\{N\text{Min}(M, x)\}$. Therefore, the variable cost for the first product is (Goyal 1984):

$$V_1(NK) = NMK_1 + N \left[\begin{array}{c} \frac{D_1 h_1}{2}\left(1 - \frac{D_1}{P_1}\right)\left(\frac{1}{N\text{Min}(M,x)}\right)^2 \\ + \frac{D_1 h_1}{2}\left(1 - \frac{D_1}{P_1}\right)\frac{1}{M-1}\left(\frac{\text{Min}(M,x) - 1}{N\text{Min}(M,x)}\right)^2 \end{array} \right] \quad (5.7)$$

The total variable cost, $Z(N, M)$, when the first product is manufactured NM times per unit of time and the second product is manufactured N times per unit of time, is:

$$Z(N,M) = N(MK_1 + K_2) + \frac{1}{N}$$
$$\times \left(\frac{\frac{Dh_1}{2}\left(1 - \frac{D_1}{P_1}\right) + \left[1 + \{\text{Min}(M,x) - 1\}^2/(M-1)\right]}{\{\text{Min}(M,x)\}^2} + \frac{D_2 h_2}{2}\left(1 - \frac{D_2}{P_2}\right) \right)$$

(5.8)

The total cost function, $Z(N, M)$, is a convex function, and there exists a local minimum value of total cost $Z(M)$ for every value of M. Therefore, a procedure must be adopted for determining the economic policy (Goyal 1984):

Step 1. Arrange the products in descending order of w^2 ratio. The second product is manufactured exactly once in every manufacturing cycle.

Step 2. (i) Determine the integer value of $M = a$ which satisfies the condition:

$$a(a+1) \geq \frac{\frac{Dh_1}{2}\left(1 - \frac{D_1}{P_1}\right)K_2}{\frac{D_2 h_2}{2}\left(1 - \frac{D_2}{P_1}\right)K_1} \geq a(a-1)$$

(ii) Determine the integer value of $M = b$ which satisfies the condition $(b+1) \geq \frac{P_2}{D_2}\left(1 - \frac{D_1}{P_1}\right) \geq b$.

(iii) Determine $\text{Min}(a, b) = y$.

Step 3.

Determine $Z(M)$ for integer values of M in the range $y \leq M \leq a$. The solution procedure is terminated at $M = M^* + 1$ if $Z(M^* + 1) \geq Z(M^*)$. Therefore the economic value of $M = M^*$.

Step 4.

Determine the economic policy as follows:

5.3 No Shortage

Table 5.3 General data for the example (Goyal 1984)

i	D_i	P_i	K_i	h_i	w^2
1	200,000	400,000	200	0.4	100
2	80,000	400,000	400	0.15	12

$$N(M^*) = \sqrt{\frac{\frac{Dh_2}{2}\left(1-\frac{D_2}{P_2}\right) + \frac{\frac{Dh_1}{2}\left(1-\frac{D_1}{P_1}\right)}{\{\text{Min}(M^*,x)\}^2}\frac{1+\{\text{Min}(M^*,x)-1\}^2}{M^*-1}}{(K_2 + K_1 M^*)}}$$

Economic production quantity for the second product is $(D_2/N(M^*))$ and for the first product $\left[\frac{D_1}{N(M^*)\text{Min}(M^*,x)}\right]$ (Goyal 1984).

During each of the remaining $m^* - 1$ runs, it is given by (Goyal 1984) $\left[\frac{D_1[\text{Min}(M^*,x)-1]}{(M^*-1)N(M^*)\text{Min}(M^*,x)}\right]$.

Example 5.1 Goyal (1984) presented an example with the data shown in Table 5.3.

Assume unit of time as 1 year, and 1 year is 250 working days.

Step 1. The w^2 ratio for the first product is 120 and for the second product is 2.
Step 2. (i) $a = 3$.
      (ii) $b = 2$.
      (iii) $\text{Min}(a, b) = 2$.
Step 3. $Z(2) = 6881.86$.
      $Z(3) = 6811.76$.
      Hence, $M^* = 3$.
Step 4. The parameters of the economic operating policy can be obtained.

5.3.2 Discrete Delivery

Wee et al. (2011) presented a multi-product single-machine EPQ model with multiple discrete deliveries and capacity and space constraints to determine the optimal period length, the optimal number of shipments, and the optimal order quantities. In order to solve the problem, they applied the extended cutting plane method, the particle swarm optimization, and harmony search algorithms.

Wee et al. (2011) assumed a manufacture sends orders to the customer and bears the transportation cost for each delivery to the customer. The customer determines the capacity of each delivery and the quantity of each shipment. The purpose of this problem is to determine the optimal replenishment period, the delivery quantity, and the number of delivery to minimize the total production inventory cost with space and capacity constraints.

According to given notations in Sect. 5.1 and the following notations presented in Table 5.4, the mathematical model is presented.

Table 5.4 The new notations for given problem

t_i	Time between two sequential shipments of each pallet for ith product
q_i	Shipment quantity for ith product
n_i^o	Number of shipments in each cycle of ith product
k_i	Capacity of pallet for ith product

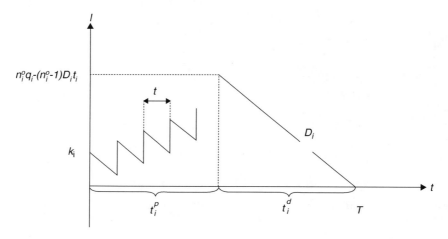

Fig. 5.2 The inventory level (Wee et al. 2011)

n_i^o, q_i, and T are the decision variables of the model. Figure 5.2 shows the inventory level of the EPQ model with discrete delivery order.

During t_i^p and t_i^d, the delivered products are produced at a constant rate (Pasandideh and Niaki 2008):

$$Q_i = n_i^o q_i \tag{5.9}$$

Also:

$$Q_i = D_i T \tag{5.10}$$

That $T = t_i^p + t_i^d$ is the summation of the production uptimes and the production downtimes. Using Eqs. (5.9) and (5.10), the total replenishment time can be modeled as follows (Wee et al. 2011):

$$T = t_i^p + t_i^d = \frac{n_i^o v_i}{D_i} \tag{5.11}$$

Since the maximum inventory level is $n_i^o q_i - (n_i^o - 1) q_i \frac{D_i}{P_i}$, t_i^d is $\frac{n_i^o q_i}{D_i} - (n_i^o - 1) \frac{q_i}{P_i}$. Moreover, the production up time is:

5.3 No Shortage

$$t_i^p = (n_i^o - 1)\frac{q_i}{P_i} \tag{5.12}$$

The total cost function consists of summary of the production, the setup, the holding, and the transportation costs as follows (Wee et al. 2011):

$$TC = CA + CP + CT + CH \tag{5.13}$$

The setup cost occurs N times per year. Therefore, the annual setup cost is $\sum_{i=1}^{n} NK_i$. Also, $N = \frac{1}{T}$, thus:

$$CA = \frac{\sum_{i=1}^{n} K_i}{T} \tag{5.14}$$

The production cost of ith product per period is $C_i Q_i$, and the annual production quantity is $NC_i Q_i$. Finally, the joint production cost is:

$$CP = \frac{\sum_{i=1}^{n} C_i n_i^o k_i}{T} = \frac{\sum_{i=1}^{n} C_i D_i T}{T} = \sum_{i=1}^{n} C_i D_i \tag{5.15}$$

Transportation cost depends on the number of shipments, and it is equal to $C_{T_i} n_i^o$ for each cycle, and the annual transportation cost is $NC_{T_i} n_i^o$. Finally, the CT in joint policy can be modeled as:

$$CT = \frac{\sum_{i=1}^{n} C_{T_i} n_i^o}{T} \tag{5.16}$$

t_i^d is built up by a collection of trapezes. The number of trapezes for product i is $m_i - 1$. If az_i^j represents the area of trapeze j of product i, the areas of trapeze 1 and 2 are:

$$az_i^1 = \left(\frac{q_i + (q_i - D_i t_i)}{2}\right) t_i = \left(\frac{2q_i - D_i t_i}{2}\right) t_i \tag{5.17}$$

$$az_i^2 = \left(\frac{(q_i - D_i t_i + q_i) + (2q_i - 2D_i t_i)}{2}\right) t_i = \left(\frac{4q_i - 3D_i t_i}{2}\right) t_i \tag{5.18}$$

Thus:

$$az_i^j = \left(\frac{2jq_i - (2j-1)D_i t_i}{2}\right) t_i \quad j = 1, \ldots, m_i - 1 \tag{5.19}$$

Finally, the area of all trapezes on the left of each cycle for ith product can be formulated as follows (Wee et al. 2011):

$$az_i^j = \sum_{j=1}^{m_i-1} \frac{2jq_it_i - 2jD_it_i^2 + D_it_i^2}{2} = \sum_{j=1}^{m_i-1} \frac{2jq_it_i}{2} - \sum_{j=1}^{m_i-1} \frac{2jD_it_i^2}{2} + \sum_{j=1}^{m_i-1} \frac{D_it_i^2}{2}$$

$$= q_it_i \sum_{j=1}^{m_i-1} j - D_it_i^2 \sum_{j=1}^{m_i-1} j + \frac{D_it_i^2}{2} \sum_{j=1}^{m_i-1} 1 \tag{5.20}$$

$$= q_it_i \frac{n_i^o(n_i^o - 1)}{2} - D_it_i^2 \frac{n_i^o(n_i^o - 1)}{2} + \frac{D_it_i^2}{2}(n_i^o - 1)$$

$$= q_it_i \frac{n_i^o(n_i^o - 1)}{2} - D_it_i^2 \frac{n_i^{o2}}{2} + D_it_i^2 \frac{n_i^o}{2} + \frac{D_it_i^2}{2}(n_i^o - 1)$$

For $k_i = P_i t_i$, one has:

$$az_i^j = \left(\frac{P_i - D_i}{2P_i^2}\right) n_i^{o2} q_i^2 + \left(\frac{D_i - P_i}{P_i^2}\right) n_i^o q_i^2 - \frac{D_i}{2P_i^2} q_i^2, \quad j$$

$$= 1, \ldots, m_i - 1 \tag{5.21}$$

The area of a triangle on the right side of each cycle of product i, (at_i), can be modeled as follows:

$$at_i = \frac{1}{2}\left(n_i^o q_i - (n_i^o - 1)t_i\right)\left(\frac{n_i^o q_i}{D_i} - (n_i^o - 1)t_i\right)$$

$$= \frac{1}{2}\left[\frac{(n_i^o q_i)^2}{D_i} - 2n_i^o q_i(n_i^o - 1)t_i + (n_i^o - 1)^2 D_i t_i^2\right] = \frac{1}{2}\left[\frac{(n_i^o q_i)^2}{D_i} - 2(n_i^o - 1)n_i^o \frac{q_i^2}{P_i} + (n_i^o - 1)^2 D_i \frac{q_i^2}{P_i^2}\right]$$

$$= \left(\frac{1}{2D_i} - \frac{1}{P_i} + \frac{P_i}{2P_i^2}\right)(n_i^o q_i)^2 + \left(1 - \frac{D_i}{P_i^2}\right) n_i^o q_i^2 + \frac{D_i}{2P_i^2} q_i^2 \tag{5.22}$$

The total areas of each cycle of product i, (a_i), is:

$$a_i = az_i + at_i = \left(\frac{P_i - D_i}{2P_i^2}\right) n_i^{o2} q_i^2 + \left(\frac{D_i - P_i}{P_i^2}\right) n_i^o q_i^2 - \frac{D_i}{2P_i^2} q_i^2$$

$$+ \left(\frac{1}{2D_i} - \frac{1}{P_i} + \frac{D_i}{2P_i^2}\right)(n_i^o q_i)^2 + \left(1 - \frac{D_i}{P_i^2}\right) n_i^o q_i^2 + \frac{D_i}{2P_i^2} q_i^2$$

$$= \left(\frac{P_i - D_i}{2P_i D_i}\right)(n_i^o q_i)^2 + \left(\frac{P_i^2 - P_i}{P_i^2}\right) n_i^o q_i^2 \tag{5.23}$$

Using $k_i = P_i t_i$ and (5.13) and assuming N periods per year yield the total annual holding cost as $Nh_i\left(\left(\frac{P_i - D_i}{2P_i D_i}\right)(n_i^o q_i)^2 + \left(\frac{P_i^2 - P_i}{P_i^2}\right) n_i^o q_i^2\right)$. Finally, the holding cost for joint production system is:

$$\text{CH} = \sum_{i=1}^{n} h_i \left(\frac{P_i - D_i}{2P_i D_i} \right) \frac{n_i^{o2} q_i^2}{T} + \sum_{i=1}^{n} h_i \left(\frac{P_i^2 - P_i}{P_i^2} \right) \frac{n_i^o q_i^2}{T} \tag{5.24}$$

Based on Eqs. (5.14)–(5.16) and (5.24) and implementing some simplifications, the total annual cost of production system can be modeled as:

$$\text{TC} = \frac{\sum_{i=1}^{n} K_i}{T} + \sum_{i=1}^{n} C_{T_i} \frac{n_i^o}{T} + \sum_{i=1}^{n} h_i \left(\frac{P_i - D_i}{2P_i D_i} \right) \frac{n_i^{o2} q_i^2}{T}$$
$$+ \sum_{i=1}^{n} h_i \left(\frac{P_i^2 - P_i}{P_i^2} \right) \frac{n_i^o q_i^2}{T} + \sum_{i=1}^{n} C_i D_i \tag{5.25}$$

For joint production systems, the total production and setup times must be smaller than the cycle length. In the model, $\sum_{i=1}^{n}(t_i^p + \text{ts}_i)$ must be smaller or equal to T. Therefore, the capacity limitation can be modeled as follows (Wee et al. 2011):

$$\sum_{i=1}^{n} (t_i^p + \text{ts}_i) \leq T \tag{5.26}$$

$$\sum_{i=1}^{n} (n_i^o - 1) \frac{v_i}{P_i} + \sum_{i=1}^{n} \text{ts}_i \leq T \tag{5.27}$$

The number of shipments must be smaller than the upper bound, and, at least, one shipment needs to be performed. One has:

$$1 \leq n_i^o \leq U_i; \quad \text{Integer}; i = 1, 2, \ldots, n \tag{5.28}$$

Finally, the complete model can be derived as follows (Wee et al. 2011):

$$\text{Min}: \text{TC} = \frac{\sum_{i=1}^{n} K_i}{T} + \sum_{i=1}^{n} C_{T_i} \frac{n_i^o}{T} + \sum_{i=1}^{n} \left(\frac{P_i - D_i}{2P_i D_i} \right) \frac{n_i^{o2} q_i^2}{T} + \sum_{i=1}^{n} \left(\frac{P_i^2 - P_i}{P_i^2} \right) \frac{n_i^o q_i^2}{T} + \sum_{i=1}^{n} C_i D_i$$

s.t.:
$$\sum_{i=1}^{n} (n_i^o - 1) \frac{q_i}{P_i} + \text{ts}_i \leq T$$
$$1 \leq n_i^o \leq U_i; \quad \text{Integer}; i = 1, 2, \ldots, n$$
$$T, v_i \geq 0; \quad i = 1, 2, \ldots, n$$
$$\tag{5.29}$$

The final model in (5.29) is a mixed integer nonlinear programming (MINLP) problem, and Wee et al. (2011) used the extended cutting plane method to solve the model. Also in order to evaluate the performance of the proposed solution method, they used the two metaheuristic algorithms including particle swarm optimization (PSO) and harmony search algorithm (HS).

Table 5.5 General data for the example (Wee et al. 2011)

Product	D_i	P_i	ts_i	K_i	C_{T_i}	h_i	C_i	U_i
1	300	5000	0.0010	500	5	2	34	10
2	350	5500	0.0015	600	7	4	32	10
3	400	6000	0.0020	700	9	6	30	10
4	450	6500	0.0025	800	11	8	28	10
5	500	7000	0.0030	900	13	10	26	10
6	550	7500	0.0035	1000	15	12	24	10
7	600	8000	0.0040	1100	17	14	22	10
8	650	8500	0.0045	1200	19	16	20	10
9	700	9000	0.0050	1300	21	18	18	10
10	750	9500	0.0055	1400	23	20	16	10
11	800	10,000	0.0060	1500	25	22	14	10
12	850	10,500	0.0065	1600	27	24	12	10
13	900	11,000	0.0070	1700	29	26	10	10
14	950	11,500	0.0075	1800	31	28	8	10
15	1000	12,000	0.0080	1900	33	30	6	10

Example 5.2 Wee et al. (2011) considered a multi-product EPQ problem with discrete deliveries and capacity constraint with 15 products. In the example, the demand rate, the production rate, and the setup time of each product are assumed to be constant for each cycle. The general data of the example are given in Table 5.5. The minimum shipment is assumed to be 1 and the maximum shipment is equal to 10. Table 5.6 shows the best results for the example using the extended cutting plane method, PSO, and HS algorithms, respectively.

Table 5.6 shows the cutting plane method has less total cost than the other methods. Furthermore, in terms of the CPU time, the computation time of the extended cutting plane method is less than the other (Wee et al. 2011).

5.3.3 Rework

Taleizadeh et al. (2012) developed an EPQ model with production capacity limitation and breakdown with immediate rework for a single-stage production system with one machine. The aim of the problem is to determine both the optimal cycle length and the optimal production quantity for each product to minimize the expected total cost (holding, production, setup, rework costs).

Proportion of defective of each product is constant in each cycle, and production rate of non-defective items is constant and is greater than the demand rate of each product. Scrap is not produced at any cycle and no defectives are produced during the rework process. Production and rework are accomplished using the same resource at the same speed, and shortage is not allowed. A real constant production

5.3 No Shortage

Table 5.6 Best results for the example cutting plane method (Wee et al. 2011)

Product	n_i^o CPM	PSO	HS	Q_i CPM	PSO	HS	T CPM	PSO	HS	TC CPM	PSO	HS
1	10	10	9	20.693	21.3	22.1	3.308	3.018	2.921	179,607	181,640	184,210
2	10	9	10	17.776	19.7	21.6						
3	10	10	8	15.580	17.4	20.8						
4	10	9	9	13.867	16.3	19.6						
5	10	10	10	12.493	14.3	17.9						
6	10	10	9	11.367	12.7	16.4						
7	10	9	10	10.427	11.4	13.2						
8	10	10	9	9.631	10.5	12.6						
9	10	9	9	8.948	9.6	11.5						
10	10	8	10	8.355	8.9	10.8						
11	10	10	7	7.836	8.1	9.1						
12	10	10	9	7.377	7.6	8.2						
13	10	10	9	6.969	6.8	7.4						
14	10	10	10	6.604	6.5	6.9						
15	10	8	10	6.276	6.1	6.1						

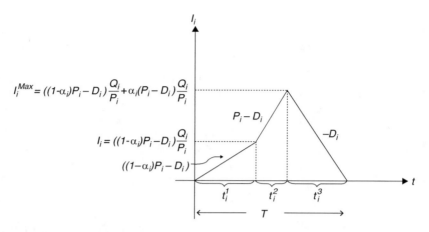

Fig. 5.3 On-hand inventory of perfect-quality items (Taleizadeh et al. 2012)

capacity limitation on a single machine in which all products are produced and nonzero set up cost are considered.

According to introduced notations in Sect. 5.1, the mathematical model is presented. Initially the problem is modeled as a single-product case, and then it is modified as a multi-product case. The basic assumption of EPQ model with rework process produced is that P_i must always be greater than or equal to the sum of demand rate D_i. Therefore, one has (Taleizadeh et al. 2012):

$$((1 - \alpha_i)P_i - D_i) \geq 0 \tag{5.30}$$

The production cycle length (see Fig. 5.3) is the summation of the production uptime, the reworking time, and the production downtime (Taleizadeh et al. 2012):

$$T = \sum_{i=1}^{n} t_i^p + t_i^r + t_i^d \tag{5.31}$$

Based on Fig. 5.3, for $i = 1, 2, \ldots, n$, one has (Taleizadeh et al. 2012):

$$t_i^p = \frac{Q_i}{P_i} \tag{5.32}$$

$$t_i^r = \alpha_i \frac{Q_i}{P_i} \tag{5.33}$$

$$t_i^d = \frac{\left(1 - \frac{D_i}{P_i} - \alpha_i \frac{D_i}{P_i}\right) Q_i}{D_i} \tag{5.34}$$

It is evident from Fig. 5.3 that (Taleizadeh et al. 2012):

5.3 No Shortage

$$I_i = ((1-\alpha_i)P_i - D_i)\frac{Q_i}{P_i} \tag{5.35}$$

$$H_i = I_i + \alpha_i(P_i - D_i)\frac{Q_i}{P_i} = ((1-\alpha_i)P_i - D_i)\frac{Q_i}{P_i} + \alpha_i(P_i - D_i)\frac{Q_i}{P_i} \tag{5.36}$$

Hence, according to Eq. (5.31), the cycle length for a single-product state is (Taleizadeh et al. 2012):

$$T = t_i^p + t_i^r + t_i^d = \frac{Q_i}{D_i} \tag{5.37}$$

or:

$$Q_i = D_i T \tag{5.38}$$

The total production cost of the system consists of setup cost, processing cost, rework cost, and inventory carrying costs. Defective items are produced in every batch and they are reworked within the same cycle. During the rework of defective items, again some processing costs and inventory holding costs are incurred for processing and holding the reworked quantities as well. The total inventory cost per year TC is:

$$\text{TC} = \overbrace{NC_iQ_i}^{\text{Production Cost}} + \overbrace{NK_i}^{\text{Setup Cost}} + \overbrace{NC_i\alpha_iQ_i}^{\text{Rework Cost}} + \overbrace{Nh_i\left[\frac{I_i}{2}(t_i^p) + \frac{I_i + H_i}{2}(t_i^r) + \frac{H_i}{2}(t_i^d)\right]}^{\text{Holding Cost of Items}} \tag{5.39}$$

The joint production policy (multi-product single machine) from Eq. (5.44) becomes:

$$\text{TC} = \sum_{i=1}^{n} C_i D_i + \frac{\sum_{i=1}^{n} K_i}{T} + \sum_{i=1}^{n} C_i \alpha_i D_i$$
$$+ \sum_{i=1}^{n} h_i \left[\frac{I_i}{2T}(t_i^p) + \frac{I_i + H_i}{2T}(t_i^r) + \frac{H_i}{2T}(t_i^d)\right] \tag{5.40}$$

Since $t_i^p + t_i^r$ are the production and rework times and ts_i is the setup time for ith product, the summation of the production, rework, and setup time for all products will be $\sum_{i=1}^{n}(t_i^p + t_i^r) + \sum_{i=1}^{n} ts_i$, and it must be smaller or equal to the period length (T). Therefore, the constraint of the model is:

$$\sum_{i=1}^{n} \left(t_i^p + t_i^r\right) + \sum_{i=1}^{n} \text{ts}_i \leq T \qquad (5.41)$$

Then, based on Eqs. (5.37), (5.38), and (5.41), one has:

$$\sum_{i=1}^{n} (1+\alpha_i) \frac{D_i}{P_i} T + \sum_{i=1}^{n} \text{ts}_i \leq T \qquad (5.42)$$

From previous equations and constraint in Eq. (5.42), one can formulate the optimization problem as:

$$\text{Min}: \ \text{TC} = \frac{\sum_{i=1}^{n} K_i}{T} + \sum_{i=1}^{n} C_i \left[[(1-\alpha_i)P_i - D_i]\frac{D_i}{2P_i} + \alpha_i[P_i - D_i]\frac{D_i}{P_i} \right] \left(1 - \frac{D_i}{P_i} - \alpha_i \frac{D_i}{P_i}\right) T^2$$

$$+ \sum_{i=1}^{n} C_i \alpha_i D_i + \sum_{i=1}^{n} C_i \left((1+\alpha_i)[(1-\alpha_i)P_i - D_i]\frac{D_i}{2P_i} + \alpha_i^2[P_i - D_i]\frac{D_i}{P_i^2} \right) T + \sum_{i=1}^{n} C_i D_i$$

$$(5.43)$$

$$\text{s.t.}: \quad T \geq \frac{\sum_{i=1}^{n} \text{ts}_i}{\left[1 - \sum_{i=1}^{n}(1+\alpha_i)\frac{D_i}{P_i}\right]} = T_{\text{Min}} \qquad (5.44)$$

In order to derive the optimal solution of the final model, a proof of the convexity of the objective function is provided. A classical optimization technique using partial derivatives is performed to derive the optimal solutions (Taleizadeh et al. 2012).

Theorem 5.1 The objective function TC in (5.43) is convex.

Proof To prove the convexity of TC $= Z$, the first and second derivatives of objective function are calculated as below (Taleizadeh et al. 2012):

$$\frac{\partial \text{TC}}{\partial T} = -\frac{\sum_{i=1}^{n} K_i}{T^2} + 2\sum_{i=1}^{n} C_i \left[[(1-\alpha_i)P_i - D_i]\frac{D_i}{2P_i} + \alpha_i[P_i - D_i]\frac{D_i}{P_i} \right] \left(1 - \frac{D_i}{P_i} - \alpha_i \frac{D_i}{P_i}\right) T$$

$$+ \sum_{i=1}^{n} C_i \left((1+\alpha_i)[(1-\alpha_i)P_i - D_i]\frac{D_i}{2P_i} + \alpha_i^2[P_i - D_i]\frac{D_i}{P_i^2} \right)$$

$$(5.45)$$

$$\frac{\partial^2 \text{TC}}{\partial T^2} = \frac{2\sum_{i=1}^{n} K_i}{T^3} + 2\sum_{i=1}^{n} C_i \left[[(1-\alpha_i)P_i - D_i]\frac{D_i}{2P_i} + \alpha_i[P_i - D_i]\frac{D_i}{P_i} \right]$$

$$\times \left(1 - (1+\alpha_i)\frac{D_i}{P_i}\right)$$

$$\geq 0 \qquad (5.46)$$

5.3 No Shortage

Table 5.7 General data for example (Taleizadeh et al. 2012)

Product	D_i	P_i	ts_i	K_i	C_i	h_i	α_i
1	400	3500	0.003	500	15	5	0.1
2	500	4000	0.004	450	12	4	0.2
3	600	4500	0.005	400	10	3	0.3
4	700	5000	0.006	350	8	2	0.4
5	800	5500	0.007	300	6	1	0.5

Table 5.8 The best results for the example (Taleizadeh et al. 2012)

Product	Uniform T_{Min}	T	T^*	Q_i	Z
1	0.1827879	0.5408731	0.5408731	216.3492	42,998.16
2				270.4366	
3				324.5239	
4				378.6112	
5				432.6985	

Since the second derivative is nonnegative, so the objective function is convex, and to solve and ensure the feasibility, the following solution procedure must be performed:

Step 1. Check for feasibility.
  If $\sum_{i=1}^{n}(1+\alpha_i)\frac{D_i}{P_i} \leq 1$ and $[(1-\alpha_i)P_i - D_i] \geq 0$, go to *Step* 2; else the problem will be infeasible.
Step 2. Calculate T using numeric method. If $T \geq 0$, go to *Step* 3; else the problem will be infeasible.
Step 3. Calculate by T_{Min} Eq. (5.44).
Step 4. If $T \geq T_{Min}$, then $T^* = T$; else $T^* = T_{Min}$.
Step 5. Calculate Q_i^* by Eq. (5.38).
Step 6. Terminate procedure.

Example 5.3 Taleizadeh et al. (2012) considered multi-product EPQ problem with breakdown and immediate rework with five products in which their general and specific data are given in Table 5.7. Table 5.8 shows the best results for the numerical example. Since the value of T is greater than T_{Min}, *Step* 4 of the procedure implies the optimality of T.

5.3.4 Auction

Nobil and Taleizadeh (2016) came up with the exact algorithm based on differentiation to solve a multi-product single-machine EPQ model for a defective production system. The defective products are reworked or are put on auction as they are.

Also, all of the products are produced by a single machine, which caused a production limit. The aim of the model is to minimize the total production costs, including holding costs, production costs, reworking costs, cost of lost profits, and setup costs. They determined the optimal production period of each product and also the percentage of the products which need to be reworked. According to given notations in Sect. 5.1 and the following notations (presented in Table 5.9), the mathematical model is presented (Nobil and Taleizadeh 2016).

Q_i, σ_i, and T are the decision variables of the model. It is evident from Fig. 5.4 that the maximum on-hand inventory of the ith item when the regular production stops is given by:

$$I_i = ((1 - \sigma_i)P_i - D_i)\frac{Q_i}{P_i} \tag{5.47}$$

Moreover, the maximum on-hand inventory when the rework stops is expressed as:

$$H_i = I_i + \sigma_i(P_i - D_i)\frac{Q_i}{P_i} \tag{5.48}$$

In Fig. 5.4, the cycle length of an item consists of three periods: the uptime production period denoted by t_i^p reworking time expressed as t_i^r, and downtime represented by t_i^d as presented in Eqs. (5.49) to (5.51)

Table 5.9 New notations for given problem (Nobil and Taleizadeh 2016)

σ_i	Proportion of produced reworkable items of ith item (decision variable)
α_i	Proportion of produced defective items of ith item ($\alpha_i = \theta_i + \sigma_i$)
θ_i	Proportion of defective items of ith item which will be auctioned
l_i	Lost profit of ith product per unit when the auction occurs
CD	Total disposal cost of all items
CC	Total warehouse construction cost

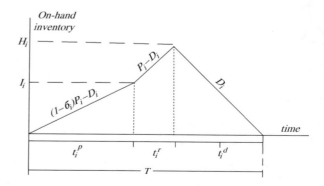

Fig. 5.4 The cycle length of the on-hand inventory (Nobil and Taleizadeh 2016)

5.3 No Shortage

$$t_i^{\text{p}} = \frac{I_i}{(1-\sigma_i)P_i - D_i} = \frac{Q_i}{P_i} \quad (5.49)$$

$$t_i^{\text{r}} = \sigma_i \frac{Q_i}{P_i} \quad (5.50)$$

$$t_i^{\text{d}} = \frac{H_i}{D_i} \quad (5.51)$$

Consequently, the length of a cycle is determined as below:

$$T_i = T = t_i^{\text{p}} + t_i^{\text{r}} + t_i^{\text{d}} = \frac{Q_i}{D_i} \quad (5.52)$$

Hence:

$$Q_i = D_i T \quad (5.53)$$

The annual total inventory cost, including the setup, production, lost profit, reworking, construction, and holding costs, is as below:

$$\text{TC} = \text{CA} + \text{CP} + \text{CD} + \text{CR} + \text{CC} + \text{CH} \quad (5.54)$$

Then the annual setup cost is:

$$\text{CA} = \sum_{i=1}^{n} NK_i = \sum_{i=1}^{n} \frac{K_i}{T} \quad (5.55)$$

The total production cost of the inventory system based on the production cost of the ith item per unit per cycle (C_i) can be obtained using Eq. (5.56):

$$\text{CP} = \sum_{i=1}^{n} NC_i Q_i = \sum_{i=1}^{n} \frac{C_i}{T}(D_i T) = \sum_{i=1}^{n} C_i D_i \quad (5.56)$$

The total lost profit is calculated as below:

$$\text{CD} = \sum_{i=1}^{n} N l_i \theta_i Q_i = \sum_{i=1}^{n} \frac{l_i \theta_i}{T}(D_i T) = \sum_{i=1}^{n} \frac{l_i(\alpha_i - \sigma_i)}{T}(D_i T)$$

$$= \sum_{i=1}^{n} l_i \alpha_i D_i - \sum_{i=1}^{n} l_i D_i \sigma_i \quad (5.57)$$

The total rework cost of the defective items is shown in Eq. (5.58):

$$\text{CR} = \sum_{i=1}^{n} NC_{R_i}\sigma_i Q_i = \sum_{i=1}^{n} \frac{C_{R_i}\sigma_i}{T}(D_i T) = \sum_{i=1}^{n} C_{R_i} D_i \sigma_i \quad (5.58)$$

Based on Fig. 5.4, the holding cost of the inventory system is obtained as:

$$\text{CH} = \sum_{i=1}^{n} \frac{h_i}{T} \left[\frac{I_i}{2}\left(\frac{Q_i}{P_i}\right) + \frac{H_i + I_i}{2}\left(\sigma_i \frac{Q_i}{P_i}\right) + \frac{H_i}{2}\left(\frac{H_i}{D_i}\right) \right]$$

$$\text{CH} = \sum_{i=1}^{n} \frac{h_i}{T} \left[\frac{I_i}{2}(t_i^p) + \frac{H_i + I_i}{2}(t_i^r) + \frac{H_i}{2}(t_i^d) \right] \quad (5.59)$$

From Eqs. (5.49) to (5.51), one has:

$$\text{CH} = \sum_{i=1}^{n} \frac{h_i}{T}$$

$$\times \left[\frac{((1-\sigma_i)P_i - D_i)(Q_i)^2}{2(P_i)^2} + \frac{\left(2 - \sigma_i 2\frac{D_i}{P_i} - \sigma_i \frac{D_i}{P_i}\right)Q_i^2}{2P_i} + \frac{\left(1 - \frac{D_i}{P_i} - \sigma_i \frac{D_i}{P_i}\right)Q_i^2}{2D_i} \right]$$

$$(5.60)$$

Finally, using Eq. (5.53) for Q_i results in:

$$\text{CH} = \sum_{i=1}^{n} \left[\left(\frac{h_i\left((D_i)(P_i)^2 + (D_i)^2(P_i) - 2(D_i)^3\right)}{2(P_i)^2} \right) T - \left(\frac{h_i\left(4(D_i)^2(P_i) - (D_i)^3\right)}{2(P_i)^2} \right) \sigma_i T + \left(\frac{h_i(D_i)^3}{2(P_i)^2} \right) \sigma_i^2 T \right]$$

$$(5.61)$$

Using Eqs. (5.55)–(5.58) and (5.61) into Eq. (5.54) results in the total inventory cost as presented in Eq. (5.62):

$$Z = TC = CA + CP + CD + CR + CC + CH$$

$$= \left\{ \begin{array}{l} \displaystyle\sum_{i=1}^{n}\frac{K_i}{T} + \sum_{i=1}^{n}(C_i + l_i\alpha_i)D_i + \sum_{i=1}^{n}D_i(C_{R_i} - l_i)\sigma_i - \sum_{i=1}^{n}\left[\frac{h_i\left(4(D_i)^2(P_i) - (D_i)^3\right)}{2(P_i)^2}\right]\sigma_i T \\ + \displaystyle\sum_{i=1}^{n}\left[\frac{h_i\left((D_i)(P_i)^2 + (D_i)^2(P_i) - 2(D_i)^3\right)}{2(P_i)^2}\right]T + \sum_{i=1}^{n}\left[\frac{h_i(D_i)^3}{2(P_i)^2}\right]\sigma_i^2 T \end{array} \right\}$$

$$(5.62)$$

The capacity of the single machine is the constraint of the EPQ inventory model. This constraint is formulated as shown in Eq. (5.63). The sum of the production,

5.3 No Shortage

rework, and setup times for all products cannot be greater than the common cycle length T. Hence:

$$\sum_{i=1}^{n} \left(t_i^p + t_i^r\right) + \sum_{i=1}^{n} \text{ts}_i \leq T \tag{5.63}$$

Substituting t_i^p and t_i^r from Eqs. (5.50) and (5.49), respectively, results in:

$$T \geq \left(\frac{\sum_{i=1}^{n} \text{ts}_i}{1 - \sum_{i=1}^{n} \frac{(1+\sigma_i)D_i}{P_i}} \right) = T_{\text{Min}} \tag{5.64}$$

The mathematical formulation of the inventory problem that aims to minimize the total inventory cost is shown as below:

$$\text{Min}: \quad Z = \sum_{i=1}^{n} \frac{K_i}{T} + \sum_{i=1}^{n} \chi_i^1 + \sum_{i=1}^{n} \chi_i^2(T) - \sum_{i=1}^{n} \chi_i^3(\sigma_i T) + \sum_{i=1}^{n} \chi_i^4(\sigma_i^2 T) + \sum_{i=1}^{n} \chi_i^5(\sigma_i)$$

s.t.

$$T \geq T_{\text{Min}}$$
$$0 \leq \sigma_i \leq \alpha_i; \quad i = 1, 2, \ldots, n$$

$$\tag{5.65}$$

where $\chi_i^1, \chi_i^2, \chi_i^3, \chi_i^4$, and χ_i^5 are the objective function coefficients.

In order to show the convexity of the nonlinear programming problem presented in Eq. (5.65), the determinant of Hessian matrix for the objective function is derived as below:

$$\text{Hessian} = \frac{2 \sum_i K_i}{T} + \left(2 \sum_i \Delta_i^2 \sigma_i + 6 \sum_i \Delta_i^3 \sigma_i^2 \right) T \tag{5.66}$$

As T is positive and all parameters are nonnegatives, the Hessian is greater than zero. Hence, taking into account that all the constraints in (5.65) are linear, the problem becomes a convex nonlinear programming problem (CNLPP). Consequently, the optimal solution can be first obtained using the derivatives. Then, the global minimum solution is determined by obtaining the optimal solution (Nobil and Taleizadeh 2016).

To find the optimal cycle length and the optimal proportion of produced reworkable for each product, the partial derivatives of the objective function Z are taken with respect to T and x_i. Setting the equations obtained by taking the derivatives, the optimal cycle length and proportion are given by:

$$T = \sqrt{\frac{\sum_i K_i - \sum_{i=1}^{n}\left(\frac{(\chi_i^5)^2}{4\chi_i^4}\right)}{\sum_i \chi_i^2}} \tag{5.67}$$

$$\sigma_i = \frac{\chi_i^3 T - \chi_i^5}{2\chi_i^4 T}; \quad i = 1, 2, \ldots, n \tag{5.68}$$

Here, a heuristic algorithm with nine steps is proposed to solve the EPQ problem under consideration.

Step 1. Calculate $\chi_i^1, \chi_i^2, \chi_i^3, \chi_i^4$, and χ_i^5 and go to *Step* 2.

Step 2. If $\sum_i K_i - \sum_{i=1}^{n}\left(\frac{(\chi_i^5)^2}{4\chi_i^4}\right) > 0$, then go to *Step* 3. Otherwise, the solution is infeasible; go to *Step* 9.

Step 3. Calculate T using Eq. (5.67) and σ_i using Eq. (5.68). Then, go to *Step* 4.

Step 4. If $\sigma_i \leq 0$, then $\sigma_i = 0$; else, if $\sigma_i \geq \alpha_i$, then $\sigma_i = \alpha_i$; then go to *Step* 5.

Step 5. If $\sum_{i=1}^{n} \frac{(1+\sigma_i)D_i}{P_i} < 1$, then go to *Step* 6. Otherwise, the solution is infeasible; go to *Step* 9.

Step 6. Obtain the lower bound of cycle length, i.e., T_{Min}, using Eq. (5.64) and go to *Step* 7.

Step 7. If $T \geq T_{\text{Min}}$, then set $T^* = T$. Otherwise, $T^* = T_{\text{Min}}$, and go to *Step* 8.

Step 8. Based on the value of T^*, calculate σ_i^*, Q_i^*, and Z^* using Eqs. (5.68), (5.53), and (5.65), respectively, and go to *Step* 9.

Step 9. Terminate the procedure.

Example 5.4 Nobil and Taleizadeh (2016) presented an example for an imperfect single-machine production system with three products. The general input data of this numerical example is shown in Table 5.10.

Then, the optimal solution is obtained based on the proposed solution procedure as follows:

Step 1. The values for $\chi_i^1, \chi_i^2, \chi_i^3, \chi_i^4$, and χ_i^5 are calculated and they are shown in Table 5.11.

Step 2. As $\sum_i K_i - \sum_{i=1}^{n}\left(\frac{(\chi_i^5)^2}{4\chi_i^4}\right) = 690.0595 > 0$, the initial feasibility is checked.

Then, go to *Step* 3.

Step 3. Using Eq. (5.67), T is obtained as:

$$T = \sqrt{\frac{\sum_i K_i - \sum_{i=1}^{3}\left(\frac{(\chi_i^5)^2}{4\chi_i^4}\right)}{\sum_i \chi_i^2}} = \sqrt{\frac{690.0595}{2728.2112}} = 0.5029$$

Moreover, based on Eq. (5.68), σ_is are calculated and are shown in Table 5.12.

5.3 No Shortage

Table 5.10 General data for example (Nobil and Taleizadeh 2016)

Item	P_i	D_i	α_i	ts_i	C_i	C_{R_i}	l_i	K_i	h_i
1	1000	300	0.10	0.010	100	50	54	3000	4
2	1100	350	0.09	0.015	90	40	42	3500	5
3	1300	320	0.08	0.020	80	30	28	4000	6

Table 5.11 The objective function coefficient (Nobil and Taleizadeh 2016)

Item	χ_i^1	χ_i^2	χ_i^3	χ_i^4	χ_i^5
1	31,620	672	666	54	−1200
2	32,823	976.2396	1025.0516	88.5847	−700
3	26,419.2	1079.9715	887.0627	58.1680	640

Table 5.12 Values of σ_i (Nobil and Taleizadeh 2016)

Item	1	2	3
σ_i	28.2595	13.6417	−3.3135

Table 5.13 The range of each σ_i (Nobil and Taleizadeh 2016)

Item	1	2	3
Range	[0–0.10]	[0–0.09]	[0–0.08]
σ_i	0.10	0.09	0.00

Step 4. The range of each σ_i is checked and they are shown in Table 5.13.
Step 5. The second feasibility condition

As $\sum_{i=1}^{n} \frac{(1+\sigma_i)D_i}{P_i} = 0.923 < 1$, the initial feasibility is checked. Then, go to *Step* 6.

Step 6. The lower bound is equal to 0.5842. Then, go to *Step* 7.
Step 7. Check the second optimality. As $T < T_{Min}$, then ($T* = T_{Min} = 0.584203359055833$), and go to *Step* 8.
Step 8. Calculate the optimal values and terminate the procedure. Based on $T^* = 0.5842$, σ_i and Q_i^* are computed and they are shown in Table 5.14. As result, the minimum annual inventory cost is $Z^* = \$110,051.754897060$.

5.3.5 Scrapped

Shafiee-Gol et al. (2016) studied pricing and production decisions in multi-product single-machine manufacturing system considering discrete delivery and rework. They solved that problem in two situations: first, when the condition is satisfied and, second, when it is not satisfied.

In this study, the finished products can only be delivered to customers if the whole lot is quality assured at the end of rework. In addition, in many circumstances, the delivery of the products is not continuous. All items produced are screened such that

Table 5.14 Values of σ_i^*, Q_i^*

	1	2	3
σ_i^*	0.10	0.09	0.00
Q_i^*	175.26	204.47	186.94

Table 5.15 New notations for given problem (Shafiee-Gol et al. 2016)

b_i	Price sensitivity of demand
φ_i	The fail portion of reworking item that scrapped
t_i^o	Preparation time for ith product
t_i^d	Time required for delivering all quality assured finished products
p	Number of fixed-quantity installments of the finished batch to be delivered to customers in each cycle, which is assumed to be a constant for all products
t_{ni}	A fixed interval of time between each installment of finished products delivered during t_i^r for ith product
$AP(T, Q)$	Total production–inventory–delivery profits per unit time

inspection cost is included in unit production cost. After the end of regular process, all defective products are reworked with an additional reworking cost. During the rework, a portion φ_i of reworked products fails and becomes scrap at an additional disposal cost C_{Si}. They considered $P_i - d_{1i} - D_i > 0$ (or $1 - x_i - D_i/P_i > 0$) to prevent occurring the shortage. When the whole lot is quality assured at the end of the rework, the finished items of each product can be delivered to customers (Shafiee-Gol et al. 2016).

In addition to notations introduced in Sect. 5.1, the presented one in Table 5.15 is used to model the on-hand problem.

The following equations can be obtained from Fig. 5.5:

$$I_i = (P_i - d_{1i})t_i^p = (P_i - d_{1i})\frac{Q_i}{P_i} = (1-x)Q_i \tag{5.69}$$

$$H_i = I_i + (P_{2i} - d_{2i})t_i^r = Q_i(1 - \varphi_i x) \tag{5.70}$$

$$t_i^p = \frac{Q_i}{P_i} = \frac{I_i}{P_i - d_{1i}} \tag{5.71}$$

$$t_i^r = \frac{xQ_i}{P_{2i}} \tag{5.72}$$

$$t_i^d = pt_{ni} = T - t_i^p - t_i^r = Q_i\left(\frac{(1-\varphi_i x)}{a_i - b_i s_i} - \frac{1}{P_i} - \frac{x}{P_{2i}}\right) \tag{5.73}$$

$$d_{1i}t_i^p = P_i x t_i^p = xQ_i \tag{5.74}$$

$$\varphi_i d_{1i} t_i^p = \varphi_i P_i x t_i^p = \varphi_i x Q_i \tag{5.75}$$

One of the necessary conditions in the multi-item production system is as follows:

5.3 No Shortage

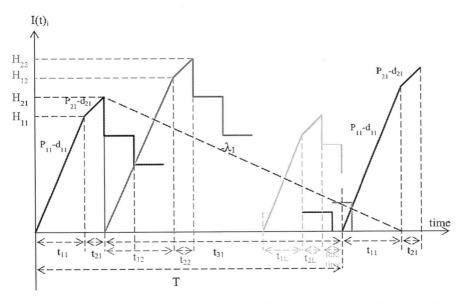

Fig. 5.5 On-hand inventory level of perfect-quality product i in the proposed multi-item production system under the common cycle time policy (Chiu et al. 2013)

$$\sum_{i=1}^{n} ts_i + \sum_{i=1}^{n} \left(t_i^p + t_i^r\right) \leq T \quad (5.76)$$

or

$$\sum_{i=1}^{n} ts_i \leq T - \sum_{i=1}^{n} \left(\frac{Q_i}{P_i} + \frac{xQ_i}{P_{2i}}\right) \quad (5.77)$$

For any given production cycle, the total production–inventory–delivery profit for all n products ($AP(T, Q)$), which is equal to sales revenue minus variable production costs, setup cost, cost of reworking defective items, disposal cost, fixed and variable delivery costs, holding cost during production uptime and rework time, and holding cost for finished goods during the delivery time, can be obtained as expressed in Eq. (5.83) (Shafiee-Gol et al. 2016).

$$\sum_{i=1}^{n} AP(\mathbf{S},\mathbf{Q}) = \sum_{i} \overbrace{s_i Q_i (1-\varphi_i x)}^{\text{revenue}} - \overbrace{C_{\text{I}i} Q_i}^{\text{inspection cost}} - \overbrace{C_{\text{R}i}(xQ_i)}^{\text{reworking cost}} - \overbrace{C_{\text{S}i}(xQ_i\varphi_i)}^{\text{disposal cost}} - \overbrace{K_i}^{\text{production setup cost}}$$

$$- \sum_{i} \overbrace{h_i \left[\frac{H_{1i}+d_{1i}t_{1i}}{2}(t_{1i}) + \frac{H_{1i}+H_{2i}}{2}(t_{2i}) + \frac{n-1}{2n}(H_{2i}t_{3i}) \right]}^{\text{holding cost}}$$

$$- \overbrace{h_{1i}\left[\frac{P_{1i}t_{2i}}{2}(t_{2i})\right]}^{\text{holding cost of reworked item}} - \sum_{i} \overbrace{C_{\text{T}i}[Q_i(1-\varphi_i x)]}^{\text{shipping cost}} - \overbrace{nK_{\text{S}i}}^{\text{fixed cost}}$$

(5.78)

Because T and every s_i and Q_i obey $T = \frac{(1-\varphi_i x)Q_i}{a_i - b_i s_i}$, $\forall i$, T and Q_i can be selected as decision variables. So:

$$s_i = \frac{a_i}{b_i} - \frac{(1-\varphi_i x)Q_i}{Tb_i} \qquad (5.79)$$

Equation (5.80) is the natural result of Eqs. (5.78) and (5.79). If constraint (5.77) is satisfied, they are optimal feasible solutions. Otherwise, Lagrangian relaxation method is used:

$$AAP(T,\mathbf{Q}) = \frac{AP(T,\mathbf{Q})}{T}$$

$$= \frac{1}{T}\left[\sum_{i=1}^{n}\left(\frac{a_i}{b_i} - \frac{(1-\varphi_i x)Q_i}{T\beta_i}\right)(1-\varphi_i x)Q_i - C_{\text{T}i}[Q_i(1-\varphi_i x)]\right.$$

$$- (nK_{\text{S}i} + K_i + C_{\text{I}i}Q_i + C_{\text{R}i}(xQ_i) + C_{\text{S}i}(xQ_i\varphi_i)) + h_{1i}\left[\frac{P_{1i}}{2}\left(\frac{xQ_i}{P_{2i}}\right)^2\right]$$

$$\left.- h_i\left[\frac{Q_i^2}{2P_{1i}} + \frac{Q_i(2-\varphi_i x - x)}{2}\left(\frac{xQ_i}{P_{2i}}\right) + \frac{n-1}{2n}Q_i(1-\varphi_i x)\left(T - \left(\frac{Q_i}{P_{1i}} + \frac{xQ_i}{P_{2i}}\right)\right)\right]\right]$$

(5.80)

In summary, the function that should be maximized is as follows:

$$AAP(T,\mathbf{Q}) = \sum_{i}\left(\frac{A_i Q_i^2}{T^2} + \frac{B_i Q_i^2}{T} + \frac{D_i Q_i}{T} + \frac{G_i}{T} + J_i Q_i\right) \qquad (5.81)$$

where:

5.3 No Shortage

$$A_i = -\frac{(1-\varphi_i x)^2}{b_i} \tag{5.82}$$

$$B_i = \frac{h_{1i}P_{1i}x^2}{2P_{2i}^2} - \frac{h_i}{2P_{1i}} - \frac{h_i x(2-\varphi_i x - x)}{2P_{2i}} + \frac{h_i(n-1)(1-\varphi_i x)}{2P_{1i}}$$
$$+ \frac{h_i(n-1)(1-\varphi_i x)x}{2P_{2i}} \tag{5.83}$$

$$D_i = \frac{a_i(1-\varphi_i x)}{b_i} + C_i + C_{Ri}x + C_{Si}x\varphi_i + C_{Ti}[(1-\varphi_i x)] \tag{5.84}$$

$$G_i = -nK_{Si} - K_i \tag{5.85}$$

$$J_i = \frac{h_i(n-1)(1-\varphi_i x)}{2n} \tag{5.86}$$

Constraint (5.77) can be regarded as (5.82). Therefore, this function should be maximized according to a linear constraint of decision variables:

$$Z - \sum_i N_i Q_i - T \leq 0 \tag{5.87}$$

From calculus, it is well known that $F(T, Q)$ is concave if and only if the following equations are satisfied:

$$\mathbf{H} = \begin{bmatrix} \dfrac{\partial^2 F(T,\mathbf{Q})}{\partial Q_1^2} & \dfrac{\partial^2 F(T,\mathbf{Q})}{\partial Q_1 \partial Q_2} & \cdots & \dfrac{\partial^2 F(T,\mathbf{Q})}{\partial Q_1 \partial Q_n} & \dfrac{\partial^2 F(T,\mathbf{Q})}{\partial Q_1 \partial T} \\ \dfrac{\partial^2 F(T,\mathbf{Q})}{\partial Q_2 \partial Q_1} & \dfrac{\partial^2 F(T,\mathbf{Q})}{\partial Q_2^2} & \cdots & \dfrac{\partial^2 F(T,\mathbf{Q})}{\partial Q_2 \partial Q_n} & \dfrac{\partial^2 F(T,\mathbf{Q})}{\partial Q_2 \partial T} \\ \vdots & \vdots & \ddots & \vdots & \vdots \\ \dfrac{\partial^2 F(T,\mathbf{Q})}{\partial Q_n \partial Q_1} & \dfrac{\partial^2 F(T,\mathbf{Q})}{\partial Q_n \partial Q_2} & \cdots & \dfrac{\partial^2 F(T,\mathbf{Q})}{\partial Q_n^2} & \dfrac{\partial^2 F(T,\mathbf{Q})}{\partial Q_n \partial T} \\ \dfrac{\partial^2 F(T,\mathbf{Q})}{\partial T \partial Q_1} & \dfrac{\partial^2 F(T,\mathbf{Q})}{\partial T \partial Q_2} & \cdots & \dfrac{\partial^2 F(T,\mathbf{Q})}{\partial T \partial Q_n} & \dfrac{\partial^2 F(T,\mathbf{Q})}{\partial T^2} \end{bmatrix} \tag{5.88}$$

$$\mathbf{H} = \begin{bmatrix} \dfrac{2K_1}{T^2} + 2\dfrac{B_1}{T} & 0 & \cdots & 0 & \dfrac{\partial^2 F(T,\mathbf{Q})}{\partial T \partial Q_1} \\ 0 & \dfrac{2K_2}{T^2} + 2\dfrac{B_2}{T} & \cdots & 0 & \dfrac{\partial^2 F(T,\mathbf{Q})}{\partial T \partial Q_2} \\ \vdots & \vdots & \ddots & \vdots & \vdots \\ 0 & 0 & \cdots & \dfrac{2K_n}{T^2} + 2\dfrac{B_n}{T} & \dfrac{\partial^2 F(T,\mathbf{Q})}{\partial T \partial Q_n} \\ \dfrac{\partial^2 F(T,\mathbf{Q})}{\partial Q_1 \partial T} & \dfrac{\partial^2 F(T,\mathbf{Q})}{\partial Q_2 \partial T} & \cdots & \dfrac{\partial^2 F(T,\mathbf{Q})}{\partial Q_n \partial T} & \dfrac{\partial^2 F(T,\mathbf{Q})}{\partial T^2} \end{bmatrix} \tag{5.89}$$

$$[Q_1 \quad Q_2 \ldots \quad Q_n \quad T] \times \mathbf{H} \times [Q_1 \quad Q_2 \ldots \quad Q_n \quad T]^T < 0 \qquad (5.90)$$

Above equation can be changed as below:

$$\sum_{i=1}^{n} \sum_{j=1}^{n} 2Q_i Q_j \frac{\partial^2 F(T,\mathbf{Q})}{\partial Q_i \partial Q_j} + 2TQ_i \frac{\partial^2 F(T,\mathbf{Q})}{\partial T \partial Q_i} + Q_i^2 \frac{\partial^2 F(T,\mathbf{Q})}{\partial Q_i^2}$$
$$+ T^2 \frac{\partial^2 F(T,\mathbf{Q})}{\partial T^2}$$
$$< 0 \qquad (5.91)$$

After calculations (Shafiee-Gol et al. 2016):

$$\sum_i Q_i^2 \left(\frac{2K_i}{T^2} + \frac{2B_i}{T} \right) + T^2 \left(\frac{2D_i Q_i}{T^3} + \frac{6K_i Q_i^2}{T^4} + \frac{2G_i}{T^3} + \frac{2B_i Q_i^2}{T^3} \right)$$
$$+ 2TQ_i \left(-\frac{D_i}{T^2} - \frac{4K_i Q_i}{T^3} - \frac{2B_i Q}{T^2} \right) < 0 \qquad (5.92)$$

and then:

$$\frac{\sum_i G_i}{T} < 0 \qquad (5.93)$$

This means that for concavity of $AAP(T, Q)$, it is enough to have $\sum_i G_i < 0$. Now, the derivative of $AAP(T, Q)$ is taken with respect to T and Q_i. And thus obtain the values of T and Q_i:

$$\frac{\partial AAP(T,\mathbf{Q})}{\partial T} = \sum_i \frac{-2K_i Q_i^2}{T^3} - \frac{B_i Q_i^2 + D_i Q_i + G_i}{T^2} = 0 \qquad (5.94)$$

$$\frac{\partial AAP(T,\mathbf{Q})}{\partial Q_i} = \frac{2A_i Q_i}{T^2} + \frac{2B_i Q_i}{T} + \frac{D_i}{T} + J_i = 0 \qquad (5.95)$$

With respect to the second term:

$$Q_i = \frac{(-D_i - J_i T)T}{2A_i + 2B_i T} \qquad (5.96)$$

By multiplying Eq. (5.94) by T^3:

5.3 No Shortage

$$\sum_i Q_i^2(2A_i + B_iT) + Q_iD_iT + G_iT = 0 \qquad (5.97)$$

After that, with replacing $Q_i = \frac{(-D_i - J_iT)T}{2A_i + 2B_iT}$ in the first Eq. (5.94):

$$\sum_i \left(B_iJ_i^2\right)T^4 + \sum_i \left(2A_iJ_i^2\right)T^3 + \sum_i \left(2A_iD_iJ_i + 4B_i^2G_i - B_iD_i^2\right)T^2$$
$$+ \sum_i (8A_iB_iG_i)T + \sum_i 4A_i^2G_i = 0 \qquad (5.98)$$

Now, the above fourth-degree equation can be solved, and the variables T and then Q_i from (5.95) can obtained. If the obtained values for the decision variables satisfied the constraint, there are optimal values. Otherwise, the Lagrangian relaxation method is used.

Example 5.5 To validate the proposed model and illustrate its various features, one numerical example is solved using MATLAB software. The information related to the example is given in Table 5.16. By solving the example, obtained values for decision variables are presented in Table 5.17. It can be easily verified that the constraint (5.77) is satisfied (0.015 < 0.2273). So, these are optimal values (Shafiee-Gol et al. 2016).

If the obtained values for the decision variables do not satisfy the constraint, the Lagrangian relaxation method should be used.

Example 5.6 In the abovementioned example, if the amount of P_i and P_{2i} change to 600, 600, 700 and 1000, 1000, 1100, respectively, the achieved values for decision variables do not satisfy the constraint:

$Q^* = [10.6352, 37.2475, 74.5102]$.
$s^* = [5006.5, 11{,}674, 17{,}508]$.
$T = 0.2098$.
The constraint: $0.015 \leqslant 0.0062$.

Because the constraint is not satisfied, they should use the Lagrangian relaxation method to find the optimal values for the variables. Because there is only one constraint, it is convenient to use the Lagrangian relaxation method and move the constraint to the objective function. Equation (5.99) is the natural result of Eqs. (5.82) and (5.85):

Table 5.16 The value of the parameters for the numerical instance (Shafiee-Gol et al. 2016)

Parameters	$i=1$	$i=2$	$i=3$
P_i	10,000	12,000	11,000
P_{2i}	1200	1400	11,000
φ_i	0.1	0.1	0.1
C_{Ii}	10	11	12
C_{Ri}	5	6	4
C_{Si}	2	2	2
C_{Ti}	0.5	0.5	0.5
K_i	15	16	12
K_{Si}	39	40	38
h_i	20	25	18
h_{1i}	35	40	30
x	0.15	0.15	0.15
β_i	0.01	0.015	0.02
D_i	100	350	700
ts_i	0.05	0.05	0.05

Table 5.17 Optimal values for variables and total profit (Shafiee-Gol et al. 2016)

Variables	$i=1$	$i=2$	$i=3$
Q_i	12.2535	42.9164	85.8503
s_i	5006.5	11,674	17,507
T		0.2417	
$AAP(T, \mathbf{Q})$		8,406,600	

$$AAP(T,u,\mathbf{Q}) = \frac{AP(T,u,\mathbf{Q})}{T}$$

$$= \left(\frac{1}{T}\right) \left(\begin{array}{c} \sum_{i=1}^{n} \left(\frac{a_i}{b_i} - \frac{(1-\varphi_i x)Q_i}{Tb_i} \right)(1-\varphi_i x)Q_i \\ - \left[\begin{array}{c} nK_{Si} + K_i + C_i Q_i + C_{Ri}(xQ_i) + C_{Si}(xQ_i\varphi_i) + C_{Ti}[Q_i(1-\varphi_i x)] \\ + h_{1i}\left[\frac{P_{1i}}{2}\left(\frac{xQ_i}{P_{2i}}\right)^2\right] + h_i \left[\begin{array}{c} \frac{Q_i^2}{2P_{1i}} + \frac{Q_i(2-\varphi_i x - x)}{2}\left(\frac{xQ_i}{P_{2i}}\right) \\ + \frac{n-1}{2n}\left(Q_i(1-\varphi_i x)\left(T - \left(\frac{Q_i}{P_{1i}} + \frac{xQ_i}{P_{2i}}\right)\right)\right) \end{array} \right] \end{array} \right] \end{array} \right)$$
$$- u\left(\sum_{i=1}^{n} ts_i - T + \sum_{i=1}^{n}\left(\frac{Q_i}{P_{1i}} + \frac{xQ_i}{P_{2i}}\right) \right)$$

(5.99)

In summary, the function that should be maximized is as follows:

$$AAP(T,u,\mathbf{Q}) = \sum_i \left(\frac{A_i Q_i^2}{T^2} + \frac{B_i Q_i^2}{T} + \frac{D_i Q_i}{T} + \frac{G_i}{T} + J_i Q_i + M_i T \cdot u + N_i \cdot u \cdot Q_i + Y_i \cdot u \right)$$

(5.100)

where:

5.3 No Shortage

$$M = 1, \quad N_i = \frac{1}{P_{1i}} - \frac{x}{P_{2i}}, \quad Y_i = \text{ts}_i, \quad Z = \sum_{i=1}^{n} \text{ts}_i \quad (5.101)$$

However, this function should be maximized according to constraint (5.87). From calculus, it is well known that $F(T, u, \mathbf{Q})$ is concave if and only if the following equations are satisfied:

$$\mathbf{H} = \begin{bmatrix} \frac{\partial^2 F(T,u,\mathbf{Q})}{\partial Q_1^2} & \cdots & \frac{\partial^2 F(T,u,\mathbf{Q})}{\partial Q_1 \partial Q_n} & \frac{\partial^2 F(T,u,\mathbf{Q})}{\partial Q_1 \partial T} & \frac{\partial^2 F(T,u,\mathbf{Q})}{\partial Q_1 \partial u} \\ \vdots & \ddots & \vdots & \vdots & \vdots \\ \frac{\partial^2 F(T,u,\mathbf{Q})}{\partial Q_n \partial Q_1} & \cdots & \frac{\partial^2 F(T,u,\mathbf{Q})}{\partial Q_n^2} & \frac{\partial^2 F(T,u,\mathbf{Q})}{\partial Q_n \partial T} & \frac{\partial^2 F(T,u,\mathbf{Q})}{\partial Q_n \partial u} \\ \frac{\partial^2 F(T,u,\mathbf{Q})}{\partial T \partial Q_1} & \cdots & \frac{\partial^2 F(T,u,\mathbf{Q})}{\partial T \partial Q_n} & \frac{\partial^2 F(T,u,\mathbf{Q})}{\partial T^2} & \frac{\partial^2 F(T,u,\mathbf{Q})}{\partial T \partial u} \\ \frac{\partial^2 F(T,u,\mathbf{Q})}{\partial u \partial Q_1} & \cdots & \frac{\partial^2 F(T,u,\mathbf{Q})}{\partial u \partial Q_n} & \frac{\partial^2 F(T,u,\mathbf{Q})}{\partial u \partial T} & \frac{\partial^2 F(T,u,\mathbf{Q})}{\partial u^2} \end{bmatrix}$$

(5.102)

$$[Q_1 \quad \cdots \quad Q_n \quad T \quad u] \times \mathbf{H} \times [Q_1 \quad \cdots \quad Q_n \quad T \quad u]^T < 0 \quad (5.103)$$

Above equation can be changed as below:

$$\sum_{i=1}^{n} \sum_{j=1}^{n} Q_i^2 \frac{\partial^2 F(T,u,\mathbf{Q})}{\partial Q_i^2} + T^2 \frac{\partial^2 F(T,u,\mathbf{Q})}{\partial T^2}$$
$$+ u^2 \frac{\partial^2 F(T,u,\mathbf{Q})}{\partial u^2} + 2Q_i Q_j \frac{\partial^2 F(T,u,\mathbf{Q})}{\partial Q_i \partial Q_j} \quad (5.104)$$
$$+ 2TQ_i \frac{\partial^2 F(T,u,\mathbf{Q})}{\partial T \partial Q_i} + 2uQ_i \frac{\partial^2 F(T,u,\mathbf{Q})}{\partial u \partial Q_i} + 2u \cdot T \frac{\partial^2 F(T,u,\mathbf{Q})}{\partial u \partial T} < 0$$

After calculations, one has:

$$\sum_i T^2 \left(\frac{2D_i Q_i}{T^3} + \frac{6A_i Q_i^2}{T^4} + \frac{2G_i}{T^3} + \frac{2B_i Q_i^2}{T^3} \right) + Q_i^2 \left(\frac{2A_i}{T^2} + \frac{2B_i}{T} \right)$$
$$+ 2TQ_i \left(-\frac{D_i}{T^2} - \frac{4A_i Q_i}{T^3} - \frac{2B_i Q}{T^2} \right) + 2N_i Q_i \cdot u + 2M_i T \cdot u < 0$$

(5.105)

and then (Shafiee-Gol et al. 2016):

$$\sum_i \frac{-2}{T}(nk_{1i}+k_i) - 2Q_i u\left(\frac{1}{p_{1i}}+\frac{x}{p_{2i}}\right) + 2uT = -45,350.40015 < 0 \quad (5.106)$$

Now, one should take the derivative of $AAP(T, u, Q)$ with respect to T, Q_i, and u and thus obtain the values of T, Q_i, and u:

$$\frac{\partial AAP(T,u,\mathbf{Q})}{\partial Q_i} = \frac{2A_i Q_i}{T^2} + \frac{2B_i Q_i}{T} + \frac{D_i}{T} + J_i + N_i \cdot u = 0 \quad (5.107)$$

$$\frac{\partial AAP(T,u,\mathbf{Q})}{\partial T} = \sum_i \frac{-2A_i Q_i^2}{T^3} - \frac{B_i Q_i^2}{T^2} - \frac{D_i Q_i}{T^2} - \frac{G_i}{T^2} + M_i \cdot u = 0 \quad (5.108)$$

$$\frac{\partial AAP(T,u,\mathbf{Q})}{\partial u} = \sum_i Y_i + M_i T + N_i Q_i = 0 \quad (5.109)$$

With respect to relation (5.107), one has:

$$Q_i = \frac{-D_i T - J_i T^2 - N_i u T^2}{2A_i + 2B_i T} \quad (5.110)$$

Next, replacing Q_i in Eqs. (5.108) and (5.109) yields to:

$$\sum_i \left(-(2A_i + B_i T)\left[\frac{D_i + J_i T + N_i u T}{2A_i + 2B_i T}\right]^2 - D_i T\left[\frac{D_i + J_i T + N_i u T}{2A_i + 2B_i T}\right] - G_i T + M_i . u T^3 \right)$$
$$= 0$$

$$(5.111)$$

$$\sum_i Y_i + M_i T - N_i T\left[\frac{D_i + J_i T + N_i u T}{2A_i + 2B_i T}\right] = 0 \quad (5.112)$$

With respect to the relation (5.112), one has:

$$u = \frac{\sum_i Y_i + M_i T - N_i T\left[\frac{D+J_i T}{2A_i + 2B_i T}\right]}{\sum_i \left[\frac{N_i^2 T}{2A_i + 2B_i T}\right]} \quad (5.113)$$

After that, by replacing it in Eq. (5.111) and solve the equation, the value of T will be obtained. Then, using Eqs. (5.113) and (5.110), the value of u and Q_i will be determined. Table 5.18 shows the optimal values of variables which also satisfy the constraint $(0.015 \leq 0.03761)$ (Shafiee-Gol et al. 2016).

Table 5.18 Optimal values for variables and total profit (Shafiee-Gol et al. 2016)

Variables	$i = 1$	$i = 2$	$i = 3$
Q_i	61.98	217.39	435.12
s_i	5020.45	11,690.36	17,521.76
T		1.226	
$AP(T, \mathbf{Q})$		8,380,581.06	

5.4 Backordering

In this subsection, five different multi-product single-machine problems with shortage backordering are presented.

5.4.1 Defective Items

Taleizadeh et al. (2010c) presented an economic production quantity model in which the production defective rate follows either a uniform or a normal probability distribution. Shortages are allowed and take backorder state, and the existence of only one machine causes a limited production capacity for the common cycle length of all products. The aim of this problem is to determine the optimal production quantity of each product such that the expected total cost is minimized. All defective items are assumed to be scrapped; i.e., no rework is allowed. The expected production rate of the scrapped items d can be expressed as $\theta = PE[X]$. Also, they assumed that there is a real constant production capacity limitation on a single machine on which all products are produced (Taleizadeh et al. 2010c).

Since the problem at hand is of multi-product type, for products $i = 1, 2, \ldots, n$, the following notations (presented in Table 5.19) and notations introduced in Sect. 5.1 are used:

The production rate P_i is always assumed to be greater than or equal to the demand rate D_i. Furthermore, the production rate of the perfect-quality items is assumed to be greater than or equal to the sum of the demand rate and the production rate of defective items; mathematically speaking, $P_i - d_i - D_i \geq 0$, or $1 - E[X_i] - D_i/P_i \geq 0$. Figure 5.6 depicts the on-hand inventory level and allowable backorder level of the EPQ model with backlogging permitted.

To model the problem, a part of the modeling procedure is adopted from Hayek and Salameh (2001). Since all products are manufactured on a single machine with a limited capacity, the cycle length for all of them is equal ($T_1 = T_2 = \cdots = T_n = T$). Then, based on Fig. 5.6, for $i = 1, 2, \ldots, n$ (Taleizadeh et al. 2010c):

$$T = \sum_{j=1}^{4} t_i^j = \frac{Q_i E(1 - X_i)}{D_i} \qquad (5.114)$$

Table 5.19 New notations for given problem (Shafiee-Gol et al. 2016)

$f_{X_i}(x_i)$: The probability distribution function of X_i
t_i^3: The permitted shortage time of the ith product
t_i^4: The time needed to satisfy all backorders in the next production of the ith product
C_S: The annual expected total scrapped item cost
$E(\cdot)$: The expected value

$$t_i^1 = \frac{I_i^1}{P_i - D_i - d_i} \tag{5.115}$$

$$I_i^1 = (P_i - D_i - d_i)\frac{Q_i}{P_i} - B_i \tag{5.116}$$

$$t_i^2 = \frac{I_i^1}{D_i} \tag{5.117}$$

$$t_i^3 = \frac{B_i}{D_i} \tag{5.118}$$

$$t_i^4 = \frac{B_i}{P_i - D_i - d_i} \tag{5.119}$$

$$(t_i^1 + t_i^4) = \frac{Q_i}{P_i} \tag{5.120}$$

The objective function of the model is the summation of the expected annual production, holding, shortage, disposal, and setup costs as:

$$Z = CP + CH + CB + CS + CA \tag{5.121}$$

The production cost per unit and the production quantity per period of the ith product are C_i and Q_i, respectively. Hence, the production cost of the ith product per period is C_iQ_i. While the total annual production cost of the ith product in a disjoint production policy is NC_iQ_i, this cost for the joint policy is $\frac{C_iQ_i}{T}$. Furthermore, based on Eq. (5.114), one has:

$$Q_i = \frac{T \times D_i}{E(1 - X_i)} = \frac{T \times D_i}{1 - E(X_i)} \tag{5.122}$$

Hence, the expected annual production cost is:

$$CP = \sum_{i=1}^{n} \frac{C_i\left[\frac{T \times D_i}{1 - E(X_i)}\right]}{T} = \sum_{i=1}^{n} \frac{C_i D_i}{1 - E(X_i)} \tag{5.123}$$

5.4 Backordering

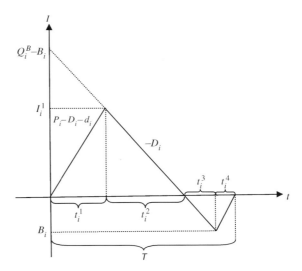

Fig. 5.6 A Production–inventory cycle (Taleizadeh et al. 2010c)

The holding cost per unit of the ith product per unit time for both the healthy and the scrapped items is h_i. According to Fig. 5.6, the total holding cost of healthy items per cycle and per year are shown in (5.124) and (5.125), respectively:

$$\sum_{i=1}^{n} h_i \left[\frac{I_i^1}{2} \left(t_i^1 + t_i^2 \right) \right] \tag{5.124}$$

$$N \sum_{i=1}^{n} h_i \left[\frac{I_i^1}{2} \left(t_i^1 + t_i^2 \right) \right] \tag{5.125}$$

However, Eq. (5.125) for the joint production policy becomes:

$$\frac{1}{T} \sum_{i=1}^{n} h_i \left[\frac{I_i^1}{2} \left(t_i^1 + t_i^2 \right) \right] \tag{5.126}$$

Finally, the expected total annual holding cost of healthy items is:

$$\sum_{j=1}^{n} h_i (P_i - d_i) \\ \times \left[\frac{(P_i - D_i - d_i) T \times D_i}{2(P_i)^2 (1 - E(X_i))^2} - \frac{B_i}{P_i (1 - E(X_i))} + \frac{(B_i)^2}{2 D_i T (P_i - D_i - d_i)} \right] \tag{5.127}$$

Since the scrapped items of each product is held until the end of its production time, similarly, based on Fig. 5.6, the total holding cost of the scrapped items per cycle and per year is shown in (5.128) and (5.129), respectively:

$$\sum_{i=1}^{n} h_i \left[\frac{d_i(t_i^1 + t_i^4)}{2} (t_i^1 + t_i^4) \right] \quad (5.128)$$

$$N \sum_{i=1}^{n} h_i \left[\frac{d_i(t_i^1 + t_i^4)}{2} (t_i^1 + t_i^4) \right] \quad (5.129)$$

Again for the joint production policy, Eq. (5.129) becomes:

$$\frac{1}{T} \sum_{j=1}^{n} h_i \left[\frac{d_i(t_i^1 + t_i^4)}{2} (t_i^1 + t_i^4) \right] \quad (5.130)$$

Hence, the expected total annual holding cost of scrapped items is:

$$\sum_{i=1}^{n} h_i d_i \left[\frac{(D_i)^2 \times T}{2(1 - E(X_i))^2 (P_i)^2} \right] \quad (5.131)$$

Finally, the expected total annual holding cost of healthy and scrapped items is:

$$\mathrm{CH} = \sum_{i=1}^{n} h_i(P_i - d_i) \left[\frac{(P_i - D_i - d_i)T \times D_i}{2(P_i)^2(1 - E(X_i))^2} - \frac{B_i}{P_i(1 - E(X_i))} + \frac{(B_i)^2}{2D_i T(P_i - D_i - d_i)} \right]$$
$$+ \sum_{j=1}^{n} h_i \left[\frac{(D_i)^2 T}{2(1 - E(X_i))^2 (P_i)^2} \right]$$

$$(5.132)$$

The shortage cost per unit of the ith product is C_{bi}, Based on Fig. 5.1, the total shortage cost per cycle and per year are shown in (5.133) and (5.134), respectively:

$$\sum_{i=1}^{n} C_{bi} \left[\frac{B_i}{2} (t_i^3 + t_i^4) \right] \quad (5.133)$$

$$N \times \sum_{i=1}^{n} C_{bi} \left[\frac{B_i}{2} (t_i^3 + t_i^4) \right] \quad (5.134)$$

Because of the joint production policy, Eq. (5.134) becomes:

5.4 Backordering

$$\frac{1}{T}\sum_{i=1}^{n} C_{bi}\left[\frac{B_i}{2}\left(t_i^3 + t_i^4\right)\right] \tag{5.135}$$

Finally, the expected total annual shortage cost is:

$$\text{CB} = \frac{1}{T}\sum_{i=1}^{n} C_{bi}\left[\frac{(P_i - d_i)(B_i)^2}{2D_i(P_i - D_i - d_i)}\right] \tag{5.136}$$

The disposal cost per unit of the scrapped item of the ith product is C_{Si}, and the quantity of scrapped items is $E(X_i)Q_i$. The expected total disposal cost per cycle is $\sum_{i=1}^{n} C_{Si}E(X_i)Q_i$. This quantity per year becomes:

$$N \times \sum_{i=1}^{n} C_{Si}E(X_i)Q_i \tag{5.137}$$

Because of the joint production policy, Eq. (5.137) becomes:

$$\frac{1}{T} \times \sum_{i=1}^{n} C_{Si}E(X_i)Q_i \tag{5.138}$$

Since $Q_i^B = \frac{T \times D_i}{E(1-X_i)} = \frac{T \times D_i}{1-E(X_i)}$, the annual expected total scrapped item cost is:

$$\text{CS} = \frac{1}{T} \times \sum_{i=1}^{n} C_{Si}E(X_i)\left[\frac{T \times D_i}{1 - E(X_i)}\right] = \sum_{i=1}^{n} \frac{C_{Si}E(X_i)D_i}{1 - E(X_i)} \tag{5.139}$$

The cost of a setup is K which occurs N times per year. So, the annual setup cost will be:

$$\text{CA} = N \times K = \frac{K}{T} \tag{5.140}$$

As a result, the objective function of the model becomes:

$$Z = CP + CH + CB + CS + CA = \sum_{i=1}^{n} \frac{C_i D_i}{1 - E(X_i)}$$

$$+ \sum_{i=1}^{n} h_i (P_i - \theta_i) \left[\frac{(P_i - D_i - d_i)T \times D_i}{2(P_i)^2 (1 - E(X_i))^2} - \frac{B_i}{P_i(1 - E(X_i))} + \frac{(B_i)^2}{2D_i T(P_i - D_i - d_i)} \right]$$

$$+ \sum_{i=1}^{n} h_i \left[\frac{(D_i)^2 T}{2(1 - E(X_i))^2 (P_i)^2} \right] + \frac{1}{T} \sum_{i=1}^{n} C_{bi} \left[\frac{(P_i - \theta_i)(B_i)^2}{2D_i(P_i - D_i - d_i)} \right] + \sum_{i=1}^{n} \frac{C_{Si} E(X_i) D_i}{1 - E(X_i)} + \frac{K}{T}$$

$$= \sum_{i=1}^{n} \left[\frac{(C_{bi} + h_i)(P_i - d_i)}{2D_i(P_i - D_i - d_i)} \right] \frac{(B_i)^2}{T} + \sum_{i=1}^{n} h_i \left[\frac{D_i[(P_i - d_i)(P_i - D_i - d_i) + D_i]}{2(P_i)^2 (1 - E(X_i))^2} \right] T$$

$$- \sum_{i=1}^{n} \left[\frac{h_i(P_i - d_i)}{P_i(1 - E(X_i))} \right] B_i + \sum_{i=1}^{n} \left[\frac{(C_i + C_{Si} E(X_i)) D_i}{1 - E(X_i)} \right] + \frac{K}{T}$$

(5.141)

To make sure that all of the n products will be produced by a single machine, a capacity limitation should be considered (Taleizadeh et al. 2010c). Since $t_i^1 + t_i^4$ and ts_i are the production time and setup time of the ith product, respectively, the summation of the total production and setup time (for all products) will be $\sum_{i=1}^{n} (t_i^1 + t_i^4) + \sum_{i=1}^{n} ts_i$ in which it should be smaller or equal to the period length (T). So the constraint of the model will be:

$$\sum_{i=1}^{n} (t_i^1 + t_i^4) + \sum_{i=1}^{n} ts_i \leq T \qquad (5.142)$$

Then:

$$T \geq \frac{\sum_{i=1}^{n} ts_i}{1 - \sum_{i=1}^{n} \frac{D_i}{P_i(1 - E(X_i))}} \qquad (5.143)$$

According to Eqs. (5.142) and (5.143), the final model becomes (Taleizadeh et al. 2010c):

$$\text{Min}: \quad Z = \frac{K}{T} + \sum_{i=1}^{n} \left[\frac{(C_{bi} + h_i)(P_i - d_i)}{2D_i(P_i - D_i - d_i)} \right] \frac{(B_i)^2}{T} - \sum_{i=1}^{n} \left[\frac{h_i(P_i - d_i)}{P_i(1 - E(X_i))} \right] B_i$$

$$+ \sum_{i=1}^{n} h_i \left[\frac{D_i[(P_i - d_i)(P_i - D_i - d_i) + D_i]}{2(P_i)^2(1 - E(X_i))^2} \right] T + \sum_{i=1}^{n} \left[\frac{(C_i + C_{Sj} E(X_i)) D_i}{1 - E(X_i)} \right]$$

s.t.:

$$T \geq \frac{\sum_{i=1}^{n} ts_i}{1 - \sum_{i=1}^{n} \frac{D_i}{P_i(1 - E(X_i))}}$$

$$B_i \geq 0 \quad \forall i, i = 1, 2, \ldots, n$$

(5.144)

5.4 Backordering

Taleizadeh et al. (2010c) proposed a solution method for the model presented in Eq. (5.144).

In order to find the optimal solution of model (5.144), they first provided a proof of the convexity of the objective function. Then, they used the derivative approach with respect to T and B_i to find the optimal point of the objective function. Furthermore, the production rate of the perfect-quality items is assumed to be greater than or equal to the sum of the demand rate and the production rate of defective items; mathematically speaking, $P_i - d_i - D_i \geq 0$, or $1 - E[X_i] - \frac{D_i}{P_i} \geq 0$ and in multi-product case should be $\sum_{i=1}^{n} \frac{D_i}{P_i(1-E[X_i])} \leq 1$. To handle the constraint, they checked the optimal solution in Eq. (5.143). If the constraint is not satisfied, then T_{Min} as the minimum value of T will be considered as the optimal point. To prove the convexity of the objective function, let us rewrite the objective function as (Taleizadeh et al. 2010c):

$$Z = \sum_{i=1}^{n} \psi_i^1 \frac{(B_i)^2}{T} - \sum_{i=1}^{n} \psi_i^2 B_i + \sum_{i=1}^{n} \psi_i^3 T + \sum_{i=1}^{n} \psi_i^4 + \frac{K}{T} \tag{5.145}$$

In which:

$$\psi_i^1 = \left[\frac{(C_{bi} + h_i)(P_i - d_i)}{2D_i(P_i - d_i - D_i)} \right] > 0 \tag{5.146}$$

$$\psi_i^2 = \left[\frac{h_i(P_i - d_i)}{P_i(1 - E(X_i))} \right] > 0 \tag{5.147}$$

$$\psi_i^3 = \left[\frac{h_i D_i[(P_i - d_i)(P_i - d_i - D_i) + D_i]}{2(P_i)^2(1 - E(X_i))^2} \right] > 0 \tag{5.148}$$

$$\psi_i^4 = \left[\frac{(C_i + C_{Si} E(X_i))D_i}{1 - E(X_i)} \right] > 0 \tag{5.149}$$

Theorem 5.2 The objective function Z in (5.143) is convex (Taleizadeh et al. 2010c).

Proof To prove the convexity of Z, one can utilize the Hessian matrix equation as:

$$[T, B_1, B_2, \ldots, B_n] \times \mathbf{H} \times [T, B_1, B_2, \ldots, B_n]^T = \frac{2K}{T} \geq 0 \tag{5.150}$$

Then, the objective function is strictly convex. The expected cost function Z is strictly convex for all nonzero T and B_i. Hence, it follows that to find the optimal production period length and the optimal level of backorder B_i, one can partially differentiate Z with respect to T and B_i and solve the resulted system of equations obtained by letting the partial derivatives be equal to zero; thus, the systems of equations become:

$$\frac{\partial Z}{\partial T} = 0 \to T = \sqrt{\frac{K}{\sum_{i=1}^{n} \psi_i^3 - \sum_{i=1}^{n} \left[\frac{(\psi_i^2)^2}{4\psi_i^1}\right]}} \quad (5.151)$$

$$\frac{\partial Z}{\partial B_i} = 0 \to B_i^* = \frac{\psi_i^2 T^*}{2\psi_i^1} = \frac{\psi_i^2}{2\psi_i^1} T^* \quad (5.152)$$

Then:

$$Q_i^* = \frac{D_j T^*}{1 - E(X_i)} = \frac{D_i}{1 - E(X_i)} \sqrt{\frac{K}{\sum_{i=1}^{n} \psi_i^3 - \sum_{i=1}^{n} \left[\frac{(\psi_i^2)^2}{4\psi_i^1}\right]}} \quad (5.153)$$

To handle the constraint, the optimal solution in Eq. (5.144) should be checked. If the constraint is not satisfied, then T_{Min} as the minimum value of T will be considered as the optimal point:

$$T_{\text{Min}} = \frac{\sum_{i=1}^{n} \text{ts}_i}{1 - \sum_{i=1}^{n} \frac{D_i}{P_i(1 - E[X_i])}} \quad (5.154)$$

To ensure the possibility and acceptability of production of all products by a machine, the steps involved in the algorithm of finding the optimal and possible values of $T^*, B_i^*, Q_i^{B^*}$ must be performed as follows:

Step 0. If $\sum_{i=1}^{n} \frac{D_i}{P_i(1-E[X_i])} \leq 1$, then go to *Step 2*; else the problem will be infeasible.
Step 1. Calculate T by Eq. (5.151).
Step 2. Calculate T_{Min} by Eq. (5.40).
Step 3. If $T \geq T_{\text{Min}}$, then $T^* = T$; else $T^* = T_{\text{Min}}$.
Step 4. Calculate B_i by Eq. (5.152).
Step 5. Calculate Q_i^* by Eq. (5.153).

Examples 5.7 and 5.8 Taleizadeh et al. (2010c) considered a multi-product inventory control problem with five products in which their general and specific data are given in Tables 5.20 and 5.21, respectively. They considered two numerical examples with uniform and normal probability distributions for X_i. The setup cost is considered $K = \$450$, and Tables 5.22 and 5.23 show the best results for the two numerical examples.

5.4 Backordering

Table 5.20 General data for Examples 5.7 and 5.8 (Taleizadeh et al. 2010c)

Item	D_i	P_i	ts_i	C_i	h_i	C_{bj}	C_{Sj}
1	200	1800	0.001	15	5	10	1
2	300	2500	0.002	12	4	8	0.8
3	400	3000	0.003	10	3	6	0.6
4	500	3500	0.004	8	2	4	0.4
5	600	4500	0.005	6	1	2	0.2

Table 5.21 Specific data for Examples 5.7 and 5.8 (Taleizadeh et al. 2010c)

	$X_i \sim U[a_i, b_i]$				$X_i \sim N[\mu_i, \sigma_i^2]$		
Product	a_i	b_i	$E[X_i]$	d_i	μ_i	σ_i^2	d_i
1	0	0.1	0.05	90	0.25	0.01	450
2	0	0.15	0.075	187.5	0.28	0.02	700
3	0	0.2	0.1	300	0.33	0.03	990
4	0	0.25	0.125	437.5	0.38	0.04	1330
5	0	0.3	0.15	675	0.42	0.05	1890

Table 5.22 The best results for Example 5.7 (uniform distribution) (Taleizadeh et al. 2010c)

	Uniform					
Product	T_{Min}	T	T^*	B_i	Q_i^*	Z
1	0.0526	0.5608	0.5608	33.02	118.06	21,614
2				48.80	181.88	
3				63.70	249.24	
4				78.21	320.46	
5				94.57	395.86	

5.4.2 Multidefective Types

Pasandideh et al. (2015) extend two works of Taleizadeh et al. (2010c, 2013a) by taking into account reworkable items of various types that require different rework rates to become perfect-quality items. Note that rework is not considered in Taleizadeh et al. (2010c), while in Taleizadeh et al. (2013a), no scrap is assumed.

This single-machine imperfect production problem assumes that perfect and imperfect-quality products are produced at certain percentages on a single machine. Furthermore, all imperfect products are classified as reworked and scrapped.

The following notations which are presented in Table 5.24 are used to model the problem on hand:

In this inventory problem, the annual constant production rate of item i in a regular production time, (P_i), is assumed to be greater than to the annual constant demand rate of product (D_i), where the annual constant imperfect production rate is $x_i P_i$. Mathematically speaking, $(1 - x_i)P_i > D_i$ or $a_i = (1 - x_i)P_i - D_i > 0$. In addition, the x_i parameter considers two types of parameters, the proportion of reworkable products (α_i^j) and the proportion of scrapped items (θ_i). After termination of the regular production, scrapped items are disposed, and the rework process starts

Table 5.23 The best results for Example 5.8 (normal distribution) (Taleizadeh et al. 2010c)

Product	Normal T_{Min}	T	T^*	B_i	Q_i^*	Z
1	0.5796	0.5777	0.5796	32.91	154.56	29,286
2				48.30	241.50	
3				61.90	346.02	
4				74.34	467.41	
5				89.27	599.57	

Table 5.24 New notations for given problem (Pasandideh et al. 2015)

i	Index of product
j	Index of defective type
α_i^j	The proportion of produced ith product with jth defective type
θ_i	The proportion of ith scrapped items
v_i^j	The ratio of the rework rate of ith item with jth defective type to the ith item production rate ($v_i^j \geq 1$)
$v_i^j P_i$	Rework rate of ith defective item type j
μ_i	Space required per unit and the maximum on-hand inventory of ith item
H_i^m	The maximum on-hand inventory of ith item
I_i	The maximum on-hand inventory of ith item, based on which the regular production process stops
H_i^j	The maximum on-hand inventory of the ith item, based on which the rework process stops for jth defective type
f_i	The unit warehouse construction cost of ith item per unit space
δ_i	Aisle space for ith item which is a percentage of its required storage space
F_i	Total required space of ith product
H_i^j	The maximum on-hand inventory of the ith item, based on which the rework process stops for jth defective type

with the $v_i^1 P_i, v_i^2 P_i, \ldots$ and $v_i^m P_i$ rates, where it is assumed no scrapped item is produced during the rework process. As the rework process of a product usually does not require more time compared to its corresponding regular production time, the rework rate is greater than or equal to the regular production rate for all products, i.e., $v_i^j \geq 1$. As a result, the rework production rate $v_i^j P_i$ of the product i is greater than or equal to the demand rate (D_i). In other words, $v_i^j P_i > D_i$ or $y_i^j = v_i^j P_i - D_i > 0$. Additionally, the following conditions are assumed to model the problem (Pasandideh et al. 2015):

- For each item, there are m types of failures that require rework.
- The number of reworkable items with percentage rework rate of α_i^1 is less than the quantity of reworkable items with the percentage rework rate of α_i^2 and so on. In other words, $\alpha_i^m \geq \alpha_i^{m-1} \geq \cdots \geq \alpha_i^1$.
- The rework rates are proportions of the regular production rate ($v_i^j P_i$).

5.4 Backordering

- The items with the percentage rework rate of α_i^2 require less processing time than the ones with the percentage rework rate of α_i^1 and so on. In other words, $v_i^m \geq v_i^{m-1} \geq \cdots \geq v_i^1 \geq 1$.

All production systems with the above conditions can benefit from the modeling and the solution procedure provided in the next sections. The proposed approach in this model enables production managers to determine the optimal period length, the lot size, and the allowable shortage of each product so as the total cost, including setup, production, warehouse construction, holding, shortage, reworking, and disposal, which is minimized (Pasandideh et al. 2015).

In order to develop an inventory model that is even more applicable to real-world inventory problems, machine capacity, limited budget, and service level requirement are imposed. Figure 5.7 shows the on-hand inventory and shortages of product i per cycle. In this figure, t_i^1 and t_i^{m+4} are the production uptimes, $t_i^2, t_i^3, \ldots, t_i^{m+1}$ are rework periods, and t_i^{m+3} and t_i^{m+3} are production downtimes. Based on what was stated in Sect. 5.2, these periods in each cycle of product i are easily obtained using Eqs. (5.155)–(5.161) as follows:

$$t_i^1 = \frac{Q_i}{P_i} - \frac{B_i}{((1-x_i)P_i - D_i)} \tag{5.155}$$

$$t_i^2 = \frac{H_i^1 - I_i}{v_i^1 P_i - D_i} = \alpha_i^1 \frac{Q_i}{v_i^1 P_i} \tag{5.156}$$

$$t_i^3 = \alpha_i^2 \frac{Q_i}{v_i^2 P_i}, \ldots, t_i^{m+1} = \alpha_i^m \frac{Q_i}{v_i^m P_i} \tag{5.157}$$

$$t_i^{m+2} = \frac{H_i^m}{D_i} \tag{5.158}$$

$$t_i^{m+3} = \frac{B_i}{D_i} \tag{5.159}$$

$$t_i^{m+4} = \frac{B_i}{(1-x_i)P_i - D_i} \tag{5.160}$$

$$T = \sum_{e=1}^{m+4} t_i^e \tag{5.161}$$

In addition, the inequality $\alpha_i = (1 - x_i)P_i - D_i > 0$ must hold in the production uptimes t_i^1 and t_i^{m+4} of a cycle. Besides, it is assumed that the time at which rework period t_i^2 starts is the time at which the regular process periods t_i^1 ends, t_i^3 after t_i^2, and so on. Moreover, scrapped items are disposed at the time the regular process stops. Then, based on Fig. 5.7, one has (Pasandideh et al. 2015):

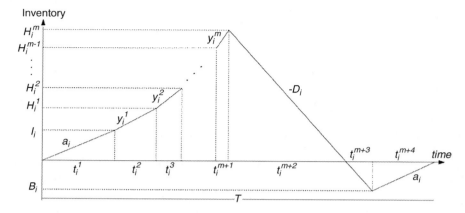

Fig. 5.7 The inventory of a product in a cycle (Pasandideh et al. 2015)

$$I_i = \alpha_i \left(\frac{Q_i}{P_i}\right) - B_i \qquad (5.162)$$

and:

$$H_i^1 = I_i + \alpha_i^1 y_i^1 \frac{Q_i}{v_i^1 P_i}; \quad y_i^1 = (v_i^1 P_i - D_i) \geq 0 \qquad (5.163)$$

therefore:

$$H_i^m = H_i^{m-1} + \alpha_i^m y_i^m \frac{Q_i}{v_i^m P_i} = I_i + \frac{Q_i}{P_i}\left(\sum_{j=1}^{m} \frac{\alpha_i^j y_i^j}{v_i^j}\right) \qquad (5.164)$$

Moreover, the cycle length is:

$$T = \frac{(1-\theta_i)Q_i}{D_i} \qquad (5.165)$$

$$Q_i = \frac{D_i T}{(1-\theta_i)} \qquad (5.166)$$

The total costs of all items TC is the sum of total setup cost CA, total production cost CP, total rework cost CR, total holding cost CH, total backorder cost CB, total disposal cost CD, and total warehouse construction cost CC of all items. In other words:

5.4 Backordering

$$\text{TC} = \text{CA} + \text{CP} + \text{CR} + \text{CH} + \text{CB} + \text{CD} + \text{CC} \tag{5.167}$$

The annual setup cost of all items is:

$$\text{CA} = \sum_{i=1}^{N} NA_i = \sum_{i=1}^{N} \frac{K_i}{T} \tag{5.168}$$

The annual production cost of all items is easily obtained as:

$$\text{CP} = \sum_{i=1}^{n} NC_i Q_i \tag{5.169}$$

For N reworks, each with a quantity of $\alpha_i^j Q_i$ and a cost of C_{Ri} per unit of item i, the annual rework cost is derived as:

$$\text{CR} = \sum_{i=1}^{n} C_{Ri}(x_i - \theta_i)\left[\frac{D_i}{(1-\theta_i)}\right] \tag{5.170}$$

Based on Fig. 5.7, the annual holding cost of the inventory system under consideration is:

$$\text{CH} = \sum_{i=1}^{n} \frac{h_i}{T}\left[\frac{I_i}{2}(t_i^1) + \frac{(I_i + H_i^1)}{2}(t_i^2) + \frac{(H_i^1 + H_i^2)}{2}(t_i^3) + \cdots + \frac{H_i^{m-1} + H_i^m}{2}(t_i^{m+1}) + \frac{H_i^m}{2}(t_i^{m+4})\right] \tag{5.171}$$

Using Fig. 5.7, the annual backorder cost of the inventory system is shown in Eq. (5.172):

$$\text{CB} = \sum_{i=1}^{n} \frac{C_{bi} B_i}{2T}\left(t_i^{m+3} + t_i^{m+4}\right) \tag{5.172}$$

For N disposals, each at a cost of C_{di} per unit of item i, the annual disposal cost of all items becomes:

$$\text{CD} = \sum_{i=1}^{n} C_{di} \theta_i \left[\frac{D_i}{(1-\theta_i)}\right] \tag{5.173}$$

As the space required per unit and the maximum on-hand inventory of ith item are μ_i and H_i^m, respectively, the required storage space of product i is $\mu_i H_i^m$, of which an additional $\delta_i \mu_i H_i^m$ is needed for its aisle space. Thus, the total required space of ith item becomes (Pasandideh et al. 2015):

$$F_i = \mu_i H_i^m (1 + \delta_i) \tag{5.174}$$

Then, the total warehouse construction cost is obtained by (Pasandideh et al. 2015):

$$\text{CC} = \sum_{i=1}^{n} f_i \mu_i H_i^m (1 + \delta_i) \tag{5.175}$$

The total cost TC reduces to (Pasandideh et al. 2015):

$$\text{TC} = Z = \sum_{i=1}^{n} \Delta_i^1 \left(\frac{B_i^2}{T}\right) + \sum_{i=1}^{n} \Delta_i^2 (T) - \sum_{i=1}^{n} \Delta_i^3 (B_i) + \sum_{i=1}^{n} \Delta_i^4 + \sum_{i=1}^{n} \Delta_i^5 \left(\frac{1}{T}\right) \tag{5.176}$$

The capacity of the single machine, the budget, and the service level of each item are the three constraints of the model described as below (Pasandideh et al. 2015). The summation of the total production, rework, and setup times for all products should be smaller than the cycle length. Therefore (Pasandideh et al. 2015):

$$\sum_{i=1}^{n} \left(t_i^1 + t_i^2 + t_i^3 + \cdots + t_i^{m-1}\right) + \sum_{i=1}^{n} \text{ts}_i \leq T \to T$$

$$\geq \frac{\sum_{i=1}^{n} \text{ts}_i + \sum_{i=1}^{n} \frac{B_i}{((1-x_i)P_i - D_i)}}{1 - \sum_{i=1}^{n} \frac{D_i}{((1-\theta_i)P_i)} \left(1 + \sum_{j=1}^{m} \frac{\alpha_i^j}{v_i^j}\right)} = T_{\min} \tag{5.177}$$

As there is a total available budget of W to produce Q_i items for each product i with a production cost of C_i, to rework $\alpha_i^j Q_i$ items for each product i with a rework cost of C_{Ri}, to dispose $\theta_i Q_i$ items for each product i with a disposal cost of C_{di}, and to construct the warehouse for all items, the budget constraint is obtained as (Pasandideh et al. 2015):

$$\sum_{i=1}^{n} \left(\begin{array}{c} C_i Q_i + C_{Ri} \alpha_i^1 + \cdots + C_{Ri} \alpha_i^m Q_i + C_{di} \theta_i Q_i + \dfrac{f_i \mu_i \alpha_i D_i (1 + \delta_i)}{P_i (1 - \theta_i)} T \\ + \dfrac{f_i \mu_i D_i (1 + \delta_i)}{P_i (1 - \theta_i)} \left(\sum_{j=1}^{m} \dfrac{\alpha_i^j y_i^j}{v_i^j} \right) T - f_i \mu_i (1 + \delta_i) B_i \end{array} \right) \leq W$$

which can be simplified to:

5.4 Backordering

$$T \leq \cfrac{W}{\sum_{i=1}^{n} \left(\cfrac{D_i}{(1-\theta_i)} \left(C_i + C_{di}\theta_i + C_{Ri}\sum_{j=1}^{m} \alpha_i^j \right) + \cfrac{f_i \mu_i D_i (1+\delta_i)}{P_i(1-\theta_i)} \left(\alpha_i + \left(\sum_{j=1}^{m} \cfrac{\alpha_i^j y_i^j}{v_i^j} \right) \right) \right)} = T^{\text{Budget}} \quad (5.178)$$

Based on the total shortage quantity and the safety factor of *i*th item, B_i and SL_i, respectively, the service level constraint is derived as (Pasandideh et al. 2015):

$$\frac{N \times B_i}{D_i} \leq SL_i \rightarrow T \geq \frac{B_i}{SL_i D_i} = T_i^{SL} \quad (5.179)$$

The final EPQ model of the problem under investigation is presented as follows (Pasandideh et al. 2015):

$$Z = \sum_{i=1}^{n} \Delta_i^1 \left(\frac{B_i^2}{T} \right) + \sum_{i=1}^{n} \Delta_i^2 (T) - \sum_{i=1}^{n} \Delta_i^3 (B_i) + \sum_{i=1}^{n} \Delta_i^4 + \sum_{i=1}^{n} \Delta_i^5 \left(\frac{1}{T} \right)$$

s.t.: $T \geq T^{\text{Min}}$

$T \leq T^{\text{Budget}}$ (5.180)

$T \geq T_i^{SL}$

$B_i \geq 0, \quad T > 0$

The objective function of the nonlinear mathematical model presented in (5.180) is convex (for more details, see Pasandideh et al. (2015)). To show this, they proved that the Hessian matrix of the objective function is positive. Besides, the constraints in Inequalities (5.177)–(5.179) all are given in linear forms, and hence all are convex as well. As a result, the mathematical model in (5.180) is a convex nonlinear programming, and the local minimum is the global solution. To find the optimal common cycle length and the optimal backorder level of each product, the partial derivatives of the objective function Z are taken with respect to T and B_i. Setting the equations obtained by taking the derivatives, the optimal cycle length and backorder level are given as (Pasandideh et al. 2015):

$$T = \sqrt{\frac{\sum_i \Delta_i^5}{\left(\sum_i \Delta_i^2 - \sum_{i=1}^{n} \frac{\left(\Delta_i^3\right)^2}{4\Delta_i^1} \right)} = \tau} \quad (5.181)$$

$$B_i = \frac{\Delta_i^3}{2\Delta_i^1} T \quad (5.182)$$

In short, the proposed eight-step solution procedure flows (Pasandideh et al. 2015):

Step 1. Initial feasibility. If $\tau > 0$, $\xi \geq 0$ and $a_i = (1 - x_i)P_i - D_i > 0$ for all products, then go to *Step* 2. Otherwise, the EPQ inventory model is infeasible; then go to *Step* 8.

Step 2. Value of the cycle length. Calculate T using Eq. (5.181). Calculate B_i using Eq. (5.182).

Step 3. Secondary feasibility. If $\sum_{i=1}^{n} \text{ts}_i - \sum_{i=1}^{n} \frac{(B_i)}{(P_i - D_i - x_i P_i)}$ and $1 - \sum_{i=1}^{n} \left\{ \frac{D_i}{P_i(1-\theta_i)} \left(1 + \sum_{j=1}^{m} \frac{\alpha_i^j}{v_i^j} \right) \right\}$ are simultaneously either positive or negative, then go to *Step* 4. Otherwise, the EPQ inventory model is infeasible. Go to *Step* 8.

Step 4. Checking the shortage level. Calculate T_i^s by Eq. (5.179), and calculate $T_{\max}^{SL}$ using $T_{\max}^{SL} = \text{Max}\{T_1^{SL}, T_2^{SL}, \ldots, T_n^{SL}\}$.

Step 5. Calculate T^{Min} using Eq. (5.177) of Pasandideh et al. (2015), calculate T^{Budget} by Eq. (5.178), calculate T_{Min}^* using $T_{\text{Min}}^* = \text{Max}\{T^{\min}, T_{\max}^{SL}\}$, and calculate T_{Max}^* using $T_{\text{Max}}^* = T^{\text{Budget}}$.

Step 6. Checking the constraints. This step involves four conditions to determine the optimal values of the decision variables as follows:

Condition 1. If $T_{\max}^* < T_{\min}^*$, then the EPQ inventory model is infeasible and then go to *Step* 8.
Condition 2. If $T_{\max}^* \geq T \geq T_{\min}^*$, then $T^* = T$ and go to *Step* 7.
Condition 3. If $T \geq T_{\max}^*$, then $T^* = T_{\max}^*$ and go to *Step* 7.
Condition 4. If $T \leq T_{\min}^*$, then $T^* = T_{\min}^*$ and go to *Step* 7.

Step 7. Finding the optimal solution. Based on the values of T^* and B^*, obtain Q_i using Eq. (5.166) and Z^* by Eq. (5.180).

Step 8. Terminating the proposed solution procedure.

Example 5.9 Pasandideh et al. (2015) considered an imperfect single-machine production system with 10 products, with the total available budget of $4,000,000,000 per period. Also, let the general data on each product be the one listed in Tables 5.25 and 5.26. Then, the optimal solution is obtained based on the proposed solution procedure as follows:

Step 1. Initial feasibility. In order to check whether the problem has a feasible solution space, all a_i, ξ_i, Δ_i^o and τ are first calculated, and they are shown in Table 5.27. As the conditions for initial feasibility hold, the problem has a feasible solution space and we go to *Step* 2.

Step 2. Value of the cycle length. Using Eq. (5.181), The cycle length is obtained as:

5.4 Backordering

Table 5.25 General data (Pasandideh et al. 2015)

Item	P_i	D_i	x_i	α_i^1	α_i^2	α_i^3	θ_i	v_i^1	v_i^2	v_i^3
1	6000	1000	0.135	0.005	0.05	0.08	0.005	2	3	4
2	6500	1100	0.12	0.005	0.045	0.07	0.005	2	3	4
3	7000	1200	0.111	0.006	0.04	0.065	0.004	2	3	4
4	7500	1300	0.102	0.006	0.036	0.06	0.003	2	3	4
5	8000	1400	0.092	0.005	0.032	0.055	0.003	2	3	4
6	8500	1500	0.082	0.005	0.027	0.05	0.003	3	4	5
7	9000	1600	0.073	0.007	0.021	0.045	0.002	3	4	5
8	9500	1700	0.063	0.007	0.016	0.04	0.002	3	4	5
9	10,000	1800	0.054	0.008	0.011	0.035	0.001	3	4	5
10	10,500	1900	0.047	0.008	0.009	0.03	0.001	3	4	5

Table 5.26 General data (continued) (Pasandideh et al. 2015)

Item	K_i	h_i	C_i	C_{Ri}	C_{di}	C_{bi}	μ_i	δ_i	f_i	ts_i	ε_i
1	400	8	46	25	20	5	2	3	50	0.0004	0.09
2	500	10	44	24	20	8	2	3	55	0.0004	0.09
3	600	12	42	23	20	11	2	3	60	0.0005	0.085
4	700	14	40	22	20	14	3	3	65	0.0005	0.085
5	800	16	38	21	16	17	3	2	70	0.0006	0.08
6	900	18	36	20	16	20	3	2	75	0.0006	0.08
7	1000	20	34	19	16	23	4	2	80	0.0007	0.075
8	1100	22	32	18	16	26	4	2	85	0.0007	0.075
9	1200	24	30	17	12	29	4	4	90	0.0008	0.07
10	1300	26	28	16	12	32	5	4	95	0.0008	0.07

$$T = \sqrt{\frac{1}{\tau} \sum_i \Delta_i^5} = \sqrt{\frac{8500}{12,613,894.7041}} = 0.0259588149281040$$

Moreover, based on Eq. (5.182), shortage quantity is calculated and they are shown in Table 5.28.

Step 3. Secondary feasibility. If $\sum_{i=1}^{n} ts_i - \sum_{i=1}^{n} \frac{(B_i)}{(P_i - D_i - x_i P_i)} < 0$ and $1 - \sum_{i=1}^{n} \left\{ \frac{D_i}{P_i(1-\theta_i)} \left(1 + \sum_{j=1}^{m} \frac{\alpha_i^j}{v_i^j}\right) \right\} < 0$, then the problem is feasible, and go to Step 4: $T^{\text{Min}} = 0.000314130978925651$.

Step 4. Checking the shortage level. Using Eq. (5.178), T_i^{SL}'s are obtained, and they are presented in Table 5.29, based on which $T_{\max}^{\text{SL}} = \text{Max}\{T_1^{\text{SL}}, T_2^{\text{SL}}, \ldots, T_n^{\text{SL}}\} = T_1^{\text{SL}} = 0.0045533$.

Step 5. Constraints' boundaries. The constraints' boundaries are calculated, and they are given in Table 5.30.

Table 5.27 Values of a_i, x_i, Δ_i^o for all products (Pasandideh et al. 2015)

Item	a_i	x_i	Δ_i^1	Δ_i^2	Δ_i^3	Δ_i^4	Δ_i^5	τ
1	4190	0.074284	12.97995	333,798.2	407.9886	49,597.99	400	12,613,895
2	4620	0.080169	20.80563	403,082.4	449.9916	51,805.03	500	
3	5023	0.077814	28.52828	478,290.4	491.9818	53,663.86	600	
4	5435	0.076581	36.27206	838,341.9	793.97	55,074.62	700	
5	5864	0.07477	44.10994	729,072.5	645.9734	56,051.96	800	
6	6303	0.07556	52.02743	836,971.3	692.996	56,611.84	900	
7	6743	0.069894	59.97305	1,265,413	979.9814	56,723.05	1000	
8	7201.5	0.070498	68.07829	1,427,115	1041.988	56,433.87	1100	

5.4 Backordering

Table 5.28 Values of B_i (Pasandideh et al. 2015)

$B_1 = 0.407971433593460$
$B_2 = 0.280723316550802$
$B_3 = 0.223835193381573$
$B_4 = 0.284110174629736$
$B_5 = 0.190078540956709$
$B_6 = 0.172883367260570$
$B_7 = 0.212088244897210$
$B_8 = 0.198659291969368$
$B_9 = 0.310626962203211$
$B_{10} = 0.369785917528195$

Table 5.29 Values of T_i^{SL} (Pasandideh et al. 2015)

TS_L^1	TS_L^2	TS_L^3	TS_L^4	TS_L^5	TS_L^6	TS_L^7	TS_L^8	TS_L^9	TS_L^{10}
0.0045	0.0028	0.0022	0.0026	0.0017	0.0014	0.0018	0.0016	0.0025	0.0028

Table 5.30 Constraints' boundaries (Pasandideh et al. 2015)

	T^{Min}	T_{Max}^{SL}	T^{Budget}	T_{Min}^*	T_{Max}^*
Value	0.000314	0.004533	0.034185	0.004533	0.042731

Step 6. Checking the constraints. As the third condition holds, i.e., $T_{Max}^* \geq T \geq T_{Min}^*$, and $T^* = T = 0.0259588$.

Step 7. Finding the optimal solution. Based on $T^* = 0.02596$, B_i^* and Q_i^* are computed, and they are shown in Table 5.31. Moreover, the minimum annual inventory cost is $Z^* = \$1,201,220.69088020$.

Step 8. Terminate the procedure solution.

5.4.3 Interruption in Manufacturing Process

An EPQ inventory model with interruption in process, scrap and rework is analyzed by Taleizadeh et al. (2014). The shortages are permitted and fully back-ordered. In this EPQ inventory model, the decision variables are cycle length and backordered quantities of each product, and the main objective is to minimize the expected total cost.

As it was discussed before, there are many real work situations in which the manufactured imperfect-quality products should be reworked or repaired with an additional cost. This model considers a manufacturing system that generates imperfect products. Furthermore, these defective products are repairable, and interruption in manufacturing process will occur. Moreover, it is assumed that there is no

Table 5.31 Values of B_i^*, Q_i^* (Pasandideh et al. 2015)

	1	2	3	4	5	6	7	8	9	10
B_i^*	0.4079	0.2807	0.2238	0.2841	0.1901	0.1729	0.2121	0.1987	0.3106	0.3698
Q_i^*	26.08	28.69	31.27	33.85	36.45	39.05	41.62	44.22	46.77	49.37

5.4 Backordering

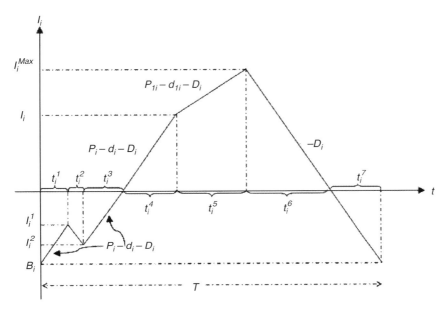

Fig. 5.8 Inventory level of perfect-quality items when interruption occurs during the shortage cycle (Taleizadeh et al. 2014)

interruption during the rework process. Since there is only a single machine, the source of regular production and rework, obviously, is the same. In this problem the following basic assumption of EPQ inventory model with random defective rate, $P - D - d > 0$ or $1 - x - \frac{D}{P} > 0$, is considered. Also, both x and θ are generated randomly between $[0, 1]$.

Owing to some realistic reasons such as finishing the raw material, regular power failure, lubrications, breakdown (corrective maintenance), re-setup for each product, preventive maintenance, predictive maintenance, and maintenance schedule, among many other relevant reasons, then an interruption in production process may occur. Consequently, according to above reasons, manufacturer needs to have an interruption during production up time.

It is important to point out that the interruption in the production process can happen when each product is being manufactured. Thus, two cases can be occurred: (1) interruption in the backorder-filling stage and (2) interruption when there are no shortages. When an interruption in the production process takes place, then the machine cannot work and should stay so until the state is changed. Figures 5.8 and 5.9 show the level of on-hand inventory of perfect-quality products in the proposed EPQ inventory model for the two proposed cases.

They assumed that certainly interruption is occurred and they can control it. This therefore would suggest us that firstly the best time of the interruption should be determined. Naturally, based on the time the interruption occurs, the inventory system cost will be different. In other words, this cost depends on the length of

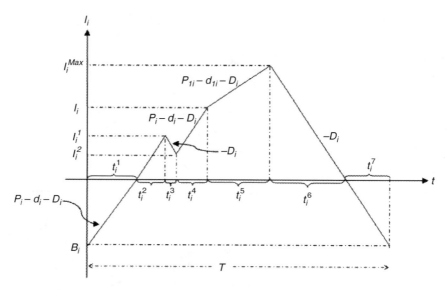

Fig. 5.9 Inventory diagram when interruption occurs during the non-shortage cycle (Taleizadeh et al. 2014)

time before interruption occurs. According to Fig. 5.8, if the interruption occurs when there are shortages, only the backordering cost and carrying cost of defective products will influence the inventory system cost. While if the interruption occurs when there are no shortages, only the carrying cost of healthy and defective products will affect the inventory system cost. So based on the time when the interruption occurs, they discussed the two following possible cases: (1) interruption when there are shortages and (2) interruption when there are no shortages.

The previous assumption simplifies the problem because one only needs to determine the optimal value for the cycle length (T). For simplicity and without loss of generality, at first, the problem is modeled for a single-product case. After, the problem is changed to a multi-product case. The basic assumption of EPQ inventory model with imperfect-quality products manufactured is that P_i must always be greater than or equal to the sum of demand rate D_i and the production rate of defective items d_i. Thus, one has:

$$P_i - d_i - D_i > 0 \quad \therefore \quad 0 \leq x_i < \left(1 - \frac{D_i}{P_1}\right) \qquad (5.183)$$

The production cycle length is the summation of the production uptime, the reworking time, the production downtime, the machine restoring time, and the shortage permitted time:

5.4 Backordering

$$T = \sum_{j=1}^{7} t_i^j \qquad (5.184)$$

If think that Fig. 5.8 shows a possible case to which manufacturer can face, then:

$$t_i^1 + t_i^3 = \frac{B_i + t_i^2 D_i}{P_i - D_i - d_i} \qquad (5.185)$$

This yields to (Taleizadeh et al. 2014):

$$t_i^3 = \frac{B_i + t_i^2 D_i}{P_i - D_i - d_i} - t_i^1 \qquad (5.186)$$

and the inventory levels I_i^1 and I_i^2 are given by:

$$I_i^1 = B_i - (P_i - d_i - D_i)t_i^1 \qquad (5.187)$$

$$I_i^2 = I_i^1 + D_i t_i^2 \qquad (5.188)$$

As a result, the shortage cost is:

$$\begin{aligned}
CS_i &= C_{bi}\left\{ \frac{B_i + I_i^1}{2} t_i^1 + \frac{I_i^1 + I_i^2}{2} t_i^2 + \frac{I_i^2 t_i^3}{2} + \frac{B_i}{2} t_i^7 \right\} \\
&= C_{bi}\left\{ \begin{array}{l} B_i t_i^1 - \frac{1}{2}(P_i - D_i - d_i)(t_i^1)^2 + B_i t_i^2 - (P_i - D_i - d_i)t_i^1 t_i^2 + \frac{1}{2}D_i(t_i^2)^2 + \frac{1}{2}B_i t_i^3 \\ -\frac{1}{2}(P_i - D_i - d)t_i^1 t_i^3 + \frac{1}{2}D_i t_i^2 t_i^3 + \frac{B_i^2}{2D_i} \end{array} \right\}
\end{aligned} \qquad (5.189)$$

After some simple algebra, one has:

$$SC_i = C_{bi}\left\{ \begin{array}{l} B_i t_i^1 - \frac{1}{2}(P_i - D_i - d_i)(t_i^1)^2 + B_i t_i^2 - (P_i - D_i - d_i)t_i^1 t_i^2 + \frac{1}{2}D_i(t_i^2)^2 + \frac{B_i^2}{2D_i} \\ +\left(\frac{1}{2}B_i - \frac{1}{2}(P_i - D_i - d_i)t_i^1\right)\left(\frac{B_i + t_i^2 D_i}{P_i - D_i - d_i} - t_i^1\right) + \frac{1}{2}D_i t_i^2\left(\frac{B_i + t_i^2 D_i}{P_i - D_i - d_i} - t_i^1\right) \end{array} \right\} \qquad (5.190)$$

Also the holding cost of the defective products (HCD) is:

$$CH_{1i}=h_{1i}\begin{Bmatrix}\frac{1}{2}P_ix_i(t_i^1)^2+P_ix_it_i^1t_i^2+P_ix_it_i^1\left(\frac{B_i+D_it_i^2}{P_i-D_i-d_i}\right)\\-P_ix_i(t_i^1)^2+\frac{1}{2}P_ix_i\left(\frac{B_i+D_it_i^2}{P_i-D_i-d_i}\right)^2+\frac{1}{2}P_ix_i(t_i^1)^2-P_ix_it_i^1\left(\frac{B_i+D_it_i^2}{P_i-D_i-d_i}\right)\end{Bmatrix}$$
$$=h_{1i}\left\{P_ix_it_i^1t_i^2+\frac{1}{2}P_ix_i\left(\frac{B_i+D_it_i^2}{P_i-D_i-d_i}\right)^2\right\}$$

(5.191)

Combining the shortage cost and holding cost of defective items, one has:

$$Z_i=C_{bi}\left\{B_it_i^2-(P_i-d_i)t_i^1t_i^2+\frac{1}{2}D_i(t_i^2)^2+\frac{1}{2}\frac{B_i^2+(D_it_i^2)^2}{P_i-D_i-d_i}+\frac{B_iD_it_i^2}{P_i-D_i-d_i}+\frac{B_i^2}{2D_i}\right\}$$
$$+h_{1i}\left\{P_ix_it_i^1t_i^2+\frac{1}{2}P_ix_i\left(\frac{B_i+D_it_i^2}{P_i-D_i-d_i}\right)^2\right\}$$

(5.192)

Then the first derivative of Z_i respect to t_i^1 gives:

$$\frac{\partial Z_i}{\partial t_i^1}=-C_{bi}((P_i-d_i)t_i^2)+h_{1i}P_ix_it_i^2=(-C_{bi}(P_i-d_i)+h_{1i}P_ix_i)t_i^2 \quad (5.193)$$

According to the above equation, it is clear that the summation of shortage cost and holding cost of defective products will be increasing on t_i^1 if:

$$C_{bi}(1-x_i)<h_{1i}x_i \quad (5.194)$$

Meaning $t_i^{1*}=0$. Otherwise, it is easy to see that the summation of shortage cost and holding cost of defective products will be decreasing on t_i^1 if $C_{bi}(1-x_i)>h_{1i}x_i$. More precisely, based on previous discussion, it implies that the t_i^1 is:

$$t_i^{1*}=\frac{B_i}{P_i-D_i-d_i} \quad (5.195)$$

In the case interruption placed when there is no shortage, as mentioned before, only the carrying cost of healthy and defective products affects the inventory system cost. According to Fig. 5.9, one has:

$$t_i^4=\frac{Q_i}{P_i}-\frac{B_i}{P_i-d_i-D_i}-t_i^2 \quad (5.196)$$

and the inventory levels I_i^1, I_i^2, I_i are given by:

5.4 Backordering

$$I_i^1 = (P_i - d_i - D_i)t_i^2 \tag{5.197}$$

$$I_i^2 = (P_i - d_i - D_i)t_i^2 - D_i t_i^3 \tag{5.198}$$

$$\begin{aligned}I_i &= I_i^2 + (P_i - D_i - d_i)t_i^4 \\ &= (P_i - D_i - d_i)t_i^2 - Dt_i^3 + (P_i - D_i - d_i)\left(\frac{Q_i}{P_i} - \frac{B_i}{P_i - D_i - d_i} - t_i^2\right)\end{aligned} \tag{5.199}$$

In this situation, since only the interruption can occur when the rework process is not started, consequently just the carrying cost during $(t_i^1 + t_i^3 + t_i^4)$, similar to the first case, should be compared. The holding cost of healthy products during $(t_i^1 + t_i^3 + t_i^4)$ is determined as:

$$\mathrm{CH}_i = h_i \left\{ \begin{array}{l} \frac{1}{2}(P_i - D_i - d_i)(t_i^2)^2 + \frac{1}{2}(2(P_i - D_i - d_i)t_i^2 - D_i t_i^3)t_i^3 \\ + \frac{1}{2}\left((P_i - D_i - d_i)\frac{Q_i}{P_i} - B_i - 2D_i t_i^3 + (P_i - D_i - d_i)t_i^2\right)\left\{\frac{Q_i}{P_i} - \frac{B_i}{(P_i - D_i - d_i)} - t_i^2\right\} \end{array} \right\} \tag{5.200}$$

This can be simplified to:

$$\mathrm{CH}_i = h_i \left\{ \begin{array}{l} (P_i - d_i)t_i^2 t_i^3 - \frac{1}{2}D_i(t_i^3)^2 + \frac{1}{2}(P_i - D_i - d_i)\frac{Q_i^2}{P_i^2} \\ -\frac{B_i Q_i}{P_i} + \frac{1}{2}\left(\frac{B_i^2 + 2D_i B_i t_i^3}{P_i - D_i - d_i}\right) - \frac{D_i Q_i t_i^3}{P_i} \end{array} \right\} \tag{5.201}$$

Also holding cost of defective products is given by:

$$\begin{aligned}\mathrm{CH}_{1i} &= h_{1i}\left\{\frac{1}{2}P_i x_i (t_i^2)^2 + P_i x_i t_i^2 t_i^3 + P_i x_i t_i^2 t_i^4 + \frac{1}{2}P_i x_i (t_i^4)^2\right\} \\ &= h_{1i} P_i x_i \left\{\frac{1}{2}(t_i^2)^2 + t_i^2 t_i^3 + t_i^2\left(\frac{Q_i}{P_i} - \frac{B_i}{P_i - D_i - d_i} - t_i^2\right) + \frac{1}{2}\left(\frac{Q_i}{P_i} - \frac{B_i}{P_i - D_i - d_i} - t_i^2\right)^2\right\}\end{aligned} \tag{5.202}$$

Yielding:

$$\mathrm{CH}_{1i} = h_{1i}\left\{P_i x_i t_i^2 t_i^3 + \frac{x_i Q_i^2}{2P_i} + \frac{P_i x_i B_i^2}{2(P_i - D_i - d_i)^2} - \frac{x_i B_i Q_i}{P_i - D_i - d_i}\right\} \tag{5.203}$$

So the summation holding costs of healthy and defective products is:

Fig. 5.10 On-hand inventory of defective items (Taleizadeh et al. 2014)

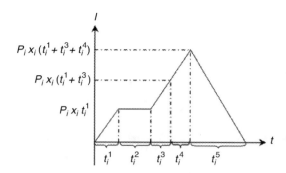

$$Z_i = h_i \left\{ (P_i - d_i) t_i^2 t_i^3 - \frac{1}{2} D_i (t_i^3)^2 + \frac{1}{2}(P_i - D_i - d_i) \frac{Q_i^2}{P_i^2} - \frac{B_i Q_i}{P_i} + \frac{1}{2} \left(\frac{B_i^2 + 2 D_i B_i t_i^3}{P_i - D_i - d_i} \right) - \frac{D_i Q_i t_i^3}{P_i} \right\}$$
$$+ h_{1i} \left\{ P_i x_i t_i^2 t_i^3 + \frac{x_i Q_i^2}{2 P_i} + \frac{P_i x_i B_i^2}{2(P_i - D_i - d_i)^2} - \frac{x_i B_i Q_i}{P_i - D_i - d_i} \right\}$$
(5.204)

Then the first derivative of Z_i in respect to t_i^2 gives:

$$\frac{\partial Z_i}{\partial t_i^2} = h_i (P_i - d_i) t_i^3 + h_{1i} \left(P x_i t_i^3 \right)$$
(5.205)

Meaning the summation of the holding costs of healthy and defective products is increasing on t_i^2. Thus, the best time for considering the interruption is $t_i^{2*} = 0$. It is worth mentioning that the best interruption time for the second case is equal to zero ($t_i^{2*} = 0$) which is as same as the best interruption time of the first case shown in Eq. (5.191). It is important to point out that the inventory behavior of these two cases is similar, and it is shown in Fig. 5.9. Thus, in the situation under which ever $C_{bi}(1 - x_i) > h_{1i} x_i$, the best value of interruption time in both cases (interruption when there are shortages and interruption when there are no shortages) is the same. Consequently, if for each product always $C_{bi}(1 - x_i) > h_{1i} x_i$ is considered, then a unique model can be used to solve the problem on hand.

Since the main issue to solve is the determination of the best time of interruption, then in both cases (i.e., when shortages are not permitted and are permitted), only the different terms of objective functions of those cases are separately studied. This is due to the fact that their same parts are independent from t_i^1 and t_i^2. They did it in this way in order to simplify the mathematical calculations. Obviously, is not necessary to calculate the total cost function because only the best time of interruption which depends on the value of t_i^1 and t_i^2 should be determined. Since t_i^1 and t_i^2 are only used in the production and interruption periods, there is no need to determine the other terms of cost function. For the situation when the best time of interruption is determined, they have showed that for both backordering and no backordering

5.4 Backordering

Fig. 5.11 On-hand inventory of scrapped items (Taleizadeh et al. 2014)

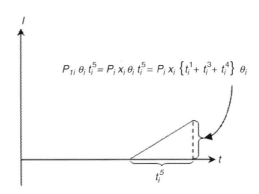

cases the time of interruption is the same. Then for this condition all terms of the total objective cost function of the inventory model must be expressed and determined (Taleizadeh et al. 2014) (Figs. 5.10 and 5.11).

According to Fig. 5.9, one has:

$$t_i^1 = \frac{B_i}{P_i - d_i - D_i} \tag{5.206}$$

$$t_i^2 = t_m \tag{5.207}$$

$$t_i^3 = \frac{I_i^1}{P_i - d_i - D_i} \tag{5.208}$$

$$t_i^4 = \frac{I_i}{P_i - d_i - D_i} \tag{5.209}$$

The total defective products manufactured during the production uptime $\left(t_i^1 + t_i^3 + t_i^4\right)$ can be computed as below:

$$d_i\left(t_i^1 + t_i^3 + t_i^4\right) = P_i x_i\left(t_i^1 + t_i^3 + t_i^4\right) = x_i Q_i \tag{5.210}$$

and the total scrap products generated during the rework process are (Taleizadeh et al. 2014):

$$d_{1i} t_i^5 = P_{1i} \theta_i t_i^5 = d_i \theta_i\left(t_i^1 + t_i^3 + t_i^4\right) = x_i Q_i \theta_i \tag{5.211}$$

Then, one has:

$$t_i^5 = \frac{x_i Q_i}{P_{1i}} \tag{5.212}$$

$$t_i^6 = \frac{I_i^{\text{Max}}}{D_i} \tag{5.213}$$

$$t_i^7 = \frac{B_i}{D_i} \tag{5.214}$$

Likewise, one has:

$$I_i^1 = D_i t_i^2 = D_i t_m \tag{5.215}$$

In order to determine the maximum level of on-hand inventory when regular production process stops, I_i, one can subtract the backordered quantity and demand during the interruption time from the total healthy manufactured products which yields:

$$I_i = (P_i - d_i - D_i)\frac{Q_i}{P_i} - B_i - D_i t_m \tag{5.216}$$

and the maximum inventory level is given by:

$$\begin{aligned} I_i^{\text{Max}} &= I_i + (P_{1i} - d_{1i} - D_i)t_i^5 \\ &= (P_i - d_i - D_i)\frac{Q_i}{P_i} - B_i - D_i t_m + (P_{1i} - d_{1i} - D_i)\frac{x_i Q_i}{P_{1i}} \end{aligned} \tag{5.217}$$

The cycle length is equal to $T = \sum_{j=1}^{7} t_i^j$ which can be verified easily summing all times as follows:

$$T = \sum_{j=1}^{7} t_i^j = \frac{Q_i}{D_i}\left[1 - \frac{d_i}{P_i} + x_i\left(1 - \frac{d_{1i}}{P_{1i}}\right)\right] \tag{5.218}$$

This yields to:

$$Q_i = \frac{D_i}{J_i} T \tag{5.219}$$

where:

$$J_i = \left[1 - \frac{d_i}{P_i} + x_i\left(1 - \frac{d_{1i}}{P_{1i}}\right)\right] \tag{5.220}$$

The total cost function per cycle is:

5.4 Backordering

$$\text{TC}'(Q_i, B_i) = \overbrace{C_i Q_i}^{\text{Production Cost}} + \overbrace{C_{Ri} x_i Q_i}^{\text{Rework Cost}} + \overbrace{C_{Si} x_i Q_i \theta_i}^{\text{Disposal Cost}} + \overbrace{K_i}^{\text{Setup Cost}}$$

$$+ \underbrace{C_{bi}\left[\frac{B_i}{2}(t_i^1 + t_i^7) + \frac{I_i^1}{2}(t_i^2 + t_i^3)\right]}_{\text{Back Ordered Cost}} + \underbrace{h_{1i}\left[\frac{P_i x_i \theta_i (t_i^1 + t_i^3 + t_i^4)}{2}(t_i^5)\right]}_{\text{Holding Cost of Scrap Product}} + \underbrace{h_i\left[\frac{I_i}{2}(t_i^4) + \frac{I_i + I_i^{\text{Max}}}{2}(t_i^5) + \frac{I_i^{\text{Max}}}{2}(t_i^6)\right]}_{\text{Holding Cost of Perfect Quality Product}}$$

$$+ \underbrace{h_i \left[\frac{P_i x_i t_i^1}{2}(t_i^1) + P_i x_i t_i^1 (t_i^2 + t_i^3 + t_i^4) + \frac{P_i x_i (t_i^3 + t_i^4)}{2}(t_i^3 + t_i^4) + P_i x_i (t_i^1 + t_i^3 + t_i^4)\frac{t_i^5}{2}\right]}_{\text{Holding Cost of Defective Product}}$$

(5.221)

Equation (5.221) can be simplified to as below (Taleizadeh et al. 2014):

$$\text{TC}'(Q_i, B_i) = \left(\frac{h_i}{2}\left(\frac{N_i}{P_i^2} + \frac{2x_i N_i}{P_i P_{1i}} + \frac{x_i^2 M_i}{(P_{1i})^2} + \frac{\left(\frac{N_i}{P_i} + \frac{x_i M_i}{P_{1i}}\right)^2}{D_i} + \frac{x_i}{P_i} + \frac{x_i^2}{P_{1i}}\right) + \frac{h_{1i} x_i^2 \theta_i}{2 P_{1i}}\right) Q_i^2$$

$$+ \left[\left(\frac{C_{bi}}{2 D_i}\right)\left(\frac{P_i - d_i}{N_i}\right) + \frac{h_i}{2}\left(\frac{1}{N_i} + \frac{1}{D_i}\right)\right] B_i^2 - \left(h_i \frac{J_i}{D_i}\right) B_i Q_i + \left[-h_i t_m J_i + C_i + C_{Ri} x_i + C_{Si} x_i \theta_i\right] Q_i$$

$$+ h_i t_m \frac{P_i}{N_i} B_i + K_i + \frac{h_i (D_i^2 + D_i N_i) t_m^2}{2 N_i} + \frac{C_{bi} D_i (t_m)^2}{2}\left(\frac{P_i - d_i}{N_i}\right)$$

(5.222)

However, Eq. (5.222) for the annual joint production policy using Eq. (5.219) becomes (Taleizadeh et al. 2014):

$$\text{TC}(T, B_i) = \frac{\text{TC}'(Q_i, B_i)}{T} = \left(\frac{h_i D_i^2}{2 J_i^2}\left(\frac{N_i}{P_i^2} + \frac{2x_i N_i}{P_i P_{1i}} + \frac{x_i^2 M_i}{(P_{1i})^2} + \frac{\left(\frac{N_i}{P_i} + \frac{x_i M_i}{P_{1i}}\right)^2}{D_i} + \frac{x_i}{P_i} + \frac{x_i^2}{P_{1i}}\right) + \frac{h_{1i} x_i^2 \theta_i D_i^2}{2 P_{1i} J_i^2}\right) T$$

$$+ \left[\left(\frac{C_{bi}}{2 D_i}\right)\left(\frac{P_i - d_i}{N_i}\right) + \frac{h_i}{2}\left(\frac{1}{N_i} + \frac{1}{D_i}\right)\right]\frac{B_i^2}{T} - h_i B_i + \left[C_i + C_{Ri} x_i + C_{Si} x_i \theta_i - h_i t_m J_i\right]\frac{D_i}{J_i}$$

$$+ \left(\frac{h_i t_m P_i}{N_i}\right)\frac{B_i}{T} + \left(K_i + \frac{h_i (D_i^2 + D_i N_i) t_m^2}{2 N_i} + \frac{C_{bi} D_i (t_m)^2}{2}\left(\frac{P_i - d_i}{N_i}\right)\right)\frac{1}{T}$$

(5.223)

Remember that the existence of only one machine in the manufacturing system results in limited production capacity. In some sense, the maximum capacity of the single machine is the only constraint of the model. This constraint is described in the following: Since $t_i^1 + t_i^3 + t_i^4, t_i^5$, and ts_i are the production uptimes, rework time and setup time of the ith product, respectively, and t_i^2 is the interruption time when machine is producing ith product, then the summation of the total production uptimes, rework time, machine repairing time, and setup time (for all products) is expressed as $\sum_{i=1}^{n}\left(t_i^1 + t_i^2 + t_i^3 + t_i^4 + t_i^5\right) + \sum_{i=1}^{n} ts_i$. Evidently, this should be smaller or equal to the period length (T). In general, a necessary condition to guarantee feasibility is (Taleizadeh et al. 2014):

$$\sum_{i=1}^{n}\left(t_i^1+t_i^2+t_i^3+t_i^4+t_i^5\right)+\sum_{i=1}^{n}\mathrm{ts}_i\leq T \quad (5.224)$$

This can be simplified to:

$$\sum_{j=1}^{5}t_i^j=t_m+Q_i\left[\frac{1}{P_i}+\frac{x_i}{P_{1i}}\right] \quad (5.225)$$

Thus:

$$\sum_{i=1}^{n}\left\{t_m+Q_i\left[\frac{1}{P_i}+\frac{x_i}{P_i^1}\right]+\mathrm{ts}_i\right\}\leq T \quad (5.226)$$

Using Eq. (5.219), one has:

$$T\geq\frac{\sum_{i=1}^{n}(\mathrm{ts}_i+t_m)}{\left(1-\sum_{i=1}^{n}\frac{D_i}{J_i}\left[\frac{1}{P_i}+\frac{x_i}{P_{1i}}\right]\right)}=T_{\mathrm{Min}} \quad (5.227)$$

Finally, from Eqs. (5.223) and (5.227), the multi-product single-machine model with scrap, rework, interruption in process, and backlogging situation becomes:

$$\mathrm{TC}(T,B_i)=\sum_{i=1}^{n}\psi_i^1\frac{B_i^2}{T}+\sum_{i=1}^{n}\psi_i^2\frac{B_i}{T}+\sum_{i=1}^{n}\psi_i^3 T-\sum_{i=1}^{n}\psi_i^4 B_i+\sum_{i=1}^{n}\psi_i^5\frac{1}{T}+\sum_{i=1}^{n}\psi_i^6$$

s.t. :
$$T\geq\frac{\sum_{i=1}^{n}(\mathrm{ts}_i+t_m)}{\left(1-\sum_{i=1}^{n}\frac{D_i}{J_i}\left[\frac{1}{P_i}+\frac{x_i}{P_{1i}}\right]\right)}=T_{\mathrm{Min}}$$
$$B_i\geq 0\quad i=1,2,\ldots,n$$

$$(5.228)$$

where:

$$\psi_i^1=\left(\frac{C_{\mathrm{bi}}}{2D_i}\right)\left(\frac{P_i-d}{N_i}\right)+\frac{h_{1i}}{2}\left(\frac{1}{N_i}+\frac{1}{D_i}\right)>0 \quad (5.229)$$

$$\psi_i^2=\frac{h_i t_m P_i}{N_i}>0 \quad (5.230)$$

5.4 Backordering

$$\psi_i^3 = \frac{h_i D_i^2}{2J_i^2}\left(\frac{N_i}{(P_i)^2} + \frac{2x_i N_i}{P_i P_{1i}} + \frac{x_i^2 M_i}{(P_{1i})^2} + \left(\frac{N_i}{P_i} + \frac{x_i M_i}{P_{1i}}\right)^2 \frac{1}{D_i} + \frac{x_i}{P_i} + \frac{x_i^2}{P_{1i}}\right)$$
$$+ \frac{h_{1i}(x_i)^2 \theta_i D_i^2}{2P_{1i} J_i^2}$$
$$> 0 \tag{5.231}$$

$$\psi_i^4 = h_i > 0 \tag{5.232}$$

$$\psi_i^5 = K_i + \frac{D_1(t_m)^2}{2N_i}(P_i - d_i)(h_i + C_{bi}) > 0 \tag{5.233}$$

$$\psi_i^6 = \frac{D_i}{J_i}(C_i + C_{Ri}x_i + C_{Si}x_i\theta_i) - h_i t_m D_i \tag{5.234}$$

It should be noted that $N_i = P_i - d_i - D_i$, $M_i = P_{1i} - d_{1i} - D_i$, and $J_i = \left[1 - \frac{d_i}{P_i} + x_i\left(1 - \frac{d_{1i}}{P_{1i}}\right)\right]$.

Theorem 5.3 The objective function $F = \text{TC}(T, B_i)$ in (5.228) is convex (Taleizadeh et al. 2014).

Proof To prove the convexity of $F = \text{TC}(T, B_i)$, one can utilize the well-known Hessian matrix equation as in (5.235). Then, according to Appendix F of Taleizadeh et al. (2014), the objective function is strictly convex. Therefore, the expected total cost function $F = \text{TC}(T, B_i)$ is strictly convex for all nonzero T and B_i. Hence, it follows that to find the optimal production period length (T) and the optimal level of backorders (B_i), one can partially differentiate $F = \text{TC}(T, B_i)$ with respect to T and B_i and solve the resulted system of equations obtained by equating the partial derivatives to zero (see Appendix G of Taleizadeh et al. (2014)):

$$[T, B_1, B_2, \ldots, B_n] \times \mathbf{H} \times [T, B_1, B_2, \ldots, B_n]^T = \frac{\sum_{i=1}^n \psi_i^5}{T} \geq 0 \tag{5.235}$$

According to Appendix G of Taleizadeh et al. (2014), the optimal solutions of decision variables are:

$$T = \sqrt{\sum_{i=1}^n \left(\frac{4\psi_i^1 \psi_i^5 - (\psi_i^2)^2}{4\psi_i^1}\right) / \sum_{i=1}^n \left(\psi_i^3 - \frac{(\psi_i^4)^2}{4\psi_i^1}\right)} \tag{5.236}$$

$$B_i = \frac{\psi_i^4 T^* - \psi_i^2}{2\psi_i^1} \tag{5.237}$$

To determine the feasible and optimal solution, the following solution procedure is proposed by Taleizadeh et al. (2014):

Table 5.32 General data for both examples (Taleizadeh et al. 2014)

Product	P_i	P_{1i}	ts_i	K_i	C_i	h_i	h_{1i}	C_{Si}	C_{bi}	C_{Ri}
1	4000	2000	0.003	500	15	5	2	3	10	1
2	4500	2500	0.004	450	12	4	2	3	8	2
3	5000	3000	0.005	400	10	3	2	3	6	3
4	5500	3500	0.006	350	8	2	2	3	4	4
5	6000	4000	0.007	300	6	1	2	3	2	5

Table 5.33 Specific data for Example 5.10 (Taleizadeh et al. 2014)

Products	D_i	t_m	$X_i \sim U[a_i, b_i]$			$d_i = P_i E[X_i]$	$\theta_i \sim U[a'_i, b'_i]$			$d_{1i} = P_{1i}E[\theta_i]$
			a_i	b_i	$E[X_i]$		a'_i	b'_i	$E[\theta_i]$	
1	600	0.003	0	0.05	0.025	100	0	0.15	0.075	150
2	700	0.003	0	0.1	0.05	225	0	0.2	0.1	250
3	800	0.003	0	0.15	0.075	375	0	0.25	0.125	375
4	900	0.003	0	0.2	0.1	550	0	0.3	0.15	525
5	1000	0.003	0	0.25	0.125	750	0	0.35	0.175	700

Step 1. Calculate $O_1 = \sum_{i=1}^{n} \frac{4\psi_i^1 \psi_i^5 - (\psi_i^2)^2}{4\psi_i^1}$, $O_2 = \sum_{i=1}^{n} \psi_i^3 - \frac{(\psi_i^4)^2}{4\psi_i^1}$, and T_{Min}.

Step 2. If $O_1 O_2 > 0$, calculate period length T using Eq. (5.234). Otherwise, there is no feasible solution; terminate the procedure.

Step 3. If T is less than T_{Min}, then $T^* = T_{Min}$, else $T^* = T$.

Step 4. Calculate B_i^* for $i = 1, 2, \ldots, n$ using Eq. (5.235).

Step 5. Calculate Q_i^* for $i = 1, 2, \ldots, n$ using Eq. (5.219).

Step 6. Terminate the procedure.

Examples 5.10 and 5.11 This section provides two numerical examples to illustrate the use of solution procedure of the presented inventory model. Consider the five product inventory control problems where the general data for both examples are given in Table 5.32. Furthermore, they considered that in both numerical examples, the defective and scrap portions (X_i and θ_i) follow a uniform distribution. Tables 5.33 and 5.34 show an additional specific data for both examples. The use of the solution procedure with Example 5.10 is illustrated as follows (Taleizadeh et al. 2014)

From *Step 1*, O_1, O_2 and T_{Min} are obtained as follows:

$$O_1 = \sum_{i=1}^{n} \frac{4\psi_i^1 \psi_i^5 - (\psi_i^2)^2}{4\psi_i^1} = 2000.155877, \quad O_2 = \sum_{i=1}^{n} \psi_i^3 - \frac{(\psi_i^4)^2}{4\psi_i^1}$$

$$= 3112.546773, \quad \text{and } T_{Min} = 0.418778052.$$

From *Step 2*, since both O_1 and O_2 are positive, then $O_1 O_2 > 0$, the period length (T) can be determined with Eq. (5.236), and its value is $T = 0.80163$.

From *Step 3*, T is not less than T_{Min}, so $T^* = T = 0.80163$.

The results of *Steps* 4 and 5 are shown in Table 5.35.

5.4 Backordering

Table 5.34 Specific data for Example 5.11 (Taleizadeh et al. 2014)

Products	D_i	t_m	a_i	b_i	$E[X_i]$	$d_i = P_i E[X_i]$	a'_i	b'_i	$E[\theta_i]$	$d_{1i} = P_{1i} E[\theta_i]$
1	630	0.005	0	0.055	0.0275	110	0	0.15	0.075	150
2	735	0.005	0	0.11	0.055	247.5	0	0.2	0.1	250
3	840	0.005	0	0.165	0.0825	412.5	0	0.25	0.125	375
4	945	0.005	0	0.22	0.11	605	0	0.3	0.15	525
5	1050	0.005	0	0.275	0.1375	825	0	0.35	0.175	700

$X_i \sim U[a_i, b_i]$, $\theta_i \sim U[a'_i, b'_i]$

Table 5.35 The best result for Example 5.10 (Taleizadeh et al. 2014)

$T_{\min}$	T	T^*	B_i	Q_i	Z
0.418778	0.80163	0.80163	135.0451	481.8815	45,364.04
			155.6826	563.9608	
			175.9271	647.3731	
			195.7637	732.4538	
			215.17	819.5579	

Table 5.36 The best result for Example 5.11 (Taleizadeh et al. 2014)

$T_{\min}$	T	T^*	B_i	Q_i	Z
1.291679	0.785506	1.291679	227.289	815.4394	48,306.31
			261.7311	954.6343	
			295.4154	1096.316	
			328.3045	1241.115	
			360.3449	1389.702	

In similar way, Example 5.11 is solved and its results are shown in Table 5.36.

5.4.4 Immediate Rework Process

Two joint production systems in a form of multi-product single machine with and without rework are studied in Taleizadeh et al. (2011) where shortage is allowed and back-ordered. For each system, the optimal cycle length and the backordered and production quantities of each product are determined such that the cost function is minimized.

Jamal et al. (2004) developed an EPQ model to determine the optimum production quantity of an item where rework is performed in two different situations. The objective function of their model was to minimize the total inventory cost of the production system under consideration. Taleizadeh et al. (2011) extended the model of Jamal et al. (2004) to be more realistic inventory control problem in which several items and several constraints are available.

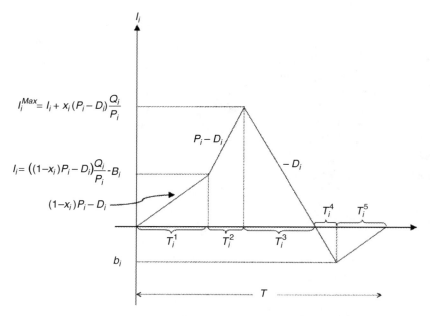

Fig. 5.12 On-hand inventory of perfect-quality items for MP-SM with rework (Taleizadeh et al. 2011)

Figure 5.12 depicts a cycle for the inventory control problem under study. In order to model the problem, a single-product problem consisting of the ith product is first developed, and then the model will extend to include several products. The basic assumption in EPQ model with rework process is that $(1 - x_i)P_i$ must always be greater than or equal to the demand rate D_i. As a result, one has:

$$(1 - x_i)P_i - D_i \geq 0 \qquad (5.238)$$

The production cycle length (see Fig. 5.12) is the summation of the production uptimes, the rework time, and the production downtimes, i.e.:

$$T = \sum_{j=1}^{5} T_i^j \qquad (5.239)$$

where T_i^1 and T_i^5 are the production uptimes (the periods in which perfect and defective items are produced), T_i^2 is the reworking time, and T_i^3 and T_i^4 are the production downtimes. In this model, a part of the modeling procedure is adopted from Jamal et al. (2004) to model the problem. Based on Fig. 5.12, one has:

5.4 Backordering

$$I_i = ((1 - x_i)P_i - D_i)\frac{Q_i}{P_i} - B_i \qquad (5.240)$$

and:

$$I_i^{\text{Max}} = I_i + x_i(P_i - D_i)\frac{Q_i}{P_i} = ((1 - x_i)P_i - D_i)\frac{Q_i}{P_i} + x_i(P_i - D_i)\frac{Q_i}{P_i} - B_i \qquad (5.241)$$

It is obvious from Fig. 5.12 that:

$$T_i^1 = \frac{Q_i}{P_i} - \frac{B_i}{(1 - x_i)P_i - D_i} \qquad (5.242)$$

$$T_i^2 = x_i \frac{Q_i}{P_i} \qquad (5.243)$$

$$T_i^3 = \frac{\left(1 - \frac{D_i}{P_i} - x_i \frac{D_i}{P_i}\right) Q_i}{D_i} - \frac{B_i}{D_i} \qquad (5.244)$$

$$T_i^4 = \frac{B_i}{D_i} \qquad (5.245)$$

$$T_i^5 = \frac{B_i}{(1 - x_i)P_i - D_i} \qquad (5.246)$$

Hence, using Eq. (5.239), the cycle length for a single-product problem is:

$$T = \sum_{j=1}^{5} T_i^j = \frac{Q_i}{D_i} \qquad (5.247)$$

or:

$$Q_i = D_i T \qquad (5.248)$$

The total cost of the system includes setup, processing, rework, shortage, and inventory carrying costs. Although the processing and rework costs are constants and do not affect the optimal solution, they are used in the objective function in order to determine the annual total cost. Defective items are produced in every batch and they are reworked within the same cycle. During the rework process, some processing and inventory holding costs are incurred for reworked quantities. Then, the total cost per year, TC(Q, B), is obtained as:

$$\text{TC}(Q, B) = \overbrace{NC_i Q_i}^{\text{Production Cost}} + \overbrace{NK_i}^{\text{Setup Cost}} + \overbrace{NC_{Ri} x_i Q_i}^{\text{Rework Cost}}$$

$$+ \overbrace{Nh_i \left[\frac{I_i}{2} (T_i^1) + \frac{I_i + I_i^{\text{Max}}}{2} (T_i^2) + \frac{I_i^{\text{Max}}}{2} (T_i^3) \right.}^{\text{Holding Cost}} + \overbrace{\left. \frac{NC_{bi} B_i (T_i^4 + T_i^5)}{2} \right]}^{\text{Shortage Cost}}$$

(5.249)

Finally, the objective function of the joint production system (MP-SM with rework) using Eq. (5.247) becomes:

$$\text{TC}(T, B) = \sum_{i=1}^{n} C_i D_i + \frac{\sum_{i=1}^{n} K_i}{T} + \sum_{i=1}^{n} C_{Ri} x_i D_i$$

$$+ \sum_{i=1}^{n} h_i \left[\frac{I_i}{2T} (T_i^1) + \frac{I_i + I_i^{\text{Max}}}{2T} (T_i^2) + \frac{I_i^{\text{Max}}}{2T} (T_i^3) \right] + \sum_{i=1}^{n} \frac{C_{bi} B_i (T_i^4 + T_i^5)}{2T}$$

(5.250)

Since the production plus rework times of the ith product is $T_i^1 + T_i^2 + T_i^5$ and the setup time is ts_i, the summation of the production, reworking, and setup time (for all products) will be $\sum_{i=1}^{n} (T_i^1 + T_i^2 + T_i^5) + \sum_{i=1}^{n} \text{ts}_i$. It is clear that this summation must be smaller or equal to the cycle length (T). Hence, the capacity constraint of the model becomes:

$$\sum_{i=1}^{n} (T_i^1 + T_i^2 + T_i^5) + \sum_{i=1}^{n} \text{ts}_i \leq T \tag{5.251}$$

Then, based on Eqs. (5.240), (5.241), and (5.244), one has:

$$\sum_{i=1}^{n} (1 + x_i) \frac{D_i}{P_i} T + \sum_{i=1}^{n} \text{ts}_i \leq T \tag{5.252}$$

From Eqs. (5.238) to (5.246), (5.248), and (5.250), the final model of the joint production system is obtained as follows:

Min : $\text{TC}(T, B)$

$$= \alpha_1 T^2 + \alpha_2 T + \frac{\alpha_3}{T} + \sum_{i=1}^{n} \alpha_{4i} \frac{B_i^2}{T} - \sum_{i=1}^{n} \alpha_{5i} B_i - \sum_{i=1}^{n} \alpha_{6i} B_i T + \sum_{i=1}^{n}$$

$$\times (C_i + x_i C_{Ri}) D_i \tag{5.253}$$

5.4 Backordering

$$\text{s.t.:} \quad T \geq \frac{\sum_{i=1}^{n} \text{ts}_i}{\left(1 - \sum_{i=1}^{n}(1+x_i)\frac{D_i}{P_i}\right)} = T_{\text{Lower}} \quad (5.254)$$

where

$$\alpha_1 = \sum_{i=1}^{n} \frac{h_i}{2} \left[\frac{((1+x_i)((1-x_i)P_i - D_i))D_i^2 + x_i(P_i - D_i)D_i^2}{P_i} \right] > 0 \quad (5.255)$$

$$\alpha_2 = \sum_{i=1}^{n} \frac{h_i}{2} \left[\frac{3((1-x_i)P_i - D_i)D_i^2}{P_i} + \frac{(P_i - (1+x_i)D_i)D_i}{P_i} \right] > 0 \quad (5.256)$$

$$\alpha_3 = \sum_{i=1}^{n} K_i > 0 \quad (5.257)$$

$$\alpha_{4i} = \frac{h_i}{2} \left[\frac{((1-x_i)P_i - 2D_i)}{((1-x_i)P_i - D_i)D_i} \right] + \frac{C_{bi}}{2} \left[\frac{(1-x_i)P_i}{((1-x_i)P_i - D_i)D_i} \right] > 0 \quad (5.258)$$

$$\alpha_{5i} = \frac{h_i}{2} \left(1 + 2x_i \frac{D_i}{P_i} + \frac{((1-x_i)P_i - D_i)}{P_i} + \frac{x_i(P_i - D_i)}{P_i} \right) > 0 \quad (5.259)$$

$$\alpha_{6i} = \frac{h_i}{2} \left[\frac{(1+x_i)D_i}{P_i} \right] > 0 \quad (5.260)$$

Note that since $\sum_{i=1}^{n}(C_i + x_i C_{Ri})D_i$ is constant, it has been removed from the objective function. In Sect. 5.4, a solution method is given for the developed model. In this case, according to Fig. 5.13, the objective function of the joint production system is obtained as follows:

$$\text{TC}(Q, B) = \overbrace{NC_iQ_i}^{\text{Production Cost}} + \overbrace{NK_i}^{\text{Setup Cost}} + \overbrace{Nh_i \left[\frac{I_i}{2}\left(T_i^1 + T_i^2\right) \right]}^{\text{Holding Cost}} + \overbrace{\frac{NC_{bi}B_i\left(T_i^3 + T_i^4\right)}{2}}^{\text{Shortage Cost}}$$

$$= \sum_{i=1}^{n} C_i D_i + \frac{\sum_{i=1}^{n} K_i}{T} + \sum_{i=1}^{n} h_i \left[\frac{I_i}{2T}\left(T_i^1 + T_i^2\right) \right] + \sum_{i=1}^{n} \frac{C_{bi}B_i\left(T_i^4 + T_i^5\right)}{2T}$$

$$(5.261)$$

where:

$$T_i^1 = \frac{Q_i}{P_i} - \frac{B_I}{(1-x_i)P_i - D_i} \quad (5.262)$$

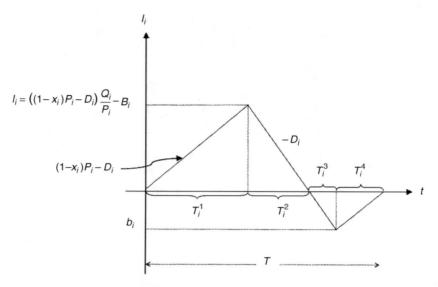

Fig. 5.13 On-hand inventory of perfect-quality items for MP-SM without rework (Taleizadeh et al. 2011)

$$T_i^2 = \frac{\left(1 - \frac{D_i}{P_i} - x_i \frac{D_i}{P_i}\right)Q_i}{D_i} - \frac{B_i}{D_i} \tag{5.263}$$

$$T_i^3 = \frac{B_i}{D_i} \tag{5.264}$$

$$T_i^4 = \frac{B_i}{(1 - x_i)P_i - D_i} \tag{5.265}$$

The capacity constraint in this case will be:

$$\sum_{i=1}^{n}\left(T_i^1 + T_i^4\right) + \sum_{i=1}^{n} \text{ts}_i \leq T \tag{5.266}$$

Hence, the final model of MP-SM EPQ without rework becomes:

$$\text{Min}: \quad \text{TC}(T,B) = \gamma_1 T + \frac{\gamma_2}{T} + \sum_{i=1}^{n} \gamma_{3i} \frac{B_i^2}{T} - \sum_{i=1}^{n} \gamma_{4i} B_i + \sum_{i=1}^{n} \gamma_{5i} \tag{5.267}$$

5.4 Backordering

$$\text{s.t.:} \quad T \geq \frac{\sum_{i=1}^{n} ts_i}{\left(1 - \sum_{i=1}^{n} \frac{D_i}{P_i}\right)} = T_{\text{Lower}} \quad (5.268)$$

where:

$$\gamma_1 = \sum_{i=1}^{n} \frac{h_i}{2} \left[\frac{(((1-x_i)P_i - D_i))D_i^2}{P_i} \right] > 0 \quad (5.269)$$

$$\gamma_2 = \sum_{i=1}^{n} K_i > 0 \quad (5.270)$$

$$\gamma_{3i} = \left(\frac{h_i + C_{bi}}{2}\right) \left[\frac{2(1-x_i)P_i}{((1-x_i)P_i - D_i)D_i}\right] > 0 \quad (5.271)$$

$$\gamma_{4i} = \frac{h_i}{2}\left(\left(1 - (1+x_i)\frac{D_i}{P_i}\right) + \frac{((1-x_i)P_i - D_i)}{P_i}\right) > 0 \quad (5.272)$$

$$\gamma_{5i} = C_i D_i + \frac{h_i}{2}\left(1 - (1+x_i)\frac{D_i}{P_i}\right)\frac{((1-x_i)P_i - D_i)D_i}{P_i^2} \quad (5.273)$$

In order to derive an optimal solution for the final model, a proof of the convexity of the objective function is first provided. Then, a classical optimization technique using partial derivatives is utilized to derive the optimal solution.

Theorem 5.4 The objective function TC(T, B) is convex (Taleizadeh et al. 2011).

Proof Using Eq. (5.274), since the Hessian matrix, results in positive values for all nonzero B_i and T, TC(T, B) = Z are convex:

$$[T, B_1, B_2, \ldots, B_n] \times \mathbf{H} \times [T, B_1, B_2, \ldots, B_n]^T$$
$$= 2\alpha_1 T^2 + \frac{2\alpha_3}{T} - 2\sum_{i=1}^{n} \alpha_{6i} B_i T > 0 \quad (5.274)$$

To find the optimal production period length T and the optimal backorder quantities b_is, partial differentiations of Z with respect to T and B_i are obtained in Eqs. (5.275) and (5.276):

$$\frac{\partial Z}{\partial T} = 2\alpha_1 T + \alpha_2 - \frac{\alpha_3}{T^2} - \frac{\sum_{i=1}^{n}\alpha_{4i}B_i^2}{T^2} - \sum_{i=1}^{n}\alpha_{6i}B_i = 0$$
$$\rightarrow 2\alpha_1 T^3 + \left(\alpha_2 - \sum_{i=1}^{n}\alpha_{6i}b_i\right)T^2 - \left(\alpha_3 + \sum_{i=1}^{n}\alpha_{4i}B_i^2\right) = 0 \quad (5.275)$$

$$\frac{\partial Z}{\partial B_i} = \frac{2\alpha_{4i}B_i}{T} - \alpha_{5i} - \alpha_{6i}T = 0 \to 2\alpha_{4i}B_i - \alpha_{5i}T - \alpha_{6i}T^2 = 0 \to B_i$$
$$= \frac{\alpha_{5i}T + \alpha_{6i}T^2}{2\alpha_{4i}} \tag{5.276}$$

Then the following solution procedure is used to solve and ensure the feasibility of the problem:

Step 1. Check for feasibility. If $(1 - x_i)P_i - D_i \geq 0$ and $\sum_{i=1}^{n}(1 + x_i)D_i/P_i < 1$, go to *Step* 2. Otherwise, the problem is infeasible.

Step 2. Find a solution point. Find the optimal solution using Eqs. (5.275) and (5.276) and by an iterative approach. To do this, start with $B_i = 0$ and insert $B_i = 0$ into Eq. (5.242). Then, the new real positive values of T are obtained. This iterative search will continue until the difference between two consecutive values of T is smaller than a given Δ.

Step 3. Check the constraint. Check the constraint based on the obtained value of T. If $T > T_{\text{Lower}}$, then $T^* = T$. Otherwise, $T^* = T_{\text{Lower}}$, and go to *Step* 4.

Step 4. Obtain the optimal solution. Based on the obtained value of T^* and using $Q_1^* = D_iT^*$, B_i^* will be derived by Eq. (5.276). Then calculate the objective function value and go to *Step* 5.

Step 5. Terminate the procedure.

The objective function of the MP-SM without rework model is convex too. To find the optimal production period length T and the optimal backorder quantities B_is, partial differentiations of Z with respect to T and B_i are obtained as are given in Eqs. (5.277) and (5.278):

$$\frac{\partial Z}{\partial T} = \gamma_1 T - \frac{\gamma_2}{T^2} - \frac{\sum_{i=1}^{n}\gamma_{3i}B_i^2}{T^2} = 0 \to T^2 = \frac{\gamma_2 + \sum_{i=1}^{n}\gamma_{3i}B_i^2}{\gamma_1} \tag{5.277}$$

$$\frac{\partial Z}{\partial B_i} = \frac{2\gamma_{3i}B_i}{T} - \gamma_{4i} = 0 \to B_i = \frac{\gamma_{4i}}{2\gamma_{3i}}T \tag{5.278}$$

By substituting Eq. (5.277) in Eq. (5.276), one has:

$$T = \sqrt{\frac{\gamma_2}{\gamma_1 - \sum_{i=1}^{n}\frac{(\gamma_{4i})^2}{4\gamma_{3i}}}} \tag{5.279}$$

$$B_i = \frac{\gamma_{4i}}{2\gamma_{3i}}T \tag{5.280}$$

In this case, the following solution procedure is used to solve and ensure the feasibility of the problem (Taleizadeh et al. 2011):

5.4 Backordering

Table 5.37 General data for Example 5.12 (Taleizadeh et al. 2011)

Product	D_i	P_i	ts_i	K_i	C_i	C_{Ri}	h_i	C_{bi}	x_i
1	100	5000	0.001	700	24	24	10	20	0.05
2	150	5500	0.002	650	22	22	9	18	0.1
3	200	6000	0.003	600	20	20	8	16	0.15
4	250	6500	0.004	550	18	18	7	14	0.2
5	300	7000	0.005	500	16	16	6	12	0.25
6	350	7500	0.006	450	14	14	5	10	0.3
7	400	8000	0.007	400	12	12	4	8	0.35
8	450	8500	0.008	350	10	10	3	6	0.4
9	500	9000	0.009	300	8	8	2	4	0.45
10	550	9500	0.01	250	6	6	1	2	0.5

Step 1. Check for feasibility. If $(1 - x_i)P_i - D_i \geq 0$ and $\sum_{i=1}^{n} \frac{D_i}{P_i} < 1$ and $\gamma_1 > \sum_{i=1}^{n} \frac{(\gamma_{4i})^2}{4\gamma_{3i}}$, go to *Step* 2. Otherwise, the problem is infeasible.

Step 2. Find a solution point. Using Eqs. (5.279) and (5.280), find the optimal solution.

Step 3. Check the constraint. Check the constraint based on the obtained value of T. If $T > T_{\text{Lower}}$, then $T^* = T$. Otherwise, $T^* = T_{\text{Lower}}$ and go to *Step* 4.

Step 4. Obtain the optimal solution. Based on the obtained value of T^* and using $Q_i^* = D_i T^*$, B_i^* will be obtained by Eq. (5.280). Then calculate the objective function value and go to *Step* 5.

Step 5. Terminate the procedure.

Examples 5.12 and 5.13 Consider two multi-product EPQ problems consisting of ten products with breakdown and capacity constraint. In these examples, the demand, production, and proportion of defective rates of each product are considered constant. Furthermore, the production rate of non-defective items is considered constant and is greater than the demand rate for each product. For each example, both the immediate rework and no rework situations are considered. Moreover, there are no scrapped or defective items during the rework process. The production and rework are accomplished using the same resource at the same speed, and shortages are allowed as backorders (Taleizadeh et al. 2011).

The general and the specific data of these examples are given in Tables 5.37 and 5.38, respectively. Tables 5.39 and 5.40 show the best results for the both examples considering immediate rework using the first solution procedure. It should be noted that a value of $\Delta = 10^{-12}$ is assumed in the solution procedure.

Using the second solution procedure, Tables 5.41 and 5.42 show the best results for both examples when rework is not performed (Taleizadeh et al. 2011).

Comparison study based on the results given in Tables 5.39, 5.40, 5.41 and 5.42 shows that lower optimum total costs are obtained for both examples in which reworking is allowed. Furthermore, the optimum cycle length obtained for the systems where rework is permitted is greater than those of the non-reworking systems (Taleizadeh et al. 2011).

Table 5.38 General data for Example 5.13 (Taleizadeh et al. 2011)

Product	D_i	P_i	ts_i	K_i	C_i	C_{Ri}	h_i	C_{bi}	x_i
1	100	5000	0.001	700	24	24	10	20	0.05
2	200	5500	0.002	650	22	22	9	18	0.1
3	300	6000	0.003	600	20	20	8	16	0.15
4	400	6500	0.004	550	18	18	7	14	0.2
5	500	7000	0.005	500	16	16	6	12	0.25
6	600	7500	0.006	450	14	14	5	10	0.3
7	700	8000	0.007	400	12	12	4	8	0.35
8	00	8500	0.008	350	10	10	3	6	0.4
9	900	9000	0.009	300	8	8	2	4	0.45
10	1000	9500	0.01	250	6	6	1	2	0.5

Table 5.39 The best results for Example 5.12 with immediate rework (Taleizadeh et al. 2011)

Product	T_{Lower}	T	T^*	Q_i	B_i	Z
1	0.1246	0.7794	0.7794	77.9442	12.9534	63,610
2				116.9163	19.4380	
3				155.8884	25.9523	
4				194.8605	32.5130	
5				233.8362	39.1414	
6				272.8047	45.7818	
7				311.7768	52.7732	
8				350.7489	60.0117	
9				389.7210	68.1049	
10				428.6931	79.7102	

Table 5.40 The best results for Example 5.13 with immediate rework (Taleizadeh et al. 2011)

Product	T_{Lower}	T	T^*	Q_i	B_i	Z
1	0.8931	0.6014	0.8931	89.3094	9.9875	101,890
2				178.6178	19.9476	
3				267.9281	29.9332	
4				357.2374	39.9866	
5				446.5468	50.1512	
6				535.8562	60.4890	
7				625.1655	71.1199	
8				714.4749	82.3392	
9				803.7843	95.09655	
10				893.936	114.3833	

5.4 Backordering

Table 5.41 The best results for Example 5.12 without rework (Taleizadeh et al. 2011)

Product	T_{Lower}	T	T^*	Q_i	B_i	Z
1	0.1294	7.2282	0.1294	12.9417	0.7356	77,215
2				19.4125	1.0236	
3				25.8834	1.2616	
4				32.3542	1.4511	
5				38.8251	1.5934	
6				45.2959	1.6893	
7				51.7668	1.7397	
8				58.2376	1.7452	
9				64.7085	1.7501	
10				71.1793	1.7589	

Table 5.42 The best results for Example 5.13 without rework (Taleizadeh et al. 2011)

Product	T_{Lower}	T	T^*	Q_i	B_i	Z
1	0.0779	3.6794	0.0779	7.7880	0.4427	127,020
2				15.5761	0.8034	
3				23.3641	1.0892	
4				31.1521	1.3051	
5				38.9402	1.4549	
6				46.7282	1.5417	
7				54.5162	1.6483	
8				62.3043	1.7173	
9				70.0923	1.7839	
10				77.8804	1.8369	

5.4.5 Repair Failure

Taleizadeh et al. (2013b) developed a multi-product single-machine EPQ model with production capacity limitation and random defective production rate and failure during repair by considering shortage backordering. The objective was to determine the optimal period lengths, backordered quantities, and order quantities. All items produced are screened, and the inspection cost per item is included in the unit production cost. All defective items produced can be reworked, and rework starts when the regular production process ends. During the regular production time, defective items may be produced randomly. The random fraction of defective items is reworked during the rework process, and complete backordering is allowed (Taleizadeh et al. 2013b). Initially the problem is modeled as a single-product case, and then it is modified as a multi-product case. The basic assumption of EPQ model with imperfect-quality items produced is that P_i must always be greater than or equal to the sum of demand rate D_i and the production rate of defective items is d_i. One has (Taleizadeh et al. 2013b):

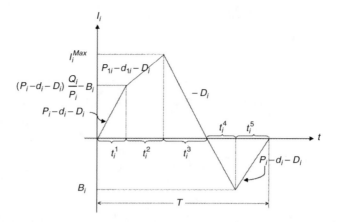

Fig. 5.14 On-hand inventory of perfect-quality items (Taleizadeh et al. 2013b)

$$P_i - d_i - D_i \geq 0 \quad \therefore 0 \leq x_i \leq \left(1 - \frac{D_i}{P_i}\right) \quad \text{or} \quad 1 - E[X_i] - \frac{D_i}{P_i} \geq 0 \quad (5.281)$$

The production cycle length (see Fig. 5.14) is the summation of the production uptime, the reworking time, the production downtime, and the shortage permitted time (Taleizadeh et al. 2013b):

$$T = \sum_{j=1}^{5} t_i^j \quad (5.282)$$

where the production uptime is t_i^1 and t_i^5, reworking time is t_i^2, and production downtime is t_i^3 and t_i^4. Also t_i^4 is the time shortage permitted, and t_i^5 is the time needed to satisfy all the backorders by the next production.

To model the problem, a part of the modeling procedure is adopted from Hayek and Salameh (2001). Since all products are manufactured on a single machine with a limited capacity, based on Fig. 5.14:

$$t_i^1 = \frac{I_i}{P_i - \lambda_i - D_i} \quad (5.283)$$

$$t_i^2 = \frac{E[X_i] \cdot Q_i}{P_{1i}} = \frac{d_i \cdot Q_i}{P_{1i} P_i} \quad (5.284)$$

$$t_i^3 = \frac{I_i^{Max}}{D_i} = Q_i \left(\frac{1}{D_i} - \frac{P_{1i} + d_i}{P_{1i} P_i} - \frac{d_{1i} d_i}{P_{1i} P_i D_i}\right) - \frac{B_i}{D_i} \quad (5.285)$$

5.4 Backordering

$$t_i^4 = \frac{B_i}{D_i} \tag{5.286}$$

$$t_i^5 = \frac{B_i}{P_i - d_i - D_i} \tag{5.287}$$

$$I_i = (P_i - d_i - D_i)\frac{Q_i}{P_i} - B_i \tag{5.288}$$

$$I_i^{\text{Max}} = I_i + (P_{1i} - d_{1i} - D_i)t_i^2 = Q_i\left(1 - \frac{D_i}{P_i} - \frac{d_{1i}\lambda_i}{P_{1i}P_i} - \frac{d_i D_i}{P_{1i}P_i}\right) - B_i \tag{5.289}$$

During rework process, the production rate of scrapped items is presented in Eqs. (5.290) and (5.291):

$$d_{1i} = P_{1i}E[\theta_i], \quad \text{where } 0 \leq \theta_i \leq 1 \tag{5.290}$$

$$d_{1i} \cdot t_i^2 = (P_{1i}E[\theta_i])\left(\frac{d_i \cdot Q_i}{P_{1i}P_i}\right) = E[\theta_i]\left(\frac{P_i E[X_i]Q_i}{P_i}\right) = E[\theta_i] \cdot E[X_i] \cdot Q_i \tag{5.291}$$

Hence, the cycle length for a single-product state is (Taleizadeh et al. 2013b):

$$T = \frac{Q_i E[1 - \theta_i x_i]}{D_i}, \quad \text{where } 0 \leq \theta_i \leq 1 \tag{5.292}$$

or:

$$Q_i = \frac{D_i T}{E[1 - \theta_i x_i]}, \quad \text{where } 0 \leq \theta_i \leq 1 \tag{5.293}$$

During the imperfect rework process, the random defective rate has a range of [0, 1], and the scrap rate has a range of $[0, Q_i E[\theta_i] E[x_i]]$. The total inventory cost per year $TC(Q, B)$ is (Taleizadeh et al. 2013b):

$$TC(Q, B) = \overbrace{NC_i Q_i}^{\text{Production Cost}} + \overbrace{NC_{Ri} E[X_i] Q_i}^{\text{Rework Cost}} + \overbrace{NC_{Si} E[X_i] Q_i E[\theta_i]}^{\text{Disposal Cost}}$$

$$+ Nh_i \overbrace{\left[\frac{I_i}{2}(t_i^1) + \frac{I_i + I_i^{\text{Max}}}{2}(t_i^2) + \frac{I_i^{\text{Max}}}{2}(t_i^3) + \frac{d_i(t_i^1 + t_i^5)}{2}(t_i^1 + t_i^5)\right]}^{\text{Holding Cost of Perfect Quality Items}}$$

$$+ \underbrace{Nh_{1i}\left[\frac{P_{1i}t_i^2}{2}(t_i^2)\right]}_{\text{Holding Cost of Imperfect Quality Items}} + \underbrace{NC_{bi}\frac{B_i}{2}(t_i^4 + t_i^5)}_{\text{Shortage Cost}} + \underbrace{NK_i}_{\text{Setup Cost}}$$

$$\tag{5.294}$$

The multi-product single-machine cost from Eq. (5.291) becomes (Taleizadeh et al. 2013b):

$$TC(Q,B) = \sum_{i=1}^{n} \frac{C_i Q_i}{T} + \sum_{i=1}^{n} \frac{C_{Ri} E[X_i] Q_i}{T} + \sum_{i=1}^{n} \frac{C_{Si} E[X_i] Q_i E[\theta_i]}{T}$$
$$+ \sum_{i=1}^{n} \frac{h_i}{T} \left[\frac{I_i}{2} (t_i^1) + \frac{I_i + I_i^{\text{Max}}}{2} (t_i^2) + \frac{I_i^{\text{Max}}}{2} (t_i^3) + \frac{d_i(t_i^1 + t_i^5)}{2} (t_i^1 + t_i^5) \right]$$
$$+ \sum_{i=1}^{n} \frac{h_{1i}}{T} \left[\frac{P_{1i} t_i^2}{2} (t_i^2) \right] + \sum_{i=1}^{n} C_{bi} \frac{B_i}{2T} (t_i^4 + t_i^5) + \sum_{i=1}^{n} \frac{K_i}{T}$$

(5.295)

Since $t_i^1 + t_i^2 + t_i^5$ are the production and rework times and ts_i is the setup time for ith product, similar to previous case, the constraint of the model is (Taleizadeh et al. 2013b):

$$\sum_{i=1}^{n} (t_i^1 + t_i^2 + t_i^5) + \sum_{i=1}^{n} ts_i \leq T \quad (5.296)$$

Then, based on Eqs. (5.283)–(5.285) and (5.288), one has (Taleizadeh et al. 2013b):

$$\sum_{i=1}^{n} \frac{D_i(P_{1i} + d_i)}{P_i P_{1i} E[1 - \theta_i X_i]} T + \sum_{i=1}^{n} ts_i \leq T \quad (5.297)$$

From Eqs. (5.283) to (5.289) and Eq. (5.293), $TC(Q,B)$ in Eq. (5.294), and constraint in Eq. (5.296), one can formulate the problem as (Taleizadeh et al. 2013b):

Min : $TC(Q,B)$

$$= \sum_{i=1}^{n} C_i^1 \frac{(B_i)^2}{T} - \sum_{i=1}^{n} C_i^2 B_i + \sum_{i=1}^{n} C_i^3 T + \sum_{i=1}^{n} C_i^4 + \sum_{i=1}^{n} \frac{K_i}{T} \quad (5.298)$$

s.t. : $T \geq \dfrac{\sum_{i=1}^{n} ts_i}{\left[1 - \sum_{i=1}^{n} \frac{D(P_{1i}+d_i)}{P_i P_{1i} E[1-\theta_i X_i]} \right]} \quad T, B_i \geq 0 \quad \forall i, i = 1, 2, \ldots, n \quad (5.299)$

where:

$$C_i^1 = \frac{C_{bi}(P_i - \lambda_i)}{D_i(P_i - d_i - D_i)} + \frac{h_i}{2(P_i - d_i - D_i)} - \frac{h_i}{2D_i} > 0 \quad (5.300)$$

5.4 Backordering

$$C_i^2 = \frac{h_i D_i}{P_i E[1-\theta_i X_i]} + \frac{h_i d_i D_i}{P_{1i} P_i E[1-\theta_i X_i]} + \frac{h_i}{E[1-\theta_i X_i]}$$
$$\times \left(1 - \frac{D_i}{P_i} - \frac{d_{1i} d_i}{P_{1i} P_i} - \frac{d_i D_i}{P_{1i} P_i}\right)$$
$$> 0 \tag{5.301}$$

$$C_i^3 = \frac{h_{1i} d_i^2 D_i^2}{2 P_{1i}(P_i)^2 (E[1-\theta_i X_i])^2} + \frac{h_i (P_i - d_i - D_i) D_i^2}{2(P_i)^2 (E[1-\theta_i X_i])^2} + \frac{h_i(P_i - d_i - D_i) d_i D_i^2}{2P_{1i}(P_i)^2(E[1-\theta_i X_i])^2}$$
$$+ h_i \left[\frac{D_i}{2E[1-\theta_i X_i]} + \frac{d_i D_i^2}{2P_i P_{1i}(E[1-\theta_i X_i])^2}\right]\left(1 - \frac{D_i}{P_i} - \frac{d_{1i}\lambda_i}{P_i P_{1i}} - \frac{d_i D_i}{P_i P_{1i}}\right) > 0$$
$$\tag{5.302}$$

$$C_i^4 = \frac{[C_i + C_{Ri} E[X_i] + C_{Si} E[X_i] E[\theta_i]] D_i}{E[1-\theta_i X_i]} + \frac{d_i D_i}{2 P_i E[1-\theta_i X_i]} > 0 \tag{5.303}$$

In order to derive the optimal solution of the final model, a proof of the convexity of the objective function is provided. A classical optimization technique using partial derivatives is performed to derive the optimal solutions (Taleizadeh et al. 2013b).

Theorem 5.5 The objective function $TC(Q, B)$ in (5.297) is convex (Taleizadeh et al. 2013b).

Proof To prove the convexity of $TC(Q, B) = Z$, the following Hessian matrix is developed (Taleizadeh et al. 2013b):

$$[T, B_1, B_2, \ldots, B_n] \times \mathbf{H} \times [T, B_1, B_2, \ldots, B_n]^T = \frac{1}{T} 2 \sum_{i=1}^{n} K_i \geq 0 \tag{5.304}$$

From Appendix 1 of Taleizadeh et al. (2013b), the objective function for all nonzero T and B_i is shown to be strictly convex. T and B_i are solved by letting the partial derivatives equal to zero. One has (Taleizadeh et al. 2013b):

$$\frac{\partial Z}{\partial T} = 0 \rightarrow T = \sqrt{\frac{\sum_{i=1}^{n} K_i}{\sum_{i=1}^{n} C_i^3 - \sum_{i=1}^{n} (C_i^2)^2 / 4 C_i^1}} \tag{5.305}$$

$$\frac{\partial Z}{\partial B_i} = 0 \rightarrow B_i = (C_i^2 / 2 C_i^1) T \tag{5.306}$$

then:

Table 5.43 General data for examples (Taleizadeh et al. 2013b)

P	D_i	P_i	P_{1i}	ts_i	K_i	C_i	h_i	h_{1i}	C_{Si}	C_{bi}	C_{Ri}
1	300	3000	2000	0.003	500	15	5	2	3	10	1
2	400	3500	2500	0.004	450	12	4	2	3	8	2
3	500	4000	3000	0.005	400	10	3	2	3	6	3
4	600	4500	3500	0.006	350	8	2	2	3	4	4
5	700	5000	4000	0.007	300	6	1	2	3	2	5

Table 5.44 Specific data for Example 5.14 (Taleizadeh et al. 2013b)

Items	$X_i \sim U[a_i, b_i]$				$\theta_i \sim U[a_i, b_i]$			
	a_i	b_i	$E[X_i]$	$d_i = P_i E[X_i]$	a_i	b_i	$E[\theta_i]$	$d_{1i} = P_{1i} E[\theta_i]$
1	0	0.05	0.025	75	0	0.15	0.075	150
2	0	0.1	0.05	175	0	0.2	0.1	250
3	0	0.15	0.075	300	0	0.25	0.125	375
4	0	0.2	0.1	450	0	0.3	0.15	525
5	0	0.25	0.125	625	0	0.35	0.175	700

$$Q_i^* = \frac{D_i}{E[1 - \theta_i X_i]} T^* \qquad (5.307)$$

The constraint below must be satisfied; otherwise, the minimum value of T will be considered as the optimal point (Taleizadeh et al. 2013b):

$$T_{\text{Min}} = \frac{1}{\left[1 - \sum_{i=1}^{n} \frac{D(P_{1i}+d_i)}{P_i P_{1i} E[1-\theta_i X_i]}\right]} \sum_{i=1}^{n} ts_i \qquad (5.308)$$

To ensure feasibility, the following solution procedure must be performed (Taleizadeh et al. 2013b):

Step 1. Check for feasibility. If $\sum_{i=1}^{n} \frac{D_i(P_{1i}+d_i)}{P_i P_{1i} E[1-\theta_i X_i]} < 1$, go to *Step* 2; else the problem will be infeasible.
Step 2. Calculate T by Eq. (5.305).
Step 3. Calculate T_{Min} by Eq. (5.308).
Step 4. If $T \geq T_{\text{Min}}$, then $T^* = T$; else $T^* = T_{\text{Min}}$.
Step 5. Calculate B_i^* by Eq. (5.304).
Step 6. Calculate Q_i^* by Eq. (5.307).
Step 7. Terminate procedure.

Examples 5.14 and 5.15 Consider a multi-product inventory control problem with five products where the general and specific data are given in Table 5.43. Specific data for Examples 5.14 and 5.15 are presented in Tables 5.44 and 5.45, respectively. They considered two numerical examples with uniform and normal probability distributions for X_i and θ_i. Tables 5.46 and 5.47 show the optimal results for the two numerical examples (Taleizadeh et al. 2013b).

5.5 Partial Backordering

Table 5.45 Specific data for Example 5.15 (Taleizadeh et al. 2013b)

Items	$X_i \sim N[\mu_i, \sigma_i^2]$ $\mu_i = E[X_i]$	σ_i^2	$d_i = P_i E[X_i]$	$\theta_i \sim N[\mu_i, \sigma_i^2]$ $\mu_i = E[\theta_i]$	σ_i^2	$d_{1i} = P_{1i} E[\theta_i]$
1	0.1	0.01	150	0.15	0.01	300
2	0.15	0.02	350	0.18	0.02	450
3	0.2	0.03	600	0.21	0.03	630
4	0.25	0.04	900	0.24	0.04	840
5	0.3	0.05	1250	0.27	0.05	1080

Table 5.46 The best results for Example 5.14 (uniform distribution) (Taleizadeh et al. 2013b)

Items	Uniform T_{Min}	T	T^*	B_i	Q_i	Z
1	0.7909	0.9183	0.9183	77.72	276	32,129
2				101.31	369.16	
3				124.49	463.49	
4				147.34	559.36	
5				169.91	657.17	

Table 5.47 The best results for Example 5.15 (normal distribution) (Taleizadeh et al. 2013b)

Items	Normal T_{Min}	T	T^*	B_i	Q_i	Z
1	1.0822	0.9060	1.0822	91.858	327.11	34,656
2				119.92	440.81	
3				147.36	558.69	
4				174.02	682.05	
5				199.78	812.37	

5.5 Partial Backordering

In this subsection, five different problems under partial backordering shortage are presented.

5.5.1 Rework

Taleizadeh et al. (2013a) developed a multi-product single-machine EPQ model with partial backordering, imperfect production, rework, budget, and service level constraints. Their objective was to minimize the joint total cost of the system subject to service level and budget constraints. Each product cycle consists of three time periods: production uptime, reworking time, and production downtime. They assumed that the total quantity of imperfect-quality items can be reworked, and no

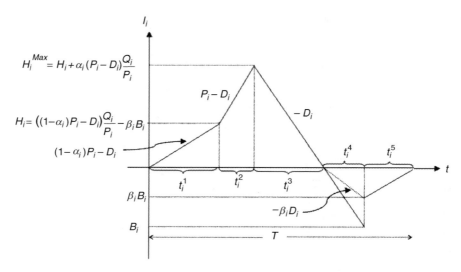

Fig. 5.15 On-hand inventory for perfect-quality items (Taleizadeh et al. 2013a)

scrap will be left at the end of the rework period. For the joint production system, capacity and budget are limited, and a fraction of the shortage is backordered (Taleizadeh et al. 2013a).

Taleizadeh et al. (2013a) extended Jamal et al.'s (2004) study by considering a more realistic inventory control problem wherein a joint production strategy with a single machine is used to produce several items under a limited capacity and service level and budget constraints.

The inventory control problem under study is shown in Fig. 5.15. A single-product problem (defined as the ith product) is initially developed in order to model the problem. The model is then further modified to extend for multiple products.

The fundamental assumption of the EPQ model with rework process is that the rate of production minus defectives must always be greater than or equal to the demand. With this, one has (Taleizadeh et al. 2013a):

$$(1 - \alpha_i)P_i - D_i \geq 0 \tag{5.309}$$

The production cycle length (see Fig. 5.15) is the sum of the production uptimes for the good and defective items, t_i^1 and t_i^5, respectively; the reworking time, t_i^2; and the production downtimes, t_i^3 and t_i^4. Therefore, one has total production cycle length of:

$$T = \sum_{j=1}^{5} t_i^j \tag{5.310}$$

5.5 Partial Backordering

As noted before, since all products are manufactured on a single machine with a limited capacity, the cycle length for all products is illustrated on Fig. 5.15. Therefore, one has:

$$t_i^1 = \frac{Q_i}{P_i} - \frac{\beta_i B_i}{(1-\alpha_i)P_i - D_i} \tag{5.311}$$

$$t_i^2 = \alpha_i \frac{Q_i}{P_i} \tag{5.312}$$

$$t_i^3 = \frac{\left(1 - \frac{D_i}{P_i} - \alpha_i \frac{D_i}{P_i}\right) Q_i}{D_i} - \frac{\beta_i B_i}{D_i} \tag{5.313}$$

$$t_i^4 = \frac{\beta_i B_i}{\beta_i D_i} = \frac{B_i}{D_i} \tag{5.314}$$

$$t_i^5 = \frac{\beta_i B_i}{(1-\alpha_i)P_i - D_i} \tag{5.315}$$

It is evident from Fig. 5.15 that:

$$H_i = ((1-\alpha_i)P_i - D_i)\frac{Q_i}{P_i} - \beta_i B_i \tag{5.316}$$

and:

$$\begin{aligned} H_i^{\text{Max}} &= H_i + \alpha_i(P_i - D_i)\frac{Q_i}{P_i} \\ &= ((1-\alpha_i)P_i - D_i)\frac{Q_i}{P_i} + \alpha_i(P_i - D_i)\frac{Q_i}{P_i} - \beta_i B_i \end{aligned} \tag{5.317}$$

Hence, from Eq. (5.310), the cycle length for a single product is:

$$T = \sum_{j=1}^{5} t_i^j = \frac{Q_i + (1-\beta_i)B_i}{D_i} \tag{5.318}$$

or:

$$Q_i = D_i T - (1-\beta_i)B_i \tag{5.319}$$

The total cost function of the model is the sum of setup, processing, rework, shortage, and inventory carrying costs. One has:

$$\text{TC} = \text{CA} + \text{CP} + \text{CR} + \text{CH} + \text{CB} + \text{CL} \tag{5.320}$$

For N setups at \$ K_i per setup, the annual setup cost is:

$$\text{CA} = \sum_{i=1}^{n} NK_i \qquad (5.321)$$

For a joint policy $N = 1/T$, one has:

$$\text{CA} = \frac{\sum_{i=1}^{n} K_i}{T} \qquad (5.322)$$

The total production cost is the summation of the product of production cost per unit and the quantity per period for all ith products which are C_i and Q_i, respectively. The annual production cost is obtained by multiplying $C_i Q_i$ with N. The cost for this joint policy is:

$$\text{CP} = \frac{\sum_{i=1}^{n} C_i Q_i}{T} \qquad (5.323)$$

The total rework cost is the summation of the product of the rework cost per unit of the ith product, and the quantities of the ith product that is to be reworked are C_{Ri} and $\alpha_i Q_i$, respectively. The annual rework cost is obtained by multiplying the total rework cost with N. The cost for this joint policy is:

$$\text{CR} = \frac{\sum_{i=1}^{n} C_{Ri} \alpha_i Q_i}{T} \qquad (5.324)$$

From Fig. 5.15, the holding cost of inventory system in independent, and joint production policies are shown in Eqs. (5.325) and (5.326), respectively. One has:

$$\text{CH} = N \sum_{i=1}^{n} h_i \left[\frac{H_i}{2} (t_i^1) + \frac{H_i + H_i^{\text{Max}}}{2} (t_i^2) + \frac{H_i^{\text{Max}}}{2} (t_i^3) \right] \qquad (5.325)$$

and:

$$\text{CH} = \frac{1}{T} \sum_{i=1}^{n} h_i \left[\frac{H_i}{2} (t_i^1) + \frac{H_i + H_i^{\text{Max}}}{2} (t_i^2) + \frac{H_i^{\text{Max}}}{2} (t_i^3) \right] \qquad (5.326)$$

The total backorder cost is the product of the backorder cost per unit of the ith item and the backorder quantity of the ith product which are C_{bi} and $\beta_i B_i$, respectively. The annual backorder cost of the system for the joint policy is:

$$\text{CB} = \frac{\sum_{i=1}^{n} C_{bi} \beta_i B_i (t_i^4 + t_i^5)}{2T} \qquad (5.327)$$

5.5 Partial Backordering

The lost sale cost of ith product per period is $\widehat{\pi}_i(1-\beta_i)B_i$. The annual lost sale cost of the system for the joint policy is:

$$CL = \frac{\sum_{i=1}^{n}\widehat{\pi}_i(1-\beta_i)B_i}{2T} \qquad (5.328)$$

where $\widehat{\pi}_i$ and $(1-\beta_i)B_i$ are the lost sale cost per unit of ith product and lost sale quantity of ith product per period, respectively. Therefore, one has (Taleizadeh et al. 2013a):

$$TC = CA + CP + CR + CH + CB + CL$$
$$= \frac{\sum_{i=1}^{n}K_i}{T} + \frac{\sum_{i=1}^{n}C_iQ_i}{T} + \frac{\sum_{i=1}^{n}C_{Ri}\alpha_i Q_i}{T} + \frac{1}{T}\sum_{i=1}^{n}h_i\left[\frac{H_i}{2}(t_i^1) + \frac{H_i + H_i^{\text{Max}}}{2}(t_i^2) + \frac{H_i^{\text{Max}}}{2}(t_i^3)\right]$$
$$+ \frac{\sum_{i=1}^{n}C_{bi}\beta_i B_i(t_i^4 + t_i^5)}{2T} + \frac{\sum_{i=1}^{n}\widehat{\pi}_i(1-\beta_i)B_i}{2T}$$

$$(5.329)$$

The objective function of the joint production system (multi-product single machine) is:

$$TC = \sum_{i=1}^{n}(C_i + \alpha_i C_{Ri})D_i + \frac{\sum_{i=1}^{n}K_i}{T}$$

$$+ h_i\left(\begin{array}{c}\frac{(D_i)^2((1-\alpha_i)P_i - D_i) + 4\alpha_i(D_i)^2((1-\alpha_i)P_i - D_i) + 2\alpha_i^2(D_i)^2(P_i - D_i)}{2(P_i)^2}\\ +(((1-\alpha_i)P_i - D_i) + \alpha_i(P_i - D_i))\left(P_i - \frac{1}{P_i} - \frac{\alpha_i}{P_i}\right)\frac{(D_i)^2}{P_i}\end{array}\right)T$$

$$- h_i\left(\begin{array}{c}\frac{((1-\alpha_i)P_i - D_i)D_i(1-\beta_i) + 2\beta_i P_i D_i}{2(P_i)^2} + \left(\begin{array}{c}\frac{((1-\alpha_i)P_i - D_i) + \alpha_i(P_i - D_i)}{P_i D_i}\\ +\left(\frac{P_i D_i - 1 - \alpha_i}{P_i}\right)\end{array}\right)\beta_i D_i\\ +\frac{(1-\beta_i)D_i(4\alpha_i((1-\alpha_i)P_i - D_i) + \alpha_i^2(P_i - D_i)) + \alpha_i\beta_i P_i D_i + \alpha_i^2 D_i(1-\beta_i)(P_i - D_i)}{(P_i)^2}\end{array}\right)B_i$$

$$- \left(h_i(((1-\alpha_i)P_i - D_i) + \alpha_i(P_i - D_i))\left(D_i - \frac{1}{P_i} - \frac{\alpha_i}{P_i}\right)\frac{(D_i(1-\beta_i))}{P_i} - \frac{(\widehat{\pi}_i - C_i - \alpha_i C_{Ri})(1-\beta_i)}{2}\right)\frac{B_i}{T}$$

$$+ \left(\begin{array}{c}\frac{(1+\alpha_i)h_i\beta_i(1-\beta_i)}{P_i} + \frac{h_i\beta_i^2}{2(1-\alpha_i)P_i - 2D_i} + \frac{h_i((1-\alpha_i)P_i - D_i)(1-\beta_i)^2}{2(P_i)^2}\\ +\frac{h_i(1-\beta_i)^2(2\alpha_i((1-\alpha_i)P_i - D_i) + \alpha_i^2(P_i - D_i))}{(P_i)^2} + \frac{C_{bi}\beta_i((1-\alpha_i)P_i - (1-\beta_i)D_i)}{2((1-\alpha_i)P_i - D_i)D_i}\\ +\frac{\beta_i(1-\beta_i)h_i(((1-\alpha_i)P_i - D_i) + \alpha_i(P_i - D_i) + P_i - (1+\alpha_i)D_i)}{P_i D_i}\\ +\frac{h_i\beta_i^2}{D_i} - h_i(((1-\alpha_i)P_i - D_i) + \alpha_i(P_i - D_i))\left(D_i - \frac{1+\alpha_i}{P_i}\right)\frac{(1-\beta_i)^2}{P_i}\end{array}\right)\frac{B_i^2}{T}$$

$$(5.330)$$

In the joint production systems having rework, the total production, rework, and setup times should be smaller than the cycle length. In our problem, $\sum_{i=1}^{n}(t_i^1 + t_i^2 + t_i^5) + \sum_{i=1}^{n} \text{ts}_i$ must be less than or equal to T. Hence, the model with capacity constraint is (Taleizadeh et al. 2013a):

$$\sum_{i=1}^{n}(t_i^1 + t_i^2 + t_i^5) + \sum_{i=1}^{n} \text{ts}_i \leq T \quad (5.331)$$

From Eqs. (5.311), (5.312), and (5.315), the capacity constraint model becomes:

$$\sum_{i=1}^{n} \frac{(1+\alpha_i)}{P_i}(D_i T - (1-\beta_i)B_i) + \sum_{i=1}^{n} \text{ts}_i \leq T \quad (5.332)$$

Since the production quantity is Q_i, the total available budget is W, and $\alpha_i Q_i$ is the number of the ith product which need rework, the budget constraint then becomes (Taleizadeh et al. 2013a):

$$\sum_{i=1}^{n} C_i Q_i + C_{Ri} \alpha_i Q_i \leq W \quad (5.333)$$

From Eqs. (5.319) and (5.333), one has:

$$\sum_{i=1}^{n}(C_i + \alpha_i C_{Ri})(D_i T - (1-\beta_i)B_i) \leq W \quad (5.334)$$

For the service level constraint, the ith product shortage quantity per period, the annual demand of the ith product, the number of periods in each year, and the safety factor of allowable shortage are B_i, D_i, N, and SL, respectively. With this, the service level constraint becomes:

$$\sum_{i=1}^{n} \frac{N \times B_i}{D_i} \leq \text{SL} \quad (5.335)$$

The service level constraint is:

$$T \geq \frac{\sum_{i=1}^{n} \frac{\lambda_{4i}}{2\lambda_{5i} D_i}}{\left(\text{SL} - \sum_{i=1}^{n} \frac{\lambda_{3i}}{2\lambda_{5i} D_i}\right)} = T_{\text{Min}}^{\text{Service Level}} \quad (5.336)$$

From Eqs. (5.330), (5.332), and (5.334), the final model of the joint production system is:

5.5 Partial Backordering

$$\text{Min}: \quad TC(T, B)$$
$$= \frac{\lambda_1}{T} + \lambda_2 T - \sum_{i=1}^{n} \lambda_{3i} B_i - \sum_{i=1}^{n} \lambda_{4i} \frac{B_i}{T} + \sum_{i=1}^{n} \lambda_{5i} \frac{B_i^2}{T} + \sum_{i=1}^{n}$$
$$\times (C_i + \alpha_i C_{Ri}) D_i \tag{5.337}$$

s.t.:

$$T \geq \frac{\sum_{i=1}^{n} \text{ts}_i - \sum_{i=1}^{n} \frac{(1+\alpha_i)}{P_i}(1-\beta_i) B_i}{\left(1 - \sum_{i=1}^{n} \frac{(1+\alpha_i) D_i}{P_i}\right)} = T_{\text{Min}}^{\text{Production}} \tag{5.338}$$

$$T \leq \frac{W + \sum_{i=1}^{n}(C_i + \alpha_i C_{Ri})(1-\beta_i) B_i}{\sum_{i=1}^{n}(C_i + \alpha_i C_{Ri}) D_i} = T_{\text{Max}}^{\text{Budget}} \tag{5.339}$$

$$T \geq \frac{\sum_{i=1}^{n} \frac{\lambda_{4i}}{2\lambda_{5i} D_i}}{\left(\text{SL} - \sum_{i=1}^{n} \frac{\lambda_{3i}}{2\lambda_{5i} D_i}\right)} = T_{\text{Min}}^{\text{Service Level}} \tag{5.340}$$

$$T, B_i \quad \forall i; i = 1, 2, \ldots, n \tag{5.341}$$

where:

$$\lambda_1 = \sum_{i=1}^{n} K_i > 0 \tag{5.342}$$

$$\lambda_2 = \sum_{i=1}^{n} h_i \left(\frac{(D_i)^2((1-\alpha_i)P_i - D_i) + 4\alpha_i(D_i)^2((1-\alpha_i)P_i - D_i) + 2\alpha_i^2(D_i)^2(P_i - D_i)}{2(P_i)^2} + (((1-\alpha_i)P_i - D_i) + \alpha_i(P_i - D_i))\left(D_i - \frac{1}{P_i} - \frac{\alpha_i}{P_i}\right)\frac{(D_i)^2}{P_i} \right) > 0 \tag{5.343}$$

$$\lambda_{3i} = h_i \left(\frac{((1-\alpha_i)P_i - D_i)D_i(1-\beta_i) + 2\beta_i P_i D_i}{2(P_i)^2} + \frac{\left(\frac{((1-\alpha_i)P_i - D_i) + \alpha_i(P_i - D_i)}{P_i D_i}\right)}{+\left(\frac{P_i D_i - 1 - \alpha_i}{P_i}\right)} \beta_i D_i \right.$$
$$\left. + \frac{(1-\beta_i)D_i(4\alpha_i((1-\alpha_i)P_i - D_i) + \alpha_i^2(P_i - D_i)) + \alpha_i \beta_i P_i D_i + \alpha_i^2 D_i(1-\beta_i)(P_i - D_i)}{(P_i)^2} \right) > 0 \tag{5.344}$$

$$\lambda_{4i} = h_i \frac{(((1-\alpha_i)P_i - D_i) + \alpha_i(P_i - D_i))}{P_i}\left(D_i - \frac{1}{P_i} - \frac{\alpha_i}{P_i}\right)D_i(1-\beta_i)$$
$$- \frac{(\widehat{\pi}_i - C_i - \alpha_i C_{Ri})(1-\beta_i)}{2}$$
$$> 0 \tag{5.345}$$

$$\lambda_{5i} = \frac{(1+\alpha_i)h_i\beta_i(1-\beta_i)}{P_i} + \frac{h_i\beta_i^2}{2(1-\alpha_i)P_i - 2D_i} + \frac{h_i((1-\alpha_i)P_i - D_i)(1-\beta_i)^2}{2(P_i)^2}$$
$$+ \frac{h_i(1-\beta_i)^2(2\alpha_i((1-\alpha_i)P_i - D_i) + \alpha_i^2(P_i - D_i))}{(P_i)^2} + \frac{C_{bi}\beta_i((1-\alpha_i)P_i - (1-\beta_i)D_i)}{2((1-\alpha_i)P_i - D_i)D_i}$$
$$+ \frac{\beta_i(1-\beta_i)h_i(((1-\alpha_i)P_i - D_i) + \alpha_i(P_i - D_i) + P_i - (1+\alpha_i)D_i)}{P_i D_i} + \frac{h_i\beta_i^2}{D_i}$$
$$- h_i(((1-\alpha_i)P_i - D_i) + \alpha_i(P_i - D_i))\left(D_i - \frac{1+\alpha_i}{P_i}\right)\frac{(1-\beta_i)^2}{P_i} > 0$$
(5.346)

Firstly, they proved the convexity of the objective function using the Hessian matrix. The roots of the objective function are then derived applying principles in differential calculus (Taleizadeh et al. 2013a).

Theorem 5.6 The objective function TC(T, S_i) in (5.337) is convex (Taleizadeh et al. 2013a).

Proof In Eq. (5.347), since the Hessian matrix is positive definite for all nonzero B_i and T, therefore TC(T, B_i) is convex (Taleizadeh et al. 2013a):

$$[T, B_1, B_2, \ldots, B_n] \times \mathbf{H} \times [T, B_1, B_2, \ldots, B_n]^T = \frac{2\lambda_1 + \sum_{i=1}^{n}\lambda_{4i}B_i}{T} > 0 \quad (5.347)$$

To derive the optimal values of the decision variables, they took the partial differentiations of TC(T, B_i) with respect to T and B_i. One has (Taleizadeh et al. 2013a):

$$\frac{\partial TC}{\partial T} = \frac{-\sum_{i=1}^{n}\lambda_{5i}B_i^2 - \lambda_1 + \sum_{i=1}^{n}\lambda_{4i}B_i}{T^2} + \lambda_2 \rightarrow T$$

$$= \sqrt{\frac{\lambda_1 - \sum_{i=1}^{n}\left(\frac{\lambda_{4i}^2}{4\lambda_{5i}}\right)}{\left(\lambda_2 - \sum_{i=1}^{n}\left(\frac{\lambda_{3i}^2}{4\lambda_{5i}}\right)\right)}} \quad (5.348)$$

$$\frac{\partial TC}{\partial B_i} = \frac{2\lambda_{5i}B_i - \lambda_{4i}}{T} - \lambda_{3i} \rightarrow B_i = \frac{\lambda_{3i}T + \lambda_{4i}}{2\lambda_{5i}} \quad (5.349)$$

where both $\lambda_1 - \sum_{i=1}^{n}\lambda_{4i}^2/4\lambda_{5i}$ and $\lambda_2 - \sum_{i=1}^{n}(\lambda_{3i}^2/4\lambda_{5i})$ should be either positive or negative simultaneously. This is to ensure that a feasible solution exists. In order to solve the above problem, they introduced the following solution procedure to ensure that all possible conditions are considered. All constraints are checked to affirm that only feasible optimal solution is obtained (Taleizadeh et al. 2013a):

5.5 Partial Backordering

Step 1. Check for initial feasibility. If $(1 - \alpha_i)P_i - D_i \geq 0$, $SL > \sum_{i=1}^{n} \lambda_{3i}/2\lambda_{5i}D_i$, and both $\lambda_1 - \sum_{i=1}^{n} \lambda_{4i}^2/4\lambda_{5i}$ and $\lambda_2 - \sum_{i=1}^{n} (\lambda_{3i}^2/4\lambda_{5i})$ are simultaneously either positive or negative, then proceed to *Step* 2. Otherwise, the problem is infeasible.

Step 2. Find a solution point. Use Eqs. (5.348) and (5.349) to calculate values for T and S_i.

Step 3. Check for secondary feasibility condition:

Condition 1. If both $\sum_{i=1}^{n} ts_i - \sum_{i=1}^{n} \frac{(1+\alpha_i)}{P_i}(1-\beta_i)B_i$ and $1 - \sum_{i=1}^{n} \frac{(1+\alpha_i)D_i}{P_i}$ are either positive or negative, go to *Step* 4. Otherwise, the problem is not feasible; then go to *Step* 6.

Condition 2. If $T_{\text{Max}}^{\text{Budget}} \leq \text{Max}\{T_{\text{Min}}^{\text{Production}}, T_{\text{Min}}^{\text{Service Level}}\}$, go to *Step* 4. Otherwise, the problem is not feasible; then go to *Step* 6.

Step 4. *Check the constraint.* Calculate $T_{\text{Min}}^{\text{Production}}, T_{\text{Max}}^{\text{Budget}}, T_{\text{Min}}^{\text{Service Level}}$ from Eqs. (5.338) to (5.340). They introduced the following conditions to arrive in determining the optimal values to be considered:

Condition 3. If $\text{Max}\{T_{\text{Min}}^{\text{Production}}, T_{\text{Min}}^{\text{Service Level}}\} \leq T \leq T_{\text{Max}}^{\text{Budget}}$, then $T^* = T$.
Condition 4. If $T \geq T_{\text{Max}}^{\text{Budget}}$, then $T^* = T_{\text{Max}}^{\text{Budget}}$.
Condition 5. If $T \leq \text{Max}\{T_{\text{Min}}^{\text{Production}}, T_{\text{Min}}^{\text{Service Level}}\}$, then $T^* = \text{Max}\{T_{\text{Min}}^{\text{Production}}, T_{\text{Min}}^{\text{Service Level}}\}$.

Step 5. Derive the optimal solution. Based on the derived value of T^*, they derived s_i^* using Eq. (5.349). For $Q_i^* = D_i T^* - (1-\beta_i)B_i^*$, the optimal values of the order quantity can be obtained. Calculate the objective function using Eq. (5.337), and then go to *Step* 6.

Step 6. Terminate the procedure.

Examples 5.16, 5.17, and 5.18 Consider a production system having imperfect production processes. All of the parameters are considered constants in each cycle. Taleizadeh et al. (2013a) assumed that the total imperfect-quality items produced are reworkable and no items will be left as scrap. The manufacturer uses the same resource for both the production and the rework processes, and due to joint production system, there is production capacity limitation and budget constraint. Shortage is allowed and a fraction of these will be backordered. Three multi-product EPQ problems with imperfect items, immediate rework, and capacity and budget constraints with partial backordering are considered for 15 products. The general and the specific data for two examples are given in Tables 5.48, 5.49, and 5.50. For the examples, it is assumed that the values for the safety factor of total allowable shortages and the available budget per period are $SL = 0.90$, $W = 400,000$; $SL = 0.90$, $W = 200,000$; and $SL = 0.99$, $W = 400,000$, for Examples 5.16, 5.17, and 5.18, respectively. The optimal results obtained using the proposed methodology is shown in Tables 5.51, 5.52, and 5.53.

In Example 5.16, since $T = 2.5619$ lies between the lower-bound $\left(\text{Max}\{T_{\text{Min}}^{\text{Production}}, T_{\text{Min}}^{\text{Service Level}}\} = 1.2673\right)$ and the upper-bound ($T_{\text{Max}}^{\text{Budget}} = 3.1557$),

Table 5.48 General data for Example 5.16 (Taleizadeh et al. 2013a)

n	D_i	P_i	ts_i	K_i	C_{Ri}	C_i	h_i	C_{bi}	$\hat{\pi}_i$	α_i	β_i
1	150	5000	0.0025	500	15	34	2	5	1	0.05	0.5
2	200	5500	0.0030	600	14	32	4	7	3	0.1	0.5
3	250	6000	0.0035	700	13	30	6	9	5	0.15	0.5
4	300	6500	0.0040	800	12	28	8	11	7	0.2	0.5
5	350	7000	0.0045	900	11	26	10	13	9	0.25	0.5
6	400	7500	0.0025	1000	10	24	12	15	11	0.05	0.6
7	450	8000	0.0030	1100	9	22	14	17	13	0.1	0.6
8	500	8500	0.0035	1200	8	20	16	19	157	0.15	0.6
9	550	9000	0.0040	1300	7	18	18	21	17	0.2	0.6
10	600	9500	0.0045	1400	6	16	20	23	19	0.25	0.6
11	650	10,000	0.0025	1500	5	14	22	25	21	0.05	0.7
12	700	10,500	0.0030	1600	4	12	24	27	23	0.1	0.7
13	750	11,000	0.0035	1700	3	10	26	29	25	0.15	0.7
14	800	11,500	0.0040	1800	2	8	28	31	27	0.2	0.7
15	850	12,000	0.0045	1900	1	6	30	33	29	0.25	0.7

Table 5.49 General data for Example 5.17 (Taleizadeh et al. 2013a)

n	D_i	P_i	ts_i	K_i	C_{Ri}	C_i	h_i	C_{bi}	$\hat{\pi}_i$	α_i	β_i
1	200	5000	0.0005	500	15	34	2	5	1	0.05	0.5
2	250	5500	0.0010	600	14	32	4	7	3	0.1	0.5
3	300	6000	0.0015	700	13	30	6	9	5	0.15	0.5
4	350	6500	0.0020	800	12	28	8	11	7	0.2	0.5
5	400	7000	0.0025	900	11	26	10	13	9	0.25	0.5
6	450	7500	0.0005	1000	10	24	12	15	11	0.05	0.6
7	500	8000	0.0010	1100	9	22	14	17	13	0.1	0.6
8	550	8500	0.0015	1200	8	20	16	19	157	0.15	0.6
9	600	9000	0.0020	1300	7	18	18	21	17	0.2	0.6
10	650	9500	0.0025	1400	6	16	20	23	19	0.25	0.6
11	700	10,000	0.0005	1500	5	14	22	25	21	0.05	0.7
12	750	10,500	0.0010	1600	4	12	24	27	23	0.1	0.7
13	800	11,000	0.0015	1700	3	10	26	29	25	0.15	0.7
14	850	11,500	0.0020	1800	2	8	28	31	27	0.2	0.7
15	900	12,000	0.0025	1900	1	6	30	33	29	0.25	0.7

therefore the condition for the feasibility of the solution algorithm is satisfied. However in Example 5.17, the obtained value is $T = 2.5619$ and is greater than both the $\text{Max}\{T_{\text{Min}}^{\text{Production}}, T_{\text{Min}}^{\text{Service Level}}\} = 1.2673$ and $T_{\text{Max}}^{\text{Budget}} = 1.6023$. And in this situation, the second condition on the third step is satisfied; therefore, they considered the upper bound as the optimal value for the cycle length which will then be $T^* = T_{\text{Max}}^{\text{Budget}} = 1.6023$. Finally in Example 5.18, since $T = 0.9386$ is smaller than both $\text{Max}\{T_{\text{Min}}^{\text{Production}}, T_{\text{Min}}^{\text{Service Level}}\} = 1.3277$ and $T_{\text{Max}}^{\text{Budget}} = 2.7035$, the third

5.5 Partial Backordering

Table 5.50 General data for Example 5.18 (Taleizadeh et al. 2013a)

n	D_i	P_i	ts_i	K_i	C_{Ri}	C_i	h_i	C_{bi}	$\hat{\pi}_i$	α_i	β_i
1	300	5000	0.0010	500	15	34	2	5	1	0.05	0.5
2	350	5500	0.0015	600	14	32	4	7	3	0.1	0.5
3	400	6000	0.0020	700	13	30	6	9	5	0.15	0.5
4	450	6500	0.0025	800	12	28	8	11	7	0.2	0.5
5	500	7000	0.0030	900	11	26	10	13	9	0.25	0.5
6	550	7500	0.0035	1000	10	24	12	157	11	0.05	0.6
7	600	8000	0.0040	1100	9	22	14	17	13	0.1	0.6
8	650	8500	0.0045	1200	8	20	16	19	157	0.15	0.6
9	700	9000	0.0050	1300	7	18	18	21	17	0.2	0.6
10	750	9500	0.0055	1400	6	16	20	23	19	0.25	0.6
11	800	10,000	0.0060	1500	5	14	22	25	21	0.05	0.7
12	850	10,500	0.0065	1600	4	12	24	27	23	0.1	0.7
13	900	11,000	0.0070	1700	3	10	26	29	25	0.15	0.7
14	950	11,500	0.0075	1800	2	8	28	31	27	0.2	0.7
15	1000	12,000	0.0080	1900	1	6	30	33	29	0.25	0.7

Table 5.51 The optimal results for Example 5.16 (Taleizadeh et al. 2013a)

n	$T_{Min}^{Production}$	$T_{Min}^{Service\ Level}$	T_{Max}^{Budget}	T	T^*	Q_i^*	B_i^*	Z^*
1	0.1159	1.2673	3.1557	2.5619	2.5619	380.10	8.33	932,400
2						505.60	13.59	
3						630.70	19.63	
4						755.40	26.30	
5						879.90	33.50	
6						1006.40	45.89	
7						1130.90	54.78	
8						1255.30	64.02	
9						1379.60	73.58	
10						1503.80	83.42	
11						1631.20	113.48	
12						1755.50	125.98	
13						1879.80	138.75	
14						2004.20	151.77	
15						2128.10	165.03	

condition on the third step is satisfied; therefore, the lower limit of the cycle length $\text{Max}\{T_{Min}^{Production}, T_{Min}^{Service\ Level}\} = 1.3277$ is the optimal value for the cycle length. All three provided examples present all the possible cases that may occur for the T (see the third step on the solution procedure). It is also important to note that if the upper bound of the cycle length T_{Max}^{Budget} is smaller than its lower-bound $\text{Max}\{T_{Min}^{Production}, T_{Min}^{Service\ Level}\}$, the problem becomes infeasible as stated as the feasibility condition provided in the first step of the proposed algorithm.

Table 5.52 The optimal results for Example 5.17 (Taleizadeh et al. 2013a)

n	$T_{\text{Min}}^{\text{Production}}$	$T_{\text{Min}}^{\text{Service Level}}$	$T_{\text{Max}}^{\text{Budget}}$	T	T^*	Q_i^*	B_i^*	Z^*
1	0.1159	1.2673	1.6023	2.5619	1.6023	237.30	6.06	473,730
2						315.50	9.89	
3						393.40	14.27	
4						471.10	19.10	
5						548.60	24.31	
6						628.10	32.09	
7						705.70	38.29	
8						783.30	44.72	
9						860.70	51.37	
10						938.10	58.21	
11						1018.50	76.71	
12						1096.10	85.13	
13						1173.60	93.72	
14						1251.10	102.48	
15						1328.50	111.40	

Table 5.53 The optimal results for Example 5.18 (Taleizadeh et al. 2013a)

n	$T_{\text{Min}}^{\text{Production}}$	$T_{\text{Min}}^{\text{Service Level}}$	$T_{\text{Max}}^{\text{Budget}}$	T	T^*	Q_i^*	B_i^*	Z^*
1	0.0857	1.3277	2.7935	0.9386	1.3277	260.70	9.74	390,040
2						325.00	13.94	
3						389.00	18.53	
4						453.00	23.43	
5						516.80	28.59	
6						583.10	35.88	
7						647.10	41.74	
8						711.10	47.76	
9						775.00	53.94	
10						838.90	60.26	
11						906.20	77.17	
12						970.30	84.76	
13						1034.40	92.47	
14						1098.40	100.30	
15						1162.40	108.26	

5.5.2 Repair Failure

A multi-product single-machine EPQ model with limited production capacity, random defective production rate, and repair failure is developed by Taleizadeh et al. (2010b). The aim of this model is to minimize the expected total annual cost by optimizing the period length, the backordered quantities, and the rework items. The

5.5 Partial Backordering

main difference between this work and what presented in Sect. 5.5.1 is considering repair failure in rework process. This model is presented in Chap. 4, Sect. 4.5.2.

5.5.3 Scrapped

Taleizadeh et al. (2010a) extended a multi-product single-machine EPQ model with the production capacity limitation and random defective production rate. The shortage was assumed to occur in combination of backorder and lost sale, and there was a limitation on the service level.

Imperfect production processes, due to process deterioration or some other factors, may randomly generate X percent of defective items at a rate d. The inspection cost per item is involved when all items are screened. All defective items are assumed to be scrapped; i.e., no rework is allowed. The expected production rate of the scrapped items d can be expressed as $d = PE[X]$ (Taleizadeh et al. 2010a).

Figure 5.16 depicts the on-hand inventory level and allowable backorder level of the EPQ model with permitted backlogging. To model the problem, they employed a part of modeling procedure used by Hayek and Salameh (2001). Then, based on Fig. 5.16, for $i = 1, 2, \ldots, n$:

$$T = \sum_{j=1}^{4} t_i^j = \frac{Q_i^B(1 - E(X_i)) + (1 - \beta_i)B_i}{D_i} \tag{5.350}$$

$$t_i^1 = \frac{I_i^1}{P_i - D_i - d_i} \tag{5.351}$$

$$I_i^1 = (P_i - D_i - d_i)\frac{Q_i}{P_i} - \beta_i B_i \tag{5.352}$$

$$t_i^2 = \frac{I_i^1}{D_i} \tag{5.353}$$

$$t_i^3 = \frac{\beta_i B_i}{\beta_i D_i} = \frac{B_i}{D_i} \tag{5.354}$$

$$t_i^4 = \frac{\beta_i B_i}{P_i - D_i - d_i} \tag{5.355}$$

$$(t_i^1 + t_i^4) = \frac{Q_i}{P_i} \tag{5.356}$$

The objective function of the model is the summation of the expected annual production, holding, shortage, disposal, and setup costs as (Taleizadeh et al. 2010a):

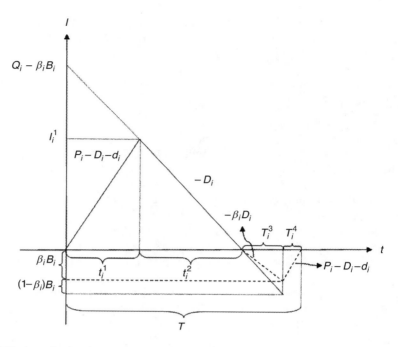

Fig. 5.16 A production–inventory cycle (Taleizadeh et al. 2010a)

$$Z = \mathrm{CP} + \mathrm{CH} + \mathrm{CB} + \mathrm{CL} + \mathrm{CS} + \mathrm{CA} \tag{5.357}$$

The production cost per unit and the production quantity per period of the ith product are C_i and Q_i, respectively. Hence, the production cost of the ith product per period is $C_i Q_i$. While the total annual production cost of the ith product in a disjoint production policy (each product is ordered separately) is $N \times C_i Q_i^B$, this cost for the joint policy (all products have a unique ordering cycle) is $\frac{C_i Q_i^B}{T}$. Furthermore, since the shortages are in combinations of backorders and lost sales, based on Eq. (5.350) one has:

$$Q_i^B = \frac{T \times D_i - (1 - \beta_i) B_i}{E(1 - X_i)} = \frac{T \times D_i - (1 - \beta_i) B_i}{1 - E(X_i)} \tag{5.358}$$

Hence, the expected annual production cost is:

$$\mathrm{CP} = \sum_{i=1}^{n} \frac{C_i \left[\frac{T \times D_i - (1-\beta_i) B_i}{1 - E(X_i)} \right]}{T} = \sum_{i=1}^{n} \frac{C_i D_i}{1 - E(X_i)} - \sum_{i=1}^{n} \left(\frac{C_i (1 - \beta_i)}{1 - E(X_i)} \right) \frac{B_i}{T} \tag{5.359}$$

5.5 Partial Backordering

The holding cost per unit of the ith product per unit time for both the healthy and the scrapped items is h_i. According to Fig. 5.16, the total holding costs of healthy items per cycle and per year are shown in (5.360) and (5.361), respectively:

$$\sum_{i=1}^{n} h_i \left[\frac{I_i^1}{2} \left(t_i^1 + t_i^2 \right) \right] \quad (5.360)$$

$$N \sum_{i=1}^{n} h_i \left[\frac{I_i^1}{2} \left(t_i^1 + t_i^2 \right) \right] \quad (5.361)$$

However, Eq. (5.361) for the joint production policy in which $N = 1/T$ becomes:

$$\frac{1}{T} \sum_{i=1}^{n} h_i \left[\frac{I_i^1}{2} \left(t_i^1 + t_i^2 \right) \right] \quad (5.362)$$

Finally, the expected total annual holding cost of healthy items is (see Appendix 2 of Taleizadeh et al. (2010a)):

$$\sum_{i=1}^{n} h_i (P_i - d_i) \left[\frac{(P_i - D_i - d_i) D_j}{2(P_i)^2 (1 - E(X_i))^2} T - \frac{[(P_i - D_i - d_i)(1 - \beta_i) + \beta_i P_i (1 - E(X_i))]}{(P_i)^2 (1 - E(X_i))^2} B_i \right.$$
$$\left. + \frac{[(P_i - D_i - d_i)^2 (1 - \beta_i)^2 + 2\beta_i (1 - \beta_i)(1 - E(X_i)) P_i (P_i - D_i - d_i) + (\beta_i P_i)^2 (1 - E(X_i))^2]}{2D_i (P_i)^2 (1 - E(X_i))^2 (P_i - D_i - d_i)} \frac{(B_i)^2}{T} \right]$$

(5.363)

Since the scrapped items of each product is assumed to be held until the end of its production time, based on Fig. 5.16, the total holding costs of the scrapped items per cycle and per year are shown in (5.364) and (5.365), respectively:

$$\sum_{i=1}^{n} h_i \left[\frac{d_i \left(t_i^1 + t_i^4 \right)}{2} \left(t_i^1 + t_i^4 \right) \right] \quad (5.364)$$

$$N \sum_{i=1}^{n} h_i \left[\frac{d_i \left(t_i^1 + t_i^4 \right)}{2} \left(t_i^1 + t_i^4 \right) \right] \quad (5.365)$$

Again, for the joint production policy, Eq. (5.365) becomes:

$$\frac{1}{T} \sum_{i=1}^{n} h_i \left[\frac{d_i \left(t_i^1 + t_i^4 \right)}{2} \left(t_i^1 + t_i^4 \right) \right] = \frac{1}{T} \sum_{i=1}^{n} h_i \left[\frac{d_i}{2} \left(\frac{Q_i}{P_i} \right)^2 \right] \quad (5.366)$$

Hence, the expected total annual holding cost of scrapped items (unit holding cost of health and scrapped items are equal) is:

$$\sum_{i=1}^{n} h_i d_i \left[\frac{(D_i)^2 T - 2D_i(1-\beta_i)B_i}{2(1-E(X_i))^2(P_i)^2} + \left(\frac{(1-\beta_i)^2}{2(1-E(X_i))^2(P_i)^2} \right) \frac{B_i^2}{T} \right] \quad (5.367)$$

Finally, the expected total annual holding cost of healthy and scrapped items is:

$$\text{CH} = \sum_{i=1}^{n} h_i(P_i - d_i) \left[\frac{(P_i - D_i - d_i)D_i}{2(P_i)^2(1-E(X_i))^2} T - \frac{[(P_i - D_i - d_i)(1-\beta_i) + \beta_i P_i(1-E(X_i))]}{(P_i)^2(1-E(X_i))^2} B_i \right.$$
$$\left. + \frac{[(P_i - D_i - d_i)^2(1-\beta_i)^2 + 2\beta_i(1-\beta_i)(1-E(X_i))P_i(P_i - D_i - d_i) + (\beta_i P_i)^2(1-E(X_i))^2]}{2(D_i)(P_i)^2(1-E(X_i))^2(P_i - D_i - d_i)} \frac{(B_i)^2}{T} \right]$$
$$+ \sum_{i=1}^{n} h_i d_i \left[\frac{(D_i)^2 T - 2D_i(1-\beta_i)B_i}{2(1-E(X_i))^2(P_i)^2} + \left(\frac{(1-\beta_i)^2}{2(1-E(X_i))^2(P_i)^2} \right) \frac{B_i^2}{T} \right]$$

$$(5.368)$$

Based on Fig. 5.16, the backordered and lost sale costs per cycle are shown in (5.369) and (5.370), respectively:

$$\sum_{j=1}^{n} C_{bi}\beta_i \left[\frac{B_i}{2} \left(t_i^3 + t_i^4 \right) \right] \quad (5.369)$$

$$\sum_{i=1}^{n} \hat{\pi}_i (1-\beta_i) B_i \quad (5.370)$$

These costs for a year become:

$$\text{CB} = N \sum_{i=1}^{n} C_{bi}\beta_i \left[\frac{B_i}{2} \left(t_i^3 + t_i^4 \right) \right] \quad (5.371)$$

$$\text{CL} = N \sum_{i=1}^{n} \hat{\pi}_i (1-\beta_i) B_i \quad (5.372)$$

Because of the joint production policy, Eqs. (5.371) and (5.372) will change to (5.373) and (5.374), respectively:

$$\text{CB} = \frac{1}{T} \sum_{i=1}^{n} C_{bi}\beta_i \left[\frac{B_i}{2} \left(t_i^3 + t_i^4 \right) \right] \quad (5.373)$$

$$\text{CL} = \frac{1}{T} \sum_{i=1}^{n} \hat{\pi}_i (1-\beta_i) B_i \quad (5.374)$$

Finally, the expected annual backordered and lost sale costs are:

5.5 Partial Backordering

$$\text{CB} = \frac{1}{T} \sum_{i=1}^{n} C_{bi}\beta_i \left[\frac{(P_i - (1-\beta_i)D_i - d_i)(B_i)^2}{2D_i(P_i - D_i - d_i)} \right] \tag{5.375}$$

$$\text{CL} = \frac{1}{T} \sum_{i=1}^{n} \widehat{\pi}_i (1-\beta_i) B_i \tag{5.376}$$

The disposal cost per unit of the scrapped item of the *j*th product is C_{Si}, and the quantity of scrapped items is $E(X_i)Q_i$. Hence, the expected total disposal cost per cycle is $\sum_{i=1}^{n} C_{Si} E(X_i) Q_i$. This quantity per year becomes:

$$N \times \sum_{i=1}^{n} C_{Si} E(X_i) Q_i \tag{5.377}$$

Because of the joint production policy, Eq. (5.377) changes to:

$$\frac{1}{T} \times \sum_{i=1}^{n} C_{Si} E(X_i) Q_i \tag{5.378}$$

Since $Q_i = \frac{T \times D_i - (1-\beta_i)B_i}{1 - E(X_i)}$, the annual expected total scrapped item cost is:

$$\begin{aligned} \text{CS} &= \frac{1}{T} \times \sum_{i=1}^{n} C_{Si} E(X_i) \left[\frac{T \times D_i - (1-\beta_i)B_i}{1 - E(X_i)} \right] \\ &= \sum_{i=1}^{n} \frac{C_{Si} E(X_i) D_i}{1 - E(X_i)} - \sum_{i=1}^{n} \left(\frac{C_{Si} E(X_i)(1-\beta_i)}{1 - E(X_i)} \right) \frac{B_i}{T} \end{aligned} \tag{5.379}$$

The cost of a setup is K which occurs N times per year. So, the annual setup is:

$$\text{CA} = N \times K = \frac{K}{T} \tag{5.380}$$

As a result, the objective function of the model becomes:

$$\text{Min } Z = CP + CH + CB + CL + CS + CA = \sum_{i=1}^{n} \frac{C_i D_i}{1 - E(X_i)} - \sum_{i=1}^{n} \left(\frac{C_i(1-\beta_i)}{1-E(X_i)} \right) \frac{B_i}{T}$$

$$+ \sum_{i=1}^{n} h_i (P_i - d_i) \left[\frac{(P_i - D_i - d_i) D_i}{2(P_i)^2 (1 - E(X_i))^2} T - \frac{[(P_i - D_i - d_i)(1-\beta_i) + \beta_i P_i (1 - E(X_i))]}{(P_i)^2 (1 - E(X_i))^2} B_i \right.$$

$$+ \left. \frac{[(P_i - D_i - d_i)^2 (1-\beta_i)^2 + 2\beta_i (1-\beta_i)(1-E(X_i)) P_i (P_i - D_i - d_i) + (\beta_i P_i)^2 (1-E(X_i))^2]}{2 D_i (P_i)^2 (1 - E(X_i))^2 (P_i - D_i - d_i)} \frac{(B_i)^2}{T} \right]$$

$$+ \sum_{i=1}^{n} h_i d_i \left[\frac{(D_i)^2 T - 2 D_i (1-\beta_i) B_i}{2(1 - E(X_i))^2 (P_i)^2} + \left(\frac{(1-\beta_i)^2}{2(1-E(X_i))^2 (P_i)^2} \right) \frac{B_i^2}{T} \right] + \frac{1}{T} \sum_{i=1}^{n} \widehat{\pi}_i (1 - \beta_i) B_i$$

$$+ \frac{1}{T} \sum_{i=1}^{n} C_{bi} \beta_i \left[\frac{(P_i - (1-\beta_i) D_i - d_i)(B_i)^2}{2 D_i (P_i - D_i - d_i)} \right] + \sum_{i=1}^{n} \frac{C_{Si} E(X_i) D_i}{1 - E(X_i)} - \sum_{i=1}^{n} \left(\frac{C_{Si} E(X_i)(1-\beta_i)}{1 - E(X_i)} \right) \frac{B_i}{T} + \frac{K}{T}$$

$$= \sum_{i=1}^{n} \left[\frac{h_i \theta_i (1-\beta_i)^2}{2(1-E(X_i))^2 (P_i)^2} + \frac{C_{bi} \beta_i (P_i - (1-\beta_i) D_i - d_i)}{2 D_i (P_i - D_i - d_i)} \right.$$

$$+ \left. \frac{[(P_i - D_i - d_i)^2 (1-\beta_i)^2 + 2\beta_i (1-\beta_i)(1-E(X_i)) P_i (P_i - D_i - d_i) + (\beta_i P_i)^2 (1-E(X_i))^2]}{2 D_i (P_i)^2 (1 - E(X_i))^2 (P_i - D_i - d_i)} \right] \frac{(B_i)^2}{T}$$

$$- \sum_{i=1}^{n} \left(\frac{h_i d_i 2 D_i (1-\beta_i)}{2(1-E(X_i))^2 (P_i)^2} + \frac{h_i (P_i - d_i)[(P_i - D_i - d_i)(1-\beta_i) + \beta_i P_i (1 - E(X_i))]}{(P_i)^2 (1 - E(X_i))^2} \right) B_i$$

$$- \sum_{i=1}^{n} \left(\frac{(C_i + \widehat{\pi}_i + C_{Si} E(X_i))(1-\beta_i)}{1 - E(X_i)} \right) \frac{B_i}{T}$$

$$+ \sum_{i=1}^{n} \left(\frac{h_i (P_i - d_i)(P_i - D_i - d_i) D_i}{2(P_i)^2 (1 - E(X_i))^2} + \frac{h_i d_i (D_i)^2}{2(1 - E(X_i))^2 (P_i)^2} \right) T + \sum_{i=1}^{n} \left[\frac{(C_i + C_{Si} E(X_i)) D_i}{1 - E(X_i)} \right] + \frac{K}{T}$$

(5.381)

The maximum capacity of the single machine and the minimum service rate are the two constraints of the model that are described in the following: Since $t_i^1 + t_i^4$ and ts_i are the production time and setup time of the ith product, respectively, the summation of the total production and setup time (for all products) will be $\sum_{i=1}^{n} (t_i^1 + t_i^4) + \sum_{i=1}^{n} ts_i$ in which it should be smaller or equal to the period length (T). So the capacity constraint of the model is:

$$\sum_{i=1}^{n} (t_i^1 + t_i^4) + \sum_{i=1}^{n} ts_i \leq T \qquad (5.382)$$

Then, based on the derivation:

$$T \geq \frac{\sum_{i=1}^{n} ts_i - \sum_{i=1}^{n} \frac{(1-\beta_i) B_i}{P_i (1 - E(X_i))}}{1 - \sum_{i=1}^{n} \frac{D_i}{P_i (1 - E(X_i))}} \qquad (5.383)$$

5.5 Partial Backordering

Since the shortage quantity of the ith product per period is B_i, the annual demand of the ith product is D_i, the number of periods in each year is N, and the safety factor of allowable shortage is SL, the service rate constraint becomes:

$$\sum_{i=1}^{n} \frac{N \times B_i}{D_i} \leq SL \qquad (5.384)$$

Finally:

$$T \geq \frac{\sum_{i=1}^{n} \frac{C_i^2}{2C_i^1 D_i}}{\left(SL - \sum_{i=1}^{n} \frac{C_i^3}{2C_i^1 D_i}\right)} = T^{SL} \qquad (5.385)$$

Based on the objective function in (5.381) and the constraints in (5.383) and (5.385), the final model becomes (Taleizadeh et al. 2010a):

$$\text{Min } Z = \sum_{i=1}^{n} \left[\frac{\hat{\pi}_i(1-\beta_i)}{2D_i} + \left(\frac{h_i d_i (1-\beta_i)^2}{2(1-E(X_i))^2 (P_i)^2}\right) + \frac{C_{bi}\beta_i(P_i - (1-\beta_i)D_i - d_i)}{2D_i(P_i - D_i - d_i)} \right.$$

$$\left. + \frac{\left[(P_i - D_i - d_i)^2(1-\beta_i)^2 + 2\beta_i(1-\beta_i)(1-E(X_i))P_i(P_i - D_i - d_i) + (\beta_i P_i)^2(1-E(X_i))^2\right]}{2D_i(P_i)^2(1-E(X_i))^2(P_i - D_i - d_i)} \right] \frac{(B_i)^2}{T}$$

$$- \sum_{i=1}^{n} \left(\frac{(C_j + C_{Si}E(X_i))(1-\beta_i)}{1-E(X_i)}\right) \frac{B_i}{T}$$

$$- \sum_{i=1}^{n} \left(\frac{h_i d_i 2D_i(1-\beta_i)}{2(1-E(X_i))^2(P_i)^2} + \frac{h_i(P_i - d_i)[(P_i - D_i - d_i)(1-\beta_i) + \beta_i P_i(1-E(\beta_i))]}{(P_i)^2(1-E(X_i))^2}\right) B_i$$

$$+ \sum_{i=1}^{n} \left(\frac{h_i\left[(P_i - d_i)(P_i - D_i - d_i)D_i + \theta_i(D_i)^2\right]}{2(P_i)^2(1-E(X_i))^2}\right) T + \sum_{i=1}^{n} \left[\frac{(C_i + C_{Si}E(X_i))D_i}{1-E(X_i)}\right] + \frac{K}{T}$$

s.t.:

$$T \geq \frac{\sum_{i=1}^{n} ts_i - \sum_{i=1}^{n} \frac{(1-\beta_i)B_i}{P_i(1-E(X_i))}}{1 - \sum_{i=1}^{n} \frac{D_i}{P_i(1-E(X_i))}}$$

$$T \geq \frac{\sum_{i=1}^{n} \frac{C_i^2}{2C_i^1 D_i}}{\left(SL - \sum_{i=1}^{n} \frac{C_i^3}{2C_i^1 D_i}\right)}$$

$$T, B_i \geq 0 \quad \forall i, i = 1, 2, \ldots, n$$

$$(5.386)$$

In order to find the optimal solution of model (5.386), they first provided a proof of the convexity of the objective function. Then, a derivative approach to find the optimal point of the objective function is presented. As stated before, for a feasibility requirement, the production rate of the perfect-quality items is assumed to be greater than or equal to the sum of the demand rate and the production rate of defective items. In multi-product model, $\sum_{i=1}^{n} \frac{D_i}{P_i(1-E[X_i])} \leq 1$. Next, the presented constraints are checked to see if the solution satisfies them. These steps will then be applied using an algorithm that is presented after the convexity proof (Taleizadeh et al. 2010a).

To prove the convexity of the objective function, let us rewrite the model as (Taleizadeh et al. 2010a):

$$\text{Min} \quad Z = \sum_{i=1}^{n} C_i^1 \frac{(B_i)^2}{T} - \sum_{i=1}^{n} C_i^2 \frac{B_i}{T} - \sum_{i=1}^{n} C_i^3 B_i + \sum_{i=1}^{n} C_i^4 T + \sum_{i=1}^{n} C_i^5 + \frac{K}{T}$$

s.t. :

$$T \geq \frac{\sum_{i=1}^{n} \text{ts}_i - \sum_{i=1}^{n} \frac{(1-\beta_i)B_i}{P_i(1-E(X_i))}}{1 - \sum_{i=1}^{n} \frac{D_i}{P_i(1-E(X_i))}}$$

$$T \geq \frac{\sum_{i=1}^{n} \frac{C_i^2}{2C_i^1 D_i}}{\left(\text{SL} - \sum_{i=1}^{n} \frac{C_i^3}{2C_i^1 D_i}\right)}$$

$$T, B_i \geq 0 \quad \forall i, i = 1, 2, \ldots, n$$

(5.387)

In which:

$$C_i^1 = \left[\frac{\hat{\pi}_i(1-\beta_i)}{2D_i} + \frac{h_i d_i (1-\beta_i)^2}{2(1-E(X_i))^2 (P_i)^2} + \frac{C_{bi}\beta_i(P_i - (1-\beta_i)D_i - d_i)}{2D_i(P_i - D_i - d_i)} \right.$$

$$\left. + \frac{\left[(P_i - D_i - d_i)^2(1-\beta_i)^2 + 2\beta_i(1-\beta_i)(1-E(X_i))P_i(P_i - D_i - d_i) + (\beta_i P_i)^2(1-E(X_i))^2\right]}{2D_i(P_i)^2(1-E(X_i))^2(P_i - D_i - d_i)} \right] > 0$$

(5.388)

$$C_i^2 = \left[\frac{(C_i + C_{Si}E(X_i))(1-\beta_i)}{1 - E(X_i)} \right] \geq 0$$

(5.389)

$$C_i^3 = \left[\frac{h_i d_i D_i (1-\beta_i)}{(1-E(X_i))^2 (P_i)^2} + \frac{h_i (P_i - d_i)[(P_i - D_i - d_i)(1-\beta_i) + \beta_i P_i (1 - E(X_i))]}{(P_i)^2 (1 - E(X_i))^2} \right]$$
$$> 0$$

(5.390)

5.5 Partial Backordering

$$C_i^4 = \left[\frac{h_i \left[(P_i - d_i)(P_i - D_i - d_i)D_i + d_i(D_i)^2 \right]}{2(P_i)^2(1 - E(X_i))^2} \right] > 0 \qquad (5.391)$$

$$C_i^5 = \left[\frac{(C_i + C_{Si}E(X_i))D_i}{1 - E(X_i)} \right] > 0 \qquad (5.392)$$

Theorem 5.7 The objective function Z in (5.387) is convex (Taleizadeh et al. 2010a).

Proof To prove the convexity of Z, one can utilize the Hessian matrix equation. Then, based on Appendix 6 of Taleizadeh et al. (2010a), the objective function is strictly convex (Taleizadeh et al. 2010a).

Since the objective function is convex, the constraints being in linear forms are convex too, the model in (5.387) is a convex nonlinear programming problem (CNLPP), and its local minimum is the global minimum. Hence, it follows that for the optimal production period length and optimal level of backorder B_j, one can partially differentiate Z with respect to T and B_j and solve the system of Eqs. (5.393) and (5.394). These equations are derived based on simultaneously letting the partial derivatives zero:

$$\frac{\partial Z}{\partial T} = 0 \rightarrow T = \sqrt{\frac{K - \sum_{i=1}^{n} \left[(C_i^2)^2 / 4C_i^1 \right]}{\sum_{i=1}^{n} \left[(C_i^3)^2 / 4C_i^1 \right]}} \qquad (5.393)$$

$$\frac{\partial Z}{\partial B_i} = 0 \rightarrow B_i^* = \frac{C_i^3 T^* + C_i^2}{2C_i^1} \qquad (5.394)$$

where to ensure feasibility both $K - \sum_{i=1}^{n} \left[(C_i^2)^2 / 4C_i^1 \right]$ and $\sum_{i=1}^{n} C_i^4 - \sum_{i=1}^{n} \left[(C_i^3)^2 / 4C_i^1 \right]$ are either positive or negative simultaneously. Then:

$$Q_i^* = \frac{D_i T^* - (1 - \beta_i) B_i^*}{1 - E(X_i)} \qquad (5.395)$$

Now:

$$T^{\text{Min}} = \frac{\sum_{i=1}^{n} \text{ts}_i - \sum_{i=1}^{n} \frac{(1-\beta_i)B_i}{P_i(1-E(X_i))}}{1 - \sum_{i=1}^{n} \frac{D_i}{P_i(1-E(X_i))}} \qquad (5.396)$$

$$T^{\text{SL}} = \frac{\sum_{i=1}^{n} \frac{C_i^2}{2C_i^1 D_i}}{\left(\text{SL} - \sum_{i=1}^{n} \frac{C_i^3}{2C_i^1 D_i}\right)} \tag{5.397}$$

where to have positive T^{SL} the constraint $\text{SL} > \sum_{i=1}^{n} \frac{C_i^3}{2C_i^1 D_i}$ should be held.

To ensure the possibility and acceptability of producing all products on a single machine and satisfying the service level constraint, the steps involved in the algorithm of finding the optimal and possible values of T^*, B_i^*, Q_i^* must be performed as follows (Taleizadeh et al. 2010a):

Step 1. If $\sum_{i=1}^{n} \frac{D_i}{P_i(1-E[X_i])} \leq 1$, $\text{SL} > \sum_{i=1}^{n} \frac{C_i^3}{2C_i^1 D_i}$, $K - \sum_{i=1}^{n} \left[\frac{(C_i^2)^2}{4C_i^1}\right] > 0$, and $\sum_{i=1}^{n} C_i^4 - \sum_{i=1}^{n} \left[\frac{(C_i^3)^2}{4C_i^1}\right] > 0$ (or the first two inequalities hold and the last two inequalities are both negative), then go to *Step* 2. Else the problem is infeasible.

Step 2. Calculate T using Eq. (5.393).
Step 3. Calculate T^{SL} by Eq. (5.397).
Step 4. Calculate B_i, $\forall\ i$, $i = 1, 2, \ldots, n$ using Eq. (5.394).
Step 5. Calculate T^{Min} by Eq. (5.396).
Step 6. If $T \geq \text{Max } \{T^{\text{Min}}, T^{\text{SL}}\}$, then $T^* = T$; else $T^* = \text{Max } \{T^{\text{Min}}, T^{\text{SL}}\}$..
Step 7. Calculate Q_i^* by Eq. (5.395), B_i^* by Eq. (5.394), and Z^* by Eq. 5.387.

In the next section, two numerical examples are given to illustrate the applications of the proposed method in cases of uniform and normal distribution functions for $f_{X_i}(x_i)$ (Taleizadeh et al. 2010a).

Examples 5.19 and 5.20 Consider a multi-product inventory control problem with five products in which their general and specific data for two examples are given in Tables 5.54 and 5.55, respectively. In Example 5.19, the probability distribution of X_i is uniform, and in Example 5.20, the distribution for X_i is normal. The setup cost is $K = \$100{,}000$, and the safety factor of total allowable shortages is $\text{SL} = 0.35$. Based on the available data of Tables 5.54 and 5.55, the problem is solved using the proposed algorithm, and the optimal results are given in Tables 5.56 and 5.57 for the uniform and normal distributions, respectively

5.5.4 Immediate Rework

Taleizadeh and Wee (2015) developed a multi-product single-machine manufacturing system with rework, production capacity constraint, and partial backordering. The defective items of n different types of products are generated x_i: $i = 1, 2, \ldots, n$ percent per cycle. So the good item quantities are $(1 - d_i)P_i$. The production and demand rates of the ith product per cycle are P_i and D_i, respectively (Taleizadeh and Wee 2015). In this production system, each cycle consists of three parts: production

5.5 Partial Backordering

Table 5.54 General data (Taleizadeh et al. 2010a)

Product	D_i	P_i	ts_i	β_i	$\hat{\pi}_i$	C_i	h_i	C_{bi}	C_{Si}
1	800	10,000	0.01	0.75	1000	500	15	350	80
2	900	11,000	0.015	0.80	900	400	12	300	70
3	1000	12,000	0.02	0.85	800	300	9	250	60
4	1100	13,000	0.025	0.90	700	200	6	200	50
5	1200	14,000	0.03	0.95	600	100	3	150	40

Table 5.55 Specific data (Taleizadeh et al. 2010a)

	$X_i \sim U[a_i, b_i]$				$X_i \sim N[\mu_i, \sigma_i^2]$			
Product	a_i	b_i	$E[X_i]$	d_i	$\mu_i = E[X_i]$	σ_i^2	d_i
1	0	0.1	0.05	500	0.25	0.01	2500
2	0	0.15	0.075	825	0.28	0.02	3080
3	0	0.2	0.1	1200	0.33	0.03	3960
4	0	0.25	0.125	1625	0.38	0.04	4940
5	0	0.3	0.15	2100	0.42	0.05	5880

uptime, rework time, and production downtime. They assumed that the total scrapped items are reworkable and no imperfect items are produced at the end of the rework process. Also, the producer has to use the same resource for production and rework processes simultaneously. Because a single machine has a limited joint production system capacity, shortage is allowed with a certain fraction of it to be backordered (Taleizadeh and Wee 2015). Figure 5.17 shows the inventory control problem under study. In the modeling, a single-product problem consisted of ith product is first developed. The fundamental assumption of an economic manufacturing model with rework process is (Taleizadeh and Wee 2015):

$$(1 - x_i)P_i - D_i \geq 0 \tag{5.398}$$

Figure 5.17 shows that T_i^1 and T_i^5 are the production uptimes for non-defective and defective items, respectively. T_i^2 is the reworking time and T_i^3 and T_i^4 are the production downtimes, respectively. Finally, the cycle length is:

$$T = \sum_{j=1}^{5} T_i^j \tag{5.399}$$

$$T_i^1 = \frac{Q_i}{P_i} - \frac{\beta_i B_i}{(1 - x_i)P_i - D_i} \tag{5.400}$$

$$T_i^2 = x_i \frac{Q_i}{P_i} \tag{5.401}$$

Table 5.56 The optimal results of Example 5.19 (uniform distribution) (Taleizadeh et al. 2010a)

Product	Uniform T_{Min}	T^{SL}	T	T^*	B_i^*	Q_i^*	Z
1	0.1578	3.0897	1.9841	3.0897	268.5	2531.2	1,625,500
2					254.6	2951.2	
3					223.6	3395.8	
4					172.6	3864.5	
5					98.9	4356.2	

Table 5.57 The optimal results of Example 5.20 (normal distribution) (Taleizadeh et al. 2010a)

Product	Normal T_{Min}	T^{SL}	T	T^*	B_i^*	Q_i^*	Z
1	0.2044	4.1222	1.6771	4.1222	349.3	4280.6	2,246,700
2					334.7	5059.8	
3					302.5	6084.9	
4					239.2	7275.1	
5					137.2	8517	

$$T_i^3 = \frac{\left(1 - \frac{D_i}{P_i} - x_i \frac{D_i}{P_i}\right) Q_i}{D_i} - \frac{\beta_i B_i}{D_i} \qquad (5.402)$$

$$T_i^4 = \frac{\beta_i B_i}{\beta_i D_i} = \frac{B_i}{D_i} \qquad (5.403)$$

$$T_i^5 = \frac{\beta_i B_i}{(1 - x_i) P_i - D_i} \qquad (5.404)$$

It is evident from Fig. 5.17 that:

$$I_i = ((1 - x_i) P_i - D_i) \frac{Q_i}{P_i} - \beta_i B_i \qquad (5.405)$$

and:

$$I_i' = I_i + x_i (P_i - D_i) \frac{Q_i}{P_i} = ((1 - x_i) P_i - D_i) \frac{Q_i}{P_i} + x_i (P_i - D_i) \frac{Q_i}{P_i} - \beta_i B_i \qquad (5.406)$$

Hence, using Eq. (5.399), the cycle length for a single-product problem is (Taleizadeh and Wee 2015):

5.5 Partial Backordering

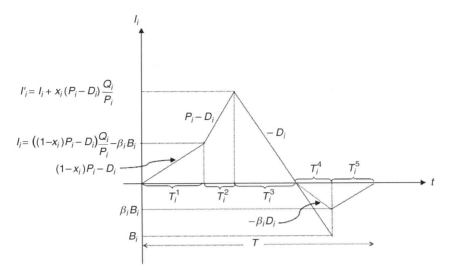

Fig. 5.17 On-hand inventory of perfect-quality items (Taleizadeh and Wee 2015)

$$T = \sum_{j=1}^{5} T_i^j = \frac{Q_i + (1-\beta_i)B_i}{D_i} \quad (5.407)$$

and the order quantity for the *i*th product is:

$$Q_i = D_i T - (1-\beta_i)B_i \quad (5.408)$$

The elements of the cost function are the setup cost, the holding cost, the processing cost, the rework cost, and the shortage cost which are expressed as (Taleizadeh and Wee 2015):

$$TC = CP + CH + CB + CL + CR + CA \quad (5.409)$$

In the following, different elements of the objective function are described (Taleizadeh and Wee 2015):

The cost of a setup is K_i which occurs N times per year. So, the annual setup cost is $\sum_{i=1}^{n} NK_i$. For a joint policy, $N = 1/T$, one has:

$$CA = \frac{\sum_{i=1}^{n} K_i}{T} \quad (5.410)$$

The production cost per unit is C_i, and the production quantity of *i*th product per period is Q_i. So, the production cost of *i*th product per period is $C_i Q_i$. The annual production cost for *i*th product is $NC_i Q_i$, and the following cost is the joint policy cost:

$$\mathrm{CP} = \frac{\sum_{i=1}^{n} C_i Q_i}{T} \tag{5.411}$$

The rework cost per unit of ith product is C_{Ri} and the quantity of ith product that needs to be reworked per period is $x_i Q_i$. So, the rework cost of ith product per period is $C_{Ri} x_i Q_i$. Hence, the rework cost for ith product per year is $NC_{Ri} x_i Q_i$, and the annual rework cost for the joint policy is:

$$\mathrm{CR} = \frac{\sum_{i=1}^{n} C_{Ri} x_i Q_i}{T} \tag{5.412}$$

From Fig. 5.17, Eqs. (5.413) and (5.414) show the inventory holding cost of the system for an independent and joint production policy, respectively:

$$\mathrm{CH} = N \sum_{i=1}^{n} h_i \left[\frac{I_i}{2} (t_i^1) + \frac{I_i + I_i'}{2} (t_i^2) + \frac{I_i'}{2} (t_i^3) \right] \tag{5.413}$$

$$\mathrm{CH} = \frac{1}{T} \sum_{i=1}^{n} h_i \left[\frac{I_i}{2} (t_i^1) + \frac{I_i + I_i'}{2} (t_i^2) + \frac{I_i'}{2} (t_i^3) \right] \tag{5.414}$$

Also, from Fig. 5.17, Eqs. (5.415) and (5.416) show the annual backordered and the lost sale costs in the joint policy production, respectively:

$$\mathrm{CB} = \frac{\sum_{i=1}^{n} C_{bi} \beta_i B_i (t_i^4 + t_i^5)}{2T} \tag{5.415}$$

$$\mathrm{CL} = \frac{\sum_{i=1}^{n} \widehat{\pi}_i (1 - \beta_i) B_i}{2T} \tag{5.416}$$

where $C_{bi} \beta_i B_i$ and $\widehat{\pi}_i (1 - \beta_i) B_i$ are the backordered and the lost sale cost of ith product per period, respectively. Consequently, one has:

$$\mathrm{TC} = \frac{\sum_{i=1}^{n} K_i}{T} + \frac{\sum_{i=1}^{n} C_i Q_i}{T} + \frac{\sum_{i=1}^{n} C_{Ri} x_i Q_i}{T} + \frac{1}{T} \sum_{i=1}^{n} h_i \left[\frac{I_i}{2} (T_i^1) + \frac{I_i + I_i'}{2} (T_i^2) + \frac{I_i'}{2} (T_i^3) \right]$$
$$+ \frac{\sum_{i=1}^{n} C_{bi} \beta_i B_i (T_i^4 + T_i^5)}{2T} + \frac{\sum_{i=1}^{n} \widehat{\pi}_i (1 - \beta_i) B_i}{2T} \tag{5.417}$$

The objective function of the joint production system (multi-product single machine) becomes:

5.5 Partial Backordering

$$
\begin{aligned}
\text{TC} = & \sum_{i=1}^{n} (C_i + C_{Ri} x_i) D_i + \frac{\sum_{i=1}^{n} K_i}{T} \\
& + \sum_{i=1}^{n} h_i \left(\frac{(D_i)^2((1-x_i)P_i - D_i) + 4d_i(D_i)^2((1-x_i)P_i - D_i) + 2x_i^2(D_i)^2(P_i - D_i)}{2(P_i)^2} \right. \\
& \left. + (((1-x_i)P_i - D_i) + x_i(P_i - D_i)) \left(D_i - \frac{1}{P_i} - \frac{x_i}{P_i} \right) \frac{(D_i)^2}{P_i} \right) T \\
& - \sum_{i=1}^{n} h_i \left(\frac{((1-x_i)P_i - D_i)D_i(1-\beta_i) + 2\beta_i P_i D_i}{2(P_i)^2} + \left(\frac{((1-x_i)P_i - D_i) + x_i(P_i - D_i)}{P_i D_i} + \left(D_i - \frac{1}{P_i} - \frac{\alpha_i}{P_i} \right) \right) \beta_i D_i \right. \\
& \left. \frac{(1-\beta_i)D_i(4x_i((1-x_i)P_i - D_i) + x_i^2(P_i - D_i)) + x_i P_i D_i \beta_i + x_i^2 D_i(1-\beta_i)(P_i - D_i)}{(P_i)^2} \right) B_i \\
& - \sum_{i=1}^{n} \left(h_i(((1-x_i)P_i - D_i) + x_i(P_i - D_i))\left(D_i - \frac{1}{P_i} - \frac{x_i}{P_i}\right) \frac{(D_i(1-\beta_i))}{P_i} - \frac{(\hat{\pi}_i - C_i - x_i C_{Ri})(1-\beta_i)}{2} \right) \frac{B_i}{T} \\
& + \sum_{i=1}^{n} \left(\frac{(1+x_i)h_i \beta_i(1-\beta_i)}{P_i} + \frac{h_i \beta_i^2}{2(1-x_i)P_i - 2D_i} + \frac{h_i((1-x_i)P_i - D_i)(1-\beta_i)^2}{2(P_i)^2} \right. \\
& + \frac{h_i(1-\beta_i)^2(2x_i((1-x_i)P_i - D_i) + x_i^2(P_i - D_i))}{(P_i)^2} + \frac{C_{bi}\beta_i((1-x_i)P_i - (1-\beta_i)D_i)}{2((1-x_i)P_i - D_i)D_i} \\
& + \frac{\beta_i(1-\beta_i)h_i(((1-x_i)P_i - D_i) + \alpha_i(P_i - D_i) + P_i - (1+x_i)D_i)}{P_i D_i} \\
& \left. + \frac{h_i \beta_i^2}{D_i} - h_i(((1-x_i)P_i - D_i) + x_i(P_i - D_i))\left(D_i - \frac{1+x_i}{P_i}\right)\frac{(1-\beta_i)^2}{P_i} \right) \frac{B_i^2}{T}
\end{aligned}
$$

(5.418)

In the joint production systems with reworks, the total production, rework, and setup times should be smaller than the cycle length. In our problem, $\sum_{i=1}^{n}\left(T_i^1 + T_i^2 + T_i^5\right) + \sum_{i=1}^{n} \text{ts}_i$ must be smaller or equal to T. Hence, the capacity constraint is:

$$\sum_{i=1}^{n}\left(T_i^1 + T_i^2 + T_i^5\right) + \sum_{i=1}^{n} \text{ts}_i \leq T \tag{5.419}$$

From Eqs. (4.400), (4.401) and (4.404), the capacity constraint model becomes:

$$\sum_{i=1}^{n} \frac{(1+d_i)}{P_i}(D_i T - (1-\beta_i)B_i) + \sum_{i=1}^{n} \text{ts}_i \leq T \tag{5.420}$$

The final model of the joint production system is (Taleizadeh and Wee 2015):

$$
\begin{aligned}
\text{Min}: \quad & \text{TC}(T, B_i) \\
& = \frac{\alpha_1}{T} + \alpha_2 T - \sum_{i=1}^{n} \alpha_{3i} B_i - \sum_{i=1}^{n} \alpha_{4i} \frac{B_i}{T} + \sum_{i=1}^{n} \alpha_{5i} \frac{B_i^2}{T} + \sum_{i=1}^{n} \\
& \times (C_i + C_{Ri} x_i) D_i
\end{aligned}
\tag{5.421}
$$

s.t. :
$$T \geq \frac{\sum_{i=1}^{n} \text{ts}_i - \sum_{i=1}^{n} \frac{(1+x_i)}{P_i}(1-\beta_i)B_i}{\left(1 - \sum_{i=1}^{n} \frac{(1+x_i)D_i}{P_i}\right)} = T_{\text{Min}}^{\text{Production}} \qquad (5.422)$$

$$T, B_i \quad \forall i; i = 1, 2, \ldots, n \qquad (5.423)$$

where

$$\alpha_1 = \sum_{i=1}^{n} K_i > 0 \qquad (5.424)$$

$$\alpha_2 = \sum_{i=1}^{n} h_i \left(\frac{(D_i)^2((1-x_i)P_i - D_i) + 4d_i(D_i)^2((1-x_i)P_i - D_i) + 2x_i^2(D_i)^2(P_i - D_i)}{2(P_i)^2} \right. \\ \left. + (((1-x_i)P_i - D_i) + x_i(P_i - D_i))\left(D_i - \frac{1}{P_i} - \frac{x_i}{P_i}\right)\frac{(D_i)^2}{P_i} \right)$$
$$> 0$$
$$(5.425)$$

$$\alpha_{3i} = h_i \left(\frac{((1-x_i)P_i - D_i)D_i(1-\beta_i) + 2\beta_i P_i D_i}{2(P_i)^2} \right. \\ \left. + \frac{(1-\beta_i)D_i(4x_i((1-x_i)P_i - D_i) + x_i^2(P_i - D_i)) + d_i P_i D_i \beta_i + x_i^2 D_i(1-\beta_i)(P_i - D_i)}{(P_i)^2} \right. \\ \left. + \left(\frac{((1-x_i)P_i - D_i) + x_i(P_i - D_i)}{P_i D_i} + \left(D_i - \frac{1}{P_i} - \frac{\alpha_i}{P_i}\right)\right)\beta_i D_i \right) > 0$$
$$(5.426)$$

$$\alpha_{4i} = \left(h_i(((1-x_i)P_i - D_i) + x_i(P_i - D_i))\left(D_i - \frac{1}{P_i} - \frac{x_i}{P_i}\right)\frac{(D_i(1-\beta_i))}{P_i} - \frac{(\widehat{\pi}_i - C_i - x_i C_{Ri})(1-\beta_i)}{2} \right) > 0 \qquad (5.427)$$

$$\alpha_{5i} = \left(\frac{(1+x_i)h_i\beta_i(1-\beta_i)}{P_i} + \frac{h_i\beta_i^2}{2(1-x_i)P_i - 2D_i} + \frac{h_i((1-x_i)P_i - D_i)(1-\beta_i)^2}{2(P_i)^2} \right. \\ \left. + \frac{h_i(1-\beta_i)^2(2x_i((1-x_i)P_i - D_i) + x_i^2(P_i - D_i))}{(P_i)^2} + \frac{C_{bi}\beta_i((1-x_i)P_i - (1-\beta_i)D_i)}{2((1-x_i)P_i - D_i)D_i} \right. \\ \left. + \frac{\beta_i(1-\beta_i)h_i(((1-x_i)P_i - D_i) + \alpha_i(P_i - D_i) + P_i - (1+x_i)D_i)}{P_i D_i} \right. \\ \left. + \frac{h_i\beta_i^2}{D_i} - h_i(((1-x_i)P_i - D_i) + x_i(P_i - D_i))\left(D_i - \frac{1+x_i}{P_i}\right)\frac{(1-\beta_i)^2}{P_i} \right) > 0$$
$$(5.428)$$

5.5 Partial Backordering

Firstly, they proved the convexity of the objective function using the Hessian matrix. The roots of the objective function are then derived using differential calculus (Taleizadeh and Wee 2015).

Theorem 5.8 The objective function TC(T, B_i) in (5.421) is convex (Taleizadeh and Wee 2015).

Proof Since the Hessian matrix is positive for all nonzero B_i and T, TC(T, B_i) is convex (Taleizadeh and Wee 2015):

$$[T, B_1, B_2, \ldots, B_n] \times \mathbf{H} \times [T, B_1, B_2, \ldots, B_n]^T = \frac{2\alpha_1 + \sum_{i=1}^{n} \alpha_{4i} B_i}{T} > 0 \quad (5.429)$$

To derive the optimal values of the decision variables, take the partial differentiations of TC(T, B_i) with respect to T and B_i:

$$\frac{\partial \text{TC}(T, B_i)}{\partial T} = \frac{-\sum_{i=1}^{n} \alpha_{5i} B_i^2 - \alpha_1 + \sum_{i=1}^{n} \alpha_{4i} B_i}{T^2} + \alpha_2 \to T$$

$$= \sqrt{\frac{\alpha_1 - \sum_{i=1}^{n} \alpha_{4i}^2 / 4\alpha_{5i}}{(\alpha_2 - \sum_{i=1}^{n} \alpha_{3i}^2 / 4\alpha_{5i})}} \quad (5.430)$$

$$\frac{\partial \text{TC}(T, B_i)}{\partial B_i} = \frac{2\alpha_{5i} B_i - \alpha_{4i}}{T} - \alpha_{3i} \to B_i = \frac{\alpha_{3i} T + \alpha_{4i}}{2\alpha_{5i}} \quad (5.431)$$

To ensure feasibility, both $\alpha_1 - \sum_{i=1}^{n} \alpha_{4i}^2 / 4\alpha_{5i}$ and $\alpha_2 - \sum_{i=1}^{n} \alpha_{4i}^2 / 4\alpha_{5i}$ should simultaneously be positive or negative. In order to solve the above problem, they introduced the following solution procedures:

Step 1. Check for feasibility. If $(1 - x_i)P_i - D_i \geq 0$, $\sum_{i=1}^{n} (1 + x_i)D_i/P_i < 1$, and both $\alpha_1 - \sum_{i=1}^{n} \alpha_{4i}^2 / 4\alpha_{5i}$ and $\alpha_2 - \sum_{i=1}^{n} \alpha_{4i}^2 / 4\alpha_{5i}$ be either positive or negative simultaneously, go to *Step* 2. Otherwise, the problem is infeasible.
Step 2. Find a solution point. Using Eqs. (5.430) and (5.431), calculate T and B_i.
Step 3. Calculate $T^* = T_{\text{Min}}^{\text{Production}}$ from Eq. (5.432). If $T \geq T_{\text{Min}}^{\text{Production}}$, then $T^* = T$; else, $T^* = T_{\text{Min}}^{\text{Production}}$.
Step 4. Derive the optimal solution. Based on the derived value of T^*, they derived B_i^*. For $Q_i^* = D_i T^* - (1 - \beta_i)B_i^*$, the optimal values of the order quantity can be obtained. Calculate the objective function and then go to *Step* 5.
Step 5. Terminate the procedure.

Examples 5.21 and 5.22 Consider a production system with production capacity limitation, imperfect production processes, immediate rework, and partial backordered quantity. The general and the specific data of these examples are given in Tables 5.58 and 5.59. The best results using the proposed methodology are shown in Tables 5.60 and 5.61 (Taleizadeh and Wee 2015). In Example 5.21,

Table 5.58 General data for Example 5.21 (Taleizadeh and Wee 2015)

Product	D_i	P_i	ts_i	K_i	C_{Ri}	C_i	h_i	C_{bi}	$\hat{\pi}_i$	x_i	β_i
1	150	5000	0.0025	500	15	34	2	5	1	0.05	0.5
2	200	5500	0.0030	600	14	32	4	7	3	0.1	0.5
3	250	6000	0.0035	700	13	30	6	9	5	0.15	0.5
4	300	6500	0.0040	800	12	28	8	11	7	0.2	0.5
5	350	7000	0.0045	900	11	26	10	13	9	0.25	0.5
6	400	7500	0.0025	1000	10	24	12	15	11	0.05	0.6
7	450	8000	0.0030	1100	9	22	14	17	13	0.1	0.6
8	500	8500	0.0035	1200	8	20	16	19	15	0.15	0.6
9	550	9000	0.0040	1300	7	18	18	21	17	0.2	0.6
10	600	9500	0.0045	1400	6	16	20	23	19	0.25	0.6
11	650	10,000	0.0025	1500	5	14	22	25	21	0.05	0.7
12	700	10,500	0.0030	1600	4	12	24	27	23	0.1	0.7
13	750	11,000	0.0035	1700	3	10	26	29	25	0.15	0.7
14	800	11,500	0.0040	1800	2	8	28	31	27	0.2	0.7
15	850	12,000	0.0045	1900	1	6	30	33	29	0.25	0.7

Table 5.59 General data for Example 5.22 (Taleizadeh and Wee 2015)

Product	D_i	P_i	ts_i	K_i	C_{Ri}	C_i	h_i	C_{bi}	$\hat{\pi}_i$	x_i	β_i
1	200	5000	0.0005	500	15	34	2	5	1	0.05	0.5
2	250	5500	0.0010	600	14	32	4	7	3	0.1	0.5
3	300	6000	0.0015	700	13	30	6	9	5	0.15	0.5
4	350	6500	0.0020	800	12	28	8	11	7	0.2	0.5
5	400	7000	0.0025	900	11	26	10	13	9	0.25	0.5
6	450	7500	0.0005	1000	10	24	12	15	11	0.05	0.6
7	500	8000	0.0010	1100	9	22	14	17	13	0.1	0.6
8	550	8500	0.0015	1200	8	20	16	19	15	0.15	0.6
9	600	9000	0.0020	1300	7	18	18	21	17	0.2	0.6
10	650	9500	0.0025	1400	6	16	20	23	19	0.25	0.6
11	700	10,000	0.0005	1500	5	14	22	25	21	0.05	0.7
12	750	10,500	0.0010	1600	4	12	24	27	23	0.1	0.7
13	800	11,000	0.0015	1700	3	10	26	29	25	0.15	0.7
14	850	11,500	0.0020	1800	2	8	28	31	27	0.2	0.7
15	900	12,000	0.0025	1900	1	6	30	33	29	0.25	0.7

since $T = 2.5619$ is greater than its lower-bound $T_{\text{Min}}^{\text{Production}} = 0.1159$, so $T^* = T = 2.5619$. In Example 5.22, $T = 2.5619$ is smaller than $T_{\text{Min}}^{\text{Production}} = 2.6247$, so $T^* = T_{\text{Min}}^{\text{Production}} = 2.6247$.

5.5 Partial Backordering

Table 5.60 The best results for Example 5.21 (Taleizadeh and Wee 2015)

Product	$T_{\text{Min}}^{\text{Production}}$	T	T^*	Q_i^*	B_i^*	TC^*
1	0.1159	2.5619	2.5619	380.10	8.33	932,400
2				505.60	13.59	
3				630.70	19.63	
4				755.40	26.30	
5				879.90	33.50	
6				1006.40	45.89	
7				1130.90	54.78	
8				1255.30	64.02	
9				1379.60	73.58	
10				1503.80	83.42	
11				1631.20	113.48	
12				1755.50	125.98	
13				1879.80	138.75	
14				2004.20	151.77	
15				2128.10	165.03	

Table 5.61 The best results for Example 5.22 (Taleizadeh and Wee 2015)

Product	$T_{\text{Min}}^{\text{Production}}$	T	T^*	Q_i^*	B_i^*	TC^*
1	2.6247	2.5619	2.6247	389.5	8.48	981,350
2				518.0	13.83	
3				646.2	19.98	
4				774.0	26.77	
5				901.6	34.10	
6				1031.2	46.79	
7				1158.8	55.86	
8				1286.2	65.28	
9				1413.6	75.03	
10				1540.8	85.07	
11				1671.3	115.88	
12				1798.7	128.65	
13				1926.0	141.69	
14				2053.3	154.99	
15				2180.5	168.55	

5.5.5 Preventive Maintenance

There are many instances in which the produced imperfect-quality items should be reworked or repaired with additional costs (Haji et al. 2009). It is assumed that during the rework process, there is no interruption. In each production run, all repairable defective items are reworked, right after the regular production process ends (Chiu and Chang 2014; Chiu 2010). They assumed that because of some controllable realistic reasons such as lubrications, re-setup for each product, programming on machine, cleaning the machine, checking the machine, and many other reasons, an interruption can happen when each product is being produced. On the other hand, they assumed that only one interruption during producing each type of product as preventive maintenance is sufficient to achieve the mentioned goal.

In Sect. 5.4.3, a multi-product single-machine economic production quantity model with preventive maintenance, scrap and rework, and full backordering is presented (Taleizadeh et al. 2014). In this section, partial backordering case from Taleizadeh (2018) is presented. He assumed that preventive maintenance can occur when the inventory level is positive or negative similar to previous case. Similar to Taleizadeh et al. (2014), generally, two cases can be studied: interruption in the backorder-filling stage (see Fig. 5.18) and interruption when there is no shortage (see Fig. 5.19). When an interruption in the production process happens, the machine cannot work and should stay until the state is changed. Figures 5.18 and 5.19 show the level of on-hand inventory of perfect-quality items in the proposed EPQ model for two proposed cases.

Since the preventive maintenance occurs in the manufacturing process, firstly the best time of the interruption should be determined. Based on the time of preventive maintenance, the total inventory cost will be different. On the other word, this cost depends on the length of time before interruption occurs. According to Fig. 5.18, if the interruption occurs when inventory level is negative, it only affects the backordering cost and carrying cost of defective items, while if the interruption occurs during positive inventory level, only the carrying cost of healthy and defective items will be affected. So based on the above explanation, about the time of interruption, the two following possible cases in determining the best time of preventive maintenance will be discussed (Taleizadeh 2018):

$$P_i - D_i - d_i \geq 0 \quad \therefore \quad 0 \leq x_i \leq \left(1 - \frac{D_i}{P_i}\right) \tag{5.432}$$

The summation of the production uptime, the production downtime, the reworking time, and the preventive maintenance length is equal to production cycle length. So one has:

$$T = \sum_{j=1}^{7} t_i^j \tag{5.433}$$

According to Fig. 5.18 showing the first case, an interruption during negative inventory level, one has:

5.5 Partial Backordering

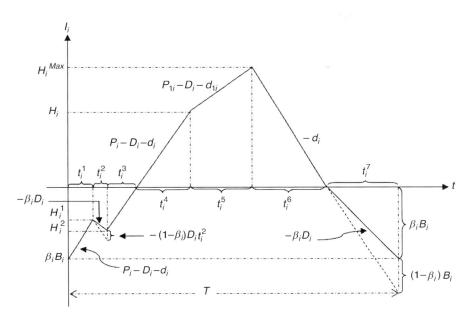

Fig. 5.18 Inventory diagram when preventive maintenance occurs during the shortage cycle (Taleizadeh 2018)

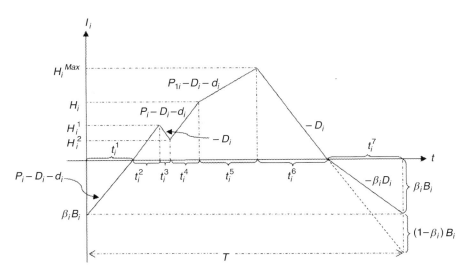

Fig. 5.19 Inventory diagram when interruption occurs during the non-shortage cycle (Taleizadeh 2018)

$$t_i^1 + t_i^3 = \frac{\beta_i B_i + t_i^2 \beta_i D_i}{P_i - D_i - d_i} \tag{5.434}$$

This gives:

$$t_i^3 = \frac{\beta_i B_i + t_i^2 \beta_i D_i}{P_i - D_i - d_i} - t_i^1 \tag{5.435}$$

In continuation:

$$H_i^1 = \beta_i B_i - (P_i - d_i - D_i) t_i^1 \tag{5.436}$$

$$H_i^2 = H_i^1 + \beta_i D_i t_i^2 \tag{5.437}$$

So the cyclic backordering cost is the area of region that is finished at t_i^3 (as shown in Fig. 5.18) multiplied by the backordering unit cost as below (Taleizadeh 2018):

$$\begin{aligned} CB_i &= C_{bi} \left\{ \frac{\beta_i B_i + H_i^1}{2} t_i^1 + \frac{H_i^1 + H_i^2}{2} t_i^2 + \frac{H_i^2 t_i^3}{2} \right\} \\ &= C_{bi} \left\{ \begin{array}{l} \beta_i B_i t_i^1 - \frac{1}{2}(P_i - D_i - d_i)(t_i^1)^2 + \beta_i B_i t_i^2 - (P_i - D_i - d_i) t_i^1 t_i^2 + \frac{1}{2}\beta_i D_i (t_i^2)^2 + \frac{1}{2}\beta_i B_i t_i^3 \\ -\frac{1}{2}(P_i - D_i - d_i) t_i^1 t_i^3 + \frac{1}{2}\beta_i D_i t_i^2 t_i^3 \end{array} \right\} \\ &= C_{bi} \left\{ \begin{array}{l} \beta_i B_i t_i^1 - \frac{1}{2}(P_i - D_i - d_i)(t_i^1)^2 + \beta_i B_i t_i^2 - (P_i - D_i - d_i) t_i^1 t_i^2 + \frac{1}{2}\beta_i D_i (t_i^2)^2 \\ + \left(\frac{1}{2}\beta_i B_i - \frac{1}{2}(P_i - D_i - d_i) t_i^1 \right) \left(\frac{\beta_i B_i + t_i^2 \beta_i D_i}{P_i - D_i - d_i} - t_i^1 \right) + \frac{1}{2}\beta_i D_i t_i^2 \left(\frac{\beta_i B_i + t_i^2 \beta_i D_i}{P_i - D_i - d_i} - t_i^1 \right) \end{array} \right\} \end{aligned}$$

$$\tag{5.438}$$

Then, Eq. (5.438) can be simplified to Eq. (5.439) as below:

$$CB_i = C_{bi} \left\{ \beta_i B_i t_i^2 - (P_i - D_i - d_i) t_i^1 t_i^2 + \frac{1}{2}\beta_i D_i (t_i^2)^2 + \frac{\beta_i}{2}(B_i + D_i t_i^2) \left(\frac{\beta_i B_i + t_i^2 \beta_i D_i}{P_i - D_i - d_i} \right) - t_i^1 t_i^2 \beta_i D_i \right\} \tag{5.439}$$

Also the lost sale cost is:

$$CL_i = \widehat{\pi}_i \left((1 - \beta_i) D_i t_i^2 + (1 - \beta_i) B_i \right) \tag{5.440}$$

and the holding cost of the imperfect items is:

5.5 Partial Backordering

$$CH_{1i} = h_{1i} \begin{Bmatrix} \frac{1}{2}P_i x_i (t_i^1)^2 + P_i x_i t_i^1 t_i^2 + P_i x_i t_i^1 \left(\frac{\beta_i B_i + t_i^2 \beta_i D_i}{P_i - D_i - d_i}\right) - P_i x_i (t_i^1)^2 \\ + \frac{1}{2}P_i x_i \left(\frac{\beta_i B_i + t_i^2 \beta_i D_i}{P_i - D_i - d_i}\right)^2 + \frac{1}{2}P_i x_i (t_i^1)^2 - P_i x_i t_i^1 \left(\frac{\beta_i B_i + t_i^2 \beta_i D_i}{P_i - D_i - d_i}\right) \end{Bmatrix}$$

$$= h_{1i} \left\{ P_i x_i t_i^1 t_i^2 + \frac{1}{2}P_i x_i \left(\frac{\beta_i B_i + t_i^2 \beta_i D_i}{P_i - D_i - d_i}\right)^2 \right\}$$

(5.441)

So the summation of backordering and lost sale costs and holding cost of imperfect items in this case for the ith product is:

$$\psi_i^1 = C_{bi} \left\{ \beta_i B_i t_i^2 - (P_i - D_i - d_i) t_i^1 t_i^2 + \frac{1}{2}\beta_i D_i (t_i^2)^2 + \frac{\beta_i}{2}(B_i + D_i t_i^2)\left(\frac{\beta_i B_i + t_i^2 \beta_i D_i}{P_i - D_i - d_i}\right) - t_i^1 t_i^2 \beta_i D_i \right\}$$

$$+ h_{1i} \left\{ P_i x_i t_i^1 t_i^2 + \frac{1}{2}P_i x_i \left(\frac{\beta_i B_i + t_i^2 \beta_i D_i}{P_i - D_i - d_i}\right)^2 \right\} + \hat{\pi}_i ((1-\beta_i) D_i t_i^2 + (1-\beta_i) B_i)$$

(5.442)

Then the first derivative of ψ_i^1 with respect to t_i^1 gives:

$$\frac{\partial \psi_i^1}{\partial t_i^1} = -C_{bi}(P_i - (1-\beta_i)D_i - d_i) t_i^2 + h_{1i} P_i x_i t_i^2$$

(5.443)

According to Eq. (5.443), the summation of backordering and lost sale costs and holding cost of imperfect items is increasing on t_i^1 if:

$$C_{bi}(P_i(1-x_i) - (1-\beta_i)D_i) < h_{1i} P_i x_i$$

(5.444)

It means $t_i^{1*} = 0$ which is meaningless, because before starting the production of each item, there is setup time during which required operations can be done and there is no need for additional preventive maintenance operation. In the other way, the summation of backordering and lost sale costs and holding cost of imperfect items is decreasing on t_i^1 if $C_{bi}(P_i(1-x_i) - (1-\beta_i)D_i) > h_{1i} P_i x_i$, meaning:

$$t_i^{1*} = \frac{\beta_i B_i}{P_i - D_i - d_i}$$

(5.445)

So the inventory control diagram of optimal case, in respect to the value of t_i^1, shown in Eq. (5.445), can be shown in Fig. 5.19. In the next section, the second possible case is studied. In the case, preventive maintenance occurs when there is no shortage; the preventive maintenance cost only affects the holding cost of healthy and imperfect items. According to Fig. 5.19:

$$t_i^4 = \frac{Q_i}{P_i} - \frac{\beta_i B_i}{P_i - D_i - d_i} - t_i^2 \qquad (5.446)$$

and:

$$H_i^1 = (P_i - D_i - d_i)t_i^2 \qquad (5.447)$$

$$H_i^2 = (P_i - D_i - d_i)t_i^2 - D_i t_i^3 \qquad (5.448)$$

$$\begin{aligned} H_i &= H_i^2 + (P_i - D_i - d_i)t_i^4 \\ &= (P_i - D_i - d_i)t_i^2 - D_i t_i^3 + (P_i - D_i - d_i)\left(\frac{Q_i}{P_i} - \frac{B_i}{P_i - D_i - d_i} - t_i^2\right) \end{aligned} \qquad (5.449)$$

In this case, since only interruption can occur when the rework process is not started, only the holding cost during $(t_i^1 + t_i^3 + t_i^4)$ similar to the first case should be investigated. The carrying cost of perfect products during $(t_i^1 + t_i^3 + t_i^4)$ is:

$$\begin{aligned} CH_i &= h_i \left\{ \frac{1}{2}(P_i - D_i - d_i)(t_i^2)^2 + \left(\frac{H_i^1 + H_i^2}{2}\right)t_i^3 + \left(\frac{H_i^2 + H_i}{2}\right)t_i^4 \right\} \\ &= h_i \left\{ \begin{array}{l} \frac{1}{2}(P_i - D_i - d_i)(t_i^2)^2 + \frac{1}{2}\left(2(P_i - D_i - d_i)t_i^2 - D_i t_i^3\right)t_i^3 \\ + \frac{1}{2}\left((P_i - D_i - d_i)\frac{Q_i}{P_i} - \beta_i B_i - 2D_i t_i^3\right)\left\{\frac{Q_i}{P_i} - \frac{\beta_i B_i}{(P_i - D_i - d_i)} - t_i^2\right\} \\ +(P_i - D_i - d_i)t_i^2 \end{array} \right\} \end{aligned}$$

$$(5.450)$$

After some factorizations:

$$CH_i = h_i \left\{ (P_i - d_i)t_i^2 t_i^3 - \frac{D_i}{2}(t_i^3)^2 + \frac{(P_i - D_i - d_i)}{2}\left(\frac{Q_i}{P_i}\right)^2 - (\beta_i B_i + D_i t_i^3)\frac{Q_i}{P_i} + \frac{\beta_i B_i}{2}\left(\frac{\beta_i b_i + 2D_i t_i^3}{P_i - D_i - d_i}\right) \right\} \qquad (5.451)$$

Also the holding cost of defective items is:

$$CH_{1i} = h_{1i}\left\{\frac{1}{2}P_i x_i(t_i^2)^2 + P_i x_i t_i^2 t_i^3 + P_i x_i t_i^2 t_i^4 + \frac{1}{2}P_i x_i (t_i^4)^2\right\}$$

$$= h_{1i} P_i x_i \left\{ \frac{1}{2}(t_i^2)^2 + t_i^2 t_i^3 + t_i^2\left(\frac{Q_i}{P_i} - \frac{\beta_i B_i}{P_i - D_i - d_i} - t_i^2\right) + \frac{1}{2}\left(\frac{Q_i}{P_i} - \frac{\beta_i B_i}{P_i - D_i - d_i} - t_i^2\right)^2 \right\} \qquad (5.452)$$

After some simplifications (Taleizadeh 2018):

5.5 Partial Backordering

$$\mathrm{CH}_{1i} = h_{1i}\left\{ P_i x_i t_i^2 t_i^3 + \frac{1}{2} x_i \frac{Q_i^2}{P_i} + \frac{1}{2} P_i x_i \left(\frac{\beta_i B_i}{P_i - D_i - d_i} \right)^2 - \frac{x_i \beta_i Q_i B_i}{P_i - D_i - d_i} \right\} \quad (5.453)$$

So the summation of holding costs of perfect and imperfect items is:

$$\psi_i^2 = h_i \left\{ (P_i - d_i) t_i^2 t_i^3 - \frac{D_i}{2}(t_i^3)^2 + \frac{(P_i - D_i - d_i)}{2}\left(\frac{Q_i}{P_i}\right)^2 - (\beta_i B_i + D_i t_i^3)\frac{Q_i}{P_i} + \frac{\beta_i B_i}{2}\left(\frac{\beta_i B_i + 2 D_i t_i^3}{P_i - D_i - d_i}\right) \right\}$$

$$+ h_{1i}\left\{ P_i x_i t_i^2 t_i^3 + \frac{1}{2} x_i \frac{Q_i^2}{P_i} + \frac{1}{2} P_i x_i \left(\frac{\beta_i B_i}{P_i - D_i - d_i} \right)^2 - \frac{x_i \beta_i Q_i B_i}{P_i - D_i - d_i} \right\} \quad (5.454)$$

Then the first derivative of ψ_i^2 in respect to t_i^2 gives:

$$\frac{\psi_i^2}{\partial t_i^2} = h_i(P_i - d_i) t_i^3 + h_{1i} P_i x_i t_i^3 \quad (5.455)$$

Equation (5.455) means that the summation of the holding costs of perfect and imperfect items are increasing on t_i^2 and the best time for considering the interruption is $t_i^{2*} = 0$. The best time for preventive maintenance of the second case, $t_i^{2*} = 0$, is as same as the best time for the preventive maintenance of the first case shown in Eq. (5.455), and the inventory control diagrams of these two cases are similar, as is shown in Fig. 5.20. So as long as $C_{bi}(P_i(1-x_i) - D_i) > h_{1i} P_i x_i$, the best times for the preventive maintenance in both cases are similar. So if we consider always for each product $C_{bi}(P_i(1-x_i) - D_i) > h_{1i} P_i x_i$, then a unique model can be used to solve the problem on hand (Taleizadeh 2018).

It is considered in Fig. 5.20 as the final inventory control diagram. So according to Fig. 5.20, one has:

$$t_i^1 = \frac{\beta_i B_i}{P_i - d_i - D_i} \quad (5.456)$$

$$t_i^2 = t_m \quad (5.457)$$

$$t_i^3 = \frac{H_i^1}{P_i - d_i - D_i} \quad (5.458)$$

$$t_i^4 = \frac{H_i}{P_i - d_i - D_i} \quad (5.459)$$

Total imperfect items produced during the production uptime, $(t_i^1 + t_i^3 + t_i^4)$, are:

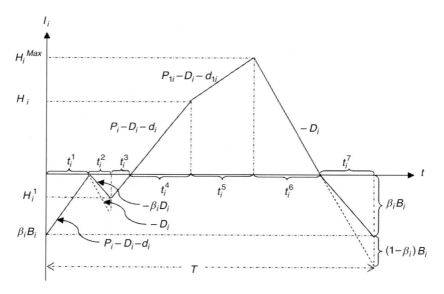

Fig. 5.20 Inventory diagram if preventive maintenance occurs at the best time (Taleizadeh 2018)

$$d_i(t_i^1 + t_i^3 + t_i^4) = P_i x_i(t_i^1 + t_i^3 + t_i^4) = x_i Q_i \tag{5.460}$$

and total scrapped items produced during the rework process are:

$$\varphi_i^1 t_i^5 = P_i^1 \theta_i t_i^5 = d_i \theta_i (t_i^1 + t_i^3 + t_i^4) = x_i Q_i \theta_i \tag{5.461}$$

So:

$$t_i^5 = \frac{x_i Q_i}{P_i^1} \tag{5.462}$$

$$t_i^6 = \frac{H_i^{Max}}{D_i} \tag{5.463}$$

$$t_i^7 = \frac{\beta_i B_i}{\beta_i D_i} = \frac{B_i}{D_i} \tag{5.464}$$

Also:

$$H_i^1 = D_i t_i^2 = D_i t_m \tag{5.465}$$

In order to determine the maximum level of on-hand inventory when regular production process stops, H_i, one can subtract the back-ordered quantity and demand during the preventive maintenance time from the total healthy produced items which gives:

5.5 Partial Backordering

$$H_i = (P_i - d_i - D_i)\frac{Q_i}{P_i} - \beta_i B_i - D_i t_m \qquad (5.466)$$

and the maximum inventory level is:

$$\begin{aligned}H_i^{\text{Max}} &= H_i + (P_{1i} - d_{1i} - D_i)t_i^5 \\ &= (P_i - d_i - D_i)\frac{Q_i}{P_i} - \beta_i B_i - D_i t_m + (P_{1i} - d_{1i} - D_i)\frac{x_i Q_i}{P_i^1}\end{aligned} \qquad (5.467)$$

Obviously, the cycle length is equal to the first case, as below:

$$T = \sum_{j=1}^{7} t_i^j = \frac{Q_i}{D_i}\left[1 - \frac{d_i}{P_i} + x_i\left(1 - \frac{d_{1i}}{P_{1i}}\right)\right] + \frac{(1-\beta_i)B_i}{D_i} \qquad (5.468)$$

This gives:

$$Q_i = \frac{D_i T - (1-\beta_i)B_i}{W_i} \qquad (5.469)$$

where:

$$W_i = \left[1 - \frac{d_i}{P_i} + x_i\left(1 - \frac{d_{1i}}{P_{1i}}\right)\right] \qquad (5.470)$$

The total cyclic cost function including production, rework, disposal, setup, backordered, holding of defective, perfect and scrapped items, and lost sale costs is (Taleizadeh 2018):

$$\begin{aligned}\text{TC}(Q_i, B_i) =\ &\overbrace{C_i Q_i}^{\text{Production Cost}} + \overbrace{C_{Ri} x_i Q_i}^{\text{Rework Cost}} + \overbrace{C_{Si} x_i q_i \theta_i}^{\text{Disposal Cost}} + \overbrace{K_i}^{\text{Setup Cost}} + C_{bi}\overbrace{\left[\frac{\beta_i B_i}{2}(t_i^1 + t_i^7) + \frac{H_i^1}{2}(t_i^2 + t_i^3)\right]}^{\text{Back Ordered Cost}} \\ &+ h_{1i}\left[\frac{P_i x_i \theta_i (t_i^1 + t_i^3 + t_i^4)}{2}(t_i^5)\right] + h_i\underbrace{\left[\frac{H_i}{2}(t_i^4) + \frac{H_i + H_i^{\text{Max}}}{2}(t_i^5) + \frac{H_i^{\text{Max}}}{2}(t_i^6)\right]}_{\text{Holding Cost of Perfect Quality Items}} + \widehat{\pi}_i\underbrace{[(1-\beta_i)D_i t_i^2 + (1-\beta_i)B_i]}_{\text{Lost Sale Cost}} \\ &+ h_i\underbrace{\left[\frac{P_i x_i t_i^1}{2}(t_i^1) + P_i x_i t_i^1(t_i^2 + t_i^3 + t_i^4) + \frac{P_i x_i (t_i^3 + t_i^4)}{2}(t_i^3 + t_i^4) + \frac{P_i x_i}{2}(t_i^1 + t_i^3 + t_i^4)t_i^5\right]}_{\text{Holding Cost of Defective Items}}\end{aligned}$$

$$(5.471)$$

After some algebra and simplifications, Eq. (5.471) can be simplified to (Taleizadeh 2018):

488 5 Multi-product Single Machine

$$\text{TC}(Q,B) = \left(\frac{h_i}{2}\left(\frac{X_i}{P_i^2}+\frac{2x_iX_i}{P_iP_i^1}+\frac{x_i}{P_i}+\frac{x_i^2}{P_i^1}+\frac{x_i^2U_i}{(P_i^1)^2}+\frac{1}{D_i}\left(\frac{X_i}{P_i}+\frac{x_iU_i}{P_i^1}\right)^2\right)+\frac{h_{1i}\theta_i(x_i)^2}{2P_i^1}\right)Q_i^2$$

$$+\left(\frac{C_{bi}\beta_iY_i}{2D_iX_i}+\frac{h_i\beta_i^2}{2}\left(\frac{1}{X_i}+\frac{1}{D_i}\right)\right)B_i^2 - h_i\beta_i\left(\frac{1}{P_i}+\frac{x_i}{P_{1i}}+\left(\frac{X_i}{P_i}+\frac{x_iU_i}{P_{1i}}\right)\right)Q_iB_i$$

$$+\left(C_i+C_{Ri}x_i+C_{Si}x_i\theta_i - h_it_m\left(\frac{D_i+X_i}{P_i}+\frac{x_i(D_i+U_i)}{P_{1i}}\right)\right)Q_i$$

$$+\left(h_i\beta_it_m\left(\frac{X_i+D_i+P_ix_i}{X_i}\right)+\widehat{\pi}_i(1-\beta_i)\right)B_i$$

$$+K_i+\frac{h_i(D_i^2+D_iX_i)t_m^2}{2X_i}+\frac{C_{bi}D_i(P_i-\varphi_i)(t_m)^2}{2X_i}+\widehat{\pi}_i(1-\beta_i)D_it_m$$

(5.472)

However, Eq. (5.472) for the annual joint production policy (multi-product single machine) using Eqs. (5.468)–(5.470) changes to:

$$\text{ATC}(T,B) = \left(\frac{h_iD_i^2}{2W_i^2}\left(\frac{X_i}{P_i^2}+\frac{2x_iX_i}{P_iP_{1i}}+\frac{x_i}{P_i}+\frac{x_i^2}{P_i^1}+\frac{x_i^2U_i}{(P_{1i})^2}+\frac{1}{D_i}\left(\frac{X_i}{P_i}+\frac{x_iU_i}{P_{1i}}\right)^2\right)+\frac{h_{1i}D_i^2\theta_i(x_i)^2}{2W_i^2P_{1i}}\right)T$$

$$+\frac{(1-\beta_i)^2}{2W_i^2}\left(h_i\left(\frac{X_i}{P_i^2}+\frac{2x_iX_i}{P_iP_{1i}}+\frac{x_i}{P_i}+\frac{x_i^2}{P_{1i}}+\frac{x_i^2U_i}{(P_{1i})^2}+\frac{1}{D_i}\left(\frac{X_i}{P_i}+\frac{x_iU_i}{P_{1i}}\right)^2\right)+\frac{h_{1i}\theta_i(x_i)^2}{P_{1i}}\right)\frac{B_i^2}{T}$$

$$+\frac{h_i\beta_i(1-\beta_i)}{W_i}\left(\frac{1}{P_i}+\frac{x_i}{P_{1i}}+\left(\frac{X_i}{P_i}+\frac{x_iU_i}{P_{1i}}\right)\right)\frac{B_i^2}{T}+\left(\frac{C_{bi}\beta_iY_i}{2D_iX_i}+\frac{h_i\beta_i^2}{2}\left(\frac{1}{X_i}+\frac{1}{D_i}\right)\right)\frac{B_i^2}{T}$$

$$+\left(\begin{array}{c}-\dfrac{D_i(1-\beta_i)h_i}{W_i^2}\left(\dfrac{X_i}{P_i^2}+\dfrac{2x_iX_i}{P_iP_{1i}}+\dfrac{x_i}{P_i}+\dfrac{x_i^2}{P_{1i}}+\dfrac{x_i^2U_i}{(P_{1i})^2}+\dfrac{1}{D_i}\left(\dfrac{X_i}{P_i}+\dfrac{x_iU_i}{P_{1i}}\right)^2\right)\\[2mm] -\dfrac{h_i\beta_iD_i}{W_i}\left(\dfrac{1}{P_i}+\dfrac{x_i}{P_{1i}}+\left(\dfrac{X_i}{P_i}+\dfrac{x_iU_i}{P_{1i}}\right)\right)-\dfrac{D_ih_{1i}\theta_i(x_i)^2(1-\beta_i)}{W_i^2P_{1i}}\end{array}\right)B_i$$

$$+\left(h_i\beta_it_m\left(\frac{X_i+D_i+P_ix_i}{X_i}\right)-\frac{(1-\beta_i)}{W_i}\left(C_i+C_{Ri}x_i+C_{Si}x_i\theta_i-h_it_m\left(\frac{D_i+X_i}{P_i}+\frac{x_i(D_i+U_i)}{P_{1i}}\right)\right)\right)\frac{B_i}{T}$$

$$+\left(K_i+\frac{h_i(D_i^2+D_iX_i)t_m^2}{2X_i}+\frac{C_{bi}D_i(P_i-d_i)(t_m)^2}{2X_i}+\widehat{\pi}_i(1-\beta_i)D_it_m\right)\frac{1}{T}+\widehat{\pi}_i(1-\beta_i)$$

$$+\frac{D_i}{W_i}\left(C_i+C_{Ri}x_i+C_{Si}x_i\theta_i-h_it_m\left(\frac{D_i+X_i}{P_i}+\frac{x_i(D_i+U_i)}{P_i^1}\right)\right)$$

(5.473)

The existence of only one machine results in limited production capacity. Since $t_i^1+t_i^3+t_i^4$, t_i^5, and ts_i are the production uptimes, rework time, and setup time of the ith product, respectively, and t_i^2 is the interruption time when the machine is producing ith product, the summation of them should be smaller or equal to the period length (T). So the capacity constraint of the model is:

5.5 Partial Backordering

$$\sum_{i=1}^{n} \left(t_i^1 + t_i^2 + t_i^3 + t_i^4 + t_i^5 \right) + \sum_{i=1}^{n} \text{ts}_i \leq T \quad (5.474)$$

using:

$$\sum_{j=1}^{5} t_i^j = t_m + Q_i \left[\frac{1}{P_i} + \frac{x_i}{P_{1i}} \right] \quad (5.475)$$

Equation (5.474) can be simplified to (Taleizadeh 2018):

$$\sum_{i=1}^{n} \left\{ t_m + Q_i \left[\frac{1}{P_i} + \frac{x_i}{P_{1i}} \right] + \text{ts}_i \right\} \leq T \quad (5.476)$$

and finally using Eq. (5.475), one has (Taleizadeh 2018):

$$T \geq \frac{\sum_{i=1}^{n} (\text{ts}_i + t_m)}{\left(1 - \sum_{i=1}^{n} \frac{D_i}{W_i} \left[\frac{1}{P_i} + \frac{x_i}{P_{1i}} \right] \right)} = T_{\text{Min}} \quad (5.477)$$

Also for the service level constraint, the shortage quantity of *i*th product per period is $B_i + D_i t_m$, the annual demand of the *i*th product is D_i, the number of periods in each year is N, and the safety factor of allowable shortages is SL. So the service level constraint is:

$$\sum_{i=1}^{n} N \left(\frac{B_i + D_i t_m}{D_i} \right) \leq 1 - \text{SL} \quad (5.478)$$

Using Eq. (5.478), one has:

$$T \geq \frac{\sum_{i=1}^{n} \frac{2 D_i \lambda_i^1 t_m - \lambda_i^2}{2 D_i \lambda_i^1}}{(1 - \text{SL}) - \sum_{i=1}^{n} \frac{\lambda_i^3}{2 D_i \lambda_i^1}} \quad (5.479)$$

The final multi-product single-machine model becomes:

$$\text{TC}(T, b_i) = \sum_{i=1}^{n} \lambda_i^1 \frac{B_i^2}{T} + \sum_{i=1}^{n} \lambda_i^2 \frac{B_i}{T} + \sum_{i=1}^{n} \lambda_i^3 T - \sum_{i=1}^{n} \lambda_i^4 B_i + \sum_{i=1}^{n} \lambda_i^5 \frac{1}{T} + \sum_{i=1}^{n} \lambda_i^6$$

s.t. :

$$T \geq \frac{\sum_{i=1}^{n}(\text{ts}_i + t_m)}{\left(1 - \sum_{i=1}^{n} \frac{D_i}{W_i}\left[\frac{1}{P_i} + \frac{x_i}{P_{1i}}\right]\right)} = T_{\text{Min}}^{\text{Capacity}}$$

$$T \geq \frac{\sum_{i=1}^{n} \frac{2D_i \lambda_i^1 t_m - \lambda_i^2}{2D_i \lambda_i^1}}{(1 - \text{SL}) - \sum_{i=1}^{n} \frac{\lambda_i^3}{2D_i \lambda_i^1}} = T_{\text{Min}}^{\text{Service Level}}$$

$$T, b_i \geq 0 \quad i = 1, 2, \ldots, n$$

(5.480)

where:

$$\lambda_i^1 = \frac{(1-\beta_i)^2}{2W_i^2}\left(h_i O_i + \frac{h_{1i} \theta_i (x_i)^2}{P_{1i}}\right) + \frac{h_i \beta_i (1-\beta_i)}{W_i} A_i$$
$$+ \left(\frac{C_{\text{bi}} \beta_i Y_i}{2D_i X_i} + \frac{h_i \beta_i^2}{2}\left(\frac{1}{X_i} + \frac{1}{D_i}\right)\right)$$
$$> 0 \tag{5.481}$$

$$\lambda_i^2 = h_i \beta_i t_m \left(\frac{X_i + D_i + P_i x_i}{X_i}\right) + \widehat{\pi}_i (1 - \beta_i) - \frac{(1-\beta_i)}{W_i} G_i \tag{5.482}$$

$$\lambda_i^3 = \frac{h_i D_i^2}{2W_i^2} O_i + \frac{h_{1i} D_i^2 \theta_i (x_i)^2}{2W_i^2 P_{1i}} > 0 \tag{5.483}$$

$$\lambda_i^4 = \frac{D_i(1-\beta_i)}{W_i^2}\left(h_i O_i + \frac{h_{1i} \theta_i (x_i)^2}{P_{1i}}\right) + \frac{h_i \beta_i D_i}{W_i} A_i > 0 \tag{5.484}$$

$$\lambda_i^5 = K_i + \frac{h_i(D_i^2 + d_i X_i) t_m^2}{2X_i} + \frac{C_{\text{bi}} D_i (P_i - d_i)(t_m)^2}{2X_i} + \widehat{\pi}_i (1-\beta_i) D_i t_m$$
$$> 0 \tag{5.485}$$

$$\lambda_i^6 = \frac{D_i}{W_i} G_i \tag{5.486}$$

It should be noted that (Taleizadeh 2018):

$$X_i = P_i - d_i - D_i, \quad U_i = P_{1i} - d_{1i} - D_i, \quad Y_i = P_i - d_i - (1-\beta_i)D_i,$$

5.5 Partial Backordering

$$W_i = \left[1 - \frac{d_i}{P_i} + x_i\left(1 - \frac{d_{1i}}{P_{1i}}\right)\right],$$

$$G_i = \left(C_i + C_{Ri}x_i + C_{Si}x_i\theta_i - h_i t_m\left(\frac{D_i + X_i}{P_i} + \frac{x_i(D_i + U_i)}{P_{1i}}\right)\right),$$

$$A_i = \frac{1}{P_i} + \frac{x_i}{P_{1i}} + \left(\frac{X_i}{P_i} + \frac{x_i U_i}{P_{1i}}\right)$$

and:

$$O_i = \frac{X_i}{P_i^2} + \frac{2x_i X_i}{P_i P_{1i}} + \frac{x_i}{P_i} + \frac{x_i^2}{P_{1i}} + \frac{x_i^2 U_i}{(P_{1i})^2} + \left(\frac{X_i}{P_i} + \frac{x_i U_i}{P_{1i}}\right)^2 D_i^{-1}.$$

Theorem 5.9 The objective function $F = \text{TC}(T, B_i)$ shown in Eq. (5.480) is convex (Taleizadeh 2018).

Proof To prove the convexity of $F = \text{TC}(T, B_i)$, one can utilize the Hessian matrix. Equation (5.487) shows the objective function is strictly convex (Taleizadeh 2018):

$$[T, B_1, B_2, \ldots, B_n] \times \mathbf{H} \times [T, B_1, B_2, \ldots, B_n]^T = \frac{\sum_{i=1}^n \lambda_i^5}{T} \geq 0 \qquad (5.487)$$

So the solutions are (Taleizadeh 2018):

$$T = \sqrt{\sum_{i=1}^n \left(\frac{4\lambda_i^1 \lambda_i^5 - (\lambda_i^2)^2}{4\lambda_i^1}\right) / \sum_{i=1}^n \left(\lambda_i^3 - \frac{(\lambda_i^4)^2}{4\lambda_i^1}\right)} \qquad (5.488)$$

$$B_i = \frac{\lambda_i^4 T - \lambda_i^2}{2\lambda_i^1} \qquad (5.489)$$

$$Q_i = \frac{D_i T - (1 - \beta_i) B_i}{1 - \frac{d_i}{P_i} + x_i\left(1 - \frac{d_{1i}}{P_{1i}}\right)} \qquad (5.490)$$

In order to obtain feasible T, the necessary condition is having both $\sum_{i=1}^n \frac{4\lambda_i^1 \lambda_i^5 - (\lambda_i^2)^2}{4\lambda_i^1}$ and $\sum_{i=1}^n \lambda_i^3 - \frac{(\lambda_i^4)^2}{4\lambda_i^1}$ either positive or negative and having both $\sum_{i=1}^n \frac{2D_i \lambda_i^1 t_m - \lambda_i^2}{2d_i \lambda_i^1}$ and $(1 - SL) - \sum_{i=1}^n \frac{\lambda_i^3}{2D_i \lambda_i^1}$ either positive or negative. So in order to obtain the feasible and optimal solution, the following solution procedure is developed by Taleizadeh (2018):

Step 1. Calculate $L_1 = \sum_{i=1}^{n} \frac{4\lambda_i^1 \lambda_i^5 - (\lambda_i^2)^2}{4\lambda_i^1}$, $L_2 = \sum_{i=1}^{n} \lambda_i^3 - \frac{(\lambda_i^4)^2}{4\lambda_i^1}$, $T_{\text{Min}}^{\text{Service Level}}$, and $T_{\text{Min}}^{\text{Capacity}}$.

Step 2. If $L_1 L_2 < 0$ or $L_2 = 0$, there is no feasible solution, so terminate the procedure. It means there is no feasible solution for the set of parameters.

Step 3. If $L_1 L_2 > 0$, meaning feasible region and optimal solution exist, calculate the period length using Eq. (5.494).

Step 4. If T is less than $\text{Max}\{T_{\text{Min}}^{\text{Service Level}}, T_{\text{Min}}^{\text{Capacity}}\}$, then $T^* = \text{Max}\{T_{\text{Min}}^{\text{Service Level}}, T_{\text{Min}}^{\text{Capacity}}\}$; else $T^* = T$.

Step 5. Calculate B_i^* for $i = 1, 2, \ldots, n$ using Eq. (5.489) and T^* obtained from *Step 4*.

Step 6. Calculate Q_i^* for $i = 1, 2, \ldots, n$ using Eq. (5.490), T^* and B_i^* obtained from *Steps 4* and *5* respectively.

Step 7. Terminate the procedure.

Examples 5.23 and 5.24 Consider a turning manufactory with only one CNC (computer numerical control) machine used to lathe metal plates to different sizes. Its customers are some factories needing metal plates in different sizes. The total demand rate of each size is deterministic, while because of the different quality of each plate, their unit costs are different. Moreover, the production rates, setup time for each product because of programming on machine, and other parameters are different too. According to the history of the factory, the manufacturer has realized that some of the customers wait to receive the product if the manufacturer could not satisfy their demands as when as they want, while some of them go to another factory to satisfy their demands. The aim of the manufacturer is to determine the best time of preventive maintenance, production and shortage quantities, and period length such that its total cost is minimized. In order to provide numerical examples for this factory, consider a production system with five products where the general data for two examples are given in Table 5.62 and the lower limit of service level is 0.9. They considered two numerical examples with uniform distribution for X_i and θ_i. Tables 5.63 and 5.64 show the specific data of both examples (Taleizadeh 2018).

Tables 5.65 and 5.66 show the best results of numerical examples. In Example 5.23, according to the solution procedure, after calculating coefficients, T should be calculated using Eq. (5.496) which is equal to 0.8642. Since $T = 0.8642 > \text{Max}\{T_{\text{Min}}^{\text{Service Level}} = 0.5965, T_{\text{Min}}^{\text{Capacity}} = 0.6745\}$, the optimal value of period length is 0.8642 (see the flowchart of the proposed solution procedure), and other decision variables should be calculated based on the optimal value of period length as shown in Table 5.65. But in Example 5.24 since $T = 0.9124 < \text{Max}\{T_{\text{Min}}^{\text{Service Level}} = 0.8954, T_{\text{Min}}^{\text{Capacity}} = 1.3945\}$, the optimal value of period length is $T^* = \text{Max}\{T_{\text{Min}}^{\text{Service Level}} = 0.8954, T_{\text{Min}}^{\text{Capacity}} = 1.3945\} = 1.3954$, and optimal order and shortage quantities should be calculated based on $T^* = 1.3954$ as shown in Table 5.66 (Taleizadeh 2018).

5.5 Partial Backordering

Table 5.62 General data for both examples (Taleizadeh 2018)

Product	P_i	P_{1i}	ts_i	K_i	C_i	h_i	h_{1i}	C_{Si}	C_{bi}	$\hat{\pi}_i$	β_i	C_{Ri}
1	10,000	5000	0.005	100	10	3	1	2	5	6	0.7	2
2	11,000	6000	0.01	200	10	3	1	2	5	6	0.7	2
3	12,000	7000	0.015	300	10	3	1	2	5	6	0.7	2
4	13,000	8000	0.02	400	10	3	1	2	5	6	0.7	2
5	14,000	9000	0.025	500	10	3	1	2	5	6	0.7	2

Table 5.63 Specific data for Example 5.23 (Taleizadeh 2018)

			$X_i \sim U[e_i, f_i]$				$\theta_i \tilde{U}[e'_i, f'_i]$		
Items	D_i	t_m	e_i	f_i	$E[X_i]$	$d_i = P_i E[X_i]$	e'_i	f'_i	$E[\theta_i]$	$d_{1i} = P_{1i} E[\theta_i]$
1	1600	0.005	0	0.01	0.005	50	0	0.05	0.0255	250
2	1700	0.005	0	0.015	0.0075	82.5	0	0.1	0.05	600
3	1800	0.005	0	0.02	0.01	120	0	0.2	0.15	1400
4	1900	0.005	0	0.025	0.0125	162.5	0	0.25	0.125	2000
5	2000	0.005	0	0.03	0.015	210	0	0.3	0.155	2700

Table 5.64 Specific data for Example 5.24 (Taleizadeh 2018)

			$X_i \sim U[e_i, f_i]$				$\theta_i \tilde{U}[e'_i, f'_i]$		
Items	D_i	t_m	e_i	f_i	$E[X_i]$	$d_i = P_i E[X_i]$	e'_i	f'_i	$E[\theta_i]$	$d_{1i} = P_{1i} E[\theta_i]$
1	2000	0.007	0	0.055	0.025	250	0	0.15	0.075	375
2	2500	0.007	0	0.1	0.05	550	0	0.2	0.1	600
3	3000	0.007	0	0.155	0.075	900	0	0.25	0.125	875
4	3500	0.007	0	0.20	0.1	1300	0	0.3	0.15	1200
5	4000	0.007	0	0.255	0.125	1750	0	0.35	0.175	1575

Table 5.65 The best results for Example 5.23 (Taleizadeh 2018)

$T_{Min}^{Capacity}$	$T_{Min}^{Service\ Level}$	T	T^*	B_i^*	Q_i^*	TC
0.6745	0.5965	0.8642	0.8642	125.2145	1354.2324	98,458.2
				141.9812	1498.1943	
				189.0294	1552.6183	
				212.7183	1654.1735	
				248.0982	1801.6512	

Table 5.66 The best results for Example 5.24 (Taleizadeh 2018)

$T_{Min}^{Capacity}$	$T_{Min}^{Service\ Level}$	T	T^*	B_i^*	Q_i^*	TC
1.3945	0.8954	0.9124	1.3945	271.5423	2452.8542	109,241.1
				293.0621	2842.3956	
				325.5423	3412.4125	
				360.4520	3965.7163	
				394.2689	4698.0254	

5.6 Conclusion

To provide a comprehensive introduction about the multi-product single-machine EPQ inventory management research status, this chapter presented the recent studies in relevant fields. The literature review framework in this chapter provided a clear overview of the multi-product single-machine study field, which can be used as a starting point for further study. This chapter presents a framework to classify different types of multi-product single machine in terms of economic production quantity problem and reviews the literature based on the framework. In this chapter, several multi-product single-machine EPQ problems are discussed in details, and the models are classified into three categories in terms of the inclusion or noninclusion of the shortage and its type. The first category includes several models in which shortage is not allowed. In the second category of models, shortage is allowed and is back-ordered. Finally, the multi-product single-machine EPQ models with partial backordering are examined.

References

Chiu, S. W. (2010). Robust planning in optimization for production system subject to random machine breakdown and failure in rework. *Computers & Operations Research, 37*, 899–908.

Chiu, Y. S. P., & Chang, H.-H. (2014). Optimal run time for EPQ model with scrap, rework and stochastic breakdowns: a note. *Economic Modelling, 37*, 143–148.

Chiu, S. W., Pai, F. Y., & Wu, W. K. (2013). Alternative approach to determine the common cycle time for a multi-item production system with discontinuous deliveries and failure in rework. *Economic Modelling, 35*, 593–596.

Eilon, S. (1985). Multi-product batch production on a single machine—a problem revisited. *Omega, 13*, 453–468.

Goyal, S. (1984). Determination of economic production quantities for a two-product single machine system. *The International Journal of Production Research, 22*, 121–126.

Hayek, P. A., & Salameh, M. K. (2001). Production lot sizing with the reworking of imperfect quality items produced. *Production Planning & Control, 12*, 584–590.

Haji, B., Haji, R., & Haji, A. (2009). Optimal batch production with rework and non-zero setup cost for rework. In *International Conference on Computers & Industrial Engineering, 2009 (CIE 2009)* (pp. 857–862). Piscataway, NJ: IEEE.

Jamal, A., Sarker, B. R., & Mondal, S. (2004). Optimal manufacturing batch size with rework process at a single-stage production system. *Computers & Industrial Engineering, 47*, 77–89.

Nobil, A. H., & Taleizadeh, A. A. (2016). A single machine EPQ inventory model for a multi-product imperfect production system with rework process and auction. *International Journal of Advanced Logistics, 5*, 141–152.

Pasandideh, S. H. R., & Niaki, S. T. A. (2008). A genetic algorithm approach to optimize a multi-products EPQ model with discrete delivery orders and constrained space. *Applied Mathematics and Computation, 195*, 506–514.

Pasandideh, S. H. R., Niaki, S. T. A., Nobil, A. H., & Cárdenas-Barrón, L. E. (2015). A multiproduct single machine economic production quantity model for an imperfect production system under warehouse construction cost. *International Journal of Production Economics, 169*, 203–214.

References

Rogers, J. (1958). A computational approach to the economic lot scheduling problem. *Management Science, 4*, 264–291.

Shafiee-Gol, S., Nasiri, M. M., & Taleizadeh, A. A. (2016). Pricing and production decisions in multi-product single machine manufacturing system with discrete delivery and rework. *Opsearch, 53*, 873–888.

Taleizadeh, A. A. (2018). A constrained integrated imperfect manufacturing-inventory system with preventive maintenance and partial backordering. *Annals of Operations Research, 261*, 303–337.

Taleizadeh, A. A., Niaki, S. T. A., & Najafi, A. A. (2010a). Multiproduct single-machine production system with stochastic scrapped production rate, partial backordering and service level constraint. *Journal of Computational and Applied Mathematics, 233*, 1834–1849.

Taleizadeh, A. A., Wee, H. M., & Sadjadi, S. J. (2010b). Multi-product production quantity model with repair failure and partial backordering. *Computers & Industrial Engineering, 59*, 45–54.

Taleizadeh, A., Jalali-Naini, S. G., Wee, H. M., & Kuo, T. C. (2013a). An imperfect multi-product production system with rework. *Scientia Iranica, 20*, 811–823.

Taleizadeh, A. A., Wee, H. M., & Jalali-Naini, S. G. (2013b). Economic production quantity model with repair failure and limited capacity. *Applied Mathematical Modelling, 37*, 2765.2774.

Taleizadeh, A. A., Cárdenas-Barrón, L. E., & Mohammadi, B. (2014). A deterministic multi product single machine EPQ model with backordering, scraped products, rework and interruption in manufacturing process. *International Journal of Production Economics, 150*, 9–27.

Taleizadeh, A. A., & Wee, H. M. (2015). Manufacturing system with immediate rework and partial backordering. *International Journal of Advanced Operations Management, 7*, 41–62.

Taleizadeh, A. A., Sadjadi, S. J., & Niaki, S. T. A. (2011). Multiproduct EPQ model with single machine, backordering and immediate rework process. *European Journal of Industrial Engineering, 5*, 388–411.

Taft, E. W. (1918). The most economical production lot. *Iron Age, 101*(18), 1410–1412.

Taleizadeh, A., Najafi, A., & Niaki, S. A. (2010c). Economic production quantity model with scrapped items and limited production capacity. *Scientia Iranica, Transaction E: Industrial Engineering, 17*, 58.

Taleizadeh, A., Cárdenas-Barrón, L., Biabani, J., & Nikousokhan, R. (2012). Multi products single machine EPQ model with immediate rework process. *International Journal of Industrial Engineering Computations, 3*, 93–102.

Taleizadeh, A. A., Sarkar, B., & Hasani, M. (2017). Delayed payment policy in multi-product single-machine economic production quantity model with repair failure and partial backordering. *Journal of Industrial and Management Optimization, 13*(5), 1–24.

Wee, H., Widyadanab, G., Taleizadeh, A., & Biabanid, J. (2011). Multi products single machine economic production quantity model with multiple batch size. *International Journal of Industrial Engineering Computations, 2*, 213–224.

Chapter 6
Quality Considerations

6.1 Introduction

Traditional economic order quantity (EOQ) models offer a mathematical approach to determine the optimal number of items a buyer should order to a supplier each time. One major implicit assumption of these models is that all the items are of perfect quality (Rezaei and Salimi 2012). However, presence of defective products in manufacturing processes is inevitable. There is no production process which can guarantee that all its products would be perfect and free from defect. Hence, there is a yield for any production process. Basic and classical inventory control models usually ignore this fact. They assume all output products are perfect and with equal quality; however, due to the limitation of quality control procedures, among other factors, items of imperfect quality are often present. So it has given researchers the opportunity to relax this assumption and apply a yield to investigate and study its impact on several variables of inventory models such as order quantity and cycle time.

As aforementioned, defective products in manufacturing processes are inevitable. A common assumption in the EPQ inventory literature with defectives is that the rework of a defective item is followed immediately after it is identified (Moussawi-Haidar et al. 2016). This assumption of continuous screening during production complicates the analysis and is not practical for most production systems, especially when the fraction of defective items is low and the production rate is high, which makes continuous screening during production very expensive. To simplify the analysis and the computation of the average inventory, it was assumed in the literature that the products are lumped into two groups, good products and to be reworked products. Some other researchers assumed that some of these defective items are sold to the secondary market with lower price.

Fig. 6.1 Categories EOQ and EPQ models

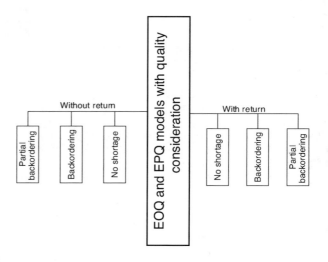

One of the assumptions in the EOQ inventory literature stated that the received items are of perfect quality. This problem has received considerable attention for the last 30 years. Porteus (1986) first considered investments for the quality improvement problem and setup cost reduction. He believed that the number of defective items in a lot depends on the probability of the production process (machine) becoming out of control, and he concluded that the quality of items can be improved by reducing lot sizes. Some researches can be found in Taleizadeh et al. (2015, 2019a, b), Mohammadi et al. (2015) Salameh & Jaber (2000), and Taleizadeh & Moshtagh (2019).

In this chapter, the EOQ and EPQ problems with quality considerations are investigated. The reviewed models are classified into two main categories including "with return" and "without return" and three subcategories related to shortage consideration. The first category includes several models which in some of them shortage is not allowed and others shortage was considered as backordering and partial backordering. Likewise, the second category consists of the same subcategories for the studies which were not considered return policy. This grouping is provided for EOQ and EPQ models separately. To provide a comprehensive introduction about the mentioned research status, in this chapter, the recent studies in relevant fields are reviewed. The literature review framework in this chapter provides a clear overview of the EOQ and EPQ models with quality considerations, which can be used as a starting point for further study. The classification is shown in Fig. 6.1.

The common notations of investigated studies are shown in Table 6.1. To integrate the presented models in this chapter similar to the previous chapters, these notations are used. The main decision variables of this field on inventory are Q and T, but in some studies other decision variables are considered too.

6.1 Introduction

Table 6.1 Notations

D	Annual demand (item/year)
h	Holding cost per unit and per unit of time ($/item/unit of time)
h_1	Holding cost of a defective item kept in inventory ($/item/unit of time)
K	Fixed order cost ($/lot)
K_S	Setup cost ($/setup)
K_d	Transport cost of defective lots back to supplier ($/lot)
P	Production rate per year (item/year)
P_1	Rework rate (item/year)
R_s	Rate of screening or inspection (item/year)
P_1	Rework rate (item/year)
C	Purchasing cost per unit of product ($/item)
C_R	Unit rework cost per item ($/item)
C_I	Inspection cost per unit of product ($/item)
C_r	Cost for returning a defective item ($/item)
C_{fa}	Cost of accepting a defective item ($/lot)
C_{fr}	Cost of rejecting a non-defective item ($/lot)
R	The cost that is paid because of the wrong rejection by the vendor ($/item)
C_R	Unit rework cost ($/item)
s	Selling price per unit of product ($/item)
v	Selling price per imperfect unit ($/item)
C_b	Backordering cost per unit of demand and per unit of time ($/item/unit of time)
g	The goodwill loss on a unit of unfilled demand ($/item)
$\hat{\pi}$	Lost sale cost per unit of unfilled demand $\hat{\pi} = s - C + g$ ($/item)
β	Rate of backordering (%)
T_i	Length of cycle (i) (year)
T_1	Length of cycle in which the inventory level is more than or equal to zero (year)
T_2	Length of cycle in which the inventory level is less than zero (year)
x	Proportion of defective items (%)
n	Size of the sample (integer)
m_1	Probability of type I error (classifying a non-defective item as defective)
m_2	Probability of type II error (classifying a defective item as non-defective)
Q	Order quantity (item)
B	The maximum backordering quantity (item)
y	Number of perfect items in a lot (item)
θ	Number of defective products in a sample of n items (item)
F	Fill rate or the percentage of cycle time that inventory level is positive (%)
TP	Total profit per cycle ($)
ETP	Expectation of total profit per cycle ($)
ATP	Annual total profit ($)
$f(x)$	Function of probability density of defective rate
$f(m_1)$	Probability density function of m_1
$f(m_2)$	Probability density function of m_2
$E[.]$	Expectation operator

6.2 Literature Review

Inventory models with imperfect-quality items have received significant attention in the literature. Muhammad and Alsawafy (2011) developed the economic order quantity model with imperfect-quality items. They considered that the incoming lot has a fraction of scrap and reworkable items, and the lot will go through a 100% inspection. Rezaei and Salimi (2012) formulated and solved a problem to determine the maximum purchasing price a buyer is willing to pay to a supplier to avoid receiving imperfect items under two conditions. First, they assumed that the buyer's selling price is independent of the buyer's purchasing price, while under the second condition, they assumed that changing the buyer's purchasing price influences the buyer's selling price and customer demand. Cheikhrouhou et al. (2018) developed an economic order quantity model for a sampling, sample quality inspection, and a returned policy of defective items. Their research filled the research gap in the literature in providing a model, which took into account the impairment loss resulting from the sample inspection. It also highlighted the strong link between order sizes, sample sizes, and lot sizes. Khan et al. (2011) determine an inventory policy for imperfect items received by a buyer. They adopted a realistic approach of screening. That is, an inspector may classify a non-defective item to be defective (type I error), and he may also classify a defective item to be non-defective (type II error). The defective items classified by the inspector and those returned from the market are accumulated and sold at a discounted price at the end of each procurement cycle.

Konstantaras et al. (2012) assume planned shortages to occur in each cycle. Two models were developed. The first model assumes an infinite planning horizon for which the optimal replenishment policy was determined. The second model assumes that the planning horizon consists of unequal cycles in each of which the percentage of imperfect-quality items reduces with every shipment following a learning curve. For this model, a closed-form solution of the total profit was derived in terms of the cost parameters and the relevant decision variables which are the replenishment points, the points where the inventory level becomes zero and the number of cycles. Hauck and Vörös (2015) considered the traditional lot-sizing problem when after arrival, the quality of each item in a lot is checked. The percentage of defective items is probability variable, and investments could be made to increase the screening rate. The variable screening rate especially implicates cases where unplanned backlogs may develop.

Aslani et al. (2017) extend an EOQ model with partial backordering when the supplier's production process has a random yield. They introduce order quantity and fill rate as decision variables. Moreover, they proposed a solution algorithm based on recursive method to solve the model and find optimal values of decision variables. In order to make improvement in yield, they investigate the effect of investing money to improve the mean and variability of yield by logarithmic functions. The results show that sometimes investment to improve the yield rate causes a reduction of costs. The word "sometimes" refers to the cost required to be invested for improvement plans. Taleizadeh and Zamani-Dehkordi (2017) define three levels of defective items in each sample according to their numbers. If this number is less than α_1, it is not necessary to inspect all the items; else if this number is between α_1 and α_2, all the items should be inspected, and if this number is more than α_2, the order is rejected and another order without any defective items is received. The rate of imperfect items in each order is p, and regarding the number of defective items, it has three levels, lower than p_1, between p_1 and p_2, and higher than p_2.

Cheng (1991) proposed a simple equation to model the relationship between unit production cost and process capability and quality assurance expenses for the EPQ problem. The optimal solution is then derived using differential calculus, which yields a simple closed-form expression for the optimal value of both production quantity and expected fraction acceptable. He also presented a sensitivity analysis of the impacts of the cost parameters on the optimal solution, followed by a discussion of the problems associated with cost estimation. Tsou et al. (2012) developed an EPQ model with continuous quality characteristic, rework and reject. The findings revealed that there is an optimal lot size, which generates minimum total cost in their model. It was also found that if the percentage of imperfect-quality items and rejected items is zero, or approaches zero, the optimal lot size of our model is equal to the classical EPQ model. In addition, it was shown that Taguchi's cost does not affect the model.

Moussawi-Haidar et al. (2016) considered the realistic case where defective items undergo quality control by the consumer or seller during the purchasing process. As soon as production is completed, a cheaper and faster screening process identifies all the defective items. They investigated two realistic cases where defective items are scrapped and when defective items are reworked. The equations to calculate the optimal total profit per unit time and order quantities were presented. Haji et al. (2009) considered an imperfect production system in which a constant percentage of defective items are produced. In their model, all the defective items produced in each cycle are reworked in the same cycle immediately after normal production ends. They assumed that a 100% inspection takes place in both the normal production and rework processes. They also assumed that type 1 errors (i.e., perfect items incorrectly rejected) and type 2 errors (i.e., imperfect items incorrectly accepted) will be committed. For this system, they obtained the optimal production quantity that minimizes the total cost of the system, which is the sum of the normal and rework processing costs, inspection costs, inventory holding costs, and inspection error costs due to inspection errors. Features of reviewed studies are given in Table 6.2.

6.3 EOQ Model with No Return

In this subsection, three models in which return policy is not considered are presented. The model development is investigated. Then, the solution procedure to solve the optimization problem is presented. Also, numerical examples have been reviewed to illustrate the implementation of the proposed method, if there is any.

Table 6.2 Features of reviewed studies

Reference	EOQ	EPQ	Shortage	Return	Partial backordering
Muhammad and Alsawafy (2011)	✓				
Cheikhrouhou et al. (2018)	✓			✓	
Khan et al. (2011)	✓			✓	
Konstantaras et al. (2012)	✓		✓		
Hauck and Vörös (2015)	✓		✓		
Aslani et al. (2017)	✓				✓
Taleizadeh and Zamani-Dehkordi (2017)	✓			✓	✓
Cheng (1991)		✓			
Tsou et al. (2012)		✓			
Moussawi-Haidar et al. (2016)		✓			
Haji et al. (2009)		✓			
Al-Salamah (2016)		✓	✓		

6.3.1 No Return Without Shortage

Rezaei and Salimi (2012) mathematically modeled the relationship between the buyer and supplier with regard to conducting the inspection, resulting in a change of the buyer's economic order quantity and purchasing price. They formulated and analyze the problem under two conditions: (1) assuming there is no relationship between the buyers' selling price, buyer's purchasing price, and customer demand and (2) assuming there is relationship between the buyers' selling price, buyer's purchasing price, and customer demand. The mathematical model of on-hand problem is presented in Sect. 2.3.5.1.

6.3.2 Two Quality Levels with Backordering

Muhammad and Alsawafy (2011) developed an economic order quantity of imperfect-quality items where the incoming lot has fractions of scrap and reworkable items. These fractions are considered to be random variables with known probability density functions. The demand is satisfied from perfect items and reworked items, whereas the scrap items are sold in a single batch at the end of the cycle with a salvage cost.

The notations which are specially used in this problem are presented in Table 6.3.

Figure 6.2 represents the model where the lot of size Q is received with purchasing price of C per unit and the fixed ordering cost K. It is assumed that each order contains a probabilistic fraction of scrap and reworkable items p_s and p_r with known probability density functions $f(p_s)$ and $f(p_r)$, respectively. Good and reworked items

6.3 EOQ Model with No Return

Table 6.3 New notations for given problem (Muhammad and Alsawafy 2011)

p_s	Percentage of scrap items (%)
p_r	Percentage of reworkable items (%)
p	Percentage of scrap and reworkable items (%)
Z_1	Inventory level after the inspection period (item)
Z_2	Inventory level after the selling of the scrap items and return reworked items (item)
Z_3	Inventory level just before receiving the reworked items (item)
Z_4	Inventory level just after receiving the reworked items (item)

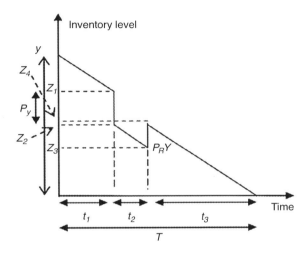

Fig. 6.2 Behavior of inventory level over time (Muhammad and Alsawafy 2011)

can be sold. On the other hand, scrap items will be sold in a batch at the end of the cycle with salvage (discount) per unit. The optimal order quantity is found by taking the difference between the total revenue and total cost, the latter of which consists of four types: procurement cost, inspection cost, rework cost, and inventory carrying cost. Revenues come from selling of good items and scrap items. Shortage is not allowed and inspection and rework processes are error-free (Muhammad and Alsawafy 2011).

Since shortage is not allowed, to avoid shortage, the number of good items is at least equal to the demand during inspection time:

$$(1 - p_s - p_r)Q \geq Dt_1 \tag{6.1}$$

Since p_s and p_r are coming from a probability density functions, they will be limited as below:

$$E(p_s) + E(p_r) \leq 1 - \frac{D}{R_s}, \quad \text{for} \quad R_s \geq D \tag{6.2}$$

The time t_1 needed to inspect the lot is:

$$t_1 = \frac{Q}{R_s} \tag{6.3}$$

The expected total revenue is the summation of sales of the good items and scrap items and it is given as:

$$E[TR(Q)] = p(1 - E(p_s))Q + vQE(p_s) \tag{6.4}$$

The expected total comprises four different costs. The first cost is the procurement cost:

$$PC(Q) = K + CQ \tag{6.5}$$

The expression for the expected total cost is (Muhammad and Alsawafy 2011):

$$E[TC(Q)] = K + CQ + C_R E(P_r)Q + C_I Q$$
$$+ h\left[\frac{E(1-p_s)QE(T)}{2} + \frac{E(p_s)Q^2}{R_s} - \frac{E(p_r^2)Q^2}{P_1}\right] \tag{6.6}$$

The expected total profit equals the expected total revenues minus the expected total cost (Muhammad and Alsawafy 2011):

$$E[TP(Q)] = E[TR(Q)] - E[TC(Q)]$$
$$= s(1 - E(p_s))Q + vQE(p_s)$$
$$- \left[K + CQ + C_R E(p_r)Q + C_I Q + h\left[\frac{E(1-p_s)QE(T)}{2} + \frac{E(p_s)Q^2}{R_s} - \frac{E(p_r^2)Q^2}{P_1}\right]\right] \tag{6.7}$$

The expected cycle period is given by:

$$E(T) = \frac{E(1-p_s)Q}{D} \tag{6.8}$$

The expected total profit per unit time is:

$$E[TPU(Q)] = \frac{E[TP(Q)]}{E(T)} \tag{6.9}$$

6.3 EOQ Model with No Return

To find the optimal order quantity, the first derivative of $E[\text{TPU}(Q)]$ is taken, set to zero, and solved for Q:

$$\frac{\partial E[\text{TPU}(Q)]}{\partial Q} = \left[\frac{1}{1-E(p_s)}\right] \times \left[\frac{KD}{Q^2} - h\left[\frac{E\left((1-p_s)^2\right)}{2} + \frac{E(p_s)D}{R_s} - \frac{E(p_r^2)D}{P_1}\right]\right] \quad (6.10)$$

From Eq. (6.10), they found the expression of the economic order quantity:

$$\text{EOQ}_2 = \left(2KD\bigg/\left\{h\left[E\left((1-p_s)^2\right) + \frac{2DE(p_s)}{R_s} - \frac{2DE(p_r^2)}{P_1}\right]\right\}\right)^{\frac{1}{2}} \quad (6.11)$$

The second derivative is equal to:

$$\frac{\partial^2 E[\text{TPU}(Q)]}{Q} = -\frac{2KD}{Q^2(1-E(p_s))} \leq 0 \quad (6.12)$$

Since the second derivative is always negative, this means that there exists a unique value of Q^* that maximizes objective function. Maximizing Eq. (6.7) is equal to minimizing Eq. (6.13) (Muhammad and Alsawafy 2011):

$$\text{EC}(Q) = \frac{1}{1-E(p_s)}\left\{\frac{KD}{Q^2} - hQ\left[\frac{E\left((1-p_s)^2\right)}{2} + \frac{E(p_s)D}{R_s} - \frac{E(p_r^2)D}{P_1}\right]\right\} \quad (6.13)$$

Example 6.1 Muhammad and Alsawafy (2011) considered an example in which $D = 50{,}000$ units/year, $C = \$25$/unit, $K = 100$/order, $h = \$5$/unit/year, $R_s = 1$ unit/min, $C_I = \$0.5$/unit, $P_1 = 0.5$ unit/min, $C_R = \$2.5$/unit, $s = \$50$/unit, and $v = \$20$/unit. They assumed the operation of the inventory model operates 8 h a day, for 365 days a year, so the annual inspection rate is (Muhammad and Alsawafy 2011):

$$R_s = 1\,\text{unit}/\text{min} \times \left(\frac{60 \times 8 \times 365\,\text{min}}{\text{year}}\right) = 175{,}200\,\text{unit}/\text{year}$$

Also, they assumed scrap and reworkable fractions, p_s and p_r, are uniformly distributed with p.d.f. as the following (Muhammad and Alsawafy 2011):

$$f(p_s) = \begin{cases} 4, & 0 \leq p_s \leq 0.25 \\ 0 & \text{otherwise} \end{cases} \qquad (6.14)$$

$$f(p_r) = \begin{cases} 12.5, & 0 \leq p_r \leq 0.08 \\ 0 & \text{otherwise} \end{cases} \qquad (6.15)$$

Then, the optimal value of Q that optimizes the expected total profit per unit time is given by (Muhammad and Alsawafy 2011):

$$\text{EOQ}_2 = \sqrt{\frac{2 \times 100 \times 50{,}000}{5\left[0.770833 + \frac{2 \times 50{,}000 \times 0.125}{175{,}200} - \frac{2 \times 50{,}000 \times 0.002133}{43{,}800}\right]}} = 1537 \, \text{units}$$

With optimal expected cost per unit time, $\text{EC}(Q) = \$7395$ (Muhammad and Alsawafy 2011).

6.3.3 Learning in Inspection with Backordering

Konstantaras et al. (2012) developed an EOQ imperfect system with learning in inspection. They assumed that 100% inspection of items is performed for each shipment, and the screening rate is faster than the demand rate. The defective items are sold at a discounted price; the fraction of defective items follows a learning curve that is either of an S-shape or of a power form learning curve. This problem is presented in Sect. 2.4.5.

6.3.4 Partial Backordering

In this subsection, the study of Aslani et al. (2017) has been presented. This problem presents an economic order quantity (EOQ) inventory model with partial backordering, where a buyer purchases its required products from a supplier. Consider an EOQ inventory model with shortage in partial backordering mode. The supplier supplies this inventory system with a production process characterized by a random yield. According to this, each lot of products produced by the supplier includes a random proportion of defective items. The yield is assumed to be a continuous random variable with known mean and variance. The buyer places its order to the supplier and receives the order immediately; in other words, there is no lead time for delivering the orders. After receiving the products, the buyer inspects all received units. Further assume that the inspection process is perfect. All rejected items in the process of the inspection are returned to the supplier without any cost to the buyer. Moreover, assume that the inspection costs of accepted items are the

6.3 EOQ Model with No Return

buyer's responsibility and the supplier pays those costs of rejected items. This can be considered a motivation granted from the supplier to the buyer in order to create a long-term relationship. They strived to model the stated problem from the buyer's point of view. In the next step, they developed two supplementary models in order to achieve improvements in the yield of the production process. One of these models is to enhance the mean of the yield, and the other one is to reduce the yield variability. If cost-effective, these supplementary models will lead to the amelioration of the yield rate. Moreover, a main result of these improvement schemes is to achieve a long-term relationship between the supplier and the buyer (Aslani et al. 2017).

The traditional EOQ model with partial backordering is proposed by Pentico and Drake (2011). The annual total cost in this model is as follows (Aslani et al. 2017):

$$\text{ATC} = \frac{K}{T} + \frac{hDF^2T}{2} + \frac{\beta C_b D(1-F)^2}{2} + \hat{\pi}D(1-\beta)(1-F) \quad (6.16)$$

In Eq. (6.16), the first term represents the purchasing cost per each year. The second term shows the holding cost per year. The third term indicates the annual costs of backordered shortage, and finally the last term indicates the cost of lost sale for unfilled demands. The optimal decision variables in the traditional EOQ + PBO model are as follows:

$$T^* = \sqrt{\frac{2K}{Dh}\left[\frac{h+\beta C_b}{\beta C_b}\right] - \frac{[(1-\beta)g]^2}{\beta h C_b}} \quad (6.17)$$

$$F^*(T) = \frac{(1-\beta)\hat{\pi} + \beta C_b T}{T(h+\beta C_b)} \quad (6.18)$$

The notations which are specially used in this problem are presented in Table 6.4.

In the traditional model, all items produced by the supplier are acceptable for the buyer (i.e., are of acceptable quality). Now assume that each lot includes a random proportion of defective items. This indicates that the supplier's production process follows random yield in producing good parts.

The notations which are specially used in this problem are presented in Table 6.4.

They showed the random yield with x, a continuous random variable with probability density function $f(x)$ and known mean and variance ($\mu \cdot \sigma^2$). Moreover, they assumed that the rate of good parts is independent of lot size. The yield of supplier's production process is defined as follows:

$$L_x \leq x \leq 1 \quad (6.19)$$

Let $y = xQ =$ the number of good parts in a received lot
$T = \frac{y}{D} =$ cycle time, time between two successive order placements
$\beta B =$ part of shortage which is backordered

Table 6.4 New notations for a given problem (Aslani et al. 2017)

μ	Mean of yield
σ^2	Variability of yield
L_x	Lower-bound μ
$\alpha(\gamma)$	Represents the investment cost required to change the companion yield parameter from γ_0 to γ
γ_0	The primary value of companion yield parameter
φ	The percentage reduction in γ for each unit of money increase in $\alpha(\gamma)$
γ_d	Desired values of companion yield
$\alpha(\gamma_d)$	Cost needed to make improvement ($/item)
σ_0^2	The initial value of variance
$\rho(\sigma^2)$	Investment cost needed to change the variance of yield rate from σ_0^2 to σ^2 ($/item)
τ	The percentage reduction in variance for each unit of money increase in $\rho(\sigma^2)$
i	Capital cost rate ($/year)
Q_{imp}	Order quantity after improvement (item)
F_{imp}	Fill rate after improvement (%)
ATC_{imp}	Annual total cost after improvement ($/year)
Q'_{imp}	Order quantity after change in variance of yield rate (item)
F'_{imp}	Fill rate after change in variance of yield rate (%)
ATC'_{imp}	Annual total cost after change in variance of yield rate ($/year)

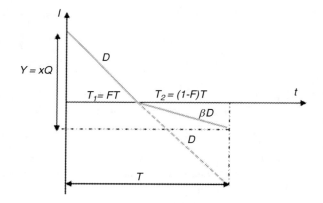

Fig. 6.3 Inventory control chart when the supplier can compensate all backordered shortage (Aslani et al. 2017)

The buyer only incurs the purchasing and inspection costs of accepted items shown with C. In order to model the problem, they considered two cases (Aslani et al. 2017):

Case I $y \geq \beta B$. In this case, the order quantity is more than the backordered shortage (see Fig. 6.3).

In this case, the profit per cycle is as follows (Aslani et al. 2017):

6.3 EOQ Model with No Return

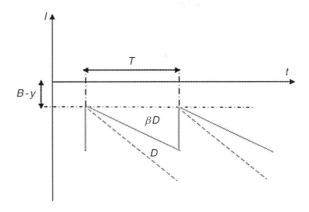

Fig. 6.4 Inventory control chart when the supplier cannot compensate all backordered shortage (Aslani et al. 2017)

$$TP_1 = sTD[F + \beta(1-F)]$$
$$- \left[K + \frac{hDF^2T^2}{2} + CDT[F + \beta(1-F)] + \frac{C_b\beta D(1-F)^2T^2}{2} \right. \\ \left. + \hat{\pi}(1-\beta)D(1-F)T \right] \quad (6.20)$$

Note that this case occurs if one has:

$$x \geq \frac{\beta B}{Q}$$

Case II $y < \beta B$. In this case, the inventory level is always negative (see Fig. 6.4). In this case, the profit per cycle can be written as follows:

$$TP_2 = s\beta DT - \left[K + C\beta DT + \frac{C_b\beta DT^2}{2} + \pi(1-\beta)DT \right] \quad (6.21)$$

Note that this case occurs if one has:

$$x < \frac{\beta B}{Q}$$

Thus, the expected average profit per cycle is calculated as follows (Aslani et al. 2017):

$$ETP = \int_{\frac{\beta B}{Q}}^{1} (TP_1)f(x)dx + \int_{L_x}^{\frac{\beta B}{Q}} (TP_2)f(x)dx \quad (6.22)$$

It must be noted that the expected average profit per cycle is impressed by how the system wants to satisfy the shortage once an order enters. If it is preferred to always satisfy all backordered shortages when an order with imperfect quality enters, then it must $\beta B/Q \leq d$. In the following, a service constraint as $\beta B/Q \leq L_x$ is imposed on the problem. This constraint ensures that the buyer will receive all backordered units during a cycle. This is constant with a supplier who is inclined to build good will and a long-lasting relationship with buyers. In other words, the supplier uses the promise of urgent fulfillment of backordered shortage as a motivation for the buyer to convince him to purchase its products. According to this, imposition of the service constraint will reduce the problem to Case I ($y \geq \beta B$).

$$\text{TP} = \int_{L_x}^{1} (\text{TP}_1) f(x) dx \tag{6.23}$$

Proposition 6.1 The function of the cycle time and the expected cycle time in this situation are as follows (Eqs. (6.24) and (6.25), respectively) (Aslani et al. 2017):

$$T = \frac{xQ}{D(F + \beta - \beta F)} \tag{6.24}$$

$$E(T) = \frac{\mu Q}{D(F + \beta - \beta F)} \tag{6.25}$$

Proof According to Fig. 6.4, one has:

$$T = T_1 + T_2 = \left(\frac{xQ - \beta D(1 - F)T}{D}\right) + (1 - F)T$$
$$= T[1 - (1 - (F + \beta) + \beta F)] \tag{6.26}$$

From Eq. (6.26), Proposition 6.1 is held.

Calculating the resulting integral function, for expected average profit per cycle (ETP), one gets (Aslani et al. 2017):

$$\text{TP} = (s - C)(F + \beta(1 - F)) \frac{\mu Q}{F + \beta - \beta F} - K - \frac{hF^2 Q^2}{2D[F + \beta - \beta F]^2} (\sigma^2 \mu^2)$$
$$- \frac{C_b \beta (1 - F)^2}{2} \frac{Q^2}{D[F + \beta - \beta F]^2} (\sigma^2 + \mu^2) - \frac{D\hat{\pi}(1 - \beta)(1 - F)\mu Q}{D(F + \beta - \beta F)} \tag{6.27}$$

The expected annual total profit is given by dividing TP by $E(T)$. Therefore,

6.3 EOQ Model with No Return

$$\text{ATP} = (s - C)D(F + \beta(1 - F))$$
$$- \left[\frac{KD(F + \beta - \beta F)}{\mu Q} + \frac{hF^2 Q}{2(F + \beta - \beta F)} \left(\frac{\sigma^2 + \mu^2}{\mu} \right) + \frac{C_b \beta (1 - F)^2 Q}{2(F + \beta - \beta F)} \left(\frac{\sigma^2 + \mu^2}{\mu} \right) + \hat{\pi}(1 - \beta)(1 - F)D \right]$$
(6.28)

We know that $\hat{\pi} = s - C + g$ and substituting $\hat{\pi}$ into the ATP, one gets:

$$\text{ATP} = (s - C)D$$
$$- \left[\frac{KD(F + \beta - \beta F)}{\mu Q} + \frac{hF^2 Q}{2(F + \beta - \beta F)} \left(\frac{\sigma^2 + \mu^2}{\mu} \right) + \frac{C_b \beta (1 - F)^2 Q}{2(F + \beta - \beta F)} \left(\frac{\sigma^2 + \mu^2}{\mu} \right) + g(1 - \beta)(1 - F)D \right]$$
(6.29)

Because demand is constant, therefore potential revenue $((s - C)D)$ is also constant. Hence, the problem of maximizing ATP can be reduced to minimizing ATC which is defined as follows in Eq. (6.30) (Aslani et al. 2017):

$$\text{ATC} = \frac{KD(F + \beta - \beta F)}{\mu Q} + \frac{hF^2 Q}{2(F + \beta - \beta F)} \left(\frac{\sigma^2 + \mu^2}{\mu} \right) + \frac{C_b \beta (1 - F)^2 Q}{2(F + \beta - \beta F)} \left(\frac{\sigma^2 + \mu^2}{\mu} \right)$$
$$+ g(1 - \beta)(1 - F)$$
(6.30)

Please note that decision variables in the problem are F and Q.

Proposition 6.2 Using classical optimization procedures for minimizing ATC, the optimal values for the decision variables can be calculated as follows (Aslani et al. 2017):

$$\begin{cases} \text{if } \beta > 1 - \sqrt{\frac{2Kh(\sigma^2 + \mu^2)}{D\mu^2 \pi^2}} : \begin{cases} a : Q^*(F) = \sqrt{\frac{2KD(F + \beta - \beta F)^2}{\left[hF^2 + C_b \beta (1 - F)^2 \right] (\sigma^2 + \mu^2)}} \\ b : F^*(Q) = \frac{\sqrt{\beta(\beta h C_b) Q (\sigma^2 + \mu^2) [\psi_3 Q - \psi_2 + \psi_1 Q]} - \beta[\psi_3 Q - \psi_2 + \psi_1 Q]}{(1 - \beta)[\psi_3 Q - \psi_2 + \psi_1 Q]} \end{cases} \\ \text{Otherwise} : \qquad c : F^* = 1, \text{ and } Q^* = \sqrt{2KD/h(\sigma^2 + \mu^2)} \end{cases}$$
(6.31)

where:

$$\psi_1 = (h + \beta C_b)(\mu^2 + \sigma^2) \tag{6.32}$$

$$\psi_2 = 2Dg(-1 + \beta)^2 \mu \tag{6.33}$$

$$\psi_3 = 2KD(-1 + \beta)^2 \tag{6.34}$$

6.3.4.1 Yield Improvement Models

In this section, making improvement in the yield rate is investigated. In the long term, one may consider the mean and the variance of the yield to be functions of capital expenditures. In this section, we seek to improve yield by investing to increase the mean of yield or to decrease its variance. Generally, doing investments to improve the yield needs cost analysis. If the investment is found affordable, then it can lead to a variety of advantages such as cost reduction and establishing a long-term relationship. When the yield improves, the total quantity of products for satisfying a constant demand decreases. Another advantage is that because of the decline in the size of each order, the inspection also decreases. Please note that we do not mean the inspection will be skipped; by "decrease in inspection," we mean because the number of products decreases, the number of inspections will decrease too; meanwhile the strategy of 100% inspection is still carried out. Elimination of the inspection is the ultimate aim, if after implementation of several improvements, the production process becomes a process without producing any imperfect item. In addition to the above points, reaching a desired situation and improving the yield to a proper state will turn the buyer into a long-term customer for the supplier.

In this section, they sought to improve yield by investing to increase the mean of yield (μ). Aslani et al. (2017) used the companion yield parameter, $\gamma = 1$ for $1 \leq \gamma \leq 1/L_x$. Please note that as μ increases from L_x to 1, γ approaches 1 from $1/L_x$. Therefore, they assumed that γ follows a logarithmic investment function as follows:

$$\alpha(\gamma) = a - bLn\gamma \tag{6.35}$$

where a and b are positive parameters given by (Aslani et al. 2017):

$$a = \frac{Ln\gamma_0}{\varphi} \quad \text{and} \quad b = \frac{1}{\varphi} \tag{6.36}$$

$\alpha(\gamma)$ represents the investment cost required to change the companion yield parameter from γ_0 to γ, as suggested by Porteus (1986), and is a strictly decreasing and convex function of γ. The target is to improve the companion yield to a desired value in order to improve the mean of yield.

Corollary 6.1 Since the yield would improve, hence the optimal order quantity and fill rate, (Q and F), would change. Substituting Q_{imp} and F_{imp} into Eq. (6.25), and considering the investment cost required to create improvements, the total cost after the improvement plan is presented in Eq. (6.37) (Aslani et al. 2017):

$$\text{ATC}_{\text{imp}}(Q_{\text{imp}}, F_{\text{imp}}) = i\alpha(\gamma) + \text{ATC}(Q_{\text{imp}}, F_{\text{imp}}) \tag{6.37}$$

Also, Aslani et al. (2017) sought to improve yield by investing to decrease the variance of yield (σ^2). In other words, they examined reducing σ^2 as variability of the

6.3 EOQ Model with No Return

yield to an acceptable value. There is a cost per year $\rho(\sigma^2)$ of reducing yield variance to an acceptable value:

$$\rho(\sigma^2) = m - nLn(\sigma^2) \tag{6.38}$$

where m and n are positive constants and $\rho(\sigma^2)$ is a convex and strictly decreasing function of σ^2 given by (Aslani et al. 2017):

$$m = \frac{Ln(\sigma_0^2)}{\tau} \text{ and } n = \frac{1}{\tau} \tag{6.39}$$

Corollary 6.2 Due to the change in variance of yield rate, it is obvious that the optimal order quantity and fill rate, (Q and F), will change. Substituting Q'_{imp} and F'_{imp} into Eq. (6.25), a new annual total cost (ATC($Q'_{imp} \cdot F'_{imp}$)) will be calculated. Finally, by considering the investment cost mandatory to create improvements, the total cost after the improvement plan will be:

$$ATC'_{imp}\left(Q'_{imp}, F'_{imp}\right) = i\alpha(\gamma) + ATC\left(Q'_{imp}, F'_{imp}\right) \tag{6.40}$$

If the condition $\beta > 1 - \sqrt{2Kh(\sigma^2 + \mu^2)/D\mu^2 C_b^2}$ is established in Proposition 6.2, the optimal functions are not closed-form; therefore, a specific procedure should be used to solve the problem and find the optimal values. To solve the problem, they proposed a solution method as follows, in Discussion 6.1:

Discussion 6.1 When $\beta > 1 - \sqrt{2Kh(\sigma^2 + \mu^2)/D\mu^2 C_b^2}$, then a recursive algorithm is implemented to find the optimal solutions. At the first step, insert $F_1 = 1$ (or any acceptable value) into the function $Q^*(F)$ (Eq. 6.31a), calculate the result, and call it Q_1. Then, at the second step, substitute the value of Q_1 into the function $F^*(Q)$ (Eq. 6.31b), and calculate F_2. At the third step, insert the value of F_2 into the function $Q^*(F)$ and calculate Q_2. Following this procedure to the next steps, they progressively approached the optimal values for Q and F. They defined two parameters which represent the appropriate time to stop the algorithm. Assume that ε_F and ε_Q are the indexes for ending the algorithm. Whenever the result of subtraction between two consecutive values of a decision variable becomes less than its index, the recursive algorithm ends. Thus, whenever $\Delta F = |F_i - F_{i-1}| < \varepsilon_F$ and $\Delta Q = |Q_i - Q_{i-1}| < \varepsilon_Q$ both take place, the algorithm will stop proceeding, and F_i and Q_i are the optimal values of decision variables. The results indicate that this algorithm is quite efficient.

Discussion 6.2 Since shortage can be written as $B = D(1 - F)T$, and also cycle time is $T = xQ/D(F + \beta - \beta F)$, it can be easily observed that the service constraint, $\beta B/Q \leq L_x$, will be satisfied if and only if the following condition is held:

$$\beta \leq L_x F/(1-F)(u-L_x), \quad (F \neq 1) \qquad (6.41)$$

As can be seen, it must be $F \neq 1$. Therefore, if the optimal value of fill rate is calculated to be $F = 1$, in order to remove this error, $F = 1 - \varepsilon$ is used where ε is an infinitesimal amount. Note that this issue does not reduce the generality of the solution algorithm. If the value of decision variables does not meet the service constraint, it means the supplier is unable to answer all backordered shortages. In these circumstances, the buyer can decide whether to continue the cooperation with this supplier or not. In this problem, the buyer is very sensitive to the service constraint, and if the supplier cannot meet the service constraint, then the buyer will switch to other suppliers (Aslani et al. 2017).

Example 6.2 Consider a company that purchases its required product from a supplier. Each received batch from this supplier includes a random proportion of defective items. This means supplier's production process works to a random yield. Assume that the yield is between 0.4 ($L_x = 0{:}4$) and 1 and is a continuous random variable. Also the mean and variance of yield are known. The mean of the yield is $\mu = 0.7$ and its variance is $\sigma^2 = 0.01$. The buyer needs 500 units of the product each year. Other data used is sequenced as $\beta = 0.5$, $g = 2$ (\$/item), $h = 4$ (\$/item/year), $K = 120$ (\$/lot), $\varepsilon_F = 0.001$ and $\varepsilon_Q = 0.1$, $C_b = 2$ (\$/unit/year) (Aslani et al. 2017).

Step 1. Is $\beta > 1 - \sqrt{2Kh(\sigma^2 + \mu^2)/D\mu^2 C_b^2}$ met?

$1 - \sqrt{\frac{2Kh(\sigma^2+\mu^2)}{D\mu^2 C_b^2}} = 0.300146 \Rightarrow \beta(= 0.5) > 1 - \sqrt{\frac{2Kh(\sigma^2+\mu^2)}{D\mu^2 C_b^2}}$

Yes.

Step 2. Calculating optimal values. Since condition $\beta > 1 - \sqrt{2Kh(\sigma^2 + \mu^2)/D\mu^2 C_b^2}$ is met, the recursive algorithm must be employed. As it can be seen in Table 6.5, at the seventh iteration, $\Delta F \leq \varepsilon_F(=0.001)$ and $\Delta Q \leq \varepsilon_Q(=0.1)$. Therefore, $F_7 = F^*$ and $Q_7 = Q^*$.

Step 3. Is $\beta \leq \frac{L_x F^*}{(1-F^*)(\mu-L_x)}$ established?

$\frac{L_x F^*}{(1-F^*)(\mu-L_x)} = 0.879859 \Rightarrow \beta(= 0.5) \leq \frac{L_x F^*}{(1-F^*)(\mu-L_x)}$

Yes.

Step 4. F^* and Q^* are optimal.

According to the solution, the optimal values are $Q^* = 379.509$, $T^* = 0.719179$, $B^* = 216.633$, and $ATC^* = 616.870$.

6.4 EOQ Model with Return

Table 6.5 The process of achieving optimal values through the recursive method (Aslani et al. 2017)

Iteration	1	2	3	4	5	6	7
F	1	0.5069	0.4206	0.4022	0.3984	0.3977	0.3975
Q	244.949	344.292	371.895	377.971	379.211	379.459	379.509
ΔF	–	0.4931	0.0863	0.0184	0.0038	0.0008	0.0001
ΔQ	–	99.343	27.603	6.076	1.240	0.248	0.05
ATC	699.854	621.702	617.115	616.880	616.870	616.870	616.870

6.4 EOQ Model with Return

In this subsection, three problems that take into account the return policy are presented. The model development is investigated. Then, the solution procedure to solve the optimization problem is presented. Also, numerical examples have been reviewed to illustrate the implementation of the proposed method, if there is any.

6.4.1 Inspection and Sampling

Cheikhrouhou et al. (2018) developed an inventory model with lot inspection policy. With the help of lot inspection, even though all products need not to be verified, still the retailer can decide the quality of products during inspection. If the retailer found that the products have imperfect quality, the products are sent back to the supplier. As it is lot inspection, misclassification errors (type I error and type II error) are introduced to model the problem.

This model emphasizes a sample inspection to avoid maximum possibility of type I and type II errors of defective items. The mathematical model analyzes two ways how to reduce defective lots. The retailer chooses whether, after inspection of defective items, they are immediately sent back to the supplier or wait for the next shipment from the supplier to send back the defective items. The model mainly stands on two major factors: order size (Q^*) and sample size (n^*), which are related to each other (Cheikhrouhou et al. 2018).

The notations which are specially used in this problem are presented in Table 6.6.

In this section, the mathematical model is introduced. Let Q represents the order quantity size of lots, and a new constant L is introduced, which represents the number of items per lots. In order to satisfy the demand D of items per year, a shipment containing several lots is received every T unit of time. In a perfect model and without defective items, the relation $D = QL/T$ is established in the stationary state, where T represents the cycle time. To calculate the total cost of the model, the following costs had to be calculated (Cheikhrouhou et al. 2018):

Table 6.6 New notations for given problem (Cheikhrouhou et al. 2018)

n	Number of items inspected per lot (integer)
K_s	Transport cost of defective lots back to the supplier (for the first subcase model) ($)
T_I	Inspection time for a shipment (time)
L	Number of items per lot (integer)
x_e	Percentage of defective items perceived by the retailer (%)
C_{lot}	Lot purchase cost ($/lot)
ATP(Q, n)	Annual total profit ($/year)

$$TC = \text{Inspection cost} + \text{Inspection error cost} + \text{Holding cost} \\ + \text{Transportation cost} + \text{Purchasing cost} + \text{Fixed order cost} \quad (6.42)$$

Upon receiving an order, an inspection based on a sample is applied. Let n represents a sample size, satisfying constraint $1 \leq n \leq L$ and $n \in N$. It is also considered that the inspection time for a shipment T_I depends on the order size and on the number of samples used for inspection. Thus, one can obtain relation $T_I = \frac{nQ}{R_s}$ and the inspection cost is (Cheikhrouhou et al. 2018):

$$\text{Inspection cost} = C_I n Q \quad (6.43)$$

As described in the model by Khan et al. (2011), their screening process is assumed to be error-free. But it is quite realistic to account for type I and type II errors committed by inspectors as it is offline inspection by the inspectors; thus, there is a chance of accepting imperfect products as perfect and rejecting perfect product as imperfect. For this reason, the increased number of samples permits to reduce the risk of falsely qualifying a lot. Depending on the sample size n, inspectors classify some non-defective lots as defectives, i.e., $(1 - x)m^n{}_1$, while some defective units as non-defectives, i.e., $x \cdot m^n{}_2$. For this reason, the percentage of defective items perceived by the retailer x_e is different from the actual one x. Thus, the fraction of defective units perceived can be obtained as (Cheikhrouhou et al. 2018):

$$E[x_e] = (1 - E[x])E[m_1]^n + E[x](1 - E[m_2]^n) \quad (6.44)$$

It is considered a cost of misclassification due to an insufficient number of samples to qualify the quality of the lot. One can assume that the number of items, which goes into type I error (false rejection), is dependent on n. The inspectors falsely reject a lot if the inspectors have a type I error, which happens on each non-defective item from the sample (Cheikhrouhou et al. 2018):

$$\text{Inspection falsely rejecting cost} = Q(1 - E[x])E[m_1]^n \quad (6.45)$$

6.4 EOQ Model with Return

The number of items, which goes into type II error (false acceptation), is dependent on n. The inspectors falsely accept a lot if they have a type II error, which happens on each defective item from the sample:

$$\text{Inspection falsely accepting cost} = QE[x]E[m_2]^n \qquad (6.46)$$

Let C_{fr} and C_{fa}, respectively, be the cost of rejecting a non-defective lot (type I error) and the cost of accepting a defective lot (type II error). In case of critical products, such as food, medical, or parts of an aircraft, the cost of acceptance ca is much more than that of a false rejection (see for reference Raouf et al. 1983). Costs of inspection error per cycle can be expressed as (Cheikhrouhou et al. 2018):

$$\text{Costs of inspection error} = C_{fa}QE[x]E[m_2]^n + C_{fr}Q(1 - E[x])E[m_1]^n \qquad (6.47)$$

Figure 6.8 represents the behavior of the inventory level. It can be noticed that the number of items that are withdrawn from the inventory is:

$$x_e QL + nQ(1 - x_e) \qquad (6.48)$$

which represents the number of defective lots plus the number of items used for the inspection in the accepted lots. Under this condition, the inventory cycle T is determined as (Cheikhrouhou et al. 2018):

$$T = \frac{QL - (xQL + nQ(1 - x_e))}{D} = \frac{Q(1 - x_e)(L - n)}{D} \qquad (6.49)$$

The remainder of the model is subdivided into two cases. The first case (Case I) considers a special transport is organized to send the defective lots back to the supplier. The model has an additional transport costs K_s, but the defective products are no longer stored in the warehouse to avoid additional holding costs. The second case (Case II) considers that for the sake of convenience, the supplier can take back the defective items in the next shipment. These two cases are the more efficient possibilities as it is logically more expensive to send a transportation cost not immediately after the screening process (because of the holding costs of defective items).

6.4.1.1 Case I: Defective Items Are Immediately Sent Back to the Supplier with an Additional Transportation Cost

As h is the holding cost per unit per item, the holding costs per cycle for the first case can be determined from Fig. 6.5 as (Cheikhrouhou et al. 2018):

Fig. 6.5 Behavior of the inventory model subcase (1) (Cheikhrouhou et al. 2018)

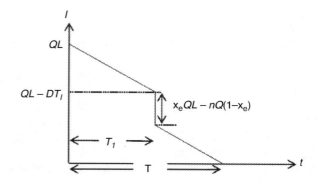

$$\text{Holding cost} = h\left(\frac{Q^2(1-x_e)^2(L-n)^2}{2D} + \frac{Q^2 nL(x_e L + n(1-x))}{R_s}\right) \quad (6.50)$$

In this case, the defective lots are sent back to the supplier with an additional cost for transportation cost K_s. The purchasing cost per cycle is determined as (Cheikhrouhou et al. 2018):

$$\text{Purchasing cost} = K_s + Q(1-x_e)C_{\text{lot}} \quad (6.51)$$

The total revenue in a cycle is:

$$\text{Total revenue} = s(QL - x_e QL - (1-x_e)nQ) = sQ((L-n)(1-x_e)) \quad (6.52)$$

The total profit is the total revenue per cycle minus the total cost per cycle divided by the cycle time and is given as follows (Cheikhrouhou et al. 2018):

$$\text{TP}_1(Q,n) = \frac{1}{T} \times \begin{bmatrix} sQ((L-n)(1-x_e)) - (K_s + Q(1-x_e)C_{\text{lot}}) - K - C_1 nQ \\ -h\left(\frac{Q^2(1-x_e)^2(L-n)^2}{2D} + \frac{Q^2 nL(x_e L + n(1-x_e))}{R_s}\right) \\ -(C_{\text{fa}} QE[x_e]E[m_2]^n + C_{\text{fr}} Q(1-E[x_e])E[m_1]^n) \end{bmatrix}$$

(6.53)

And the annual total profit is:

6.4 EOQ Model with Return

$$\text{ATP}_1(Q, n) = \frac{D}{Q(1 - x_e)(L - n)}$$

$$\times \begin{bmatrix} sQ((L-n)(1-x_e)) - (K_s + Q(1-x_e)C_{lot}) - K - C_1 nQ \\ -h\left(\dfrac{Q^2(1-x_e)^2(L-n)^2}{2D} + \dfrac{Q^2 nL(x_e L + n(1-x_e))}{R_s}\right) \\ -(C_{fa} QE[x_e]E[m_2]^n + C_{fr} Q(1-E[x_e])E[m_1]^n) \end{bmatrix}$$

(6.54)

For maximization of the profit, by taking partial derivatives with respect to Q and n, one can obtain (Cheikhrouhou et al. 2018):

$$\frac{\partial \text{ATP}_1(Q,n)}{\partial Q} = \frac{(K+K_s)D}{Q^2(1-E[x_e])(L-n)} - h\frac{(1-E[x_e])(L-n)}{2}$$
$$- h\frac{DLn(E[x_e]L + n(1-E[x_e]))}{R_s(1-E[x_e])(L-n)}$$

(6.55)

And:

$$\frac{\partial \text{ATP}_1(Q,n)}{\partial n} = -\frac{DC_{lot}}{(L-n)^2} - \frac{C_1 DL}{(1-E[x_e])(L-n)^2} - \frac{(K+K_s)D}{Q(1-E[x_e])(L-n)^2}$$
$$- \frac{DC_{fa} E[x]E[m_2]^n}{(1-E[x_e])(L-n)}\left[\frac{1}{(L-n)} + \ln(E[m_2])\right]$$
$$- \frac{DC_{fr}(1-E[x])E[m_1]^n}{(1-E[x_e])(L-n)}\left[\frac{1}{(L-n)} + \ln(E[m_1])\right] + h\frac{Q(1-E[x_e])}{2}$$
$$- h\frac{DLQ(E[x_e]L^2 + 2nL(1-E[x_e]) - n^2(1-E[x_e]))}{R_s(1-E[x_e])(L-n)^2}$$

(6.56)

For sufficient condition, the second-order derivative can be obtained as follows: All terms of Hessian matrix are obtained as follows:

$$\frac{\partial^2 \text{ATP}_1(Q,n)}{\partial Q^2} = \frac{2(K+K_s)D}{Q^3(1-E[x_e])(L-n)} < 0 \qquad (6.57)$$

$$\frac{\partial^2 \text{ATP}_1(Q,n)}{\partial n \partial Q} = \frac{(K+K_s)D}{Q^2(1-E[x_e])(L-n)^2} + \frac{h(1-E[x_e])}{2}$$
$$- h\frac{DL(E[x_e]L^2 + (2Ln - n^2)(1-E[x_e]))}{R_s(1-E[x_e])(L-n)^2}$$

(6.58)

$$\frac{\partial^2 \text{ATP}_1(Q,n)}{\partial n^2} = -\frac{2DC_{\text{lot}}}{(L-n)^3} - \left(\frac{2(K+K_s)D + 2C_1DLQ}{Q(1-E[x_e])(L-n)^3}\right) - \frac{2hDQL^3}{R_s(1-E[x_e])(L-n)^3}$$
$$- \frac{DC_{\text{fa}}E[x]E[m_2]^n}{Q(1-E[x_e])(L-n)} \times \left[\frac{2}{(L-n)^2} + \frac{\ln(E[m_2])}{(L-n)} + \ln(E[m_2])^2\right]$$
$$- \frac{DC_{\text{fr}}(1-E[x])E[m_1]^n}{(1-E[x_e])(L-n)} \times \left[\frac{2}{(L-n)^2} + \frac{\ln(E[m_1])}{(L-n)} + \ln(E[m_1])^2\right] < 0$$
(6.59)

Proposition 6.3 For $n < L$ and for R_s more than 1, the expected profit function $\text{TP}_1(y, n)$ is concave. Thus, there always exist an order size y and a sample size n which maximize the total profit.

Proof See Appendix 2 of Cheikhrouhou et al. (2018).

According to Proposition 6.2, it could be stated that the profit function is always concave and has a solution, which maximizes the total profit.

6.4.1.2 Case II: The Supplier Will Take Back the Defective Lots in the Next Shipment

In this case, the holding costs are increased by an additional term corresponding to the holding costs of the defective items (gray area in Fig. 6.6) (Cheikhrouhou et al. 2018):

$$\text{Holding cost} = h\left(\frac{Q^2(1-x_e)^2(L-n)^2}{2D} + \frac{Q^2L(x_eL + n(1-x_e))}{R_s}\right)$$
$$+ h_1\left[\frac{Q^2x(1-x_e)(L-n)^2}{D} + \frac{Qx_eL(xL + (L+n))}{R_s}\right]$$
(6.60)

The supplier takes back the defective items in the next shipment. Thus, there is not any additional cost for transportation. The purchase cost per cycle is determined as (Cheikhrouhou et al. 2018):

$$\text{Purchase cost} = Q(1-x_e)C_{\text{lot}}$$
(6.61)

The total profit $\text{TP}(Q, n)$ is determined in the same way and is given as (Cheikhrouhou et al. 2018):

6.4 EOQ Model with Return

Fig. 6.6 Behavior of the inventory for Case II (Cheikhrouhou et al. 2018)

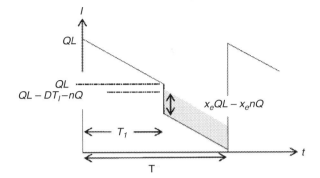

$$TP_2(Q,n) = \begin{pmatrix} sQ((L-n)(1-x_e)) - h\left(\dfrac{Q^2(1-x_e)^2(L-n)^2}{2D} + \dfrac{Q^2L(x_eL+n(1-x_e))}{R_s}\right) \\ -h_1\left[\dfrac{Q^2x(1-x_e)(L-n)^2}{D} + \dfrac{Qx_eL(xL+(L+n))}{R_s}\right] - K - C_1nQ - Q(1-x_e)C_{\text{lot}} \end{pmatrix}$$
(6.62)

And the annual total profit is:

$$ATP_2(Q,n) = \dfrac{D}{Q(1-x_e)(L-n)}$$
$$\times \begin{pmatrix} sQ((L-n)(1-x_e)) - K - C_1nQ - Q(1-x_e)C_{\text{lot}} \\ -h\left(\dfrac{Q^2(1-x_e)^2(L-n)^2}{2D} + \dfrac{Q^2L(x_eL+n(1-x_e))}{R_s}\right) \\ -h_1\left[\dfrac{Q^2x(1-x_e)(L-n)^2}{D} + \dfrac{Qx_eL(xL+(L+n))}{R_s}\right] \end{pmatrix}$$
(6.63)

A similar statement on Proposition 6.2 can be proposed regarding the existence of a maximum for the $ATP_2(Q, n)$ function (Cheikhrouhou et al. 2018).

As the solutions are dependent on each other, an algorithm for the numerical study is required. Here a procedure in order to find the optimal solution of the problem is presented by Cheikhrouhou et al. (2018).

Step 1. Find the maximum values of $ATP_1(Q, n)$ and $ATP_2(Q, n)$ by solving the zero value of gradients, using Newton–Raphson's method.

Step 2. Compare both cases and select the more profitable one. If the two cases give profits, then more profitable case can be accepted and less profitable case can be rejected. Otherwise, Step 3 must be applied for both cases.

Step 3. Let (Q^*, n^*) be the optimal values obtained corresponding to the optimal total profit function. The aim is to obtain an integer value of n and Q corresponding to convenient practical cases. Compute the values of ATP($\lfloor y \rfloor$, $\lfloor n \rfloor$), ATP($\lfloor y \rfloor$ + 1, $\lfloor n \rfloor$), ATP($\lfloor y \rfloor$, $\lfloor n \rfloor$ + 1), and ATP($\lfloor y \rfloor$ + 1, $\lfloor n \rfloor$ + 1), and finally choose between the four results with the twin value which gives the highest profit.

Example 6.3 Cheikhrouhou et al. (2018) used NBBARY algorithm to obtain the optimal order size and sample size. They assumed D = 50,000 units/year, L = 300 unit, K = 3000 \$/order, K_s = 5000 \$, h = 2 \$/unit/unit of time, h_1 = 5 \$/unit/unit of time, R_s = 5840 units/year, s = 8 \$/units, C_I = 15 \$/units, C_{lot} = 500 \$/lot, C_{fa} = 900,000 \$/lot, C_{fr} = 700 \$/lot, $x \sim U(0, 0.04)$, $m_1 \sim U(0, 0.05)$, and $m_2 \sim U(0, 0.1)$. In order to calculate the expected value of x,

$$E[x] = \int_a^b xf(x)dx = \int_a^b x\frac{1}{b-a}dx = \frac{b+a}{2} = \frac{0.04+0}{2} = 0.02 \quad (6.64)$$

Following the same process, the values of $E[m_1]$ = 0.025 and $E[m_2]$ = 0.05 are obtained (Cheikhrouhou et al. 2018). By numerical experiment, a maximum profit of 248,291 occurs for an order size Q of 47.04 and a sample size n of 2.48. The second case has a maximum profit of 278,924 for an order size of 37.45 and a sample size of 2.79. The second case is consequently the model to adopt as in these conditions it is more profitable to keep defective lots in inventory until the next shipment. After analyzing the four possible values of integer parameters, the optimum combination of decision variables reveals to be (Q^*, n^*) = (2, 37) with a profit of 278,924 (Cheikhrouhou et al. 2018).

6.4.2 Inspection Error

Khan et al. (2011) have focused on an optimal production/order quantity that takes care of imperfect processes. An imperfect inspection process (Raouf et al. 1983) is utilized to describe the defective proportion of the received lot. That is, the inspector may commit errors while screening. The probability of misclassification errors is assumed to be known. The inspection process would consist of three costs: (a) cost of inspection, (b) cost of type I errors, and (c) cost of type II errors.

The notations which are specially used in this problem are presented in Table 6.7.

Consider a lot of size Q being delivered to the buyer. It is assumed that each lot contains a fixed proportion x of defective items. An inspector screens out the defective items from the lot with fixed rate of misclassifications. That is, a proportion m_1 of non-defective items are classified to be defective and a proportion m_2 of defective items are classified to be non-defective. It is assumed that the probability density functions, $f(x)$, $f(m_1)$, and $f(m_2)$, are known. It is also assumed that the items that are returned from the market are stored with those that are classified as defective by the inspector. They are all sold as a single batch at the end of each cycle at a

6.4 EOQ Model with Return

Table 6.7 New notations for a given problem (Cheikhrouhou et al. 2018)

B_1	The batch classified as defective by the inspector (item)
B_2	The batch of returned units from the market (item)
Z_1	Inventory level (item)

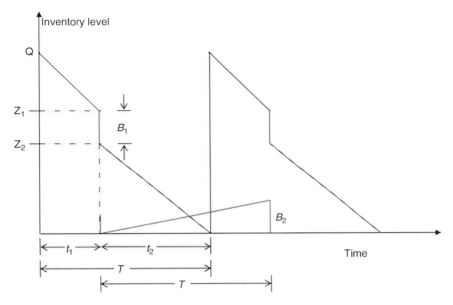

Fig. 6.7 Behavior of the inventory level over time (Khan et al. 2011)

discounted price. The behavior of the inventory level is illustrated in Fig. 6.7 (Khan et al. 2011).

The screening and consumption of the inventory continue until time t_1, after which all the defectives (B_1) are withdrawn from the inventory as a single batch and are sold to the secondary market. The consumption process continues at the demand rate until the end of cycle time T. Due to inspection error, some of the items used to fulfill the demand would be defective. These defective items are later returned to the inventory and are shown in Fig. 6.7 as B_2. To avoid shortages, it is assumed that the number of non-defective items is at least equal to the adjusted demand, that is, the sum of the actual demand and items that are replaced for the ones returned (xm_2Q) from the market over T. Thus (Khan et al. 2011):

$$Q - Q(1-x)m_1 - Qx(1-m_2) \geq DT + xm_2Q \quad (6.65)$$

$$Q(1-x) - Qx(1-m_2) \geq DT \quad (6.66)$$

$$Q(1-x)(1-m_2) \geq DT \quad (6.67)$$

Consider now the different cases of misclassifications that an inspection process can have. There are four possibilities in such an inspection process. These are as

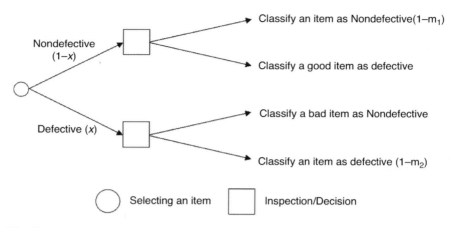

Fig. 6.8 Four possibilities in the inspection process (Khan et al. 2011)

follows: Case I, a non-defective item is classified as non-defective; Case II, a non-defective item is classified as defective; Case III, a defective item is classified as non-defective; and Case IV, a defective item is classified as defective. This scenario is depicted in Fig. 6.8 (Khan et al. 2011).

The items in batch B_2 are returned from the market at the rate Qxm_2/T and are taken from the inventory with batch B_1. Therefore, the revenue from salvaging $B = B_1 + B_2$ items is given by (Khan et al. 2011):

$$vQ(1-x)m_1 + vQx \qquad (6.68)$$

The revenue from selling the good items is computed as (Khan et al. 2011):

$$sQ(1-x)(1-m_1) + sQxm_2 \qquad (6.69)$$

The total profit per cycle can now be written as the difference between the total revenue and total cost per cycle, that is (Khan et al. 2011):

$$\text{TP}(Q) = s\overbrace{Q(1-x)(1-m_1) + sQxm_2}^{\text{Revenue from good items}} + \overbrace{vQ(1-x)m_1 + vQx}^{\text{Salvage revenue}}$$
$$- \left[\underbrace{K}_{\text{Fixed cost}} + \underbrace{CQ}_{\text{Purchasing cost}} + \underbrace{C_IQ}_{\text{Inspection cost}} + \underbrace{C_R(1-x)m_1}_{\text{Rework cost}} + \underbrace{\frac{h}{2}\left\{\left(\frac{2}{R_s} - \frac{D}{R_s^2} + \frac{\psi^2}{D}\right)Q^2 + Qxm_2T\right\}}_{\text{Holding cost}} \right]$$
(6.70)

where $\psi = 1 - \frac{D}{x} - (m_1 + x) + x(m_1 + m_2)$. It can be demonstrated that this expected annual profit follows a concave function. Thus, the optimal order size

6.4 EOQ Model with Return

that represents the maximum annual profit is determined by setting the first derivative equal to zero and solving for Q to get (Khan et al. 2011):

$$Q^* = \sqrt{\frac{2KD}{hE[x]E[m_2](1-E[x])(1-E[m_1]) + hD\left(\frac{2}{R_s} - \frac{D}{R_s^2} + \frac{E[\psi^2]}{D}\right)}} \quad (6.71)$$

Example 6.4 Consider a production system that replenishes the buyer's orders instantly. This system is not perfect, i.e., it produces some defective items. The inspection process that screens out the defective items is also imperfect. The probability density functions for the fraction of defective items and the inspection errors are mostly taken from the history of a supplier/machine and workers (Khan et al. 2011).

$D = 50,000$ item/year, $C = 25$ \$/unit, $K = 100$ \$/order, $s = 50$ \$/item, $v = 20$ \$/item, $R_s = 175,200$ item/year, $C_1 = 0.5$ \$/item, $h = 5$ \$/item/unit of time, $C_{fa} = 500$ \$/item, and $C_{fr} = 100$ \$/item.

$$f(x) = \begin{cases} 25, & 0 \le x \le 0.05 \\ 0 & \text{otherwise} \end{cases}, f(m_1) = \begin{cases} 25, & 0 \le m_1 \le 0.05 \\ 0 & \text{otherwise} \end{cases}, \text{ and } f(m_2) = \begin{cases} 25, & 0 \le m_2 \le 0.05 \\ 0 & \text{otherwise} \end{cases}.$$

Since $E(m_1) = E(m_1) = E(x) = 0.02$, substituting the above values in Eq. (6.71), the optimal value of order size is obtained as $Q^* = 1455$ units and $\psi = 1,095,090$ \$/year (Khan et al. 2011).

6.4.3 Different Defective Quality Levels and Partial Backordering

In this subsection, the mathematical model of Taleizadeh and Zamani-Dehkordi (2017) has been presented. This problem considers all the situations that have happened in reality. One of these situations is the presence of the defective items in each received lot, and the other situation is being the group of customers that do not wait to fulfill their requirements from the vendor and choose another one to get their orders, so the proportion of the backordered items becomes lost sales. This study considers both mentioned situations simultaneously to model the inventory system.

The notations which are specially used in this problem are presented in Table 6.8.

There are some assumptions that should be considered to model the inventory system to access more efficient and usable results. One of these assumptions is inspection. There are some defective items in each order that a firm or enterprise receives, and these items should be recognized and managers should choose a particular decision according to their inspection strategy. They considered an

Table 6.8 New notations for a given problem (Taleizadeh and Zamani-Dehkordi 2017)

x_1	Minimum level of defective percent in lot size (%)
x_2	Maximum level of defective percent in lot size (%)
α_1	Minimum level of defective percent in sample size (%)
α_2	Maximum level of defective percent in sample size (%)
TP_i	The total profit of case i, $i = 1, 2, 3$ ($/year)

inspection approach that is explained in the following. In this approach, three levels for the number of defective items in each sample are considered. According to the number of defective items that are in an order, they would decide what to do with the received order. According to the mentioned explanation, they defined three levels for the number of defective items in each sample; if this number is less than α_1, it is not necessary to inspect all the items; else if this number is between α_1 and α_2, all the items should be inspected, and if this number is more than α_2, the order is rejected and another order without any defective items is received. They defined these three levels because they can make the situation clearer for the managers. For example, the managers know if the number of defective items in a random sample chosen from each lot is lower than the particular number; it is beneficial that they ignore the inspection process and subsequently the inspection cost is eliminated. On the other hand, if the number of defective items in a sample would be between two particular numbers, it is beneficial that the inspection process is considered, though it leads to inspection cost. Finally, if the number of defective items in a sample becomes greater than the particular number, it is beneficial to return the lot to the supplier, because many defective items may exist in each lot (according to the chosen sample). Therefore, it may affect their prestige if these lots are delivered to their customers (Taleizadeh and Zamani-Dehkordi 2017).

The rate of imperfect items in each order is x, and regarding the number of defective items, it has three levels, lower than x_1, between x_1 and x_2, and more than x_2. Another assumption is related to the customers. There are two types of customers. The first one is the customers that do not change the vendor that they have chosen before to fulfill their demand, even though they know that they should sometimes wait more than the regular time for their order. On the other hand, the second one is the customers that are not patient enough to wait to fulfill their demand; these customers prefer to receive their orders by the other vendors because they do not want to wait for the previous vendor or their demand is critical, so their demand should be fulfilled soon. Because of the second one, a particular proportion of backordered items become lost sales so partial backordering instead of full backordering is applied (Taleizadeh and Zamani-Dehkordi 2017).

According to the mentioned explanations, they had three situations regarding the number of defective items in each order. In the first situation, the number of defective items that are in the sample chosen from the lot is more than the upper line which is α_2, so the buyer rejects the lot and receives another order without any defective items. In this situation, it is beneficial for the managers to return the lot to the supplier, because they conclude there are many defective items in each lot and it

6.4 EOQ Model with Return

affects their prestige badly if these lots are delivered to their customer. In the second situation, the number of defective items is between α_1 and α_2, so the buyer inspects all the items because the mangers know that if the number of defective items in a sample becomes between two particular numbers, it is beneficial that the inspection process is considered, though it leads to inspection cost. The defective items are separated from the perfect items. Then, the perfect items are sold at the particular price within the cycle, and the defective items are sold after the inspection time at the lower price than the perfect items. In the third situation, the number of defective items that are in the sample is lower than the lower line α_1, so the buyer does not intend to the inspection process and prefer to return the lot to the supplier. All the items are sold with the particular price that is determined for the perfect items. After selling the items, customers can return the defective items and get the particular amount of money instead of it. Each of these situations has a particular probability that affects the formulation of the total revenue function. In the following, these situations are precisely described and the process is shown in Fig. 6.9 (Taleizadeh and Zamani-Dehkordi 2017).

6.4.3.1 Case I: $\theta > \alpha_2$

The first case is the one that the number of imperfect products in a chosen sample is more than the upper limit that is α_2 and determined by the buyer and the supplier. In this situation, all the items are rejected, and a new order is received that has not any imperfect items. The revenue of this case in a cycle is computed as the following: $sD[F + \beta(1 - F)]$, the amount of money that is received by selling products per unit of time; $\frac{nC_1}{T}$, the cost of inspection of a sample per unit of time; $\frac{K}{T}$, the ordering cost per unit of time; $CD[F + \beta(1 - F)]$, cost of buying products per unit of time; $\frac{hDTF^2}{2}$, holding cost per unit of time; $gD(1 - \beta)(1 - F)$, shortage cost that is related to lost sales per unit of time; and $\frac{\beta C_b DT(1-F)^2}{2}$, shortage cost that is related to backorders per unit of time. Also there is a situation that buyers reject the lot wrongly; in other words, the buyer rejects the lot while the number of defective items is less than α_2, so the particular amount of money is considered for this situation that the buyer should pay to the supplier. The probability of this situation is computed as $\psi = P(x \leq x_2 | \theta \geq \alpha_2)$, and the penalty that the buyer should pay per cycle is computed as $\frac{\psi C_{fr}}{T}$. So the total revenue per unit of time is obtained from the formulations above (Taleizadeh and Zamani-Dehkordi 2017):

$$TP_1 = sD[F + \beta(1 - F)]$$
$$- \begin{bmatrix} \frac{K}{T} + CD[F + \beta(1-F)] + \frac{hDTF^2}{2} + \frac{nC_1}{T} \\ +gD(1-\beta)(1-F) + \frac{\beta C_b DT(1-F)^2}{2} + \frac{\psi C_{fr}}{T} \end{bmatrix} \quad (6.72)$$

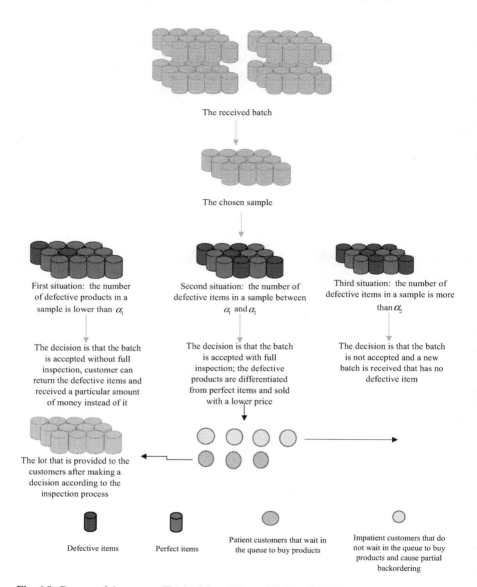

Fig. 6.9 Process of the system (Taleizadeh and Zamani-Dehkordi 2017)

6.4.3.2 Case II: $\alpha_1 \leq \theta \leq \alpha_2$

The second case is the one that the number of imperfect products in a chosen sample is between the lower limit and upper limit (α_1 and α_2); in this situation, all the items should be inspected to separate the defective items from the perfect items. Then he sells the perfect items with the particular price within the cycle and sells the defective items after the inspection time with the lower price than the perfect items. The revenue of this case is computed as the following: $sD(1-x)[F+\beta(1-F)]$, the amount of money that is received per unit of time by selling the perfect units that are $(1-x)$ percentage of all the units that are purchased; $vxDF$, the amount of money that is received per unit of time by selling the imperfect items that are x percentage of all the units that are purchased; $\frac{K}{T}$, the ordering cost per unit of time; $CD[F+\beta(1-F)]$, cost of buying products per unit of time; C_IDF, cost of screening all units per unit of time (cost of inspection of the sample is computed in this part of total cost); $\frac{hD(1-x)^2TF^2}{2}$, holding cost of the perfect units per unit of time; hxD^2TF^2/R_S, holding cost of the imperfect units per unit of time; and $gD(1-\beta)(1-F)$, shortage cost that is related to lost sales per unit of time. So the total revenue per unit of time is obtained from the formulation below (Taleizadeh and Zamani-Dehkordi 2017):

$$TP_2 = sD(1-x)[F+\beta(1-F)] + vxDF$$
$$- \begin{bmatrix} K/T + CD[F+\beta(1-F)] + C_IDF + hD(1-x)^2TF^2/2 \\ +hxD^2TF^2/R_S + gD(1-\beta)(1-F) + \beta C_bDT(1-F)^2/2 \end{bmatrix} \quad (6.73)$$

6.4.3.3 Case III: $\theta \leq \alpha_1$

Finally, the third situation is the one that the number of imperfect products in a chosen sample is lower than the lower limit (α_1) that is expected. In this situation, none of the items is inspected, and all the products are given to the customers as their orders with the price that is considered for the perfect products. Customers can receive the particular amount of money instead of the defective items that they return to the vendor. The revenue of this situation is computed as the following: $sD[F+\beta(1-F)]$, the amount of money that is received per unit of time by selling the perfect units that are $(1-x)$ percentage of all the units that are purchased; K/T, the ordering cost per unit of time; $CD[F+\beta(1-F)]$, cost of buying products per unit of time; $hD(1-x)^2TF^2/2$, holding cost of the perfect units per unit of time; nC_I/T, the cost of inspection of a sample per unit per time; hxD^2TF^2/R_S, holding cost of the imperfect units per unit of time; $gD(1-\beta)(1-F)$, shortage cost that is related to lost sales per unit of time; and $\beta C_bDT(1-F)^2/2$, shortage cost that is related to backorders per unit of time. The total number of returned item is computed as $(x + x^2 + x^3 + \cdots)$, and we know that $\lim_{m \to \infty}(x + x^2 + x^3 + \cdots + x^m) = x/(1-x)$, so we have $C_rDFx/(1-x)$ as a return cost of imperfect units per unit of time that the customers return them to a vendor. So the total revenue per unit of time is obtained from the formulation above:

$$TP_3 = sD(1-x)[F + \beta(1-F)] + vxDF$$
$$- \left[\begin{array}{l} \dfrac{K}{T} + CD[F + \beta(1-F)] + \dfrac{nC_1}{T} + \dfrac{hD(1-x)^2 TF^2}{2} \\ + \dfrac{hxD^2 T\varphi^2}{x} + C_r DF \dfrac{x}{1-x} + gD(1-\beta)(1-F) + \dfrac{\beta C_b DT(1-F)^2}{2} \end{array} \right] \quad (6.74)$$

In order to obtain the optimal values of decision variables, Taleizadeh and Zamani-Dehkordi (2017) first derived the expected value of the total profit for each case. Each case is done with a certain probability computed according to the intervals related to the number of defective items in each sample and subsequently the numbers of defective items in each lot. So according to these probabilities and related case for each one, the total profit per year that includes all three cases is (Taleizadeh and Zamani-Dehkordi 2017):

$$ETP_1 = sD[F + \beta(1-F)]$$
$$- \left[\begin{array}{l} \dfrac{K}{T} + CD[F + \beta(1-F)] + \dfrac{hDTF^2}{2} + \dfrac{nC_1}{T} \\ + gD(1-\beta)(1-F) + \dfrac{\beta C_b DT(1-F)^2}{2} + \dfrac{\psi C_{fr}}{T} \end{array} \right] \quad (6.75)$$

For the second case, one has:

$$TP_2 = sDE_2(1-x)[F + \beta(1-F)] + vE_2(x)DF$$
$$- \left[\begin{array}{l} \dfrac{K}{T} + CD[F + \beta(1-F)] + C_1 DF + \dfrac{hDE_2(1-x)^2 TF^2}{2} \\ + \dfrac{hE_2(x)D^2 TF^2}{R_S} + gD(1-\beta)(1-F) + \dfrac{\beta C_b DT(1-F)^2}{2} \end{array} \right] \quad (6.76)$$

And finally for the third case:

$$ETP_3 = sD(E_3(1-x))[F + \beta(1-F)] + v(E_3(x))DF$$
$$- \left[\begin{array}{l} \dfrac{K}{T} + CD[F + \beta(1-F)] + \dfrac{nC_1}{T} + \dfrac{hD\left(E_3\left((1-x)^2\right)\right)TF^2}{2} + C_r DF\left(E_3\left(\dfrac{x}{1-x}\right)\right) \\ + gD(1-\beta)(1-F) + \dfrac{\beta C_b DT(1-F)^2}{2} \end{array} \right]$$
$$(6.77)$$

To access a general optimum order quantity, Taleizadeh and Zamani-Dehkordi (2017) combined the three cases, so the expected for x is calculated in three cases as below. For the first case, they had formulated the expectation of the number of defective products in an order received by the vendor, while the number of defective

6.4 EOQ Model with Return

items in the sample is more than α_2, $\overline{E_1}(x)$ is computed as below (Taleizadeh and Zamani-Dehkordi 2017):

$$\overline{E}_1(x) = \Pr(\theta \geq \alpha_2 \cap x \leq x_1)E_{x \leq x_1}[x] + \Pr(\theta \geq \alpha_2 \cap x_1 \leq x \leq x_2)E_{x_1 \leq x \leq x_2}[x]$$
$$+ \Pr(\theta \geq \alpha_2 \cap x_2 \geq x)E_{x \geq x_2}[x] \quad (6.78)$$

For the second case, because the number of defective products in the sample is between α_1 and α_2, one will have (Taleizadeh and Zamani-Dehkordi 2017):

$$\overline{E}_2(x) = \Pr(\alpha_1 \leq \theta \leq \alpha_2 \cap x \leq x_1)E_{x \leq x_1}[x] + \Pr(\alpha_1 \leq \theta \leq \alpha_2 \cap x_1 \leq x \leq x_2)E_{x_1 \leq x \leq x_2}[x]$$
$$+ \Pr(\alpha_1 \leq \theta \leq \alpha_2 \cap x_2 \geq x)E_{x \geq x_2}[x] \quad (6.79)$$

And if the number of defective products in the random sample is lower than α_1, one will have (Taleizadeh and Zamani-Dehkordi 2017):

$$\overline{E}_3(x) = \Pr(\theta \leq \alpha_1 \cap x \leq x_1)E_{x \leq x_1}[x] + \Pr(\theta \leq \alpha_1 \cap x_1 \leq x \leq x_2)E_{x_1 \leq x \leq x_2}[x]$$
$$+ \Pr(\theta \leq \alpha_1 \cap x_2 \geq x)E_{x \geq x_2}[x] \quad (6.80)$$

Finally, the total revue is:

$$\text{ETP} = \Pr(\theta \geq \alpha_2)(sD[F + \beta(1-F)]) + \Pr(\alpha_1 \leq \theta \leq \alpha_2)(sD(E_2(1-x))[F + \beta(1-F)])$$
$$+ \Pr(\theta \leq \alpha_1)(sD(\overline{E}_3(1-x))[F + \beta(1-F)] + v(\overline{E}_3(x))DF + v(\overline{E}_2(x))DF$$
$$- \left[\frac{K}{T} + CD[F + \beta(1-F)] + \frac{hDTF^2}{2} + \frac{C_1n}{T} + gD(1-\beta)(1-F) + \frac{\beta C_b DT(1-F)^2}{2} + \frac{\psi C_{fr}}{T}\right]$$
$$- \left[\begin{array}{c}\frac{K}{T} + CD[F + \beta(1-F)] + C_1 DF + \frac{hD\left(\overline{E}_2\left((1-x)^2\right)\right)TF^2}{2} + \frac{h(\overline{E}_2(x))D^2TF^2}{R_S} \\ + gD(1-\beta)(1-F) + \frac{\beta C_b DT(1-F)^2}{2}\end{array}\right]$$
$$- \left[\begin{array}{c}\frac{K}{T} + CD[F + \beta(1-F)] + \frac{C_1n}{T} + \frac{hD\left(\overline{E}_3\left((1-p)^2\right)\right)TF^2}{2} + C_r DF\left(\overline{E}_3\left(\frac{x}{1-x}\right)\right) \\ + gD(1-\beta)(1-F) + \frac{\beta C_b DT(1-F)^2}{2}\end{array}\right]$$
$$(6.81)$$

Taleizadeh and Zamani-Dehkordi (2017) proved that the ETP is a concave function. To access optimum F, they set T as constant and set $\frac{\partial \text{ETP}}{\partial F} = 0$ and then derived:

$$F^*(T) = \frac{-B_1 - B_2(T)}{2A_1(T)} \tag{6.82}$$

where B_1, $B_2(T)$, and $A_1(T)$ are computed as below:

$$B_1 = \Pr(\theta \geq \alpha_2)(s - s\beta - C + C\beta + g(1-\beta))D + \Pr(\alpha_1 \leq \theta \leq \alpha_2)(sD(E_2(1-x)) \\ - sD(E_2(1-x))\beta + v(E_2(x))D - CD + CD\beta - C_1D + gD(1-\beta)) \tag{6.83}$$

$$B_2(T) = -[\Pr(\theta \geq \alpha_2) + \Pr(\alpha_1 \leq \theta \leq \alpha_2) + \Pr(\theta \leq \alpha_1)]\beta C_b DT \tag{6.84}$$

$$A_1(T) = -\Pr(\theta \geq \alpha_2)\left(\frac{hDT}{2} + \frac{\beta C_b DT}{2}\right) - [\Pr(\alpha_1 \leq \theta \leq \alpha_2) + \Pr(\theta \leq \alpha_1)]\frac{\beta C_b DT}{2} \\ - \Pr(\alpha_1 \leq \theta \leq \alpha_2)\frac{hD\left(E_2\left((1-x)^2\right)\right)T}{2} - \Pr(\theta \leq \alpha_1)\frac{hD\left(E_3\left((1-x)^2\right)\right)T}{2} \tag{6.85}$$

Moreover, to get the optimum T by solving $\frac{\partial \text{ETP}}{\partial T} = 0$, they obtained:

$$T^* = \frac{-B_1^2 - 4\left[\Pr(\theta \geq \alpha_2)\left(\frac{hD}{2} + \frac{\beta C_b D}{2}\right) - (\Pr(\alpha_1 \leq \theta \leq \alpha_2) + \Pr(\theta \leq \alpha_1))\frac{\beta C_b D}{2}\right]C_3}{2((-(\Pr(\theta \geq \alpha_1) + \Pr(\alpha_1 \leq \theta \leq \alpha_2))\beta C_b D - \Pr(\theta \leq \alpha_1)\beta C_b DT)^2} \\ \frac{-\Pr(\alpha_1 \leq \theta \leq \alpha_2)\frac{hD\left(E_2\left((1-x)^2\right)\right)}{2} - \Pr(\theta \leq \alpha_1)\frac{hD\left(E_3\left((1-x)^2\right)\right)}{2}}{} \\ -8C_2\left(\Pr(\theta \geq \alpha_2)\left(-\frac{hD}{2} - \frac{\beta C_b D}{2}\right) + \Pr(\theta \leq \alpha_1)\right)\left(-\frac{hD\left(E_3\left((1-p)^2\right)\right)}{2} - \frac{\beta C_b D}{2}\right) \tag{6.86}$$

where:

$$C_2 = -\frac{\beta C_b D}{2}(\Pr(\theta \geq \alpha_2) + \Pr(\alpha_1 \leq \theta \leq \alpha_2) + \Pr(\theta \leq \alpha_1)) \tag{6.87}$$

$$C_3 = -(K + nC_1 + \psi C_{\text{fr}})(\Pr(\theta \geq \alpha_2)) - K(\Pr(\alpha_1 \leq \theta \leq \alpha_2)) \\ - (K + nC_1)\Pr(\theta \leq \alpha_1)) \tag{6.88}$$

Example 6.5 Taleizadeh and Zamani-Dehkordi (2017) provided some numerical results according to their real case. They considered $n = 20$, $\alpha_1 = 1$, and $\alpha_2 = 4$, and also the buyer and the vendor should define the maximum limit that is considered as

6.4 EOQ Model with Return

0.15, and the buyer defines the minimum limit as 0.06. According to this information (Taleizadeh and Zamani-Dehkordi 2017),

$$E(x) = \frac{1}{b-a}\int_a^b x\,dx = \frac{b^2-a^2}{2(b-a)} = \frac{b+a}{2} \qquad (6.89)$$

$$E(x^2) = \frac{1}{b-a}\int_a^b x^2\,dx = \frac{b^3-a^3}{3(b-a)} = \frac{b^2+a^2+ab}{3} \qquad (6.90)$$

$$E\left(\frac{x}{1-x}\right) = \frac{1}{b-a}\int_a^b \frac{x}{1-x}\,dx = \frac{(\ln(1-a)+a)-(\ln(1-b)+b)}{b-a} \qquad (6.91)$$

$$E[x \leq 0.06] = \frac{1}{0.06-0}\int_0^{0.06} x\,dx = \frac{0.06^2 - 0^2}{2(0.06-0)} = \frac{0.06+0}{2} = 0.03 \qquad (6.92)$$

$$E[0.06 \leq x \leq 0.15] = \frac{1}{0.15-0.06}\int_{0.06}^{0.15} x\,dx = \frac{0.15^2 - 0.06^2}{2(0.15-0.06)}$$
$$= \frac{0.15+0.06}{2} = 0.105 \qquad (6.93)$$

$$E[0.15 \leq x \leq 0.25] = \frac{1}{0.25-0.15}\int_{0.15}^{0.25} x\,dx = \frac{0.25^2 - 0.15^2}{2(0.25-0.15)}$$
$$= \frac{0.25+0.15}{2} = 0.20 \qquad (6.94)$$

A sample of 20 products is chosen randomly, and the probabilities of different situations according to the number of defective items in the chosen selection are obtained (Taleizadeh and Zamani-Dehkordi 2017):

(a) If the number of defective products θ is more or equal to $\alpha_2 = 4$, the buyer rejects the order and receives another lot instead of it that does not have any defective items.
(b) If the number of defective products θ is equal to 2 or 3, all the items should be inspected.
(c) If the number of defective products θ is lower or equal to $\alpha_1 = 1$, no item is inspected.

The probability of being θ defective items in a sample is calculated as (Taleizadeh and Zamani-Dehkordi 2017):

Table 6.9 The probability of different situations (Taleizadeh and Zamani-Dehkordi 2017)

	$b \leq 0.0600$	$0.0600 \leq b \leq 0.1500$	$0.1500 \leq b \leq 0.2500$
$\theta \leq 1.00$	0.796	0.186	0.018
$1.00 \leq \theta \leq 4.00$	0.237	0.460	0.304
$\theta \geq 4.00$	0.004	0.032	0.963

Table 6.10 The optimal values

β	F^*	T^*	ETP
0.1	0.74	0.1454	75,559
0.2	0.63	0.0787	28,321
0.3	0.49	0.559	7393.1
0.4	0.32	0.0441	1972.0
0.5	0.23	0.0366	514.72

$$f(\theta; n, \lambda) = \binom{n}{\theta} x^\theta (1-x)^{n-\theta} \qquad (6.95)$$

For instance, they calculated the probability of the situation that $\theta \leq 1$ and $x = 0.01$ (Taleizadeh and Zamani-Dehkordi 2017):

$$f(1; 20, 0.010) = \Pr(\theta = 0) + \Pr(\theta = 0)$$
$$= \binom{20}{0} 0.010^0 (1 - 0.010)^{20} + \binom{20}{1} 0.010^1 (1 - 0.010)^{19} = 0.983 \qquad (6.96)$$

The probabilities of different situations are provided in Table 6.9 (Taleizadeh and Zamani-Dehkordi 2017).

Also the other parameters are $D = 50$ units per year, $v = 20$ \$/unit, $x = 1$ unit/min, $C_I = 0.5$ \$/unit, $C_r = 15$ \$/unit, $C_{fr} = 70$ \$/unit, $s = 50$ \$/unit, $C = 25$ \$/unit, $C_b = 1$ \$/unit, $g = 2$ \$/unit, and $K = 10$ \$/order. The results are presented in Table 6.10.

6.5 EPQ Model Without Return

In this subsection, two models of EPQ models without return policy are presented. Quality assurance and screening are the main topics of these models.

6.5 EPQ Model Without Return

Table 6.11 New notations for a given problem (Cheng 1991)

ε	Expected fraction of products found to be acceptable in a production run (%)
$P(\varepsilon) = a/(1-\varepsilon)^b$	Unit cost of production ($/item)

6.5.1 Quality Assurance Without Shortage

In this subsection, the study of Cheng (1991) has been presented. His problem considers an EPQ model with imperfect production processes and quality-dependent unit production cost. Also, discussion of the procedure for determining the optimal solution is presented. Finally, the problems associated with cost estimation are addressed. Consider the case where a company employs a production process with a certain level of capability to manufacture a single product, the quality of which is monitored by some quality assurance programmer. The capability of the process and the effectiveness of the quality assurance programmer depend on a great variety of factors such as production technology, machine capability, jigs and fixtures, work methods, use of online monitoring devices, skill level of the operating personnel, and inspection, maintenance, and replacement policies. A high level of process capability in conjunction with a stringent quality assurance system will result in products of an acceptable level of quality being more consistently produced, hereby reducing the subsequent costs of scrap and rework of substandard products and wasted materials and labor hours. There is also an array of intangible benefits resulting from adopting advanced production technology and improving quality assurance, which include improvement in yields, process flexibility, product variety, development and delivery lead times, profit margin, market share, and customer goodwill (Cheng 1991).

However, superior process capability and effective quality management can be achieved only through substantial investment in plant, machinery, equipment, and employee training. Evidently this will increase the production overhead that will inevitably be apportioned to the individual products by the costing system, so that a higher unit cost of production will result. In addition, the unit direct cost of production, which consists of three cost components (direct labor, direct material, and direct expenses), also goes up. This is so because a more capable process calls for more skillful labor and more expensive tools, which will result in higher direct labor and direct expenses. Furthermore, a more effective quality assurance programmer means the use of better-quality material and more stringent quality assurance policies, which will increase the direct material and direct expenses (Cheng 1991). The notations which are specially used in this problem are presented in Table 6.11.

While it is hard to quantify accurately the various intangible benefits, an increase in production overhead will, as stated above, invariably push up the unit cost of production. Thus, it seems rational to hypothesize that the unit (tangible) cost of production is an increasing function of process capability and quality assurance expenses.

However, this is not to deny that the increased (tangible) costs can often be offset by the increased (intangible) benefits as witnessed by such manufacturers as Ford,

Firestone, Toyota, Xerox, and many others who have invested heavily in modern production technologies and quality assurance systems. They report that for every dollar they spend in quality control, they save two dollars in total (intangible) production costs and costs related to product returns (an intangible benefit) (Cheng 1991):

$$\text{Total relevant cost per production run} = K_S + p(\varepsilon)Q + \frac{h\varepsilon^2 Q^2}{2D} \quad (6.97)$$

where $a, b > 0$ are constant real numbers chosen to provide the best fit of the estimated cost function. The objective is to minimize the total relevant cost per unit time subject to the constraint that the expected fraction of products found to be acceptable in a production run cannot exceed 100%. So the function which should be minimized is as below (Cheng 1991):

$$C(Q, \varepsilon) = \frac{\text{Total relevant cost per production run}}{\left(\frac{Q\varepsilon}{D}\right)} = \frac{D}{Q\varepsilon}\left[K_S + p(\varepsilon)Q + \frac{h\varepsilon^2 Q^2}{2D}\right]$$

s.t. $0 \leq \varepsilon \leq 1$

$$(6.98)$$

To solve the constrained minimization problem, Cheng (1991) took the first partial derivative of $C(Q, \varepsilon)$ with respect to Q and ε and proved that the following equations are the optimal values of decision variables:

$$Q^* = (1 + b)\sqrt{2K_S D/h} \quad (6.99)$$

$$\varepsilon^* = \frac{1}{1 + b} \quad (6.100)$$

6.5.2 Quality Screening and Rework Without Shortage

Moussawi-Haidar et al. (2016) explicitly integrated the inspection time into the economic production model with rework and demonstrated the significant effect that the inspection time has on the results. They considered a manufacturing process with random supply and a screening process conducted during and at the end of production. They analyzed two scenarios for dealing with the defective items produced: selling at a discount and reworking. For each scenario, the demand during production is met using non-defective items only. The expected profit functions are developed using the renewal reward theory, and closed-form expressions for the optimal production lot size are derived. This problem is considered in Sect. 4.3.4.

6.6 EPQ Model with Return

In this subsection, two models of EPQ models with return policy are presented. Continuous quality characteristic, inspection errors, and 100% quality screening are the main topics of these models.

6.6.1 Continuous Quality Characteristic Without Shortage

Tsou et al. (2012) extended the traditional EPQ model by considering imperfect-quality items and continuous quality characteristic. In many manufacturing industries such as textiles, if products do not enjoy some precise and rather complete dimension and quality properties, they are sold as second or third rate. However, sometimes, it is possible to disguise the defect through some rework (reprocessing) such as re-dyeing or surface repairing. In case the defect is not repairable, e.g., an item lacks minimum required durability, the product is discarded as useless.

It must be noted that this study assumes an item produced follows a general distribution pattern. As shown in Fig. 6.1, the produced items are divided into three categories: they are perfect, imperfect, or defective. An item is perfect if the quality characteristic is inside the internal specification limits. An item is imperfect if the quality characteristic goes beyond the internal specification limits and yet inside the external specification limits. An item is defective if its quality characteristics depart from external specification limits (see Fig. 6.10) (Tsou et al. 2012).

It is assumed that perfect items are kept in stock and sold at a price, imperfect items are sold at a discounted price when imperfection is identified, and defective item can be reworked or rejected. If the quality characteristic of an item is above the upper specification level (USL), it can be reworked. But an item must be rejected if its quality characteristic is below the lower specification level (LSL). For example, if the quality characteristic is length of a wire above the USL after the process, it can be reworked, but if it is low after the process, it must be rejected (Tsou et al. 2012).

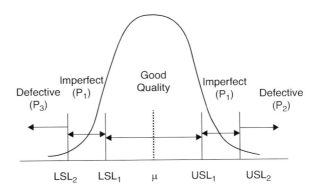

Fig. 6.10 The distribution of product quality characteristic (Tsou et al. 2012)

Table 6.12 New notations for a given problem (Tsou et al. 2012)

LSL_1	Lower specification limit for perfect items
LSL_2	Lower specification limit for imperfect items
USL_1	Upper specification limit for perfect items
USL_2	Upper specification limit for imperfect items
μ	Target quality characteristic
p_1	Percentage of imperfect-quality items (%)
p_2	Percentage of reworked items (%)
p_3	Percentage of rejected items (%)
x	Random variable which represents the actual value of the quality characteristic (%)
$f_0(x)$	Qualitative probability density function (general distribution)
$f_1(x)$	Qualitative probability density function for an item that has been reworked (general distribution)

After the rework process, each item is assumed to be good. The distribution of reworked item is assumed to follow a general distribution within the internal specification limits. A perfect-quality item can be sold at a price, $s, with a discount of Taguchi's cost of poor quality (COPQ) as its major quality characteristic departs from the target value. The imperfect-quality items could be used in low-end production situations or sold to a particular purchaser at a discounted price, $v, also with a discount of Taguchi's COPQ as its major quality characteristic departs from the target value. Quality characteristics outside the external specification limits are considered to be defective and will be scrapped directly. In the next section, the mathematical model is developed (Tsou et al. 2012).

The model developed in this study can be used to determine the optimal lot size in different manufacturing companies, textile industries, or ornamental products manufacturing industries; it must be borne in mind that this model suffers from some limitations, which are basically associated with the underpinning assumptions. Moreover, the model is limited to the deterministic (non-stochastic) models of cost parameters, similar production, and rework rates (Tsou et al. 2012).

The notations which are specially used in this problem are presented in Table 6.12.

The study centers on the following assumptions: A process produces a single product in a batch size of Q; storage and withdrawals are uniform and continuous; the demand rate for the product is deterministic and constant; and the factor of shortages or backorders is ignored. Figure 6.10 shows the distribution of product quality with different quality characteristic ranges. Different ranges for product quality distribution have been listed in Table 6.13. Figure 6.11 is the roadmap for our decision-making to handle products with different quality levels (Tsou et al. 2012).

Tsou et al. (2012) assumed that the product quality distribution follows the general distribution function, $f_0(x)$. After the rework process, a defective item is assumed to be of good quality such that the actual value of the quality characteristics

6.6 EPQ Model with Return

Table 6.13 Different ranges for product quality distribution

$LSL_1 \leq x \leq USL_1$	Item is good and kept in stock after inspection and sold at a full price with a discount of Taguchi's COPQ
$LSL_2 \leq x \leq LSL_1$ or $USL_1 \leq x \leq USL_2$	Item is imperfect and sold at a lower price with a discount of Taguchi's COPQ
$USL_2 \leq x$	Defective items can be reworked instantaneously at a cost, and then kept in stock. After the rework process, item is assumed to be of good quality with distribution, $f_1(x)$
$x \leq LSL_2$	Item is defective and not rework able. So, it is rejected with a cost

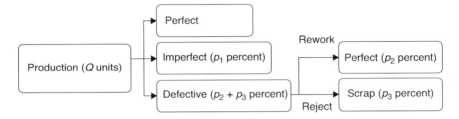

Fig. 6.11 Roadmap for decision-making on imperfect and defective products (Tsou et al. 2012)

has a general distribution with probability density function, $f_1(x)$, and quality characteristic value between LSL_1 and USL_1:

$$\int_{-\infty}^{+\infty} f_0(x)dx = 1 \tag{6.101}$$

$$\int_{LSL_1}^{USL_1} f_1(x)dx = 1 \tag{6.102}$$

A perfect-quality item sold at a price, $s, with a discount of Taguchi's COPQ as its quality characteristic departs from the target value μ. The imperfect item could be sold to a low-end buyer at a discounted price, $v, and a discount of Taguchi's COPQ as its major quality characteristic departs from the target value. Figure 6.12 illustrates the behavior of the inventory level per cycle. As shown in Fig. 6.12, the Tp requires that Q items go to the storage if the production rate is P and without producing the imperfect and defective production. From Fig. 6.12, TpD items are sold, thus decreasing the final warehouse. Defective products have the similar effect on the final inventory of stock (Tsou et al. 2012).

From the definition above, p_1 and p_2 can be defined as (Tsou et al. 2012):

Fig. 6.12 The behavior of the inventory level per cycle (Tsou et al. 2012)

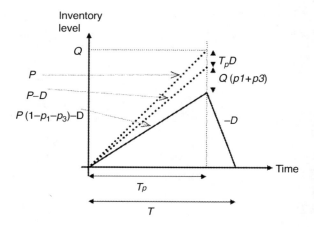

$$p_1 = \int_{LSL_2}^{LSL_1} f_0(x)dx + \int_{USL_1}^{USL_2} f_0(x)dx \qquad (6.103)$$

$$p_2 = \int_{USL_2}^{+\infty} f_0(x)dx \qquad (6.104)$$

$$p_3 = \int_{-\infty}^{LSL_2} f_0(x)dx \qquad (6.105)$$

The proportion of perfect item is:

$$1 - (p_1 + p_3) \qquad (6.106)$$

The cycle time, T, can be calculated as:

$$Q(1 - p_1 - p_3) = TD \rightarrow T = \frac{Q(1 - p_1 - p_3)}{D} \qquad (6.107)$$

And the production time, T_p, can be calculated as:

$$Q = TP \rightarrow T_p = \frac{Q}{P} \qquad (6.108)$$

They defined the total revenue and the total cost per cycle as TR(Q) and TC(Q), respectively. The total profit per cycle, TP(Q), is the total revenue per cycle minus the total cost per cycle. It is given as:

6.6 EPQ Model with Return

$$TP(Q) = TR(Q) - TC(Q) \tag{6.109}$$

The total revenue per cycle, TR(Q), is the total sales volume of product within the specification limits. Therefore, one has (Tsou et al. 2012):

$$\begin{aligned} TP(Q) &= \int_{LSL_1}^{USL_1} f_0(x)Q(s - L_0(x))dx + \int_{LSL_2}^{LSL_1} f_0(x)Q(v - L_1(x))dx \\ &+ \int_{USL_1}^{USL_2} f_0(x)Q(v - L_1(x))dx + \int_{USL_2}^{+\infty} f_0(x)dx \int_{LSL_1}^{USL_1} f_1(x)Q(s - L_0(x))dx \\ &= Q \Bigg[s(1 - p_1 - p_3) + vp_1 - \int_{LSL_1}^{USL_1} f_0(x)L_0(x)dx - \int_{LSL_2}^{LSL_1} f_0(x)L_1(x)dx \\ &\quad - \int_{USL_1}^{USL_2} f_0(x)L_1(x)dx - p_2 \int_{LSL_1}^{USL_1} f_1(x)L_0(x)dx \Bigg] \end{aligned} \tag{6.110}$$

The total revenue per unit time can be written as (Tsou et al. 2012):

$$TRY(Q) = TR(Q)/T$$

$$TP(Q) = D \Bigg[s(1 - p_1 - p_3) + vp_1 - \int_{LSL_1}^{USL_1} f_0(x)L_0(x)dx - \int_{LSL_2}^{LSL_1} f_0(x)L_1(x)dx \\ - \int_{USL_1}^{USL_2} f_0(x)L_1(x)dx - p_2 \int_{LSL_1}^{USL_1} f_1(x)L_0(x)dx \Bigg] / (1 - p_1 - p_3) \tag{6.111}$$

It can be seen that total revenue per unit time for any distribution is independent of Q. Therefore, the maximization of total profit per unit time is the same as minimization of the total cost per unit time. The total cost per cycle can be found as (Tsou et al. 2012):

$$TC(Q) = CQ + C_R Q p_2 + K_S + C_1 Q + h\frac{1}{2}Q\left((1 - p_1 - p_3) - \frac{D}{P}\right)T \tag{6.112}$$

The total cost per unit time can be written as (Tsou et al. 2012):

$$\text{TCY}(Q) = \text{TC}(Q)/T$$
$$\text{TCY}(Q) = (CQ + C_R Q p_2 + 1)\frac{D}{1-p_1-p_3} + \frac{K_S D}{Q(1-p_1-p_3)} + h\frac{1}{2}Q\left(1-p_1-p_3-\frac{D}{P}\right)$$
(6.113)

By differentiating and equating $d(\text{TCY}(Q))/dQ = 0$, the optimal lot size, Q^*, which generates the minimum expected total cost, can be obtained by (Tsou et al. 2012):

$$Q^* = \sqrt{\frac{2DK_S}{h(1-p_1-p_3-\frac{D}{P})(1-p_1-p_3)}}$$
(6.114)

The second derivatives of Eq. (6.113) are positive for all positive Q, which implies that there exists a unique Q^* that minimizes Eq. (6.113).

Example 6.6 To verify the usefulness of the previous model, Tsou et al. (2012) considered a production system with the following parameters: $K_S = \$125/\text{cycle}$; $C = \$0.1/\text{unit}$; $C_R = \$0.05/\text{unit}$; $h = \$15/\text{unit/year}$; $D = 15{,}000$ units/year; $P = 20{,}000$ units/year; $C_I = \$0.02/\text{unit}$; $p_1 = 15\%$; and $p_2 = 10\%$ and $p_3 = 5\%$ (Tsou et al. 2012). Using Eqs. (6.114) and (6.113), the optimal lot size and the minimum total relevant cost are $Q^* = 2500$ and $\text{TCY}(Q^*) = 4219$.

6.6.2 Inspections Errors Without Shortage

Haji et al. (2009) provided a framework to integrate the existence of products with imperfect-quality items, inspection errors, rework, and scrap items into a single economic production quantity (EPQ) model. To achieve this objective, a suitable mathematical model is defined, and the optimal production lot size that minimizes the total cost is obtained. Sensitivity analysis is carried out for this model. The sensitivity analysis results indicate that the model is very sensitive to defective proportions and type 1 errors of inspection. Nowadays, given the progress made, inspection errors are often ignored. But the findings of this study show that the values of EPQ and total cost are very sensitive to type I error of inspection. If the existence of such errors is ignored, then the obtained results will differ considerably from the optimal outcome. This will impose additional costs to the system.

Haji et al. (2009) considered a production system in which imperfect-quality items are passed to the rework process; after the rework process, they are classified as either good items or scraps. These important issues must be addressed when dealing with imperfect production and rework processes:

- Imperfect-quality items must be separated so that they are not passed to stock.

6.6 EPQ Model with Return

Table 6.14 New notations for a given problem (Haji et al. 2009)

β	Proportion of defects in the production process in each cycle (%)
α	Proportion of scraps in the rework process in each cycle (%)
e_1	Proportion of good items that are incorrectly rejected in each cycle (%)
e_2	Proportion of bad items that are incorrectly accepted in each cycle (%)
Q	Net batch quantity needed per cycle to satisfy demand (item)
Q_i	Input batch quantity required to be processed per cycle (item)
Q_r	The number of reworkable items (item)
v_1	The cost per imperfect item incorrectly accepted in the production process ($/item)
v_2	The cost per perfect item incorrectly rejected in the rework process ($/item)
v_3	The cost per scrap item incorrectly accepted in the rework process ($/item)

- Errors may be committed while screening products during the production and rework processes to separate imperfect-quality items and scraps. Imperfect-quality items or scraps may be incorrectly accepted (type II error), and good items may be incorrectly rejected (type I error).

Assuming the rework process, scrap production, and inspection errors (type I and type II errors), a suitable mathematical model is defined, and then the optimal production lot size that minimizes the total system cost is obtained. The notations which are specially used in this problem are presented in Table 6.14.

Consequently, these two operations involve two types of costs: setup cost and processing cost. Since the production rate (P) is greater than the demand rate (D), inventory is accumulated during the production period. Having enough inventory sometimes helps satisfy demand during the period wherein production is stopped due to various reasons. They further assumed that the net production rate, after excluding the defective items, is constant and greater than the demand rate.

Figure 6.13 shows that inventory increases in the first phase of production, i.e., during the normal production time t_p at rate R_1, which is equal to:

$$R_1 = P[(1-\beta)(1-e_1) + \beta e_2] - D \tag{6.115}$$

At the end of this phase, the rework of defective items starts. During the rework phase, inventory increases at rate R_2, which is equal to:

$$R_2 = P[(1-\alpha)(1-e_1) + \alpha e_2] - D \tag{6.116}$$

Since Q_i stands for the input quantity in each cycle, then the required processing time for this quantity (the production time), t_p, is:

$$t_p = \frac{Q_i}{P} \tag{6.117}$$

It is assumed that the quality of the output of the production process is not perfect. At each cycle of the production process, a fixed fraction β of non-conforming items

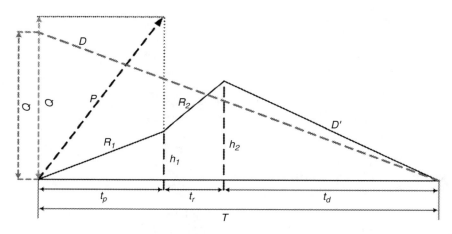

Fig. 6.13 A comparison of inventories with defective and non-defective products (Haji et al. 2009)

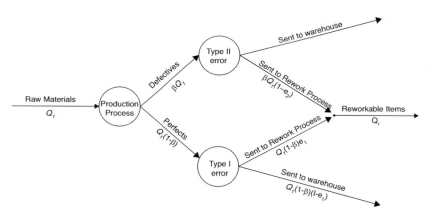

Fig. 6.14 The product flow diagram in the regular production process (Haji et al. 2009)

may be produced. Therefore, the number of non-conforming items produced at the production process is equal to βQ_i and the number of perfect-quality items at the end of the production process is $(1 - \beta)Q_i$. The perfect items at the production process are inspected and put in the inventory to be used when necessary, and the non-conforming items are screened for the rework process. It is assumed that raw materials and input products are of perfect quality. Since the inspection process is error-prone, types I and II errors may be committed. That is, non-conforming items may be incorrectly accepted and conforming items may be incorrectly rejected. From type I errors, the amount of correctly accepted items passed to inventory is $(1 - \beta) Q_i (1 - e_1)$. From type II errors, the quantity of incorrectly accepted items is $\beta Q_i e_2$ (see Figs. 6.14 and 6.15). Hence, the number of items recognized as good items, Q, is:

6.6 EPQ Model with Return

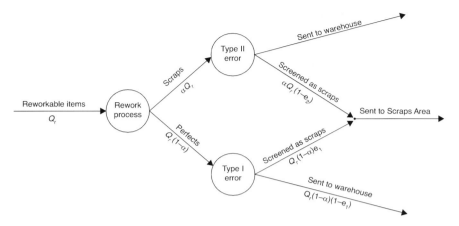

Fig. 6.15 The product flow diagram in the rework process (Haji et al. 2009)

$$Q = Q_i(1-\beta)(1-e_1) + Q_i\beta e_2 \tag{6.118}$$

The number of reworkable items (denoted by Q_r) consists of the number of incorrectly rejected good items and correctly accepted defective items for rework, computed as:

$$Q_r = [\beta(1 - e_1 - e_2) + e_1]Q_i \tag{6.119}$$

The total system cost $\mathrm{TC}(Q)$ can be obtained as follows:

$$\mathrm{TC}(Q) = \overbrace{\frac{K_S D}{Q}}^{\text{Setup cost}} + \overbrace{\frac{CD}{\varphi}}^{\text{Production cost}} + \overbrace{C_R \omega \frac{D}{\varphi}}^{\text{Rework cost}} + \overbrace{C_I(1+\omega)\frac{D}{\varphi}}^{\text{Inspection cost}} + \overbrace{\frac{D}{\varphi}\gamma}^{\text{Inspection error cost}}$$

$$+ \overbrace{\frac{hDQ}{2P\varphi}\left[\frac{\omega}{\varphi}(L-M) + \frac{P}{D}(L+M\omega) - 1 - \omega\right]}^{\text{Holding cost}}$$

$$\tag{6.120}$$

where:

$$\lambda = (\alpha + \beta)\left(e_1 e_2 + e_1^2 - e_1\right) + \alpha\beta\left(2e_1 + 2e_2 - 2e_1 e_2 - e_1^2 - e_2^2 - 1\right) + 1 - e_1^2 \tag{6.121}$$

$$\varphi = \lambda - e_2(\beta + \alpha\omega) \tag{6.122}$$

$$L = (1-\beta)(1-e_1) + \beta e_2 \tag{6.123}$$

Table 6.15 A comparison of the results of changing the β values (Haji et al. 2009)

B	Q^*	TC (Q^*)
0.05	115.74	100,070.18
0.1	122.36	101,871.91
0.15	131.02	103,660.92
0.2	142.79	105,431.03
0.25	159.83	107,172.30
0.3	187.18	108,866.79
0.36	259.35	110,778.05
0.37	283	111,071.71

$$M = (1-\alpha)(1-e_1) + \alpha e_2 \tag{6.124}$$

$$\gamma = v_1 \beta e_2 + v_2 \omega (1-\alpha) e_1 + v_3 \omega \alpha e_2. \tag{6.125}$$

Haji et al. (2009) showed that TC(Q) is a convex function of Q. Therefore, setting the first derivative of objective function equal to zero will yield an optimum value as below:

$$Q^* = \sqrt{\frac{2K_S P \varphi}{h\left[\frac{\omega}{\varphi}(L-M) + \frac{P}{D}(L+M\omega) - 1 - \omega\right]}} \tag{6.126}$$

Example 6.7 Haji et al. (2009) presented an example with the following data: $D = 3000$, item/year; $K_S = 60$, \$/setup; $P = 5000$, item/year; $h = 80$, \$/item/year; $\alpha = 0.01$; $\beta = 0.05$; $e_1 = 0.05$; $e_2 = 0.01$; $v_1 = 10$, \$/item; $v_2 = 8$, \$/item; $v_3 = 7$, \$/item; $C = 30$, \$/item; $C_R = 10$, \$/item; and $C_I = 1$, \$/item.

The results (see Table 6.15) show that the model is very sensitive to parameters e_1 and β, whereas it is much less sensitive to parameters α and e_2. In this case, the changes in TC are directly related to the changes in all parameters. The changes in Q are directly related to the changes in β and e_1, whereas these changes are inversely related to the changes in e_2.

6.7 Conclusion

In this chapter, 11 inventory models with quality factors are presented. These models consider the traditional lot-sizing problem when after arrival, the quality of each item in a lot is checked. All models are categorized in four sections about EOQ and EPQ models with and without returns. Learning in inspections, sampling plans, inspection errors, different quality levels, quality assurance, quality screening, continues quality characteristics and different types of shortages are the main topics investigated and differ models.

References

Al-Salamah, M. (2016). Economic production quantity in batch manufacturing with imperfect quality, imperfect inspection, and destructive and non-destructive acceptance sampling in a two-tier market. *Computers & Industrial Engineering, 93*, 275–285.

Aslani, A., Taleizadeh, A. A., & Zanoni, S. (2017). An EOQ model with partial backordering with regard to random yield: Two strategies to improve mean and variance of the yield. *Computers & Industrial Engineering, 112*, 379–390.

Cheikhrouhou, N., Sarkar, B., Ganguly, B., Malik, A. I., Batista, R., & Lee, Y. H. (2018). Optimization of sample size and order size in an inventory model with quality inspection and return of defective items. *Annals of Operations Research, 271*, 445–467.

Cheng, T. C. E. (1991). EPQ with process capability and quality assurance considerations. *Journal of the Operational Research Society, 42*(8), 713–720.

Haji, A., Sikari, S. S., & Shamsi, R. (2009). The effect of inspection errors on the optimal batch size in reworkable production systems with scraps. *International Journal of Product Development, 10*(1–3), 201–216.

Hauck, Z., & Vörös, J. (2015). Lot sizing in case of defective items with investments to increase the speed of quality control. *Omega, 52*, 180–189.

Khan, M., Jaber, M. Y., & Bonney, M. (2011). An economic order quantity (EOQ) for items with imperfect quality and inspection errors. *International Journal of Production Economics, 133*(1), 113–118.

Konstantaras, I., Skouri, K., & Jaber, M. Y. (2012). Inventory models for imperfect quality items with shortages and learning in inspection. *Applied Mathematical Modelling, 36*(11), 5334–5343.

Mohammadi, B., Taleizadeh, A. A., Noorossana, R., & Samimi, H. (2015). Optimizing integrated manufacturing and products inspection policy for deteriorating manufacturing system with imperfect inspection. *Journal of Manufacturing Systems, 37*, 299–315.

Moussawi-Haidar, L., Salameh, M., & Nasr, W. (2016). Production lot sizing with quality screening and rework. *Applied Mathematical Modelling, 40*(4), 3242–3256.

Muhammad, A., & Alsawafy, O. (2011). Economic order quantity for items with two types of imperfect quality. *An International Journal of Optimization and Control: Theories & Applications (IJOCTA), 2*(1), 73–82.

Pentico, D. W., & Drake, M. J. (2011). A survey of deterministic models for the EOQ and EPQ with partial backordering. *European Journal of Operational Research, 214*, 179–198.

Porteus, E. L. (1986). Optimal lot sizing, process quality improvement and setup cost reduction. *Operations Research, 34*(1), 137–144.

Raouf, A., Jain, J. K., & Sathe, P. T. (1983). A cost-minimization model for multicharacteristic component inspection. *IIE Transactions, 15*, 187–194.

Rezaei, J., & Salimi, N. (2012). Economic order quantity and purchasing price for items with imperfect quality when inspection shifts from buyer to supplier. *International Journal of Production Economics, 137*(1), 11–18.

Salameh, M. K., & Jaber, M. Y. (2000). Economic production quantity model for items with imperfect quality. *International Journal of Production Economics, 64*(1), 59–64.

Taleizadeh, A. A., & Zamani-Dehkordi, N. (2017). Economic order quantity with partial backordering and sampling inspection. *Journal of Industrial Engineering International, 13*, 331–345.

Taleizadeh, A. A., Tavassoli, S., & Bhattacharya, A. (2019a). An inventory system for a deteriorating product with inspection policy under multiple prepayments and delayed payment and linked-to order quantity. *Annals of Operations Research, 287*, 403–437.

Taleizadeh, A. A., & Moshtagh, M. S. (2019). A consignment stock scheme for closed loop supply chain with imperfect manufacturing processes, lost sales, and quality dependent return: Multi levels Structure. *International Journal of Production Economics, 217*, 298–316.

Taleizadeh, A. A., Yadegari, M., & Sana, S. S. (2019b). Production models of multiple products using a single machine under quality screening and reworking policies. *Journal of Modelling in Management, 14*(1), 232–259.

Taleizadeh, A. A., Kalantari, S. S., & Cárdenas-Barrón, L. E. (2015). Determining optimal price, replenishment lot size and number of shipments for an EPQ model with rework and multiple shipments. *Journal of Industrial & Management Optimization, 11*(4), 1059–1071.

Tsou, J.-C., Hejazi, S. R., & Barzoki, M. R. (2012). Economic production quantity model for items with continuous quality characteristic, rework and reject. *International Journal of Systems Science, 43*(12), 2261–2267.

Chapter 7
Maintenance

7.1 Introduction

The role of the equipment condition in controlling quality and quantity is well known (Ben-Daya and Duffuaa 1995). Equipment must be maintained in top operating conditions through adequate maintenance programs. Despite the strong link between maintenance production and quality, these main aspects of any manufacturing system are traditionally modeled as separate problems. Few attempts have been made to integrate them in a single model that captures their underlying relationships.

There exists an extensive literature addressing the issue of production planning and an equally broad literature tackling maintenance planning questions. Production planning models seek typically to balance the costs of setting up the system with the costs of production and material holding, while maintenance models attempt typically to balance the costs and benefits of sound maintenance plans in order to optimize the performance of the production system. In both domains, issues of production modeling and maintenance modeling have experienced an evident success both from theoretical and applied viewpoints.

In this chapter, the EPQ problems considering maintenance policy under different situations are presented. The presented models are classified into three main categories. The first category includes several models in which shortage is not allowed. The second and third categories include models in which shortage was considered as backordering and partial backordering. To provide a comprehensive introduction about the mentioned research status, the studies in relevant fields are reviewed. The classification is shown in Fig. 7.1.

The common notations of presented models are shown in Table 7.1. To integrate the chapter, the presented notations in Table 7.1 are used for all models. The main decision variables of this field on inventory are Q and T, but in some studies, other decision variables are considered too.

Fig. 7.1 Category of presented EPQ models

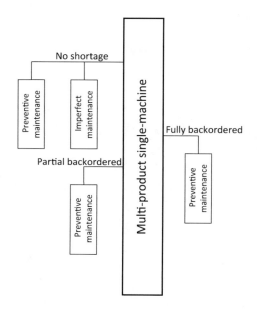

7.2 No Shortage

7.2.1 Preventive Maintenance

Ben-Daya and Makhdoum (1998) considered a production process producing a single item. A production cycle begins with a new system which is assumed to be in an in-control state, producing items of acceptable quality. However, after a period of time in production, the process may shift to an out-of-control state. The elapsed time for the process to be in the in-control state, before the shift occurs, is a random variable assumed to follow a general distribution with increasing hazard rate. The process is inspected at times $t_1, t_2, \ldots, t_n$ to assess its state. In fact, the output quality of the product is monitored by an $\bar{x}$-control chart. PM activities are coordinated with quality control inspection and are carried out periodically at a subset of the above time epochs according to one of the three PM policies. The production cycle ends either with a true alarm signaling that the system is out of control or after m inspection intervals, whichever occurs first. The number m is a decision variable. The process is then restored to the in-control state and to as good as new condition by maintenance and/or replacement. The usual assumptions of the classical EPQ model apply here. In particular, the demand is constant and continuous, and all demands must be met (Ben-Daya and Makhdoum 1998). The specific notations which are used for this problem is presented in Table 7.2.

7.2 No Shortage

Table 7.1 Notations

D	Demand rate (item/year)
h	Holding cost per unit and per unit of time ($/unit/unit of time)
K	Fixed order cost ($/order)
K_S	Setup cost ($/setup)
P	Production rate (item/year)
C	Purchasing cost per unit of product ($/unit)
C_{PM}	Cost of preventive maintenance ($/period)
C_{PM}^m	Maximum cost of preventive maintenance ($/period)
y_j	Actual age of the process right before the jth PM (time)
w_j	Actual age of the process right after the jth P (time)
L_j	Length of the ith sampling (inspection) interval (time)
t_j	Time at the end of jth interval or at jth inspections (time)
$f(t)$	Probability density function of the time to shift distribution
$F(t)$	Cumulative time to shift distribution function $\overline{F}(t) = 1 - F(t)$
$r(t)$	Hazard function $r(t) = f(t)/\overline{F}(t)$
m	Number of inspection interval or inspection undertaken during each production run (integer)
η	Imperfectness factor
C_I	Inspection cost ($/unit)
C_b	Backordering cost per unit of demand and per unit of time ($/unit/unit of time)
B_n	The maximum backordering quantity in units for the nth cycle (item)
T	Length of production cycle (time)
Q	Order quantity (item)
n	Sample size (integer)
$f(t)$	Probability density function of the time to shift distribution
ETC	Expected total cost ($)

It is obvious that the shift pattern to the out-of-control state of a preventively maintained system is not the same as that of a system which is not maintained. The effect of PM on the system is modeled as follows: whenever a PM is carried out, the age of the system is reduced by a certain amount. This is equivalent to a reduction in the shift rate of the system to the out-of-control state (Ben-Daya and Makhdoum 1998).

The joint production, maintenance, and quality model is based on the following assumptions:

1. The duration of the in-control period is assumed to follow an arbitrary probability distribution, $f(t)$, having an increasing hazard rate $r(t)$ and a cumulative distribution function $F(t)$.
2. The process is inspected at times $t_1, t_2, \ldots$, to determine its state, and the output quality of the product is monitored by an $\bar{x}$-control chart. The inspection interval

Table 7.2 Specific notations (Ben-Daya and Makhdoum 1998)

w_1	Warning limit coefficient of the xi-control chart
r_{max}	Failure rate threshold beyond which PM activities should be performed
k	Control limit coefficient of the xi-control chart
Z_T^{max}	The expected time to perform the highest level of a PM (time)
Z_1	The expected time to perform a PM (time)
Z_2	The expected time to repair the process if a failure is detected (time)
C_f	Cost per false alarm (\$/alarm)
C_a	Cost to locate and repair the assignable cause (\$)
C_{in}	Quality cost per unit time while producing in control (\$/unit of time)
C_{out}	Quality cost per unit time while producing out of control (\$/unit of time)
C_S	Cost incurred by producing a non-conforming item (\$/unit)
$C_R(\tau)$	Restoration cost (\$)
K_n	Fixed sampling cost (\$/sample)
C_n	Cost per unit sample (\$/unit of sample)
α	Pr (exceeding control limits one process in control)
β	Pr (not exceeding control limits one process out of control)
n_{PM}	Expected number of PMs during the production cycle (integer)
γ_j	Fraction used to compute the reduction in the process age at time γ_j (%)
p_i	The conditional probability that the process shifts to the out-of-control state during the jth interval given that it was in the in-control state at the beginning of the jth interval
$I(t)$	The function of the inventory level (item)
E (PM)	Expected preventive maintenance cost per production cycle (\$)
E (HC)	Expected inventory holding cost per inventory cycle (\$)
E (QC)	Expected quality control cost per production cycle including the repair cost (\$)
ETC	Expected total cost (\$)

lengths $L_1, L_2, \ldots$, are chosen such that the integrated hazard rate over each interval is the same.

3. Quality inspection activities are carried out at the end of each interval. If a PM criterion is met at the end of the interval, production ceases for an amount of time Z_1 to carry out preventive maintenance activities.
4. If any inspection shows that the state of the process is out of control, production ceases until the accumulated on-hand inventory is depleted to zero. If the process is found to be in control, production continues until the next sampling is due or the predetermined level of inventory is accumulated.
5. The production cycle begins with a new system and ends either with a true alarm or after a specified number m of inspection intervals, whichever occurs first. In other words, if no true alarm is observed by time t_{m-1}, then the cycle is allowed to continue for an additional time L_m. At time t_m, necessary maintenance work is carried out. Therefore, there is no cost of sampling and charting during the mth sampling interval. The process is brought back to the in-control state by repair

7.2 No Shortage

and/or replacement. Thus, a renewal occurs at the end of each cycle. This type of renewal process (Ross 2013) has the property that the expected cost per unit time can be expressed as the ratio of the expected cost per cycle to the expected length of the cycle.

Other important assumptions can be found in Ben-Daya and Makhdoum (1998). They developed the main cost functions corresponding to the three PM policies. For each policy, they derived the quality control, inventory holding, and preventive maintenance costs.

Policy 7.1 The process is inspected at times $t_1, t_2, \ldots$, to assess the state of the process. At times $t_n, t_{2n}, t_{3n}, \ldots$, where n is a decision variable, the process is shut down, and both quality control inspections and PM tasks are carried out in parallel. Note that $\{t_n, t_{2n}, t_{3n}, \ldots\}$ is a subset of $\{t_1, t_2, t_3, \ldots\}$. The expected total cost consists of the setup cost K_S, the expected quality control cost $E(QC)$, the expected inventory holding cost $E(HC)$, and the expected preventive maintenance cost $E(PM)$.

Ben-Daya and Makhdoum (1998) derived expressions for both the expected quality control cost per cycle $E(QC)$ and the expected cycle length $E(T)$. The expected production cycle length includes (1) the expected time for inspection intervals when the process is in control, (2) the expected time for detecting the presence of an assignable cause, (3) the preventive maintenance time, and (4) the repair time. The expected quality control cost includes the expected cost of operating while in control with no alarm, the expected cost of false alarm, the expected cost of operating while out of control with no alarm, the repair cost, and the cost of sampling minus the salvage value for working equipment of age t.

If PM is not carried out, it is assumed that the time for quality inspection is negligible. However, if PM activities are performed, then it is assumed that the expected time to perform a PM is proportional to the PM level and is given by:

$$Z_1 = Z_1^{\max}\left(C_{PM}/C_{PM}^m\right) \tag{7.1}$$

Theorem 7.1 In Policy 7.1, $E(T)$ is given by (Ben-Daya and Makhdoum 1998):

$$E(T) = Z_2 + \sum_{j=1}^{m} L_j(1-p_i) + \sum_{j=1}^{m-1} Z_{1_{j-1}} \prod_{i=1}^{j-1}(1-p_i) + \beta \sum_{j=1}^{m-1} p_j \prod_{i=1}^{j-1}(1-p_i)$$
$$\times \left[\sum_{i=j+1}^{m-1}\left(l_j - Z_{1_{i-1}}\right)\beta^{i-j-1} + l_m\beta^{m-j-1}\right]$$

(7.2)

where:

$$Z_{1_j} = \begin{cases} Z_1 & j = n, 2n, 3n, \ldots \\ 0 & \text{otherwise and for } j = m \end{cases} \quad (7.3)$$

The quality control cost $E(\text{QC})$ is given by:

$$\begin{aligned} E(\text{QC}) = & (K_n + C_n n) \left\{ 1 + \sum_{j=1}^{m-2} \prod_{i=1}^{j} (1 - p_i) + \beta \sum_{j=1}^{m-2} p_j \prod_{i=1}^{j} (1 - p_i) \sum_{i=0}^{m-j-2} \beta^i \right\} \\ & + (C_{\text{in}} - C_{\text{out}}) \sum_{j=1}^{m} \frac{p_j \prod_{i=0}^{j-1} (1 - p_i)}{F(y_j) - F(w_{j-1})} \int_{w_{j-1}}^{y_j} t f(t) dt + (C_{\text{out}} - C_{\text{in}}) \sum_{j=1}^{m} y_j p_j \prod_{i=1}^{j-1} (1 - p_i) \\ & + C_{\text{in}} \sum_{j=1}^{m} L_j \prod_{i=1}^{j-1} (1 - p_i) + C_a \left[\sum_{j=1}^{m} p_j \prod_{i=1}^{j-1} (1 - p_i) + \prod_{i=1}^{m} (1 - p_i) \right] \\ & + C_{\text{out}} \beta \left[\sum_{j=1}^{m-1} p_j \prod_{i=1}^{j-1} (1 - p_i) \sum_{i=j+1}^{m} L_i \beta^{i-j-1} \right] + \alpha C_f \sum_{j=1}^{m-1} \prod_{i=1}^{j} (1 - p_i) - \prod_{i=1}^{m} (1 - p_i) L(t_m) \end{aligned}$$
$$(7.4)$$

The proof of this theorem is given in Appendix A of Ben-Daya and Makhdoum (1998).

The total expected inventory holding cost, $E(\text{HC})$, is defined as:

$$E(\text{HC}) = h \int_0^T I(t) dt \quad (7.5)$$

The integral in (7.5) is determined by computing the expected area $E(A)$ under the function $I(t)$. Hence:

$$E(\text{HC}) = h \int_0^T I(t) dt = h E(A) \quad (7.6)$$

In order to compute $E(A)$, the expression of the inventory levels at times $t_j + Z_{1j}$ is required and is given by the following lemma:

Lemma 7.1 Let I_j be the inventory level at time $t_j + Z_{1j}$, $j = 1, 2, \ldots, m-1$, I_m be the inventory level at time t_m, and then:

$$I_j = I_{j-1} + (P - D)L_j - D Z_{1_j} \quad j = 1, 2, \ldots, m \quad (7.7)$$

where Z_{1j} is given by (7.3), $I_0 = 0$, and if $I_j < 0$ it is set equal to zero. The proof of this lemma is clear from Fig. 7.2.

The expected area $E(A)$ under the function $I(t)$ is given by the following theorem:

7.2 No Shortage

Fig. 7.2 Inventory levels, where t_{PMj} is the time at which jth PM is performed (Ben-Daya and Makhdoum 1998)

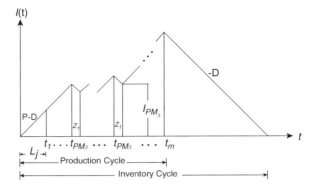

Theorem 7.2 Let:

$$U_j = \begin{cases} [2I_{j-1} + (P-D)L_j]\dfrac{L_j}{2} + \dfrac{[I_{j-1} + (P-D)L_j]^2}{2D} & \text{if } I_j = 0 \\ [2I_{j-1} + (P-D)L_j]\dfrac{L_j}{2} + [I_{j-1} + (P-D)L_j + I_j]\dfrac{Z_{1j}}{2} & \text{if } I_j > 0 \end{cases} \text{ for } j$$

$$= 1, 2, \ldots, m$$

(7.8)

where Z_{1j} is given by (7.3). Let $B_j = I_j^2/2D, j = 1, 2, \ldots, m$. Then $E(A)$ is given by:

$$E(A) = \sum_{j=1}^{m} U_j \prod_{i=1}^{j-1}(1-p_i) + (1-\beta)\sum_{j=1}^{m-1} B_j p_j \prod_{i=1}^{j-1}(1-p_i)$$
$$+ \beta \sum_{j=1}^{m-1} p_j \prod_{i=1}^{j-1}(1-p_i)\left[\sum_{i=j+1}^{m}\beta^{i-j-1}U_i + \beta^{m-j}B_m\right] + B_m \prod_{i=1}^{m-1}(1-p_i)$$

(7.9)

The proof of this theorem is given in Appendix B of Ben-Daya and Makhdoum (1998).

In many preventive maintenance models, the system is assumed to be as good as new after each preventive maintenance action. However, a more realistic situation is one in which the failure pattern of a preventively maintained system changes. One way to model this is to assume that, after PM, the failure rate of the system is somewhere between as good as new and as bad as old. This concept is called imperfect maintenance and was introduced by many authors (Nakagawa 1980; Pham and Wang 1996). It can be assumed that the failure rate of the equipment is decreased after each PM. This amounts to a reduction in the age of the equipment. In this problem, Ben-Daya and Makhdoum (1998) assumed that the reduction in the age of the equipment is a function of the cost of preventive maintenance C_{PM}. Let $y_k(w_k)$ denote the effective age of the equipment right before (right after) the kth PM. Let:

$$\gamma_k = \eta^{k-1}\frac{C_{\text{PM}}}{C_{\text{PM}}^m} \tag{7.10}$$

where $0 < \eta \leq 1$. Linear and nonlinear relationships between age reduction and PM cost are considered. In the linear case:

$$w_k = (1 - \gamma_k)y_k \tag{7.11}$$

and in the nonlinear case:

$$w_k = \left(1 - \gamma_k^\varepsilon\right)y_k, \quad 0 < \varepsilon \leq 1 \tag{7.12}$$

Note that the effective age of the equipment at time t_j is given by:

$$y_1 = l_1 \tag{7.13}$$
$$y_j = w_{j-1} + l_j, \quad j = 1, 2, \ldots, m \tag{7.14}$$

The expected PM cost is given by the following theorem:

Theorem 7.3 The expected maintenance cost during a complete cycle is given by:

$$E(\text{PM}) = \sum_{j=1}^{m-1} C_{\text{PM}_j} \prod_{i=1}^{j}(1-p_i) + \beta \sum_{j=1}^{m-1} C_{\text{PM}_j} p_j \prod_{i=1}^{j}(1-p_i)$$
$$+ \beta \sum_{j=1}^{m-2} E(M_j) p_j \prod_{i=1}^{j}(1-p_i) \tag{7.15}$$

where:

$$C_{\text{PM}_j} = \begin{cases} C_{\text{PM}} & j = n, 2n, \ldots \\ 0 & \text{otherwise} \end{cases} \tag{7.16}$$

and $E(M_j)$ is the residual expected PM cost after time t given that the process is at an out-of-control state at t_j. Note that the expected number of PMs per production cycle is given by:

7.2 No Shortage

$$n_{PM} = \begin{cases} E(PM)/C_{PM} & \text{if } C_{PM} \neq 0 \\ 0 & \text{otherwise} \end{cases} \quad (7.17)$$

The proof of this theorem is given in Appendix C of Ben-Daya and Makhdoum (1998).

Policy 7.2 The process is inspected at times $t_1, t_2, \ldots$, to assess the state of the process. PM activities are performed only at those intervals at which the failure rate of the process reaches a preset threshold. At those intervals, the process is shut down, and both quality control inspections and PM tasks are carried out in parallel (Ben-Daya and Makhdoum 1998).

Under the assumptions described before and the requirements of Policy 7.2, the expected cycle length $E(T)$ is the same as (7.1) except that Z_{1j} is redefined as follows:

$$Z_{1_j} = \begin{cases} Z_1 & \text{if } r(t_j) \geq r_{max} \\ 0 & \text{otherwise} \end{cases} \quad (7.18)$$

where $r(t_j)$ is the failure rate at time t_j and r_{max} is the present threshold. The quality control cost $E(QC)$ per cycle is given by (7.4). The total expected inventory holding cost, $E(HC)$, is defined by (7.5), and the inventory levels are given by the following lemma:

Lemma 7.2 Let I_j be the inventory level at time $t_j + Z_{1j}, j = 1, 2, \ldots, m-1, I_m$ be the inventory level at time t_m, and then:

$$I_j = I_{j-1} + (P-D)L_j - DZ_{1_j} \quad j = 1, 2, \ldots, m \quad (7.19)$$

where Z_{1j} is given by (7.3), $I_0 = 0$, and if $I_j < 0$ it is set equal to zero.
The proof of this lemma is clear from Fig. 7.2.

The expected area $E(A)$ under the function $I(t)$ is the same as (7.9) except that U_j is redefined as follows:

$$U_j = \begin{cases} [2I_{j-1} + (P-D)L_j]\dfrac{L_j}{2} + \dfrac{[I_{j-1} + (P-D)L_j]^2}{2D} & \text{if } I_j = 0 \\ [2I_{j-1} + (P-D)L_j]\dfrac{L_j}{2} + [I_{j-1} + (P-D)L_j + I_j]\dfrac{Z_{1_j}}{2} & \text{if } I_j > 0 \end{cases} \text{ for } j$$

$$= 1, 2, \ldots, m$$

(7.20)

Note that Z_{1j} is given by (7.18) (Ben-Daya and Makhdoum 1998). The expected maintenance cost during a complete cycle is the same as (7.15) except that C_{PM_j} is redefined as follows:

$$C_{PM_j} = \begin{cases} C_{PM} & \text{if } r(t_j) \geq r_{max} \\ 0 & \text{otherwise and for } j = m \end{cases} \quad (7.21)$$

The expected number of PMs n_{PM} during the production cycle is given by (7.17). Note that the age of the equipment at time t_j right before PM is given by (7.14) and right after the PM is given by:

$$w_j = \begin{cases} (1 - \gamma_j)y_j & \text{if } r(t_j) \geq r_{max} \\ y_j & \text{otherwise} \end{cases} \quad (7.22)$$

Policy 7.3 The process is inspected at times $t_1, t_2, \ldots$, to assess the state of the process. PM activities are performed only at those intervals at which two consecutive values of sample means fall in the warning zone. At those intervals, the process is shut down, and both quality control inspections and PM tasks are carried out in parallel. Under the assumptions described before and the requirements of Policy 7.3, the expected cycle length $E(T)$ is the same as (7.1) except that Z_{1j} is replaced by $p_w^2 Z_{1j}$, where p_w is the probability that a sample mean falls in the warning zone (Ben-Daya and Makhdoum 1998).

The quality control cost $E(QC)$ per cycle is exactly the same as (7.4). The total expected inventory holding cost, $E(HC)$, is defined by (7.5), and the inventory levels are given by the following lemma.

Lemma 7.3 Let I_j be the inventory level at time $t_j + Z_{1j}, j = 1, 2, \ldots, m - 1$, I_m is the inventory level at time t_m, and then:

$$I_j = (1 - p_w^2)[I_{j-1} + (P - D)L_j] - p_w^2[I_{j-1} + (P - D)L_j - DZ_1], \quad j = 1, 2, \ldots, m - 1 \quad (7.23)$$

$$I_m = I_{m-1} + (P - D)l_m \quad (7.24)$$

where $I_0 = 0$ and if $I_j < 0$ is equal to zero. The proof of this lemma is clear from Fig. 7.2.

The expected area at a given inspection interval can be computed as (the probability of performing PM) × (inventory area if PM is performed) + (the probability of not performing PM) × (inventory area if PM is not performed):

7.2 No Shortage

$$U_j = \begin{cases} p_w^2\left([2I_{j-1}+(P-D)L_j]\dfrac{L_j}{2}+\dfrac{[I_{j-1}+(P-D)L_j]^2}{2D}\right) \\ +(1-p_w^2)\left([2I_{j-1}+(P-D)L_j]\dfrac{L_j}{2}\right) & \text{for } I_j=0,\ j=1,2,\ldots,m-1 \\[4pt] \left([2I_{j-1}+(P-D)L_j]\dfrac{L_j}{2}+[I_{j-1}+(P-D)L_j-I_j]\dfrac{Z_1}{2}\right) \\ +(1-p_w^2)\left([2I_{j-1}+(P-D)L_j]\dfrac{L_j}{2}\right) & \text{for } I_j>0,\ j=1,2,\ldots,m-1 \\[4pt] [2I_{j-1}+(P-D)L_j]\dfrac{L_j}{2} & \text{for } j=m \end{cases}$$

(7.25)

The expected maintenance cost during a complete cycle is given by:

$$E(\text{PM}) = C_{\text{PM}} n_{\text{PM}} \tag{7.26}$$

where n_{PM} is the expected number of PMs. Since no PM is performed at time t_m and the probability of performing PM at time t_j, $j = 1, 2, \ldots, m-1$ is p_w^2, then n_{PM} is given by:

$$n_{\text{PM}} = (m-1)p_w^2 \tag{7.27}$$

Again the expected age of the equipment at a given inspection interval can be computed as (the probability of performing PM) × (age of the equipment if PM is performed) + (the probability of not performing PM)× (age of the equipment if no PM is performed). So, the age of the equipment at time t_j right before PM is given by (7.14) and right after the PM is given by:

$$w_j = p_w^2\left(1 - \eta^{j-1}\dfrac{C_{\text{PM}}}{C_{\text{PM}}^{\text{m}}}\right)y_j + (1-p_w^2)y_j \tag{7.28}$$

7.2.1.1 Integrated Model and Solution Method

The expected total cost per unit time is given by (Ben-Daya and Makhdoum 1998):

$$\text{ETC} = \dfrac{K_S + E(\text{HC}) + E(\text{QC}) + E(\text{PM})}{E(T)} \tag{7.29}$$

The expression of $E(T)$ is given by:

$$E(T) = \frac{P}{D}(E(T) - n_{\text{PM}}Z_1) \qquad (7.30)$$

Next, they discussed the problem of solving the above integrated production, quality, and maintenance models to obtain the optimal values for the decision variables. Also, they discussed the way by which the frequency of sampling should be regulated and the optimization procedure used to determine the optimal design parameter values and the optimal preventive maintenance effort (Ben-Daya and Makhdoum 1998).

The problem is to determine simultaneously the optimal production run time and hence the optimal EPQ, the optimal preventive maintenance level, and the optimal design parameters of the $\bar{x}$-control chart, namely, $L_1, L_2, \ldots, L_m$, the sample size n, and the control limit coefficient k. In addition, the optimal values of 1, $r_{\max}$, and warning limit coefficient for Policies 7.1–7.3, respectively, will be also determined simultaneously with the abovementioned decision variables. For a Markovian shock model, a uniform sampling scheme provides a constant integrated hazard over each interval. Banerjee and Rahim (1988) extended this fact to non-Markovian shock models by choosing the length of sampling intervals such that the integrated hazard over each interval is the same for all intervals, that is (Ben-Daya and Makhdoum 1998):

$$\int_{t_j}^{t_{j+1}} r(t)dt = \int_0^{t_1} r(t)dt \qquad (7.31)$$

Since the failure rate is reduced at the end of each interval because of PM activities, condition (7.31) becomes:

$$\int_{w_{j-1}}^{y_j} r(t)dt = \int_0^{l_1} r(t)dt, \quad j = 2, \ldots, m \qquad (7.32)$$

If the time that the process remains in the in-control state follows a Weibull distribution, that is, its probability density function is given by:

$$f(t) = \lambda v t^{v-1} e^{-\lambda t^v}, \quad t > 0, \ v \geq 1, \ \lambda > 0 \qquad (7.33)$$

Then using (7.32), the length of the sampling intervals $L_j, j = 2, \ldots, n$ can be determined recursively as follows:

$$L_j = \left[(w_{j-1})^v + L_1^v\right]^{1/v} - w_{j-1}, \quad j = 2, \ldots, m \qquad (7.34)$$

The problem is to determine the values of the decision variables m, n, k, L_1, and C_{PM}, which define the PM level, that minimize the expected total cost ETC. Recall that the age of the equipment after a PM is reduced proportional to the PM cost C_{PM}.

7.2 No Shortage

The cost function is minimized using the pattern search technique of Hooke and Jeeves (1961).

Example 7.1 Ben-Daya and Makhdoum (1998) presented an example with parameter values $Z_1^{max} = 0.1$, $Z_2 = 1.0$, $K_n = \$2.0$, $C_n = \$0.5$, $C_f = \$500$, $C_a = \$1100$, $C_{in} = \$50$, $C_{out} = \$950$, $\delta = 0.5$ (δ is the amount of shift in the mean when the process is out of control, measured in standard deviations), $L(t_m) = 100e^{-tm}$, $D = 1400$ units, $P = 1500$ units, $h = \$0.1$, $K_S = \$20$, $\eta^{j-1} = 0.99$, $\lambda = 0.05$, and $\upsilon = 2$.

They used a computer program coded in Fortran implementing the Hooke and Jeeves (1961) procedure and run on a 486 PC to obtain the results presented below. The results of Policy 7.1 with L set equal to 1 obtained for different PM levels corresponding to $C_{PM}^m = \$300$ for both cases are summarized in Tables 7.3 and 7.4.

These results illustrate clearly the trade-offs between PM levels and quality control costs. The increase in PM level yields reductions in quality control costs. With no PM, the quality control cost is \$400.1. With a PM level of \$100, the quality control cost is reduced to \$276.1 and \$235.9 for both Case I and Case II, respectively. The optimum PM level when $C_{PM}^m = \$300$ is obtained when $C_{PM} = \$300$ (linear case), leading to a quality cost of \$183.9 and an overall cost of \$367.6 much less than without PM (\$429.7). One might also notice that at low values of C_{PM}, more reductions in both quality and overall costs are obtained for Case II than that for Case I. This is because for the same cost of PM, Case II yields much more reduction in the age of the equipment than Case I (Ben-Daya and Makhdoum 1998).

PM activities also affect the economic production quantity (EPQ). As a matter of fact, it is noticeable from Tables 7.2 and 7.3 that PM does affect the production cycle (t_m) since for higher PM levels we have longer production cycles. This is because when PMs are performed during the production cycle, quality control costs are reduced and hence longer procedure cycle will be still feasible. Some different numerical examples are provided by Ben-Daya and Makhdoum (1998) for other policies.

7.2.2 Imperfect Preventive Maintenance

Ben-Daya (2002) developed an integrated model for the joint determination of economic production quantity and preventive maintenance (PM) level for an imperfect process having a general deterioration distribution with increasing hazard rate. The effect of PM activities on the deterioration pattern of the process is modeled using the imperfect maintenance concept. It is assumed that after each PM, the age of the system is reduced proportional to the PM level.

Consider a production process producing a single item. A production cycle begins with a new system which is assumed to be in an in-control state, producing items of acceptable quality. However, after a period of time in production, the process may shift to an out-of-control state. The elapsed time for the process to be in the in-control

Table 7.3 Case I: Effect of PM level for linear improvement ($Q^* =$ EPQ) (Ben-Daya and Makhdoum 1998)

C_{PM}	m	N	k	L_1	α	$1-\beta$	t_m	QC	HC	PM	Q^*	ETC
0	26	49	2.51	1.88	0.012	0.8387	4.3	400.1	25.4	0	6502	429.7
100	37	56	2.51	1.54	0.0121	0.8909	7.3	294.7	24.7	107.2	10,539	429.3
200	38	64	2.49	1.54	0.0128	0.9344	10.3	238.6	17	152.3	14,717	409.7
300	29	70	2.47	1.56	0.0136	0.9569	13.4	199.9	13.3	166.8	18,987	381.6

7.2 No Shortage

Table 7.4 Case II: Effect of nonlinear PM level for nonlinear improvement (Ben-Daya and Makhdoum 1998)

C_{PM}	m	n	k	L_1	α	$1-\beta$	t_m	QC	HC	PM	Q^*	ETC
0	26	49	2.51	1.88	0.012	0.8387	4.3	400.1	25.4	0	6502	429.7
100	24	53	2.53	1.28	0.0114	0.8667	9.5	235.9	31.6	103.4	13,719	373
200	22	64	2.49	1.48	0.0129	0.9349	11.2	218.2	23.5	138.2	16,065	381.6
300	29	70	2.47	1.56	0.0136	0.9569	13.4	199.9	13.3	166.8	18,987	381.6

Table 7.5 Specific notations (Ben-Daya 2002)

α	Percentage of non-conforming units produced when the process is in the out-of-control state (%)
τ_i	Detection delay during interval i (time)
t_j	Time of the jth PM, $t_j = \sum_{i=1}^{j} L_j$ (time)
N_i	Number of non-conforming items produced (t_{i-1}, t_i) (item)
p_j	The conditional probability that the process shifts to the out-of-control state during the time interval (t_{j-1}, t_j) given that the process was in control at time t_{j-1}
ETC	Expected total cost (%)

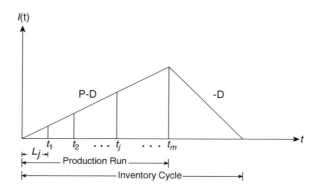

Fig. 7.3 Inventory cycle; the jth PM is performed at time t_j (Ben-Daya 2002)

state, before the shift occurs, is a random variable assumed to follow a general distribution with increasing hazard rate. The process is inspected at times $t_1, t_2, \ldots, t_m$ to assess its state, and at the same time, PM activities are carried out. The production cycle ends either when the system is out of control or after m inspection intervals whichever occurs first. The number m is a decision variable. The process is then restored to the in-control state and to the as good as new condition by maintenance and/or replacement. The usual assumptions of the classical EPQ model apply here. In particular, the demand is constant and continuous, and it must be met (Ben-Daya 2002).

The specific notations which are used for this problem are presented in Table 7.5.

The total expected cost per cycle consists of the setup cost, inventory holding cost, PM cost, inspection cost, cost of producing non-conforming items, and restoration cost. Before deriving these costs, let us determine the expected production cycle length (Ben-Daya 2002). The expected inventory cycle length is given by (see Fig. 7.3):

$$E(\text{CT}) = \frac{P}{D} E(T) \tag{7.35}$$

where $E(T)$ is the expected production run length which is given by the following lemma:

7.2 No Shortage

Table 7.6 Associated probabilities (Ben-Daya 2002)

State	Probability	Expected residual cost
In control	$1 - p_1$	$E(T_1)$
Out of control	p_1	0

Lemma 7.4 The expected production cycle is given by (Ben-Daya 2002):

$$E(T) = \sum_{j=1}^{m} L_j \prod_{i=1}^{j-1} (1 - p_i) \tag{7.36}$$

Proof Let $E(T_j)$ be the expected residual time in the cycle beyond t_j given that the process was in control at time t_j, $E(T_0) = E(T)$. Then:

$$p_j = \frac{F(y_j) - F(w_{j-1})}{\overline{F}(w_{j-1})} \tag{7.37}$$

In order to find the expression of $E(T)$, consider the possible states of the process at the end of the first interval (at time $t_1 = l_1$). For each possible state, the expected residual time in the cycle and the associated probabilities are presented in Table 7.6.

Consequently, $E(T) = L_1 + (1 - p_1)E(T_1)$. Similarly, for $j = 1, 2, \ldots, m - 2$, one has $E(T_j) = L_{j+1} + (1 - p_{j+1})E(T_{j+1})$. Note that $E(T_{m-1}) = L_m$; therefore (Ben-Daya 2002):

$$E(T) = \sum_{j=1}^{m} L_j \prod_{i=1}^{j-1} (1 - p_i) \tag{7.38}$$

In this model, the setup cost is K_S, and the inventory holding cost is $hE(T)^2(P - D)P/2D$. As mentioned earlier, Ben-Daya (2002) used the concept of imperfect maintenance. After each PM, the age of the system is somewhere between as good as new and as bad as old depending on the level of PM activities. The reduction in the age of the equipment is a function of the cost of preventive maintenance. Let $y_k(w_k)$ denote the effective age of the equipment right before (after) the kth PM. Similar to previous model of Ben-Daya and Makhdoum (1998), Ben-Daya (2002) used Eqs. (7.10), (7.11), (7.13), and (7.14) including $\gamma_k = \eta^{k-1} \frac{C_{PM}}{C_{PM}^m}$, $w_k = (1 - \gamma_k)y_k$, $y_1 = l_1$, and $y_j = w_{j-1} + l_j$, $j = 2, \ldots, m$.

Now, let us turn to the problem of determining the expected preventive maintenance cost. Since the inspection is error-free and after each inspection PM activities are carried out, the expected PM cost per cycle, E(PM), is given by the following lemma:

Lemma 7.5 The expected PM cost per cycle is given by (Ben-Daya 2002):

$$E(\text{PM}) = C_{\text{PM}} \sum_{j=1}^{m-1} \prod_{i=1}^{j} (1 - p_i) \quad (7.39)$$

It is assumed that no PM is carried out at the end of the last interval. The proof of this lemma is similar to that of Lemma 7.4.

The expected number of inspections is equal to the number of PMs in addition to one inspection at the end of the cycle. Hence, inspection cost is $E(\text{IC}) = C_I(n_{\text{PM}} + 1)$.

The inspection cost is separated from PM so that alternative PM policies can be considered. Here, a PM is carried out with each inspection. An alternative policy would be to inspect the system at $l_1, l_1 + l_2, \ldots$ but perform PM only at a subset of these time epochs. The reader is referred to Ben-Daya and Duffuaa (1995) for alternative PM policies (Ben-Daya 2002).

The expected number of non-conforming items during the jth interval is given by (Ben-Daya 2002):

$$E(N_j) = \int_{w_{j-1}}^{y_j} \alpha(y_i - t) P f_c(t) dt$$

$$= \alpha P \left(y_j [F_c(y_j) - F_c(w_{j-1})] - \int_{w_{j-1}}^{y_j} t f_c(t) dt \right) \quad (7.40)$$

where:

$$f_c = f / \overline{F}(w_{j-1}) \quad (7.41)$$

$$F_c = F / \overline{F}(w_{j-1}) \quad (7.42)$$

The total expected number of non-conforming items per production run is given by the following lemma:

Lemma 7.6

$$E(N) = \sum_{j=1}^{m} p_j E(N_j) \prod_{i=1}^{j-1} (1 - p_i) \quad (7.43)$$

Proof Let $E(B_j)$ be the expected residual number of non-conforming items produced in the cycle beyond t_j given that the process was in control at time t_j, $E(N) = E(B_o)$ (Ben-Daya 2002).

In order to find the expression of $E(N)$, consider the possible states of the process at the end of the first interval (at time $t_1 = l_1$). For each possible state, the expected residual time in the cycle and the associated probabilities are presented in Table 7.7.

7.2 No Shortage

Table 7.7 Associated probabilities (Ben-Daya 2002)

State	Probability	Expected residual number of non-conforming items
In control	$1 - p_1$	$E(B_1)$
Out of control	p_1	$E(N_1)$

Consequently:

$$E(N) = p_1 E(N_1) + (1 - p_1) E(B_1) \tag{7.44}$$

Similarly, for $j = 1, 2, \ldots, m-2$, one has

$$E(B_j) = p_{j+1} E(N_{j+1}) + (1 - p_{j+1}) E(B_{j+1}) \tag{7.45}$$

Note that $E(B_{m-1}) = p_m E(N_m)$. Therefore:

$$E(N) = \sum_{j=1}^{m} p_j E(N_j) \prod_{i=1}^{j-1} (1 - p_i) \tag{7.46}$$

The total cost of producing non-conforming items per unit time is given by:

$$E(\mathrm{DC}) = \alpha C_S P \sum_{j=1}^{m} p_j \prod_{i=1}^{j-1} (1 - p_i)$$

$$\times \left\{ y_j [F_c(y_j) - F_c(w_{j-1})] - \int_{w_{j-1}}^{y_j} t f_c(t) dt \right\} \tag{7.47}$$

Since the restoration cost depends on the detection delay, the restoration cost during the jth interval is given by:

$$E(\mathrm{RC}) = \int_{w_{j-1}}^{y_j} \tau(y_i - t) f_c(t) dt$$

$$= (C_{R0} + C_{R1} y_j) \left([F_c(y_j) - F_c(w_{j-1})] - C_{R1} \int_{w_{j-1}}^{y_j} t f_c(t) dt \right) \tag{7.48}$$

because the restoration cost changes linearly with the detection delay as below:

$$C_R(y_j - t) = C_{R0} + C_{R1}(y_j - t) \tag{7.49}$$

where C_{R0} and C_{R1} are some constants. The restoration cost per cycle is given by the following lemma:

Lemma 7.7 The expected restoration cost per cycle is given by (Ben-Daya 2002):

$$E(\text{RC}) = \sum_{j=1}^{m} p_j \prod_{i=1}^{j-1} (1 - p_j)$$
$$\times \left[(C_{R0} + C_{R1} y_j) \left[F_c(y_j) - F_c(w_{j-1}) \right] - C_{R1} \int_{w_{j-1}}^{y_j} t f_c(t) dt \right] \quad (7.50)$$

The proof of this lemma is similar to that of Lemma 7.6.

Let $E(\text{QC}) = E(\text{DC}) + E(\text{RC})$ denote quality-related costs. To solve the problem, recall that the expected total cost per unit time is given by (Ben-Daya 2002):

$$\text{ETC} = \frac{K_S + E(\text{IC}) + E(\text{HC}) + E(\text{QC}) + E(\text{PM})}{E(\text{CT})} \quad (7.51)$$

where K_S, $E(\text{IC})$, $E(\text{HC})$, $E(\text{QC})$, $E(\text{PM})$, and $E(\text{CT})$ are the setup cost, inspection cost, inventory holding cost, quality-related costs (cost of non-conforming items and restoration cost), PM cost, and the expected inventory cycle length, respectively. Next, Ben-Daya (2002) discussed the problem of solving the above integrated model to obtain the optimal values for the decision variables. Also, they would have discussed the way by which the inspection frequency should be regulated and the optimization procedure used to determine the optimal solution. The problem is to determine simultaneously the optimal lengths of the inspection intervals, namely, l_1, l_2, ..., l_m; the optimal PM level, C_{PM}; and the number of inspections, m. For a Markovian shock model, a uniform inspection scheme provides a constant integrated hazard over each interval. Banerjee and Rahim (1988) extended this fact to non-Markovian shock models by choosing the length of inspection intervals such that the integrated hazard over each interval is the same for all intervals, as presented by Ben-Daya and Makhdoum (1998) in Eqs. (7.31) and (7.32), and Eqs. (7.33) and (7.44) are used too.

The problem is to determine the values of the decision variables m, L_1, and C_{PM}, which define the PM level, that minimize the expected total cost ETC. Recall that the age of the equipment after a PM is reduced in proportion to the PM cost C_{PM}. The cost function is minimized using the pattern search technique of Hooke and Jeeves (1961). However, due to the characteristics of the cost function, some modifications to the standard method have to be made to account for the inherent integrality constraint on the number of inspections. The optimal value of $m \geq 2$ is determined by the inequalities $\text{ETC}(m - 1) \geq \text{ETC}(m)$ and $\text{ETC}(m) \leq \text{ETC}(m + 1)$.

Example 7.2 Ben-Daya (2002) presented numerical examples to illustrate important aspects of the developed integrated model. The process shift mechanism is assumed to follow a Weibull distribution. The Weibull scale and shape parameters are $\lambda = 5$ and $v = 2.5$, respectively. The following data are used for other parameters: $D = 500$ units, $P = 1000$ units, $h = \$0.5$, $K_S = \$150$, $C_S = \$20$, $C_I = \$10$, $C_{R0} =$

Table 7.8 Effect of PM on quality and total cost (Ben-Daya 2002)

α	C_{PM}	M	Q	E(QC)	ETC
0.2	0	9	387	33.22	315.36
	0.5	10	692	22.41	265.09
	1	5	747	19.68	255.8
0.4	0	8	350	36.66	346.81
	0.5	10	685	24.74	276.26
	1	5	713	21.55	265.09

$10, $C_{R1} = 0.15$, and ? $= 0.99$. The relationship between PM cost, C_{PM}, and the improvement in the age of the system will be used to investigate the effect of PM level on both quality control-related costs and total expected costs. The results obtained for different PM levels corresponding to $C_0^{PM} = 20$ are summarized in Table 7.8, where Q is the lot size.

These results illustrate clearly the trade-offs between PM levels and quality-related costs. The increase in PM level yields reductions in quality control costs. With no PM, the quality-related costs amount to $33.22 and $36.66 for $\alpha = 0.2$ and 0.4, respectively. With a PM level of 50%, the quality-related costs are reduced to $22.41 ($24.74). The optimum PM level when $C_{PM}^m = 20$ is obtained when $C_{PM} = $20, leading to quality-related costs of $19.68 ($21.55) and an overall cost of $255.80 ($265.09) much less than without PM ($315.36 ($346.81)).

7.2.3 Imperfect Maintenance and Imperfect Process

Sheu and Chen (2004) developed an integrated model for the joint determination of both economic production quantity and level of preventive maintenance (PM) for an imperfect production process. This process has a general deterioration distribution with increasing hazard rate. The effect of PM activities on the deterioration pattern of the process is modeled using the imperfect maintenance concept.

The production operation system is considered to produce a single item. A production cycle begins with a production system assumed to be in an in-control state: that is, the system produces items of acceptable quality. However, after a period of time in production, the process may shift to an out-of-control state. The elapsed time for the process in the in-control state, before the shift occurs, is a random variable that is assumed to follow a general distribution with increasing hazard rate. The process is inspected at times $t_1, t_2, \ldots, t_m$ the state of the production system whether it is kept in the in-control state or not. At the same time, PM activities are carried out, but the production system has to stop. The production cycle will be stopped either when the system is transferred to the type II out-of-control state or after the mth inspection, whichever occurs first. The process is then restored to the in-control state and to the as good-as-new condition by a complete repair or replacement if necessary (Sheu and Chen 2004).

Table 7.9 Specific notations (Ben-Daya 2002)

θ	Probability of a type II out-of-control state when the system is out of control
γ_k	Imperfectness coefficient at the kth PM
α_1	Non-conforming rate with type I out-of-control state
α_2	Non-conforming rate with type II out-of-control state
C_{mr}	Cost of minimal repair by type I out-of-control state ($)
$N_j^{(\mathrm{I})}$	Produced non-conforming items within (t_{j-1}, t_j) due to the type I out-of-control state (item)
$N_j^{(\mathrm{II})}$	Produced non-conforming items within (t_{j-1}, t_j) due to the type II out-of-control state (item)
n_{mr}	The expected number of minimal repairs per production cycle (integer)
n_{PM}	The expected number of PMs per production cycle (integer)

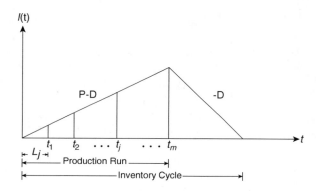

Fig. 7.4 Inventory cycle; the jth PM is performed at time t_j (Sheu and Chen 2004)

The specific notations which are used for this problem is presented in Table 7.9. The total expected cost per cycle can be presented as below:

$$\mathrm{TC} = \text{setup cost} + \text{holding cost} + \text{PM cost} + \text{inspection cost} \\ + \text{quality-related cost (nonconforming items cost} + \text{restoration cost)} \tag{7.52}$$

Before deriving these costs, let us determine the expected production cycle length. The expected inventory cycle length is given by (see Fig. 7.4) (Sheu and Chen 2004):

$$E(\mathrm{CT}) = \frac{P}{D} E(T) \tag{7.53}$$

where $E(T)$ is the expected production run length which is given by the following lemma (Sheu and Chen 2004):

7.2 No Shortage

Lemma 7.8 The expected production cycle is given by (Sheu and Chen 2004):

$$E(T) = \sum_{j=1}^{m} L_j \prod_{i=1}^{j-1} (1 - \theta p_i) \tag{7.54}$$

Proof The related proof can be performed similar to Lemma 7.4.

In this model, the setup cost is K_S, and the inventory holding cost is $hE(T)^2(P-D)P/2D$ similar to the previous presented models. For preventive maintenance cost and inspection cost, Sheu and Chen (2004) followed Ben-Daya and Makhdoum (1998) and Ben-Daya (2002) and used $\gamma_k = \eta^{k-1}\frac{C_{PM}}{C_{PM}^m}$, $w_k = (1 - \gamma_k)y_k$, $y_1 = l_1$, and $y_j = w_{j-1} + l_j$, $j = 2, \ldots, m$ in their model. Then to calculate the expected PM cost per cycle, $E(PM)$, they presented Lemma 7.9.

Lemma 7.9 The expected PM cost per cycle is given by (Sheu and Chen 2004):

$$E(PM) = C_{PM} \sum_{j=1}^{m-1} \prod_{i=1}^{j} (1 - p_i) + (1 - \theta)C_{mr} \sum_{j=1}^{m-1} p_j \prod_{i=1}^{j-1} (1 - \theta p_i) \tag{7.55}$$

It is assumed that no PM is carried out at the end of the last interval. The proof of this lemma is similar to that of Lemmas 7.4 and 7.8.

Remark 7.1 n_{PM} can be explained as the expected number of PMs per production cycle as below:

$$n_{PM} = \sum_{j=1}^{m-1} \prod_{i=1}^{j} (1 - \theta p_i) \tag{7.56}$$

Remark 7.2 n_{mr} can be explained as the expected number of minimal repairs per production cycle.

$$n_{mr} = (1 - \theta) \sum_{j=1}^{m-1} p_j \prod_{i=1}^{j-1} (1 - \theta p_i) \tag{7.57}$$

Moreover, the expected number of inspections is equal to the number of PMs in addition to one inspection at the end of the cycle. Hence, $E(IC) = C_I(n_{PM} + 1)$ (Sheu and Chen 2004).

In order to calculate the production cost of non-conforming items, Lemma 7.10 is presented by Sheu and Chen (2004).

Lemma 7.10
$$E\left(N_j^{(\mathrm{I})}\right) = \int_{w_{j-1}}^{y_j} \alpha_\mathrm{I} P(y_j - t) \frac{(1-\theta)f(t)\left[\overline{F}(t)\right]^{-\theta}}{\left[\overline{F}(w_{j-1})\right]^{1-\theta}} dt \qquad (7.58)$$

Proof For more detailed description, see Sheu and Chen (2004).

Lemma 7.11
$$E\left(N_j^{(\mathrm{II})}\right) = \int_{w_{j-1}}^{y_j} \alpha_\mathrm{II} P(y_j - t) \frac{\theta f(t)\left[\overline{F}(t)\right]^{\theta-1}}{\left[\overline{F}(w_{j-1})\right]^{\theta}} dt \qquad (7.59)$$

Proof For more detailed description, see Sheu and Chen (2004).

The expected number of non-conforming items during the *j*th interval is given by $E\left(N_j^{(\mathrm{I})}\right)$ and $E\left(N_j^{(\mathrm{II})}\right)$. The total expected number of non-conforming items per production run is given by the following lemma (Sheu and Chen 2004):

Lemma 7.12
$$E(N) = \sum_{j=1}^{m} \left[(1-\theta)E\left(N_j^{(\mathrm{I})}\right) + \theta E\left(N_j^{(\mathrm{II})}\right)\right] p_j \prod_{i=1}^{j-1} (1 - \theta p_i) \qquad (7.60)$$

where:

$$E\left(N_j^{(\mathrm{I})}\right) = \int_{w_{j-1}}^{y_j} \alpha_\mathrm{I} P(y_j - t) \frac{(1-\theta)f(t)\left[\overline{F}(t)\right]^{-\theta}}{\left[\overline{F}(w_{j-1})\right]^{1-\theta}} dt \qquad (7.61)$$

$$E\left(N_j^{(\mathrm{II})}\right) = \int_{w_{j-1}}^{y_j} \alpha_\mathrm{II} P(y_j - t) \frac{\theta f(t)\left[\overline{F}(t)\right]^{\theta-1}}{\left[\overline{F}(w_{j-1})\right]^{\theta}} dt \qquad (7.62)$$

Proof For more detailed description, see Sheu and Chen (2004).

The total cost of producing non-conforming items per unit time is given by Sheu and Chen (2004):

$$E(\mathrm{DC}) = C_\mathrm{S} E(N)$$
$$= C_\mathrm{S} \sum_{j=1}^{m} \left[(1-\theta)E\left(N_j^{(\mathrm{I})}\right) + \theta E\left(N_j^{(\mathrm{II})}\right)\right] p_j \prod_{i=1}^{j-1} (1 - \theta p_i) \qquad (7.63)$$

To derive the restoration cost per production cycle, the following lemma is presented by Sheu and Chen (2004).

Lemma 7.13 The expected restoration cost per production cycle is given by Sheu and Chen (2004):

7.2 No Shortage

$$E(\text{RC}) = \sum_{j=1}^{m} \theta p_j \prod_{i=1}^{j-1} (1 - \theta p_i)$$

$$\times \left((C_{R0} + C_{R1} y_j) \left[1 - \left(\frac{\overline{F}(y_j)}{\overline{F}(w_{j-1})} \right)^{\theta} \right] - r_1 \int_{w_{j-1}}^{y_j} t \frac{\theta f(t) [\overline{F}(t)]^{\theta-1}}{[\overline{F}(w_{j-1})]^{\theta}} \, dt \right) \quad (7.64)$$

where C_{R0} and C_{R1} are similar to what is presented in the work of Ben-Daya (2002).

Proof For more detailed description, see Sheu and Chen (2004).

Remark 7.3 The quality-related cost $E(\text{QC}) = E(\text{DC}) + E(\text{RC})$ (Sheu and Chen 2004).

Remark 7.4 There are several special cases in the integrated model. For example, when $m = \theta = 1$, this is the classical economic production quantity model. Another case was considered by Ben-Daya (2002) when $\theta = 1$ (Sheu and Chen 2004).

The expected total cost is composed of the setup cost, inspection cost, inventory holding cost, quality-related costs (i.e., cost of non-conforming items, restoration cost), and PM cost. For a renewal reward process (Ross 1996), one has the expected total cost per expected cycle length as follows (Sheu and Chen 2004):

$$\text{ETC} = \frac{K_S + E(\text{IC}) + E(\text{HC}) + E(\text{QC}) + E(\text{PM})}{E(\text{CT})} \quad (7.65)$$

The problem is to determine simultaneously the optimal lengths of the inspection intervals, namely, $L_1, L_2, \ldots, L_m$; the optimal PM level, C_{PM}; and the number of inspections, m (Sheu and Chen 2004). For a Markovian shock model, Sheu and Chen (2004) used Eqs. (7.31)–(7.34) of Ben-Daya and Makhdoum (1998). As mentioned before, the problem is to determine the values of the decision variables m, L_1, and C_{PM}, which define the PM level, that minimize the expected total cost ETC. Recall that the age of the equipment after a PM is reduced in proportion to the PM cost C_{PM}. The cost function is minimized using the pattern search technique of Hooke and Jeeves (1961). However, due to the characteristics of the cost function, some modifications to the standard method have to be made to account for the inherent integrality constraint on the number of inspections. The optimal value of $m \geq 2$ is determined by the inequalities $\text{ETC}(m - 1) \geq \text{ETC}(m)$ and $\text{ETC}(m) \leq \text{ETC}(m + 1)$. Therefore, the optimal value m^* and L_1^* can be obtained by the following procedure if the level of PM is determined (Sheu and Chen 2004).

Step 1. Estimate m_0, the maximum number of inspections undertaken during each production run, either from historical records or from the condition of production.

Step 2. For $m = 1$, one can search an optimal value l_1 subject to the expected total cost ETC_1.

Step 3. One has the expected total cost from ETC_2 to $\text{ETC}m_0$ by repeating *Step 2* for $m = 2, 3, \ldots, m_0$.

Step 4. The optimal values l_1 and m satisfy the following condition:

$$\text{ETC}(L_1^*, m^*; C_{\text{PM}}) = \text{Min}\{\text{ETC}_j, \ j = 1, \ldots, m_0\}$$

Example 7.3 Sheu and Chen (2004) presented numerical examples to illustrate important aspects of the developed integrated model. The process shift mechanism is assumed to follow a Weibull distribution similar to the previous cases with parameters $\lambda = 5$ and $\nu = 2.5$. They considered $\theta = 0.1$, $K_S = \$150$, $D = 500$ units, $P = 1000$ units, $h = \$0.5$, $C_S = \$20$, $C_I = \$10$, $C_{R0} = \$10$, $C_{R1} = \$0.15$, ▢ $= 0.99$, $\alpha_I = 0.2$, $\alpha_{II} = 0.4$, $C_{\text{PM}}^m = 20$, and $C_{\text{mr}} = 10$ (Sheu and Chen 2004). The results for no PM under the setup cost $K_S = \$150$ is $(L_1^*, m^*, Q^*, E(\text{TC}), C_{\text{PM}}) = (0.308, 3, 472.42, 300.90, 0)$, and the optimal value of PM cost obtained is 20 leading to the expected total cost amount $(L_1^*, m^*, Q^*, E(\text{TC}), C_{\text{PM}}) = (0.261, 3, 768.03, 262.63, 20)$.

7.2.4 Aggregate Production and Maintenance Planning

Aghezzaf et al. (2007) developed an aggregate production and maintenance planning problem in which a set of items must be produced in lots on a capacitated production system throughout a specified finite planning horizon. They assumed that the production system is subject to random failures and that any maintenance action carried out on the system, in a period, reduces the system's available production capacity during that period. The objective is to find an integrated lot-sizing and preventive maintenance strategy of the system that satisfies the demand for all items over the entire horizon without backlogging and which minimizes the expected sum of production and maintenance costs.

At the aggregate planning level, the only work they were aware of is that of Weinstein and Chung (1999). They proposed a three-part model to evaluate an organization's maintenance policy. In their approach, an aggregate production plan is first generated, then a master production schedule is developed to minimize the weighted deviations from the goals specified at the aggregate level, and finally work center-loading requirements are used to simulate equipment failures during the planning horizon. We have used several experiments to test the significance of various factors for maintenance policy selection. These factors include the category of maintenance activity, maintenance activity frequency, failure significance, maintenance activity cost, and aggregate production policy. The fundamental difference between Weinstein–Chung's approach and ours lies in the fact that our model takes, explicitly, into consideration the reliability parameters of the system at the early stage of the planning process, that is, when the aggregate plan is to be developed. As

7.2 No Shortage

Table 7.10 Specific notations (Ben-Daya 2002)

$H = N\tau$	Planning horizon (time)
N	Number of periods (integer)
τ	Each period fixed length (time)
D_{it}	Demand rate of *ith* item should be satisfied in each period t (item/period)
$C_{\max}$	Production system nominal capacity (item)
L_p	Capacity units consumed for each planned preventive (item)
L_r	Capacity units consumed for each unplanned maintenance (item)
ρ_i	The processing time for each unit of product i (time)
K_{it}	Fixed cost of producing item i in period t (\$/setup)
C_{it}	Variable cost of producing one unit of item i in period t (\$/item)
h_{it}	Variable cost of holding one unit of item i by the end of period t (\$/unit/unit of time)
C_r	Cost to carry out a corrective maintenance action ($C_{PM} < C_r$) (\$)
Q_{it}	Quantity of item i produced in period t (item)
I_{it}	Inventory of item i at the end of period t (item)
y_{it}	Binary variable (y_{it} equals to 1 if item i is produced in period t and 0 otherwise)
T	Preventive maintenance cycle (time)

a consequence, the chances that their model performs better at the simulation phase are higher (Aghezzaf et al. 2007).

The specific notations which are used for this problem is presented in Table 7.10.

They are given a planning horizon including N periods of fixed length and a set of products P to be produced during this planning horizon. They assumed that the production system has a known nominal capacity and that each maintenance action consumes a certain percentage of this capacity. Thus, they assumed that each planned preventive and unplanned maintenance action consumes, respectively, $L_p = aC_{\max}$ and $L_r = bC_{\max}$ capacity units (with $0 \leq a \leq b \leq 1$). Note that the assumption $a \leq b$ may be justified by the fact that more capacity resources may be consumed in case of a random failure since some offline activities for repair must in this case be accomplished online. Finally, it is assumed that the failure probability density function $f(t)$ and the cumulative distribution function $F(t)$ of the production system are known. They let $r(t)$ be the failure rate of a system at time t. It is well known that $r(t)$ is given by Aghezzaf et al. (2007):

$$r(t) = \frac{f(t)}{1 - F(t)} \qquad (7.66)$$

The maintenance policy suggests to replace the production system at predetermined instances $T = k\tau, 2k\tau, 3k\tau, \ldots$ and to carry out a minimal repair whenever an unplanned failure occurs. All maintenance actions are supposed to be perfectly performed (Aghezzaf et al. 2007):

Minimize $\sum_{t\in H}\sum_{i\in P}(K_{it}y_{it} + C_{it}Q_{it} + h_{it}I_{it}) + \frac{N\tau}{T}\left(C_{PM} + C_r\int_0^T r(t)dt\right)$ (7.67)

s.t.

$Q_{it} + I_{it-1} - I_{it} = D_{it}$ for $t \in H, i \in P$ (7.68)

$Q_{it} \le \left(\sum_{s\in H, s\ge t} D_{is}\right) y_{it}$ for $t \in H, i \in P$ (7.69)

$\sum_{i\in P} \rho_i Q_{it} \le C(t)$ for $t \in H$ (7.70)

$Q_{it}, I_{it}, T \ge 0;\quad y_{it} \in \{0,1\}$ for $t \in H, i \in P$ (7.71)

where the function $C(t)$ defines the available capacity in each period t. This capacity $C(t)$ is given by (Aghezzaf et al. 2007):

$$C(t) = C_{max} - L_p - L_r \int_0^\tau r(u + (t-1)\tau)du \quad (7.72)$$

if the preventive maintenance takes place in period t. For other period t, the capacity $C(t)$ is given by:

$$C(t) = C_{max} - L_r \int_0^\tau r(u + (t-1)\tau)du \quad (7.73)$$

To solve the above mathematical programming problem (PPM), the length of the planning horizon H is assumed, and the length of the preventive maintenance cycle T is given in multiples of the basic planning period duration τ (i.e., $H = N\tau$ and $T = k\tau$). Let $n_I = [N/k]$ if the ratio N/k is integer and $n_I = t[N/k] + 1$ otherwise (where $[N/k]$ is the highest integer smaller or equal than N/k).

The maintenance and planning model (PPM) can now be rewritten as follows (Aghezzaf et al. 2007):

Minimize $Z(k) = \sum_{n=1}^{n_I}\left(C_{PM} + \sum_{t=(n-1)k+1, t\le N}^{nk} C_r \int_0^\tau r(u + (t-(n-1)k-1)\tau)\right.$

$\left. + \sum_{i\in P}(K_{it}y_{it} + C_{it}Q_{it} + h_{it}I_{it})\right)$

(7.74)

s:t: : Constraints Eqs: (7.72) and (7.73), and :

7.2 No Shortage

Table 7.11 Products' periodic demand (Aghezzaf et al. 2007)

Period	D_{1t}	D_{2t}	Period	D_{1t}	D_{2t}
1	2	3	5	2	3
2	3	2	6	3	2
3	2	3	7	2	3
4	3	2	8	3	2

$$\sum_{i \in P} p_I Q_{it} \leq C(t) = \begin{cases} C_{\max} - L_P - L_\tau \int_0^\tau r(u + (t-1)\tau) & \text{if } t = (n-1)k + 1 \\ C_{\max} - L_\tau \int_0^\tau r(u + (t-1)\tau) & \text{if } t = (n-1)k + 2 \leq t \leq nk \end{cases}$$

(7.75)

The decision variables remain, for each product i and each period t, Q_{it}, I_{it}, and y_{it} together with the variable k which defines the optimal length of the preventive maintenance cycle T ($T = k\tau$). To determine the optimal values of production plan and the length T of the maintenance cycle, the following procedure has been used (Aghezzaf et al. 2007).

Step 1. Based on the value of k, determine n_1 and the corresponding maintenance cost function terms, and then determine the available capacities $C^k(t)$ in period t.

Step 2. Solve the resulting pure production planning problem (PPM$_r$ for fixed values of k) using any selected algorithm. In this case, the production planning problem using the mixed integer solver of CPLEX is solved.

Step 3. Compare the resulting values $Z(k)$, and select for the optimal preventive maintenance period size of the value k^* such that $Z(k^*) = \min_k \{Z(k)\}$.

The production plan associated with $Z(k^*)$ is selected as the final production plan (Aghezzaf et al. 2007).

Example 7.4 Aghezzaf et al. (2007) presented an example in which planning horizon composed of 8 production periods, each with an available maximal capacity of $C_{\max} = 15$. Two products are to be produced in lots so that the demands are satisfied. They assumed $h_{it} = 2$, $C_{it} = 5$, and $h_{it} = 2$ for $i = 1, 2$. Also Table 7.11 shows the setup, production, and holding costs for each product and the periodic demands of each product, respectively.

Table 7.12 shows the optimal plan for the two products without taking into account the capacity lost in maintenance (assuming that the system will not fail and does not require any preventive maintenance). The total cost for the optimal production plan is equal to 417 (Aghezzaf et al. 2007).

Now, one considers the preventive maintenance model with minimal repair at failure as the selected maintenance strategy of the production system with the following parameters: the cost of a preventive maintenance action is set to $C_{PM} = \$28$, and the cost of minimal repair action at failure is given by $C_r = \$35$. Aghezzaf et al. (2007) assumed that the system lifetime is distributed according to Gamma distribution with the parameters $G(\alpha = 2, \lambda = 1)$.

Table 7.12 Optimal production plan with maximum capacity (Aghezzaf et al. 2007)

		Q_{it}		I_{it}		y_{it}	
		Products i					
Periods t		1	2	1	2	1	2
1		5	10	3	7	1	1
2		0	0	0	5	0	0
3		7	0	5	2	1	0
4		0	0	2	0	0	0
5		0	10	0	7	0	1
6		8	0	5	5	1	0
7		0	0	3	2	0	0
8		0	0	0	0	0	0

Table 7.13 Specific notations (Salameh and Ghattas 2001)

Q	Buffer stock level (item)
T	Running time of the production unit per cycle (time)
t	Preventive maintenance time per cycle (time)
P	Buffer replenishment rate (units/unit time) (item/year)
D	Consumption rate from the buffer during t (units/year)
h	Buffer holding cost ($/unit/unit time)
$B(t)$	Shortage per preventive maintenance cycle t (item)

7.3 Backordering

7.3.1 Preventive Maintenance

Preventive maintenance, an essential element of the just-in-time structure, induced the idea of this problem. Performing regular preventive maintenance results in a shutdown of the production unit for a period of time to enhance the condition of the production unit to an acceptable level. During such interruption, a just-in-time buffer is needed so that normal operations will not be interrupted. The optimum just-in-time buffer level is determined by trading off the holding cost per unit per unit of time and the shortage cost per unit of time such that their sum is minimum (Salameh and Ghattas 2001).

The specific notations which are used for this problem is presented in Table 7.13.

The main goal of Salameh and Ghattas (2001) was to determine the optimum just-in-time buffer level to withstand the regular preventive maintenance interruption. The following assumptions are applied throughout this problem:

- The just-in-time buffer is not subject to deterioration or obsolescence.
- The regular preventive maintenance guarantees that the probability of a breakdown of the production unit during T is approximately zero.
- Before the beginning of any normal preventive maintenance, the just-in-time buffer is Q.
- T is large enough compared with t, so that during any time period T, buffer replenishment starts from a zero level.

7.3 Backordering

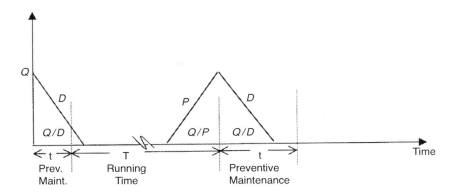

Fig. 7.5 Buffer level behavior (Salameh and Ghattas 2001)

- Unused buffer inventory during t is depleted to zero during the next cycle T.

The behavior of the system is depicted in Fig. 7.5. Defining $I(t)$ to be the average inventory level during the period $(t + T)$, then:

$$I(t) = \left[\frac{Q^2}{2D} + \frac{Q^2}{2P}\right]\left[\frac{1}{t+T}\right] \qquad (7.76)$$

Therefore, the expected average inventory per cycle of length $(t + T)$ and hence per unit of time will be:

$$E[I(t)] = \left[\frac{P+D}{PD}\right]\frac{Q^2}{2}E\left(\frac{1}{t+T}\right) = \left[\frac{Q^2}{2D} + \frac{Q^2}{2P}\right]\int_0^\infty \frac{f(t)}{t+T}dt \qquad (7.77)$$

where:

$$E\left(\frac{1}{t+T}\right) = \int_0^\infty \frac{f(t)}{t+T}dt$$

Therefore, the average buffer carrying cost will be:

$$\frac{hQ^2(D+P)}{2PD}E\left(\frac{1}{t+T}\right)$$

If the buffer supply time Q/D is less than the preventive maintenance time, then the stock out time will be $t - Q/P$; otherwise, the stock out time will be zero, i.e.:

$$\text{stock out time} = \begin{cases} 0 & t \leq \frac{Q}{D} \\ t - \frac{Q}{D} & t > \frac{Q}{D} \end{cases}$$

Therefore, the shortage per preventive maintenance cycle can be expressed as:

$$B(t) = \begin{cases} 0 & t \leq \frac{Q}{D} \\ D\left(t - \frac{Q}{D}\right) = Dt - Q & t > \frac{Q}{D} \end{cases} \qquad (7.78)$$

The expected shortage per preventive maintenance cycle t is:

$$E[B(t)] = D \int_{Q/D}^{\infty} \left(t - \frac{Q}{D}\right) f(t) dt \qquad (7.79)$$

The expected cycle length is:

$$E(t + T) = T + E(t) \qquad (7.80)$$

Therefore, the expected number of cycles per unit time is $1/T + E(t)$, and the expected number of units short per unit time will be:

$$\frac{E[B(t)]}{E(t+T)} = \frac{D}{T + E(t)} \int_{S/D}^{\infty} \left(t - \frac{Q}{D}\right) f(t) dt \qquad (7.81)$$

Hence, the storage cost per unit time is:

$$C_b \frac{D}{T + E(t)} \int_{Q/D}^{\infty} \left(t - \frac{Q}{D}\right) f(t) dt$$

Let TCU(S) be the expected total cost per unit time. TCU(S) can be expressed as:

$$\text{TCU}(Q) = \frac{hQ^2(P+D)}{2PD} E\left(\frac{1}{T+t}\right) + \frac{C_b D}{T + E(t)} \int_{Q/D}^{\infty} \left(t - \frac{Q}{D}\right) f(t) dt \qquad (7.82)$$

Differentiating Eq. (7.82) with respect to the buffer stock level Q yields to:

$$\frac{d\text{TCU}(Q)}{dQ} = \frac{hQ}{PD}(P+D) E\left(\frac{1}{T+t}\right) - \frac{C_b D}{T + E(t)} \int_{Q/D}^{\infty} \left(t - \frac{Q}{D}\right) f(t) dt \qquad (7.83)$$

7.3 Backordering

$$\frac{d^2 \text{TCU}(Q)}{dQ^2} = \frac{hQ}{PD}(P+D)E\left(\frac{1}{T+t}\right) + \frac{C_b D}{D(T+E(t))} f\left(\frac{Q}{D}\right) \quad (7.84)$$

$d^2\text{TCU}(Q)/dQ^2 \geq 0$ for all $Q \geq 0$. Hence, TCU(Q) admits a unique minimum at $Q = Q^*$, where:

$$\left.\frac{d\text{TCU}(Q)}{dQ}\right|_{Q=Q^*} = 0$$

or:

$$\int_{Q/D}^{\infty} f(t)dt = \frac{hQ(P+D)(T+E(t))}{C_b PD} E\left(\frac{1}{T+t}\right) \quad (7.85)$$

Example 7.5 To illustrate the use of the model developed in previous section, consider a situation where $T = 30$ days, $P = 15{,}000$ units/year, $D = 30{,}000$ units/year, $h = \$28.00$/unit/year, and $C_b = \$1.00$/unit. Preventive maintenance time t is uniformly distributed over the interval $[a, b]$, with $a = 1$ day and $b = 3$ days. Therefore:

$$f(t) = \begin{cases} 1/2 & 1 \leq t \leq 3 \\ 0 & \text{otherwise} \end{cases}$$

The total cost function in Eq. (7.82) reduces to:

$$\text{TCU}(S) = \frac{h(P+D)}{2PD(b-a)} \ln\left(\frac{T+b}{T+a}\right) Q^2 + \frac{2C_b D}{(2T+a+b)(b-a)}$$

$$\times \left(\frac{b^2}{2} - \frac{b}{D}Q + \frac{Q^2}{2D^2}\right) \quad (7.86)$$

The optimal value of Q that minimizes the above equation is $Q^* = 170.44$ units, with $\text{TCU}(Q^*) = \$671.31$. For $Q = 0$ (no just-in-time buffer), the expected total annual cost is $\text{TCU}(0) = \$2109.38$. The minimum just-in-time buffer level that reduces the probability of stock out during preventive maintenance to zero can be obtained by setting Eq. (7.79) equal to zero and solving for S. In this example, this minimum value of Q is 250 units. The corresponding expected annual cost is $\$984.70$. The above results are summarized in Table 7.14.

Table 7.14 Results of example

	Q	TCU(Q)
No buffer	0	$2109.38
Optimal buffer level	170.44	$671.31
Minimum buffer such that the probability of stock out is zero	250	$984.70

7.4 Partial Backordering

7.4.1 Preventive Maintenance

Taleizadeh (2018) developed a multi-product single-machine economic production quantity model with preventive maintenance and scraped and rework process. Shortages are permitted and a fraction of them is backlogged. Capacity and service level are limitations of the production system. It is assumed that preventive maintenance can be performed when the inventory level is positive or negative. Indeed, two different scenarios are modeled, and according to the comparisons between their costs, a new scenario according to the best time of preventive maintenance is investigated and modeled. The aim of this problem is to determine the best time for preventive maintenance, production, and backordered quantities of each item and common cycle length, such that the expected total cost is minimized. The more details of this study can be found in Sect. 5.5.5 of Chap. 5.

7.5 Conclusion

In this chapter, six inventory models with maintenance are presented. These models consider the traditional production problem in which preventive maintenance is considered. All models are categorized in three sections about the type of shortages including no shortage, backordering, or partial backordering. Preventive maintenance, imperfect preventive maintenance, imperfect maintenance and production, aggregate production, and maintenance planning are the main topics investigated and differ models in this chapter.

References

Aghezzaf, E. H., Jamali, M. A., & Ait-Kadi, D. (2007). An integrated production and preventive maintenance planning model. *European Journal of Operational Research, 181*(2), 679–685.

Ben-Daya, M. (2002). The economic production lot-sizing problem with imperfect production processes and imperfect maintenance. *International Journal of Production Economics, 76*(3), 257–264.

Banerjee, P. K., & Rahim, M. A. (1988). Economic design of control charts under Weibull shock models. *Technometrics, 30*(4), 407–414.

References

Ben-Daya, M., & Duffuaa, S. O. (1995). Maintenance and quality: the missing link. *Journal of Quality in Maintenance Engineering, 1*(1), 20–26.

Ben Daya, M., & Makhdoom, M. (1998). Integrated production and quality model under various preventive maintenance policies. *Journal of the Operational Research Society, 49*, 840–853.

Hooke, R., & Jeeves, T. A. (1961). "Direct Search'" solution of numerical and statistical problems. *Journal of the ACM (JACM), 8*(2), 212–229.

Nakagawa, T. (1980). A summary of imperfect preventive maintenance policies with minimal repair. *RAIRO-Operations Research, 14*(3), 249–255.

Pham, H., & Wang, H. (1996). Imperfect maintenance. *European Journal of Operational Research, 94*(3), 425–438.

Ross, S. M. (2013). *Applied probability models with optimization applications*. Chelmsford: Courier Corporation.

Ross, S. M. (1996). *Stochastic processes* (2nd ed.). New York: Wiley.

Salameh, M. K., & Ghattas, R. E. (2001). Optimal just-in-time buffer inventory for regular preventive maintenance. *International Journal of Production Economics, 74*, 157–161.

Sheu, S. H., & Chen, J. A. (2004). Optimal lot-sizing problem with imperfect maintenance and imperfect production. *International Journal of Systems Sciences, 35*, 69–77.

Taleizadeh, A. A. (2018). A constrained integrated imperfect manufacturing-inventory system with preventive maintenance and partial backordering. *Annals of Operations Research, 261*, 303–337.

Weinstein, L., & Chung, C. H. (1999). Integrating maintenance and production decisions in a hierarchical production planning environment. *Computers & Operations Research, 26*(10–11), 1059–1074.

Index

A
Aggregate planning, 574
Aggregate production, 574, 575, 582
Algebraic manipulation, 121
Annual rework cost, 452
Annual total inventory, 387
Arithmetic–geometric mean inequality method, 186–188
Auction, 385–390
Average profit, 116

B
Backorder case, 33
Backorder cost (BC), 94, 98, 101, 145
Backordering
    and rework (*see* Rework and backordering)
    backorder costs, 261
    cost function, 267
    defective products, 268
    EPQ model, 261
    imperfect quality and inspection, 70–74
    imperfect-quality items, 109–112
    inventory average, 263
    learning in inspection, 101–109
    mathematical expression, 263, 267
    multiple quality characteristic screening, 74–82
    notations, 70, 71
    partial (*see* Partial backordering)
    partial derivatives, 266
    preventive maintenance, 578–580
    rejection, 82–87
    second partial derivatives, 265
    solution, 270
    total cost function, 264
Backordering cost (BC), 90, 101, 102, 109
    and lost sale costs, 113
Backordering level, 274
Backordering, multi-product single-machine EPQ model
    defective items
        annual production cost, 402
        annual shortage cost, 405
        data, 408, 409
        joint production policy, 403–405
        normal distribution, 410
        objective function, 402, 405, 407
        optimal production quantity, 401
        production cost per unit, 402
        production–inventory cycle, 403
        production rate, 401, 407
        scrapped, 401, 404
        shortages, 401
        total production and setup time, 406
        uniform distribution, 409
    interruption, manufacturing process (*see* EPQ inventory model with interruption, manufacturing process)
    multidefective types (*see* Multidefective types)
    repair failure, 443, 444, 446–448
    rework proces, 433–441
Backordering period, 118
Backorders, 30, 34, 36
Batches
    imperfect-quality, 13
    maximum purchasing price, 38
    multiple, 13
    received, 37

© Springer Nature Switzerland AG 2021
A. A. Taleizadeh, *Imperfect Inventory Systems*,
https://doi.org/10.1007/978-3-030-56974-7

Batches (*cont.*)
  rejection of defective supply batches, 16–18
Behavior of inventory level, 93
Beta-binomial distribution, 51
Binomial distribution, 49, 51
Buy and repair options, 56–61
Buyer makes decisions, 50
Buyer's purchasing price, 500
Buyer's selling price, 37, 39–42, 500

C
Capacity constraint model, 321, 384, 439, 447, 454, 466, 475, 488
Classical EOQ model, 244
Commodity, 62
Commodity flow, 62, 68, 69
Complete backordering, 124
Computer numerical control (CNC), 492
Continuous delivery, EPQ models with scrap
  cyclic inventory cost, 158
  expected annual cost, 158
  fully backordered
    arithmetic–geometric mean inequality method, 186–189
    designer window treatments manufacturer, 199, 200, 202
    random breakdown, 196, 197, 200, 204, 207, 208
    random defective rate, 189–191, 193–199
  HGA *vs*. GAMS software, 168
  imperfect production process, 155
  imperfect-quality items, 156
  manufactured product, 159
  MINLP problem, 164, 166
  multi-product and multi-machine, 160–164
  on-hand inventory, 158
  optimal production quantity, 159
  probability density function, 158
  production downtime, 157
  production rate, perfect-quality items, 156
  production uptime, 157
  proposed HGA, 165, 168
  solution procedures, 157
  with partial backordered shortage, 223–230
Convex nonlinear programming problem (CNLPP), 389, 469
Corrective maintenance, 575
Cost-efficiency, 125
Cost equation, 292
Cost function, 63
Cumulative number, 28
Cutting plane method, 380, 381
Cycle duration, 100

Cycle time, 93, 96
Cyclic backordering cost, 482
Cyclic inventory-related cost, 84

D
Defective
  and defective delivery occurrences, 82
  and deterioration costs, 20
  fraction, 74
  and good items, 18, 24
  HC, 25, 27, 78
  inventory level, 19, 57, 60, 76, 77
  learning curve, 101
  non-inspected items, 53
  number and portion, 75
  percentage, 7, 70, 71
  produced items, 17
  random proportion, 10
  supply batches, 82–87
  uninspected items, 50
Defective items, 311, 313, 383, 435, 443, 501
  cost expression, 257
  on-hand inventory, 255, 256
  production, 255, 258
  reworked, 259
Defective products, 370, 385, 392
  decision variables, 268
  inventory level, 269
  notation and analysis, 271
  optimal backordering level, 276
  optimal total cost, 273
  production–inventory model, 269
  production time, 272
  total cycle time, 271
Demand rate, 1
Deterioration, 8, 10, 18, 19, 23
Different values, 107
Discrete delivery, EPQ models with scrap
  machine breakdown
    buyer's cost, 177, 178, 180, 181
    buyer's inventory level, 173
    cost of transportation, 173
    expected production–inventory cost per cycle, 180
    on-hand inventories, perfect and defective vendor items, 173
    optimal run time, 185
    random variable, 173
    stochastic breakdown and multiple shipments, 173
    total cost function, 185
    vendor's cost, 170, 173–176, 179
    with and without breakdowns, 175, 181, 182, 184

Index 587

machine failures, 185
multi-delivery
   algebraic approach, 170
   finished items, 166
   integer number, 172
   optimal number of shipments, 172
   optimal replenishment lot size, 172
   random scrap rate, 172
   randomness of scrap rate, 169
   supplier's inventory holding during delivery time, 168
   total holding costs, 169
   total production–inventory–delivery costs per cycle, 166
   produced items, 166
   vendor–buyer integrated EPQ model with scrap, 166
Discrete shipments, 153, 154
Disposal cost, 18
Distribution function, 50, 87, 294

E

Economic frequency, 373
Economic lot size model, 1
Economic manufacturing quantity (EMQ), 312
Economic operating policy, 375
Economic order quantity (EOQ) model, 2, 4, 50, 147
   assumptions, 498
   buy and repair options, 56–61
   cost function, 62
   formulation, 7
   holding and ordering costs, 153
   inspection shifts from buyer to supplier, 37–42
   inventory models, 153
   learning in inspection, 506
   maintenance, 16–18
   partial backordering, 506–510
   quality levels with backordering, 502–505
   reworkable items, 153
   robust production–inventory model, 153
   sampling inspection plans, 43–50
   scrapped items, 153, 155
   traditional models, 497
   with no return without shortage, 502
Economic policy, 374
Economic production quantity (EPQ), 92, 99, 324, 561
   breakdown, 1
   economic lot size model, 1
   inventory models, 2, 3
   manufacturing system, 2
   multi-product single machine, 3, 4

production model, 367
rework process, 3
single machine, 367
single-product single-stage manufacturing system, 367
traditional model, 1
End-of-cycle backorders, 84
End-of-cycle inventory, 82
Entropy cost, 64–65, 68, 69
Entropy EOQ
   with screening, 65–70
   without screening, 61–65
EOQ imperfect system, 506
EOQ model with return
   defective quality levels and partial backordering, 525–530, 533, 534
   Hessian matrix, 519
   holding costs per cycle, 517
   holding costs, defective items, 520
   imperfect inspection process, 522–525
   inspection and sampling, 515–517
   NBBARY algorithm, 522
   purchase cost per cycle, 520
EPQ inventory model with interruption, manufacturing process
   annual joint production policy, 429
   backlogging situation, 430
   backorder-filling stage, 421
   data, 432, 433
   decision variables, 419, 431
   defective products, 419
      holding cost, 423–426
      shortage cost, 424
   imperfect products, 419, 422
   inventory level, perfect-quality items, 421
   inventory system cost, 421
   no shortages, 421, 422, 424, 426
   objective function, 431
   on-hand inventory
      defective items, 426
      scrapped items, 427
   optimal solution, 431
   production capacity, 429
   production cycle length, 422
   random defective rate, 421
   reasons, 421
   rework process, 425
   shortage cost, 423
   shortages, 419, 422, 426
   total cost function per cycle, 428
   total defective products, 427
   total objective cost function, 427
   total scrap products, 427
EPQ inventory models, 2, 3
EPQ inventory system, 329

EPQ model, 235, 248
　categories, 236
　classical, 237
　EOQ, 237, 238
　GP, 238
　notations, 236, 237
　Porteus's model, 238
EPQ model with return
　continuous quality characteristic without shortage, 537–542
　inspections errors without shortage, 542–546
　sensitivity analysis, 542
EPQ model without return
　quality assurance without shortage, 535, 536
　quality screening and rework without shortage, 536
Equipment condition, 549
Equivalent number, 28, 29
Error type I, 500, 501, 515, 516, 522
Error type II, 500, 515–517, 522
Expectation value, 124
Expected cost, 82
Expected cycle time, 36
Expected profit per cycle, 37
Expected total cost (ETC), 305
Expected total profit (ETP), 240
Expected total profit per time unit, 38
Extended cutting plane method, 379, 380

F

Failure rate, 555, 557, 560, 575
Fill rate, 118, 123
Finite planning horizon model, 106
Finite production model, 307
First-order condition, 123
First-order derivatives, 85, 106
Fixed backorder cost, 261
Fixed cost, 87, 88
Fixed lifetime, 8
Fixed setup, 89, 94, 97
Fixed transportation cost, 9, 56, 57, 75, 87, 88, 126, 128

G

General and specific data, 324
Geometric programming (GP) approach, 238
Geometric random variable, 84
Geometric series, 45
Goodwill cost, 10, 18, 23–27, 72, 77, 81

H

Harmony search algorithm (HS), 379
Heat flow, 62
Hessian matrix, 54, 91, 92, 95, 96, 111, 112, 117, 311, 316, 322, 330, 407, 431, 447, 456, 469, 491
Holding cost (HC), 10, 11, 13, 15, 18, 20, 23–28, 30, 31, 35, 36, 51, 53, 57, 58, 63, 77, 78, 87, 88, 90, 94, 98, 101, 102, 114, 127–129, 131, 134, 137, 140
　order quantities, 27
　with learning effects, 27
　without learning effects, 27
Hypergeometric distribution, 51

I

Imperfect EOQ model
　categories, 7, 8
　deterioration, 8
　HC, 10
　imperfect quality, 10–15
　inventory control purposes, 9
　JIT manufacturing environment, 10
　lifetime constraints, 8
　notations, 7, 9
　obsolescence, 8
　perishability, 8
　probability distribution, 10
　random proportion, 10
　real manufacturing environment, 10
　vendor–buyer inventory policy, 10
Imperfect inventory system, 56
Imperfect item sales, 242
　differentiating and equating, 243
　inspection process, 239
　inventory level per cycle, 240, 241
　production period, 243
Imperfect maintenance
　expected PM cost per cycle, 571
　expected production cycle length, 570
　inspection intervals, 573
　integrated model, 573
　inventory holding cost, 571
　non-conforming items, 572
　pattern search technique, 573
　PM activities, 569
　process shift mechanism, 574
　and production process, 569
　renewal reward process, 573
　restoration cost per production cycle, 572
　specific notations, 570

Index

Imperfect preventive maintenance (PM)
   associated probabilities, 565, 567
   classical EPQ model, 564
   expected cost, 564, 565
   expected inventory cycle length, 564
   expected restoration cost per cycle, 568
   inspection cost, 566
   inventory holding cost, 565
   non-conforming items, 566
   process shift mechanism, 568
   production cycle, 561, 564
   quality control costs, 569
   restoration cost, 567
   specific notations, 564
Imperfect production processes, 223
Imperfect production system, 236
Imperfect products, 409
   partial backordering
      cycle time, 128, 136
      decision variables, 140
      demand, 127
      demand and inspection rates, 128
      expected total profit, 140
      fraction of cycle time, 134
      HC, 128, 129, 131, 134, 137
      identified and studied, 127
      optimal cycle time, 128
      optimal policy, 140
      optimal values, decision variables, 135
      process failure, 126
      profit function, 137
      repair cost, 134
      repair process, 128
      SC, 129, 132, 134, 137
      screening period, 127
      total cost, 127
      TP, 129, 132, 135
      uniform distribution, 140
      variable cost, 127
Imperfect quality, 7, 10, 18, 25, 56
   expected profit, 11, 12
   fixed cost, 13
   fixed ordering cost, 11
   and inspection, 70–74
   inventory level, 11
   optimal value, 14
   perfect and imperfect inventory levels, 13, 15
   perfect-quality item, 11
   probability, 82
   screening process, 11
Imperfect rework process, 324
   defective items, 251

   EPQ model, 249
   notations, 250, 312
   optimal production, 253
   perfect-quality items, 314
   practical production process, 249
   production uptime, 250
   scrap items, 252
   solution procedures, 249
Imperfect-quality EMQ model, 313
Imperfect-quality items, 69, 235, 239, 246, 252, 308
In control, 553, 565–567
Inspection
   with backordering, 506
   buyer's responsibility and supplier pays, 507
   economic production model, 536
   errors, 542, 543
   imperfect inspection process, 522
   and rework processes, 503
   and sampling, 515, 516
Inspection cost, 47, 166, 193, 223
Inspection cost per time unit, 90
Inspection errors, 501, 517, 522, 523, 525, 537, 542, 543
Inspection process, 239
Inspection/screening period, 60
Instantaneous replenishment
   acceptance sampling plan, 50
   AOQL, 55, 56
   average inventory per cycle, 52
   beta-binomial distribution, 51
   binomial distribution, 49
   Hessian matrix, 54
   integer nonlinear program, 54
   inventory behavior, 52
   inventory-related costs, 51, 52
   notations, 50, 51
   parameters, 55
   probability of acceptance, 55
   quality-related cost, 53
   renewal–reward theorem, 52
   solution method, 55
   varying, 55, 56
Integer nonlinear program, 54
Integrated model, 303
Integrated procurement–production–inventory system, 208–216
Inventories
   control systems, 1
   EPQ inventory models, 2
   inventory order quantity models, 2
   managing inventories, 1

Inventories (*cont.*)
  variables, 4
Inventory average, 287
Inventory control problem, 331, 450
Inventory cost analysis, 247
Inventory cost per cycle, 252
Inventory cycle, 88, 97, 302
Inventory cycle length, 46
Inventory level, 129, 132, 134, 136, 142, 143
  behavior, 11, 51, 52, 70, 71, 88, 89, 93, 97, 101, 103, 109, 134
  cycle time, 128
  defective items, 76, 77
  differential equation, 114
  dynamics, 51
  FTD, 131
  graphic representation, 119
  maximum, 97, 111
  over time, 24
  perfect and imperfect, 15
  problem on hand, 24
  repair option, 57, 60
  repaired products, 87, 127
  replenishment cycle, 75, 76
  starting and ending, 82
  variation, 19, 104
  zero, 11, 30, 93
Inventory level, EPQ model, 376
Inventory management, 76
Inventory model, 358, 500
Inventory order quantity models, 2
Inventory-related costs, 51, 52
Investment cost, 508, 512, 513

J

Joint production policy, 383
Joint production system, 321
Just-in-time (JIT) manufacturing environment, 10

L

Lagrangian function, 39–41
Lagrangian relaxation method, 394, 397
Learning curve, 101
Learning effects
  EOQ model, 26
  HC, 23–28
  in inspection with backorders, 34
  in inspection with lost sales, 31
  notations, 29
  transfer of learning, 28–37

Linear backorder cost, 261
Lost profit, 387
Lost sale cost, 30, 31, 34, 62, 113, 125, 128, 129, 227, 453
Lower specification level (LSL), 537

M

Machine repair, 154, 173, 179, 200
Maintenance, 4, 16–18
  aggregate production and maintenance planning, 574
  imperfect maintenance and process (*see* Imperfect maintenance)
  imperfect preventive maintenance, 561
  preventive (*see* Preventive maintenance (PM))
  production and quality, 549
  production planning models, 549
Maintenance and planning model (PPM), 576
Maintenance planning, 549, 574, 575, 582
Maintenance policy, 575
Management systems, 113
Manufacturing processes, 4
  defective products, 497
Manufacturing system, 357
Marketing, 1
Markovian shock model, 568
Material cost, 64
Materials, 1, 2
Mathematica 6.0, 106
Mathematical equation, 287, 291
Mathematical expressions, 98
Mathematical model, 382, 386
Mathematical modeling and analysis, 235
MATLAB, 397
Maximum limit/minimum limit, 49
Maximum purchasing price, 42
Mean of yield, 508, 512
Mixed integer nonlinear programming (MINLP), 379
Multidefective types
  annual backorder cost, 413
  annual constant production rate, 409
  annual holding cost, 413
  annual production cost, 413
  backorder level, 415
  budget, 414
  capacity, single machine, 414
  cycle length, 412
  EPQ model, 415
  imperfect single-machine production system, 416

Index

inequalities, 411, 415
inventory and shortages, product, 411, 412
inventory model, 411
objective function, 415
optimal cycle length, 415
parameter types, 409
perfect and imperfect-quality products, 409
rework production rate, 410
rework rates, 409
reworks, 413
scrapped items, 411
service level, 414
shortage quantity, 417
space, 413
total shortage quantity/safety factor, 415
total warehouse construction cost, 414
Multi-delivery, 166, 168

See also Discrete delivery, EPQ models with scrap

Multi-delivery policy
  costs per cycle, 333
  defective items, 332, 334
  on-hand inventory, 332
  notations, 335
  optimal production, 335
  partial rework
    defective items, 339
    EPQ model, 337
    non-conforming items, 339
    production quantity, 339
    proposed model, 337
  and quality assurance, 332
  rework, 332
Multi-delivery single machine
  algebraic derivation, 344
  defective rate, 343
  non-conforming items, 340
  notations, 342
  on-hand perfect-quality inventory, 341
Multi-item production system, 392, 393
Multi-product and multi-machine, 160–164
Multi-product case, 382
Multi-product inventory control problem, 231
Multi-product model, 468
Multi-product single-machine EPQ model
  backordering (*see* Backordering, multi-product single-machine EPQ model)
  capacity, 368
  categories, 367, 368
  classification, 367
  decision variables, 370
  defective products, 370

features, reviewed studies, 372
mathematical model, 371
non-conforming items, 370
no shortage
  auction, 385–391
  discrete deliveries, 375, 377, 379, 380
  rework, 380, 382, 383, 385
  scrapped, 391, 392, 394–397, 399, 400
  simple model, 371, 373–375
optimal cycle length, 370
optimal production quantity, 370
partial backordering shortage (*see* Partial backordering shortage, multi-product single-machine EPQ model)
preventive maintenance, 371
production capacity limitation, 370, 371
production quantities, products, 368
random defective production rate, 371
service level, 370, 371
shortages, 370
space, 368
total annual cost, 371
total budget, 370
trade credit policy, 371
Multi-product single-machine production system, 223–230
Multi-product two-machine
  fabrication costs, 349
  notations, 345
  on-hand inventory level, 346
  parameters, 351
  quality screening, 345
Multi-products single manufacturing system, 3

N

Negative inventory, 113
Non-conforming item, 552, 564, 566, 567, 570–572
Non-defective items, 310, 380, 441
Nonlinear programming model (NLPP), 70
Non-Markov shock model, 305
Non-shortage cycle, 481

O

Objective function, 384, 385, 389, 391, 467
Obsolescence, 8
On-hand inventory, 246
Optimal cost, 64
Optimal decision variables, 79
Optimal lot size, 501
Optimal order quantity, 50

Optimal PM level, 568, 573, 577, 578
Optimal policy, 61, 123, 125
Optimal production, 248, 249, 312, 351, 560
Optimal replenishment policy, 500
Optimal solution, 501
Optimal values, 244, 309
Optimization software, 165
Order quantity, 500
Ordering cost, 11, 15, 51, 53, 102
Ordering cycle, 11, 13, 37, 120, 121
Out of control, 550, 552, 553, 564, 565, 567

P
Partial backordering, 506–510, 582
    economic manufacturing model, 318
    EMQ model, 318
    EOQ model
        imperfect products (*see* Imperfect products)
        imperfect-quality products, 112–118
        replacement, imperfect products, 142–147
        screening, 118–126
    joint production systems, 321
    modeling procedure, 318
    non-defective and defective items, 318
    partial differentiations, 323
    policy production, 320
    production and demand rates, 318
    production capacity limitation, 323
    production cost, 319
    single product problem, 319
Partial backordering shortage, multi-product single-machine EPQ model
    preventive maintenance, 479–481, 483–492
    repair failure, 460
    rework, 449–452, 454–457, 459, 470, 471, 473–475, 477
    scrapped, 461–467, 469, 470
Partial derivatives, 447
Particle swarm optimization (PSO), 379
Penalty cost
    shortage, 78
Penalty cost purchasing cost, 102
Perfect-quality item, 70, 245, 250, 275
Perishability, 8, 9
Pharmaceutical companies, 190
Physical thermodynamic system, 61
Planning horizon, 574–577
Positive inventory, 294
Practical production systems, 300

Preventive maintenance (PM), 371, 479–481, 483–492
    age reduction and PM cost, 556
    computer program coded in Fortran, 561
    cost functions, 553
    expected maintenance cost, 556, 557, 559
    expected production cycle length, 553
    expected shortage, 580
    imperfect maintenance, 555
    imperfect process, 561
    integrated model and solution method, 559, 560
    inventory levels, 554, 555
    joint production, 551
    just-in-time structure, 578
    maintenance, 551
    minimal repair, 577
    optimal preventive maintenance level, 560
    optimum just-in-time buffer level, 578
    out of control, 550
    partial backordering, 582
    PM effect, 551
    PM level for linear improvement, 562
    production cycle, 550, 552
    quality control cost, 554, 558
    quality control inspections and PM tasks, 557
    quality inspection activities, 552
    quality model, 551
    specific notations, 550, 552, 578
    system changes, 555
    total expected inventory holding cost, 557
Price–demand relationship, 39–41
Probability, 82
Probability density function, 43, 47, 49, 50, 70, 80, 81, 109, 112, 117, 315, 339
Probability distribution, 10, 239, 260
Probability function, 23
Product
    handling, 9
    transportation, 9
Product cycle, 449
Product deterioration, 8
Product flows, 2
Production cost, 94, 97
Production cycle length, 325, 450
Production–inventory cycle, 225, 403
Production inventory model, 295
Production–inventory situation, 238
Production operation system, 569
Production period, 93
Production planning, 549

Index   593

Production process, 239
Production rate, 2, 401
Production system, 61, 304
Products
    categorization, 7
Profit function, 100, 357
Proposed manufacturing problem, 357
Purchase cost, 118

Q
QCU, 53
Quality
    continuous quality characteristic, 537–540
    defective levels and partial backordering, 525–527
    economic order quantity model, 500
    EPQ models, 501, 537
    imperfect-quality items, 500
    inventory models, 500
    investments, 498
    quality assurance without shortage, 535, 536
    quality screening and rework without shortage, 536
    with backordering, 502–505
Quality-related cost, 53
Quality screening
    defective items, 255
    notations, 255
    production and screening, 255
    production quantity model, 254

R
Random defective rates, 344
    backorders, 283
    beta distribution, 289
    cost equation, 284, 292
    cyclic backordering, 283
    mathematical equation, 282
    maximum inventory level, 280
    production process, 307
    production time, 280
    uniform distribution, 277
Random variable, 71, 82, 85, 92, 99
Real manufacturing environment, 23
Rejection
    defective supply batches, 82–87
Renewal reward theorem, 52, 72, 79, 85, 115, 247, 252
Repair, 552, 575, 577

Repair cost, 87, 134, 552, 553
Repair failure, 371, 443, 444, 446–448, 460, 461
    defective items, 325
    material, 324
    modeling procedure, 325
    notations, 328
    production capacity, 324
    random defective rate, 328
    rework process, 325
    traditional EPQ model, 324
Replenishment cycle, 76, 103, 104
Replenishment policy, 107
Restoration cost, 564, 567, 568, 572, 573
Return policy, 501
Rework, 3
Rework and backordering
    avoiding interruptions, 87
    BC, 90
    distribution function, 87
    fixed cost, 87, 88
    HC, 87, 90
    Hessian matrix, 91, 92
    inventory cycle, 88
    production cycle, 87
    production period, 87, 88
    random variable, 87
    repair cost, 87
    total cost, 87
    total profit, 91
    variable cost, 87
Rework policy, 296
    and preventive maintenance
        inventory control systems, 300
        inventory system, 300
        linear and nonlinear relationships, 304
        non-conforming items, 301
        production cycle, 300, 301, 304
        profit function, 306
        quantity, 303
    EPQ model, 244, 298
    inventory cost per cycle, 246
    parameter values, 300
    perfect-quality items, 245
    production rate, 245
    regular production process, 245
    setup cost, 299
Rework process, 313, 501, 503, 538, 542–545
Rework process and scraps
    backordering, 294
    EPQ model, 293
    optimal solutions, 296

Rework process and scraps (*cont.*)
   production cycle time, 294
   total cost function, 295
Reworkable items, 357, 500, 502, 503, 538, 545

S
Salameh and Jaber's model, 66
Sales revenue per time unit, 89
Sample sizes, 500, 515, 516, 522
Sampling, 515–517
Sampling inspection plans
   buyer draws, 43
   decisions, 43
   economic order quantity, 43
   notations, 43, 44
   $p<p_0$, 45–47
   $p>p_1$, 44, 46, 47
   $p_0<p<p_1$, 44, 45, 47
   probability density function, 43
Schedule shortage period, 272
Scrap, 2, 3, 380
Scrap and rework costs, 235
Scrap items, 315, 401, 404, 411
   categories, 155
   defective items, 153
   EPQ model, 153
   no shortage
      continuous delivery (*see* Continuous delivery, EPQ models with scrap)
      cyclic inventory cost, 158
      discrete delivery (*see* Discrete delivery, EPQ models with scrap)
      manufactured product, 159
      multi-product and multi-machine, 160, 162–164
      steady production rate, 155
   steady production rate, 155
Screening
   entropy EOQ with, 65–70
   entropy EOQ without, 61–65
   multiple quality characteristics, 74–82
   partial backordering, 118–126
Screening cost, 102, 118
Screening process, 10, 11, 51, 65, 70, 74, 75, 81, 82, 109, 118, 123, 128
   deterioration and demand, 19
   HC, 18
   input parameters, 22, 23
   inventory variation of EOQ model, 19, 20
   monotony constraint, 21
   notations, 19

numerical results, 83
optimal results, 22, 23
optimal values, 22
parameters, 82
unconstrained problem, 21
Screening rate, 81
Screening time, 101
Second law of thermodynamics, 62
Second partial derivatives, TP, 103
Second-order derivatives, 86
Second-order Hessian matrix, 284
Second-order partial derivatives, 79
Second-order sufficient conditions (SOSC), 111
Selling price, 70, 156, 193
Service level, 454, 461, 470, 489
Service level constraint
   on EPQ model with random defective rate, 217–222
Shipment cost, 15, 26, 28, 101, 104–106, 109
Shipment decisions
   machine utilization, 352
   multi-delivery policy, 353
   on-hand inventory, 352
   renewal reward theorem, 355
   rework process, 352
Shortage cost (SC), 73, 129, 132, 134, 137
Shortage period, 118, 123
Shortage quantity, 96
Shortages, 2, 3, 401
Single-item production process, 300
Single-machine production system, 390
Single-product case, 382
Single-product single-stage manufacturing system, 3
Solution procedure, 361
S-shaped logistic learning curve, 26
Start-of-cycle inventory, 82, 84
Stock-out period, 29, 113
Supply chain, 2, 7

T
Taylor series expansion, 81
Temperature changes, 62
Thermodynamic law, 68
Thermodynamic system, 62, 68
Time-varying demand
   and product deterioration, 8
Total annual holding cost, 378
Total backorder cost, 452
Total cost function, 72, 374, 377, 451
Total costs (TC), 24, 87, 91, 98, 102, 104, 120

Index 595

Total cost per cycle, 58
Total forgetting, 33
Total inventory cost, 274, 314, 383, 388
Total lost profit, 387
Total production cost, 383, 386, 387, 452
Total production–inventory–delivery profit, 393
Total profit (TP), 25, 91, 92, 94–96, 98–101, 105, 110, 121, 129, 132, 135, 144, 145
Total profit functions, 46
Total profit per cycle (TP), 32
Total profit per time unit (TPU), 25, 26, 38
Total revenue (TR), 24, 72, 101, 104, 110, 120
Total revenue per cycle, 58, 66
Total rework cost, 387, 452
Total scrap items, 251
Total variable cost, 374
Transportation cost, 57, 88, 100, 127, 128, 139, 377
Triangular distribution, 285
Type I and II errors, 50

U
Uniform distribution, 80, 100, 112, 117, 146
Uniform probability density function, 48
Upper specification level (USL), 537

V
Variability of yield, 500, 508
Variable cost, 87
Variable transformation approach, 120
Vendor–buyer inventory policy, 10

W
Waste, 535
Weibull distribution, 306, 560, 568, 574
Weinstein–Chung's approach, 574

Y
Yield improvement models, 512, 513

Printed in the United States
by Baker & Taylor Publisher Services